P9-EDT-829

MODERN DIGITAL DESIGN

McGraw-Hill Series in Electrical Engineering

Consulting Editor
Stephen W. Director, *Carnegie-Mellon University*

Circuits and Systems
Communications and Signal Processing
Control Theory
Electronics and Electronic Circuits
Power and Energy
Electromagnetics
Computer Engineering
Introductory
Radar and Antennas
VLSI

Previous Consulting Editors

Ronald N. Bracewell, Colin Cherry, James F. Gibbons, Willis W. Harman,
Hubert Heffner, Edward W. Herold, John G. Linvill, Simon Ramo,
Ronald A. Rohrer, Anthony E. Siegman, Charles Susskind, Frederick E. Terman,
John G. Truxal, Ernst Weber, and John R. Whinnery

Computer Engineering

Consulting Editor
Stephen W. Director, *Carnegie-Mellon University*

Bartee: *Digital Computer Fundamentals*
Bell and Newell: *Computer Structures: Readings and Examples*
Garland: *Introduction to Microprocessor System Design*
Gault and Pimmel: *Introduction to Microcomputer-Based Digital Systems*
Givone: *Introduction to Switching Circuit Theory*
Givone and Roesser: *Microprocessors/Microcomputers: Introduction*
Hamacher, Vranesic, and Zaky: *Computer Organization*
Hayes: *Computer Organization and Architecture*
Kohavi: *Switching and Finite Automata Theory*
Lawrence-Mauch: *Real-Time Microcomputer System Design: An Introduction*
Levine: *Vision in Man and Machine*
Peatman: *Design of Digital Systems*
Peatman: *Design with Microcontrollers*
Peatman: *Digital Hardware Design*
Ritterman: *Computer Circuit Concepts*
Sandige: *Modern Digital Design*
Sze: *VLSI Technology*
Taub: *Digital Circuits and Microprocessors*
Wiatrowski and House: *Logic Circuits and Microcomputer Systems*

MODERN
DIGITAL
DESIGN

Richard S. Sandige

Department of Electrical Engineering
The University of Wyoming

McGraw-Hill Publishing Company

New York St. Louis San Francisco Auckland Bogotá Caracas
Hamburg Lisbon London Madrid Mexico Milan Montreal
New Delhi Oklahoma City Paris San Juan São Paulo
Singapore Sydney Tokyo Toronto

This book was set in Times Roman by Publication Services.
The editors were Alar E. Elken and John M. Morriss;
the production supervisor was Friederich W. Schulte.
The cover was designed by Joseph Gillians.
Project supervision was done by Publication Services.
R. R. Donnelley & Sons Company was printer and binder.

MODERN DIGITAL DESIGN

1 2 3 4 5 6 7 8 9 0 DOC DOC 9 5 4 3 2 1 0

ISBN   0-07-054857-9

Library of Congress Cataloging-in-Publication Data

Sandige, Richard S.
    Modern digital design / Richard S. Sandige.
        p.   cm.—(McGraw-Hill series in electrical engineering.
    Computer engineering)
    Includes bibliographies and index.
    ISBN 0-07-054857-9
    1. Logic circuits—Design and construction.   2. Logic design.
    I. Title.   II. Series.
    TK7868.L6S26   1990   68808
    621.39′5—dc20   89-12401

ABOUT THE AUTHOR

Richard S. Sandige is a professor in the Department of Electrical Engineering at The University of Wyoming. Dr. Sandige is a former employee of Hewlett Packard, where he was involved in the design and manufacture of computer workstations. He is the author or co-author of three other books: *Electronic Testing and Troubleshooting*, *DC-10 Laboratory Manual*, and *Digital Concepts Using Standard Integrated Circuits*. Dr. Sandige received his Ph.D. in Electrical Engineering from Texas A&M University.

CONTENTS

2   Number Systems, Number Representations, and Codes

3   Minimizing Functions Using Maps

6 Obtaining Realizable Logic Diagrams Using SSI Devices

7 Implementing Logic Functions Using MSI and Programmable Devices

Part III Sequential Logic Design

8 Sequential Logic Circuits and Bistable Memory Devices

9 Synchronous Sequential Logic Circuit Design

10 Asynchronous Sequential Logic Circuit Design

PREFACE

Electrical Engineering, Computer Science, and Computer Engineering curricula each provide one or more courses for the study of digital logic. This text is designed to provide a beginning course or courses in digital logic design, perhaps at the sophomore or junior level. After mastering the material in this text, the student will possess all the necessary tools and concepts for pursuing advanced studies in the areas of switching theory and finite automata theory of logical machines.

Each chapter begins with an introduction and provides the student with intended instructional goals. Many explicit examples are provided throughout each chapter. IEEE standard logic symbols are used throughout the text to introduce the student to these new logic symbols. Each chapter concludes with a list of references and a large set of problems.

The text is divided into three parts: Part I, Fundamental Concepts; Part II, Combinational Logic Design; and Part III, Sequential Logic Design. In Part I fundamental concepts are presented in Chapters 1 through 4 to provide a mathematical basis from which to study the remaining chapters. These topics consist of switching algebra and standard boolean functions; number systems, number representations, and codes; minimizing functions using maps; and additional minimization techniques. These fundamental topics are required to provide a firm foundation for the study of logic design and prepare the student for designing the combinational and sequential circuit presented in Parts II and III.

Part II, consisting of Chapters 5 through 7, provides an in-depth presentation of combinational logic design with a top-down design approach. The topics include top-down design process for gate level combinational logic design, obtaining realizable logic diagrams using SSI devices, and implementing logic functions using MSI and programmable devices. These chapters provide beginning students with a strong foundation from which to extend their knowledge in the real world of digital design.

The presentation in Part III, Chapters 8 through 10, represents a rather complete discussion of bistable devices and both synchronous and asynchronous sequential logic design. Topics include sequential logic circuits and bistable memory devices, synchronous sequential logic circuit design, and asynchronous sequential logic circuit design. The discussion in these chapters provides a well-rounded presentation for beginning students to learn sequential logic design from both the classical point of view and the more modern software design tools point of view.

In the real world of digital design, programmable devices are being used in increasing numbers in almost every design. An introduction to programmable logic devices (PLDs) in a first course provides the foundation for using these versatile devices in later courses. The software programs necessary for using programmable devices are readily available for IBM PC XT/AT and IBM PS/2 compatible machines. Schematic capture programs are also available, and the reference sections in Chapters 6, 7 , and 9 provide sources for available schematic capture and programmable device software design tools. In the last section of Chapter 9, state machine design examples are presented using the software package PLDesigner by Minc Incorporated. The logic equation, truth table, state machine, and waveform entry methods provided by PLDesigner makes this one of the most versatile and advanced software packages available for designing with programmable logic devices.

The Appendix consists of four sections: Overview of IEEE Std 91-1984 (Explanation of Logic Symbols), Digital Logic Devices, Selected Data Sheets, and Digital Circuit Types.

The sequence of the material in this text has been chosen with great care so that the text can be used as a one-quarter course, a two-quarter or one-semester course, or a full-year course (three quarters or two semesters). In some instances reference is made to material presented earlier in the book. The reader may thus be referred to a previous section, table, or figure so the ideas or contents of these references can be used in the current discussion.

As mentioned earlier, this text is suited for students in an Electrical Engineering, Computer Science, or Computer Engineering curriculum. If the text is used for a one-quarter course, one possible order of presentation is as follows: All of Chapter 1; Section 2-3; all of Chapter 3 except Sections 3-5 and 3-7; all of Chapter 5 up through Section 5-5 (choose selected designs in Section 5-4 and 5-5); all of Chapter 6 except Sections 6-4, 6-5, and 6-7; all of Chapter 7 except Sections 7-4, 7-6, and 7-8; all of Chapter 8 except Section 8-4 and Section 8-6 (however, one needs to cover grated D latches and edge-triggered flip-flops in Section 8-4); and all of Chapter 9 except Section 9-5. Sections that are not presented in class can be assigned as outside reading or, in the case of Sections 7-8 and 9-5, outside computer activity.

If the text is used for a one-semester or two-quarter course, it is recommended that the following material be presented: All of Chapter 1; Sections 2-3 and 2-7 (and sections 2-5 and 2-6 for a two-quarter course only); all of Chapter 3 except Section 3-5; Sections 4-2, 4-3, 4-4, and 4-6; all of Chapter 5 (choose selected designs in Section 5-5); all of Chapter 6 except Sections 6-4 and 6-5; all of Chapter 7 except Sections 7-4 and 7-8; all of Chapter 8; all of Chapter 9 except Section 9-5; and for a two-quarter course only, Sections 10-2 and 10-3. Sections 7-8 and 9-5 should be assigned as outside computer activity. Other sections can be assigned as outside reading.

For those curricula that allow a two-semester or three-quarter sequence, the entire text can be covered to provide an in-depth presentation, including programmable logic and the software programs that support these devices. The guidelines provided above are recommendations; they can be modified to meet the needs of individual curricula.

The trend in digital design in many areas, such as the computer graphics area, is to perform as much functionality as possible in hardware and as little as possible in software. The necessary speed required for performing graphics with high resolution, for example, requires faster hardware devices (fast propagation delays) and fewer software routines (fewer instruction lines). This requires engineers to use hardware as much as possible and software sparingly. PLDs are used in practically every design to reduce PC board real estate and speed up the design cycle. Software packages such as PALASM and PLDesigner are used as productivity tools to shorten the design time of engineers. Schematic design packages such as Schema, OrCAD, P-CAD, and FutureNet are also used as productivity tools to assist the designer in drawing detailed logic diagrams and in some cases as schematic entry for PLD and gate arrays designs. Software design tools are a necessary and important part of modern digital design, as shown through the use of some of these tools in the text.

A special thanks to my very good friends at Hewlett Packard: Jim McLucas, Tom Thrasher, and Dave Klink. These two engineers and one computer scientist helped make this a better book by providing excellent recommendations and suggestions. I would also like to thank my daughter Heidi for her proofreading and editorial suggestions, which have made this book much more readable. Many thanks to the individuals associated with the following companies who provided helpful suggestions, comments, and even hardware and software donations.

Hewlett Packard Company
Minc Incorporated
Texas Instruments Incorporated
Advanced Micro Devices, Inc.
Signetics Company
National Semiconductor Corporation
DATA I/O Corporation
LOGICAL Devices, Inc.
OrCAD Systems Corporation
ALDEC Company
Phase Three Logic, Inc.

ACCEL Technologies, Inc.

EXEL Microelectronics, Inc.

International CMOS Technology, Inc.

Motorola Inc.

Cypress Semiconductor Corporation

Intel Corporation

Altera Corporation

Samsung Semiconductor Inc.

Visionics Corporation

RETNEL Systems

Pistohl Electronic Tool Co.

Omation, Inc.

Xilinx Inc.

Personal CAD Systems Inc.

Wyle Laboratories

I would also like to extend thanks and appreciation to my editor, Alar Elken, who constantly set deadlines that were realistic and achievable. Without those deadlines the book would surely have been a year late. Thanks also to my son Michael for writing the C program that helped me meet my deadline with the book's index.

Special thanks are in order for the following reviewers, who contributed by providing their technical expertise and many recommendations for the sequence and depth of the material presented in the book.

P. I. P. Boulton, University of Toronto

Tae-Sang Chung, University of Kentucky

John D. Dixon, University of North Dakota

Christopher Druzgalski, California State University–Long Beach

Doug Ernie, University of Minnesota

Paul T. Hulina, Pennsylvania State University

David J. Johnson, University of Washington

Edwin C. Jones, Jr., Iowa State University

Charles R. Kime, University of Wisconsin–Madison

Phil Noe, Texas A&M University

Paule D. Stigall, University of Missouri–Rolla

The reviewers and especially the author feel confident that this book will fill a need in the education of prospective engineers and computer scientists.

Richard S. Sandige
Laramie, Wyoming

MODERN DIGITAL DESIGN

PART
I

FUNDAMENTAL
CONCEPTS

CHAPTER
1

SWITCHING ALGEBRA AND STANDARD BOOLEAN FUNCTIONS

1-1 INTRODUCTION AND INSTRUCTIONAL GOALS

The study of engineering and computer science should be built on a foundation of mathematics. We start this chapter with that idea in mind by presenting the mathematics associated with the study of digital design. A mathematical model called switching algebra or two-valued Boolean algebra is presented to provide a foundation for the development of techniques and procedures for the design and analysis of both combinational circuits and sequential circuits.

This chapter is packed with ideas, definitions, and concepts that are used either directly or indirectly in designing digital products on a daily basis.

Our discussion begins with Section 1-2, A Mathematical Model, where a foundation is laid for a mathematical model for two-valued switching circuits. Section 1-3, The Algebra of Logic, continues with an emphasis on Huntington's first set of postulates which provides our definition of Boolean algebra. The concept of logic functions are then presented in Section 1-4, Digital Logic Functions, where a comparison is made between digital design and digital analysis. In Section 1-5, Introduction to Logic Symbols, the IEEE logic symbols are presented and discussed. Section 1-6, Boolean Algebra Theorems, offers a means for logic reduction or simplification, and a number of useful theorems are introduced along with several methods for proving theorems. In Section 1-7, Minimizing Boolean Functions Algebraically, we present algebraic minimization techniques for reducing or simplifying Boolean functions.

The remaining sections in this chapter are considered mandatory for a reasonably good foundation of what is in the ensuing chapters. Section 1-8, Canonical or Standard Forms for Boolean Functions, is required to understand the design process. Functions are often expressed in either a standard sum of products form or a standard product of sums form. A strong presentation is provided in this section relating the sum of products and product of sums forms to the definitions for minterms and maxterms. The concept of functional completeness is also presented and illustrated by circuit diagrams.

Section 1-9, Specifying Designs Using Logic Descriptions, and Section 1-10, Number of Different Functions for *n* Independent Variables, rounds out our discussions. Knowing the various ways designs may be specified and realizing the vast number of different functions possible with just a few variables equips the reader with the tools to move on.

After studying this chapter, you should be prepared to

1. Write the postulates and theorems of Boolean algebra.
2. Prove Boolean algebra theorems using Boolean postulates, Venn diagrams, and perfect induction.
3. Analyze drawings for switching circuits that use either logic symbols or switches, and obtain their Boolean functions.
4. Draw logic diagrams that represent Boolean functions using both switches and logic symbols.
5. Expand Boolean functions into their canonical or standard forms.
6. Obtain minimum Boolean functions using algebraic reduction.
7. Write functions specified in truth tables in standard sum of products and standard product of sums forms.
8. Convert between standard sum of products and standard product of sums forms.
9. List the different sets of functionally complete hardware devices.
10. Obtain truth tables and corresponding Boolean functions given the design specifications.
11. Obtain logic statements and corresponding Boolean functions given the design specifications.
12. Determine the number of different output functions possible for a given number of independent input variables.

1-2   A MATHEMATICAL MODEL

A switching circuit can be symbolically represented with inputs on the left and outputs on the right by a block diagram as shown in Fig. 1-1*a*. This representation

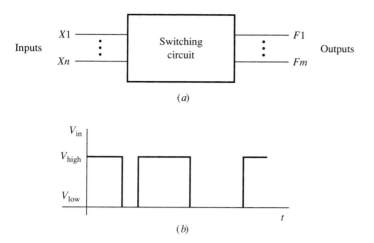

FIGURE 1-1
(*a*) Block diagram of a switching circuit, (*b*) typical ideal switching waveform.

implies that a switching circuit can be constructed that contains a discrete number of inputs from $X1$ to Xn as well as a discrete number of outputs from $F1$ to Fm. The inputs $X1$ to Xn are each driven by a switching waveform (or signal) that typically varies between two distinct voltage values or states as illustrated in Fig. 1-1*b*. Switching circuits are constructed from a wide variety of circuit elements ranging from slow magnetically actuated relay contacts to modern, high-speed gallium arsenide transistors. Gallium arsenide is a semiconductor material used to design transistors.

Relay switching circuits are reminiscent of older telephone exchange circuits while gallium arsenide transistors are being implemented in the fastest mainframe computers. Between these two types of circuits lies a wealth of currently available silicon integrated circuit technologies. Silicon is by far the most commonly used semiconductor material for designing transistors today. Two of the more popular technologies are transistor-transistor logic (TTL) and complementary metal oxide semiconductor logic (CMOS). Digital designs today typically utilize switching circuit-building blocks using TTL and CMOS perhaps more than any other type of technology. It is interesting to note that Silicon Valley, a stretch of landscape south of San Francisco in California, rose to prominence because of the many companies in that location that produced silicon circuits.

Circuits driven with continuous waveforms are referred to as analog circuits. For example, an analog filter circuit like the low-pass filter circuit illustrated in Fig. 1-2*a* is driven by a continuous waveform such as the periodic triangular waveform shown in the Fig. 1-2*b*. A low-pass filter such as this is a circuit designed to allow a band of low frequencies from zero to some upper cutoff frequency to pass through it, while preventing higher frequencies from passing through it. The continuous input signal V_{in} is a variable that takes on a multitude of values since its value varies in infinitesimally small steps from 0 to $+V_{max}$.

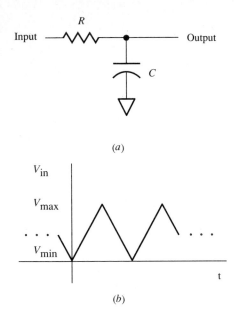

(a)

(b)

FIGURE 1-2
(a) Analog filter circuit, (b) triangular waveform.

Though signals that are associated with analog circuits are continuous in nature, signals that are associated with digital circuits have a two-step appearance.

Because of the two-step or two-state nature of digital circuits, digital circuits have many advantages over analog (continuous) circuits. For example, they have greater reliability than analog circuits because when devices are operated with only two states they simply operate more dependably. Due to the two-valued binary nature of digital circuits they may be designed from the outset to possess a specified accuracy; therefore, a given circuit or system has a specified accuracy even over varying environmental conditions. Two-state digital circuits use the binary number system and are referred to by many different names such as binary circuits, switching circuits, digital switching circuits, logic circuits, digital logic circuits, and digital systems.

One of the few places where analog circuits outperform digital circuits is in their output response time. The output signal of an analog circuit that has infinitesimally small input steps responds to each input step in a continuous manner. Digital systems are designed to depend on two-valued waveforms that are regulated at a sampling rate based on the frequency of a periodic enabling device (commonly called the system clock). Circuits that operate using sampling, or time slices, have a slower output response time than analog circuits because operations can not be performed faster than the time dictated by the period of the system clock ($T = 1/f$, where T is the period and f is the frequency or speed of the system clock).

Switching circuits can be roughly divided into two classes: combinational logic circuits and sequential logic circuits. A combinational logic circuit has the property that its outputs are totally determined by its external inputs. The majority of sequential logic circuits depend on a system clock and memory devices to preserve past inputs. Sequential logic circuits will be presented in Chapter 8.

Switching circuits in general have switching variables that can exist in only two states. Realizable electronic switching circuits and analog circuits have inherent delays. Switching circuit delays are called propagation delays, that is, the time interval between a change in an output switching waveform as a result of a change in an input switching waveform. The first class of switching circuits we will discuss are combinational circuits. The speed or response of combinational circuits is limited only by the propagation delay time of the devices being used. What is the fastest known switch, and how fast is it? The Josephson junction is the fastest known switch. The operation of this device depends on superconductivity and the tunneling of electrons across an insulating barrier. This extremely fast switch is capable of changing state in as little as 6 picoseconds (6 ps), that is, six trillionths of a second. Gallium arsenide switching devices have been operated at 10 ps, while the fastest experimental silicon logic devices switch on and off in 30 ps. Compare this to the popular bipolar silicon TTL (transistor-transistor logic) switching devices, that is, off-the-shelf components, which have a switching speed generally greater than 1 nanosecond (1 ns).

1-2-1   Two-Valued Switching Circuits

A practical example of a two-valued switching circuit is shown in Fig. 1-3. This circuit contains switches that can exist in only two states: open or closed. When the appropriate group of switches is closed in Fig. 1-3 the light bulb is turned on; otherwise, the light bulb is turned off. Two switching variables are associated with each switch, the input control variable X and the output transmission variable T. The input variable is associated with the mechanical action that must be initiated to close an open switch or open a closed switch, while the output variable is associated with data transmission across the two terminals of the switch, that is, electrical continuity. When there is no continuity across the terminals, an open circuit occurs across the terminals and the output transmission variable T is equal to 0. When continuity exists across the terminals, the terminals are connected together and the output transmission variable T is equal to 1. It is customary

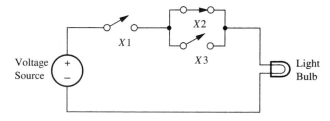

FIGURE 1-3
Switching circuit realized with switches.

only to show the input variables and not the transmission variables when drawing switches, as illustrated in Fig. 1-3.

In this book we will deal exclusively with two-valued switching circuits. The mathematics of switching circuits deals with the logical relations between the input variables and the output variables rather than with the type of circuit technology being used. The mechanical and electrical nature of various switching circuit technologies is best left to your design laboratory.

1-3 THE ALGEBRA OF LOGIC

An English mathematician, George Boole, in 1854 published a book entitled *An Investigation of the Laws of Thought*. In his book, Boole introduced the idea of examining the truth or falsehood of complicated statements via a special algebra of logic. Statements or variables in Boole's algebra were only allowed to take on two values: those that were completely false, which may be assigned the value 0, and those that were completely true, which may be assigned the value 1. Thus a switching, digital, or logic variable in an abstract sense can take on only two values, a logic 0 and a logic 1.

There are two major types of mechanical switches; toggle and push-button (or momentary) switches. A toggle switch such as a light switch on the wall has the property of 'memory.' Memory devices will be discussed in Chapter 8. Push-button or momentary switches such as push-button telephone and computer keyboard switches do not have memory. This is the class of switches we will discuss in this chapter. These switches are analogous to transistor switches that are used to design today's modern switching circuits.

Figure 1-4*a* shows a push-button or momentary switch that is normally open. We will refer to this type of switch as a positive switch. Using Boole's idea of an algebra of logic, a positive switch has a transmission variable of 0 (no continuity) if the input variable is 0 since a 0 implies no action is taken to close the switch, and a transmission variable of 1 (continuity exists) if the input variable is 1 since a 1 implies an action is taken to close the switch. Switches $X1$ and $X3$ shown in Fig. 1-3 are normally open, positive switches.

A push-button switch that is normally closed is drawn in a closed position as shown in Fig. 1-4*b*. This type of switch will be referred to as a negative switch since its operation is opposite to that of a positive switch. A negative switch has a transmission variable that is 1 (continuity exits) if the input variable is 0 since no action is taken to open the switch. A negative switch has a transmission variable of 0 (no continuity) if the input variable is 1 since an action is taken to open the switch. Switch $X2$ shown in Fig. 1-3 is a normally closed, negative switch.

In 1938, a research assistant named Claude Shannon in the department of electrical engineering at Massachusetts Institute of Technology provided the first applications of the principles developed by Boole to the design of electrical switching circuits. Shannon's published paper, "A Symbolic Analysis of Relay and Switching Circuits," was an abstract of a thesis presented at MIT for the degree of master of science.

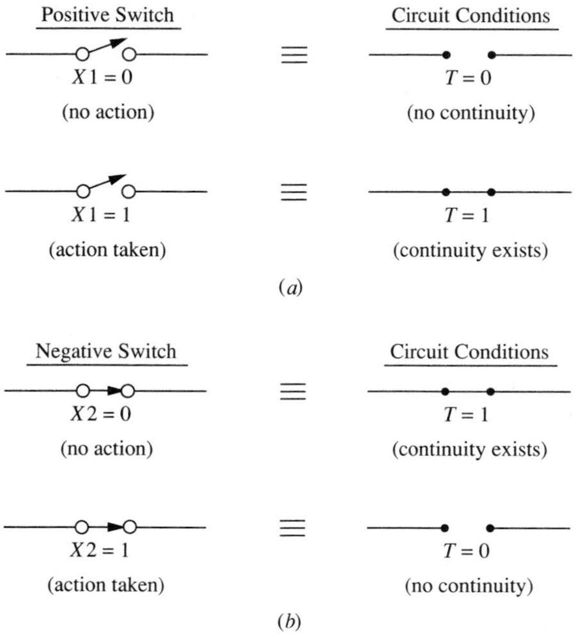

Positive Switch | Circuit Conditions

$X1 = 0$ $\equiv$ $T = 0$

(no action)   (no continuity)

$X1 = 1$ $\equiv$ $T = 1$

(action taken)   (continuity exists)

(*a*)

Negative Switch | Circuit Conditions

$X2 = 0$ $\equiv$ $T = 1$

(no action)   (continuity exists)

$X2 = 1$ $\equiv$ $T = 0$

(action taken)   (no continuity)

(*b*)

FIGURE 1-4
Circuit conditions, (*a*) a positive switch, (normally open, push-button or momentary switch) (*b*) a negative switch (normally closed, push-button or momentary switch).

The interesting thing about Shannon's paper was his demonstration that a mathematical technique was available to express all types of switching circuits. Using ordinary algebraic expressions, analog circuits could often be reduced or simplified mathematically. Shannon showed that Boolean algebraic expressions can be used in a similar fashion to mathematically reduce or simplify switching circuits. Logic designers today use Boolean algebra to design the switching circuit components of modern digital computers in the same manner as analog designers use conventional mathematics to design operational amplifiers for modern audio and communications equipment.

Modern switching circuits found in today's personal computers, hand-held calculators, traffic light controllers, talking toys, jet airliners, and modern super-computers predominantly utilize switching circuitry constructed from transistors. In a similar manner a bipolar (either PNP or NPN) transistor and a metal oxide semiconductor (MOS) transistor (either N channel or P channel) each has the same characteristic as a momentary switch. Consider the circuit conditions shown in Fig. 1-5; when the bipolar transistor or MOS transistor is turned off (its input control variable is 0), then its output transmission variable is 0 (no continuity), and when either transistor is turned on (its input control variable is 1), then its output transmission variable is 1 (continuity exists).

It is not important that you understand how transistors operate. It is only important that you understand that when they operate as two-state devices they operate in a straightforward manner just like a switch operates, only much faster.

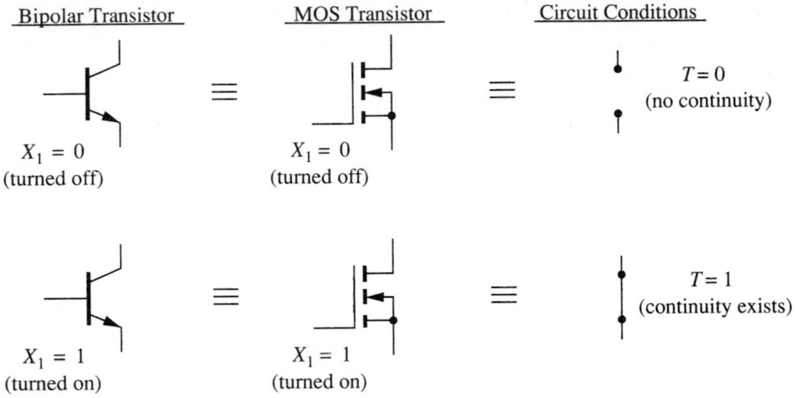

Bipolar Transistor		MOS Transistor		Circuit Conditions

$X_1 = 0$
(turned off)

$X_1 = 0$
(turned off)

$T = 0$
(no continuity)

$X_1 = 1$
(turned on)

$X_1 = 1$
(turned on)

$T = 1$
(continuity exists)

FIGURE 1-5
Circuit conditions for bipolar and MOS transistors (these types act just like a positive switch).

Not only are transistors faster, they require very little power, and are much more reliable than their switch or relay counterparts. Switching circuits constructed using transistors vary from individually packaged transistors for high-power applications to integrated circuit packages using thousands of transistors on a single silicon chip like the Intel 80386 or the Motorola 68030 microprocessors.

1-3-1 Defining a Boolean Algebra Based on Huntington's Postulates

When a new mathematics is introduced, certain self-evident mathematical statements or propositions are stated without proof and are called postulates, axioms, premises, or maxims. Any additional propositions may then be proven using the given set of postulates. These propositions are called theorems. A theorem, like a postulate, is a rule concerning a relationship between Boolean variables. Our first task is to introduce a sufficient set of postulates that define a two-valued Boolean algebra (actually several different sufficient sets of postulates are possible). Simple circuits using switches can be used to provide us with physical insight into what the two-valued postulates of Boolean algebra may represent. Except for the special terminology connecting it with digital circuits, switching algebra or Boolean algebra, as it is more commonly called, is identical to the algebra of propositional logic considered as an abstract algebra, and the algebra of sets.

For a detailed comparison of each of these algebras refer to *The Principles of Switching Circuits*, by H. E. Edwards in the list of references.

In 1904 E. V. Huntington presented a paper entitled "Sets of Independent Postulates for the Algebra of Logic." Huntington's postulates provide the basic propositions for defining a many-valued Boolean algebra of class MB = $\{0, 1, 2, \ldots, n\}$. Switching algebra is based on Huntington's postulates for a

Boolean algebra of class $B = \{0, 1\}$ and two undefined binary operations + and $\times$ (today $\bullet$ is used instead of $\times$ since $\times$ is used as a variable). Many different sets of independent postulates have been published, but Huntington's "first set" of postulates is the most consistently used for defining two-valued Boolean algebra. Two-valued Boolean algebra also uses variables; these variables can only take on values that are in class B. Huntington's postulates for a two-valued Boolean algebra of class $B = \{0, 1\}$ are listed in Fig. 1-6.

It should be observed that the Postulates P2 through P5 are presented as dual pairs. Each member of a dual pair (such as $X + 0 = X$) can be obtained from the other member of the pair ($X \bullet 1 = X$) by interchanging the elements 0 and 1 (often called the identity elements) and the binary operators + and $\bullet$. This is called the principle of duality and shows that valid relationships exist between the variables and operators of the postulates when the identity elements (0 and 1) and binary operators (+ and $\bullet$) are interchanged.

To understand Huntington's first set of postulates we will first present a set of definitions for the operators used in the postulates and then show that they satisfy the postulates. The postulates will then be expressed using switches, and finally we will express the postulates in terms of logic symbols.

The commutative properties (P3a and P3b) expressed by $X + Y = Y + X$ and $X \bullet Y = Y \bullet X$, and the distributive property (P4b) expressed by $X \bullet (Y + Z) = X \bullet Y + X \bullet Z$ should already be somewhat familiar since these properties have the same form in ordinary algebra (symbolic algebra for real numbers). Some

---

Postulates for a Two-Valued Boolean Algebra of Class $B = \{0, 1\}$

P1a: If X and Y are in B, then $X + Y$ is in B.
P1b: If X and Y are in B, then $X \bullet Y$ is in B.

P2a: There is an element 0 such that $X + 0 = X$
     for every variable X.
P2b: There is an element 1 such that $X \bullet 1 = X$
     for every variable X.

P3a: $X + Y = Y + X$.   (Commutative with respect to +)
P3b: $X \bullet Y = Y \bullet X$.   (Commutative with respect to $\bullet$)

P4a: $X + Y \bullet Z = (X + Y) \bullet (X + Z)$. (+ is distributive over $\bullet$)
P4b: $X \bullet (Y + Z) = X \bullet Y + X \bullet Z$.   ($\bullet$ is distributive over +)

P5:   For every variable X there is a variable $\overline{X}$
    (called the complement of X) such that
    a: $X + \overline{X} = 1$ and
    b: $X \bullet \overline{X} = 0$.

P6:   There are at least two distinct elements in B.

FIGURE 1-6
Huntington's first set of postulates.

of the properties in Huntington's postulates are unique to Boolean algebra. To effectively use the properties of Huntington's postulates they must be learned.

1-3-2 OR, AND, and Complement Operator Definitions

As you may have observed, the binary operators $+$ and $\cdot$ are not defined in Huntington's postulates. The binary operators and the complement operator are defined in the three separate tables illustrated in Figs. 1-7. Each operator can be shown to satisfy the Boolean algebra Postulates P1 through P6. Each table in Fig. 1-7 is called a table of combinations, and provides a list of the values for all possible combinations of the independent variables (X and/or Y) that are used in defining each operator. Since each input variable can exist as either a 1 or a 0, for two variables, a table of combinations of X and Y, or truth table as it is more often called, only contains four (2×2) rows.

The reason a table of combinations is also called a truth table is because the table was often generated using true for logic 1 and false for logic 0, hence the term truth table emerged. Today truth tables are practically always generated using the elements 1 and 0 in the table. In Fig. 1-7a a row entry in column $X + Y$ is 1 when one or more variables X or Y is 1, and 0 otherwise. In Fig. 1-7b a row entry in the column $X \cdot Y$ is 1 when both variables X and Y are 1, and 0 otherwise.

The $+$ or OR operator, as it is more commonly called, performs the OR operation (logical addition, or logical sum) of the two variables X and Y. When we consider 0 and 1 as numbers and the $+$ symbol to be the arithmetic addition operator, then the OR operation performs arithmetic addition for three out of the four rows of the OR operator truth table.

The $\cdot$ or AND operator performs the AND operation (logical multiplication, or logical product) of the two variables X and Y. It is purely coincidental that if we consider 0 and 1 to be numbers and the $\cdot$ symbol to be the arithmetic multiplication operator, then the AND operation performs arithmetic multiplication for all four rows of the AND operator truth table.

Analogous arithmetic operations are simply used to illustrate the probable origin of the $+$ and $\cdot$ operator symbols for the OR and AND binary operators.

In Fig. 1-7c, the $^-$ or complement (inversion, negation, or NOT) operator performs the complement operation (logical complement) of the single variable X. The complement operator performs the 1's complement of 0 and 1 if we consider

X	Y	$X + Y$
0	0	0
0	1	1
1	0	1
1	1	1

(a)

X	Y	$X \cdot Y$
0	0	0
0	1	0
1	0	0
1	1	1

(b)

X	$\overline{X}$
0	1
1	0

(c)

FIGURE 1-7
Operator definitions, (a) $+$ or OR operator, (b) $\cdot$ or AND operator, (c) $^-$ or complement operator.

them as numbers (the 1's complement of 0 is 1 while the 1's complement of 1 is 0.) Binary arithmetic (among others) uses the 1's complement in performing subtraction, and this topic will be pursued in Chapter 2. The entry in column $\overline{X}$ is 1 when variable X is 0, and 0 when the variable X is 1. Standard usage encourages using first the overbar $^-$ for the logical complement, then depending on the availability of the printing media being used either the symbol $\neg$ or the tilde $\sim$ preceding the variable being complemented as $\neg X$ or $\sim X$. The logical complement is sometimes represented by the apostrophe symbol ' following the variable being complemented as X'.

The OR, AND, and complement operators can also be defined by writing all possible combinations of the identity elements with each respective operator as illustrated below:

Definition of OR operator	Definition of AND operator	Definition of complement operator
$0 + 0 = 0$	$0 \cdot 0 = 0$	$\overline{0} = 1$
$0 + 1 = 1$	$0 \cdot 1 = 0$	$\overline{1} = 0$
$1 + 0 = 1$	$1 \cdot 0 = 0$	
$1 + 1 = 1$	$1 \cdot 1 = 1$	

Definition 1-1. A Boolean expression is a constant such as 1 or 0, a single Boolean variable such as X, Y, or Z, or several constants and/or variables used in combination with one or more binary operators or complement operators such as $X + Y \cdot Z$, $X + 1$, *or* $X \cdot \overline{X}$.

Since X is a Boolean expression we can also say that $\overline{X}$ is a Boolean expression. In switching theory X and $\overline{X}$ are customarily termed literals: each occurrence of an uncomplemented or complemented variable is a literal. We can also see from the Boolean postulates that certain Boolean expressions such as $X \cdot 1 = X$ or $X \cdot (Y + Z) = X \cdot Y + X \cdot Z$ are equivalent. Using the equal sign in this manner signifies a Boolean equation just as it does in ordinary algebra.

Product terms and sum terms are often used when discussing Boolean expressions. Expressions such as $X \cdot Y$, and $X \cdot \overline{Y} \cdot Z$ are called product terms since these expressions only contain literals that are ANDed together, that is, products are formed. Expressions such as $\overline{X} + Y$, and $\overline{X} + \overline{Y} + Z$ are called sum terms. In this case literals are only ORed together, hence the name sum terms.

Definition 1-2. The hierarchy or order of precedence of the binary operators and the complement operator are: complement ($^-$), AND ($\bullet$), then OR ($+$). The complement operator has the most precedence, while the OR operator has the least. Parentheses may be used as in ordinary algebra to circumvent operator hierarchy.

For example, the order in which the expression $X \cdot Y + Z$ is evaluated is first X AND Y then $(X$ AND $Y)$ OR Z. In other words, the product of X and Y is first formed and then summed with Z. Parentheses may not be needed if you understand the order of precedence of the operators. The order in which the expression $X \cdot Y + X \cdot Z$ is evaluated is X AND Y, X AND Z then $(X$ AND $Y)$ OR $(X$ AND $Z)$. In this case the product terms $X \cdot Y$ and $X \cdot Z$ are first formed then these product terms are summed. When parentheses are used, the effect is either to make a complex expression less ambiguous, or to establish a desired order of precedence of the operators. As in ordinary algebra, Boolean expressions inside parentheses are always evaluated first. Both of these examples form a final expression that is referred to as a sum of products (SOP) form of expression. The form of the expression on the right side of postulate P4a, $(X + Y) \cdot (X + Z)$, may be referred to as a product of sums (POS) form of expression for obvious reasons (see Fig. 1-6).

Recall the theorem of mathematical induction which states that if you can show that a proposition is true for two variables, and the truth of the proposition for k variables can be shown to imply the truth of the proposition for $k + 1$ variables, then the proposition must be true for any number of variables. If k is small, we can easily test the proposition to see if it holds for each and every case. If the proposition does hold for each and every case, then we can say the proposition holds perfectly or by perfect induction. For Boolean algebra this means that we can verify that a proposition holds by perfect induction if the proposition contains a small number of variables and every case can be verified. The following example illustrates the usefulness of the principle of perfect induction, that is, verifying all possible cases.

To use the OR, AND, and complement operators as part of our two-valued Boolean algebra, we must first show that they each satisfy the mathematical premise on which we have based our Boolean algebra (Huntington's Postulates). This can be done using the principle of perfect induction as illustrated in the following example.

Example 1-1. Show that the following OR, AND, and complement operators satisfy the Boolean algebra postulates P1 through P6.

$0 + 0 = 0$	$0 \cdot 0 = 0$	$\bar{0} = 1$
$0 + 1 = 1$	$0 \cdot 1 = 0$	$\bar{1} = 0$
$1 + 0 = 1$	$1 \cdot 0 = 0$	
$1 + 1 = 1$	$1 \cdot 1 = 1$	

Solution Postulate P1 is satisfied because all elements being operated on are contained in $B = \{0, 1\}$.

P2a is satisfied by observing that $0 + 0 = 0$ and $1 + 0 = 1$ are true. P2b is satisfied by observing that $0 \cdot 1 = 0$ and $1 \cdot 1 = 1$ are true.

P3a and P3b are satisfied by observing that the order of the elements for the OR and AND operations are interchangeable.

We can show that P4a is satisfied by constructing a truth table containing three variables as shown in Fig. E1-1a. The OR, AND, and complement operators are used in the construction of the table. The parentheses used on the right side of P4a are necessary to circumvent operator hierarchy. By evaluating the expression on the left side of P4a and also the expression on the right side of P4a using all 8 (2 × 2 × 2) possible combinations of the variables X, Y, and Z we can see by perfect induction that P4a is satisfied.

Next we construct a truth table containing the two possible values of the variable X as shown in Fig. E1-1b. Filling in the table for the expression on the left side of P5a and the expression on the right side of P5a we can see that both expressions evaluate to 1, proving by perfect induction that P5a is satisfied.

The verification of P4b and P5b can also be proven by perfect induction, but these are left as exercises for the student.

P6 is satisfied by inspection since element 0 and element 1 are not equal.

As we have demonstrated in this example the OR, AND, and complement operators do indeed satisfy the Boolean Postulates P1 through P6 and therefore qualify as operators for our two-valued Boolean algebra. These operators are the primary operators used throughout this text.

X	Y	Z	$Y \cdot Z$	$X + Y \cdot Z$	$X + Y$	$X + Z$	$(X + Y) \cdot (X + Z)$
0	0	0	0	0	0	0	0
0	0	1	0	0	0	1	0
0	1	0	0	0	1	0	0
0	1	1	1	1	1	1	1
1	0	0	0	1	1	1	1
1	0	1	0	1	1	1	1
1	1	0	0	1	1	1	1
1	1	1	1	1	1	1	1

Left side   =   Right side
of P4a   of P4a

(a)

X	$\overline{X}$	$X + \overline{X}$	1
0	1	1	1
1	0	1	1

Left   =   Right
side   side
of   of
P5a   P5a

(b)

FIGURE E1-1

1-4 DIGITAL LOGIC FUNCTIONS

An easy way to visualize Postulates P2 through P5 is in terms of circuits that utilize switches as shown in Fig. 1-8. A 0 is represented by two adjacent terminals that are never connected, while a 1 is represented by two adjacent terminals that are always connected. An OR operation is performed by combining items (switches and/or adjacent terminals) in parallel as indicated by the first circuit for P2a. An AND operation is performed by combining items (switches and/or adjacent terminals) in series as shown by the first circuit in P2b.

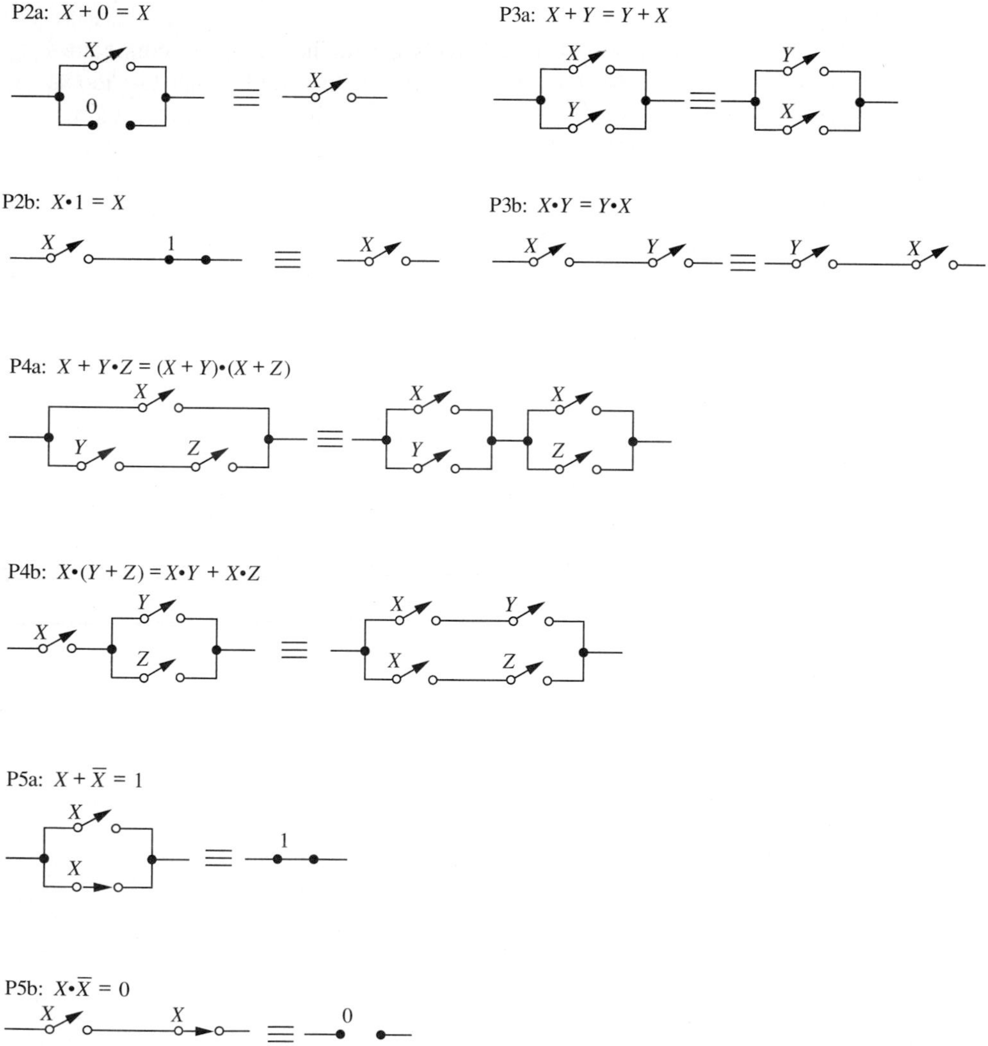

P2a: $X + 0 = X$

P3a: $X + Y = Y + X$

P2b: $X \cdot 1 = X$

P3b: $X \cdot Y = Y \cdot X$

P4a: $X + Y \cdot Z = (X + Y) \cdot (X + Z)$

P4b: $X \cdot (Y + Z) = X \cdot Y + X \cdot Z$

P5a: $X + \overline{X} = 1$

P5b: $X \cdot \overline{X} = 0$

FIGURE 1-8
Boolean algebra Postulates P2 through P5 represented by switches.

To analyze the circuits in Fig. 1-8, remember that for a positive switch you must push its button to make continuity, while a negative switch has continuity *until* you push its button. For the circuits representing P3a (both circuits have a parallel connection and therefore represent an OR operation), it should be obvious that continuity does not exist across either circuit when neither switch X nor switch Y is pushed. It should be understood when we say either "a switch is pushed" or "push a switch," we are referring to pushing the push button associated with the switch. If switch X is pushed in either circuit, but not switch Y, then continuity exists across both circuits. If switch Y is pushed in either circuit, but not switch X, continuity also exists across both circuits. Pushing both switches X and Y of course provides continuity across both circuits.

For the first circuit representing P5b (this circuit has a series connection and therefore represents an AND operation), since both switches are labeled X they must be pushed simultaneously. When neither switch is pushed, the positive switch remains open while the negative switch remains closed and no continuity exists across the circuit. When both switches are pushed, the positive switch closes and the negative switch opens, and again no continuity exists across the circuit. By analyzing each set of circuits in this manner one can easily begin to understand not only how switching circuits that are represented by switches can perform the expressions represented by the Boolean algebra postulates, but one should also begin to better understand what the postulates mean in terms of simple switching circuits.

This type of analysis is similar to the one that Shannon used for switching circuits when he proposed Boolean algebra as a mathematical tool. By closely observing the algebraic expressions in Fig. 1-8 and their corresponding circuit representations, it should be clear that the expressions with fewer literals always result in a simpler circuit, that is, a circuit with less switches.

> **Definition 1-3.** A Boolean function (or Boolean equation) is a dependent variable such as F set equal to a Boolean expression such as $X + \bar{Y}$ where X and Y are independent variables. The functional notation $F(X, Y)$ is also used in Boolean algebra in the same way it is used in ordinary algebra. It is standard practice to capitalize all variables. A function may be assigned a value of 0 or 1 for each of the possible 2^n combinations of binary values for n independent variables.

Postulates P2 through P5 are repeated again in Fig. 1-9 using logic symbols. Here a 0 is represented by a signal line that is always inactive (a logic 0), while a 1 is represented by a signal line that is always active (a logic 1). The symbol for the OR operation is the OR element, the symbol for the AND operation is the AND element, and the symbol for the complement operation is the Inverter. Graphic symbols such as the OR element, AND element, and the Inverter represent Boolean functions or the physical devices that carry out Boolean functions.

The graphic symbols shown in Fig. 1-9 for the OR element and the AND element only have two inputs (inputs are drawn on the left side of the symbols).

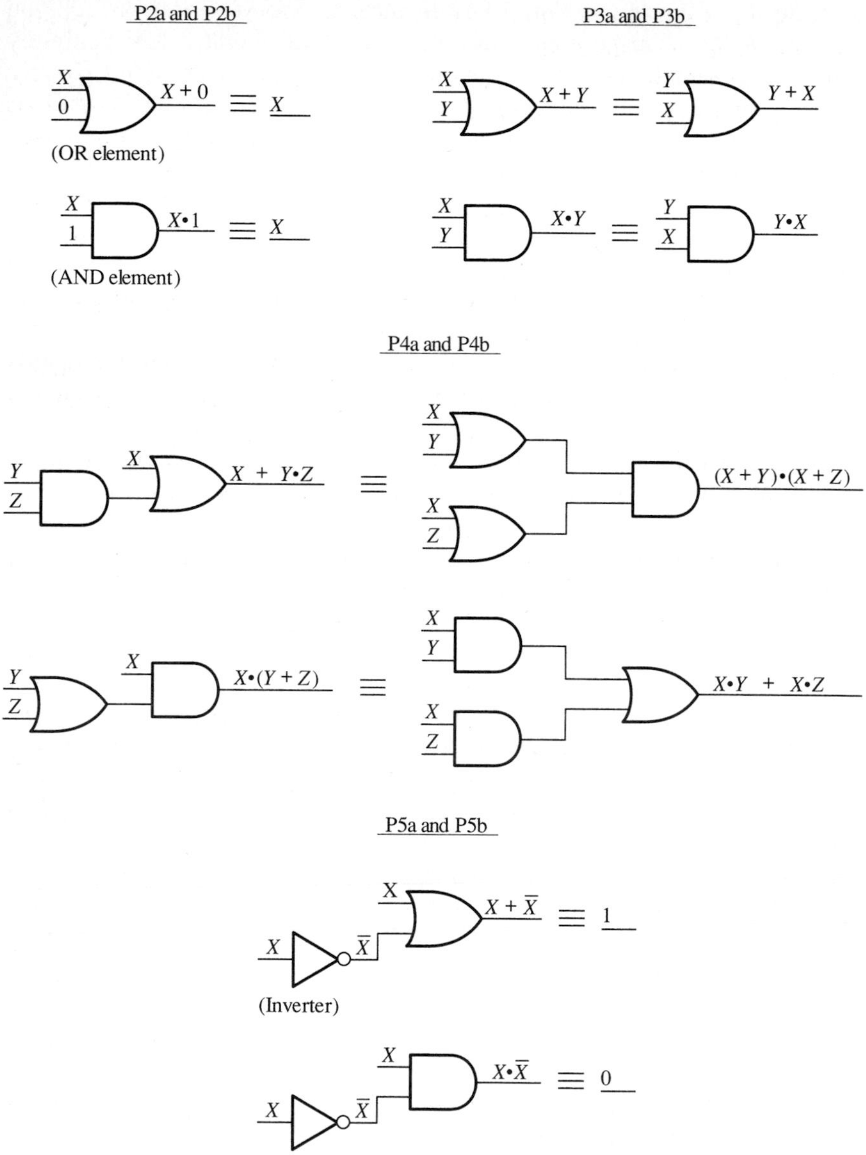

P2a and P2b

P3a and P3b

$X + 0 \equiv X$

(OR element)

$X \cdot 1 \equiv X$

(AND element)

$X + Y \equiv Y + X$

$X \cdot Y \equiv Y \cdot X$

P4a and P4b

$X + Y \cdot Z \equiv (X + Y) \cdot (X + Z)$

$X \cdot (Y + Z) \equiv X \cdot Y + X \cdot Z$

P5a and P5b

(Inverter)

$X + \overline{X} \equiv 1$

$X \cdot \overline{X} \equiv 0$

FIGURE 1-9
Boolean algebra Postulates P2 through P5 represented by logic symbols.

Physical devices can be designed with more than two inputs, and we will have occasion to use certain graphic symbols with more than two inputs. The first symbol or circuit shown for P3a is an OR element that performs the OR operation on its inputs X and Y to obtain $X + Y$ at its output. The first symbol for P3b is an AND element that performs the AND operation on its inputs X and Y to obtain

$X \cdot Y$ at its output. In both cases the order of labeling the inputs is unimportant as seen by the equivalent circuits.

When an output of one of the logic symbols acts as an input for another logic symbol as illustrated by the first circuit of P4a, the logic operation performed first (the AND operation in this case) has precedence over the OR operation (the operation performed second), and the final output expression is $X + Y \cdot Z$. In the first circuit of P4b, the OR operation is the one performed first and has precedence over the AND operation which is performed second. The output for this circuit is $X \cdot (Y + Z)$ which has to be written with parentheses to circumvent the normal order of precedence of the operators. For the first circuit of P5a in Fig. 1-9, the X input is first complemented by the Inverter and then ORed with the uncomplemented X input to achieve the output $X + \overline{X}$. With a little practice you should be able to verify (or obtain) the output expressions for each of the circuits in Fig. 1-9.

By using symbols, we can study the logic characteristics of devices without learning specific knowledge of their internal circuitry. In this text we present the logic symbols in the context of electrical application; however, most of the symbols may also be applied to diversified mechanical, electromechanical, electro-optical, pneumatic, or hydraulic systems.

Look at Fig. 1-9 and observe that for postulate P4a the diagram using logic symbols for the expression with the smallest number of literals $X + Y \cdot Z$ results in a circuit with fewer symbols than the circuit for the equivalent expression $(X + Y) \cdot (X + Z)$. This is also true for the circuits for Postulates P2a, P2b, P4b, P5a, and P5b. In order that functions may be realized with fewer symbols it is common practice in the design of switching circuits to try to obtain the simplest expressions possible for the Boolean functions, that is, obtain expressions with the smallest number of literals.

Over the course of the next few chapters we will demonstrate not only how to obtain mathematical expressions that represent various switching circuits, but we will also illustrate how to reduce or simplify those mathematical expressions into forms that represent simpler circuits. By accomplishing a valid reduction of a mathematical expression that represents a switching circuit, we effectively reduce the amount of circuitry required to realize or implement the mathematical expression.

Consider the following Boolean function where F is the dependent variable and X and Y are the independent variables.

$$F(X, Y) = X + \overline{Y}$$

The expression on the right side of the equal sign represents the exact manner in which the function F is related to X and Y. Logic designers use a number of different methods to represent Boolean functions. In Fig. 1-10a we show the truth table for the function $F = X + \overline{Y}$. The truth table was obtained by substituting all four possible combinations for the input variables into the expression on the right side of the function and solving for the dependent variable F. A logic circuit for the function $F = X + \overline{Y}$ is shown in Fig. 1-10b using logic symbols. The two independent inputs X and Y are shown on separate input lines and the dependent

$F = X + \bar{Y}$

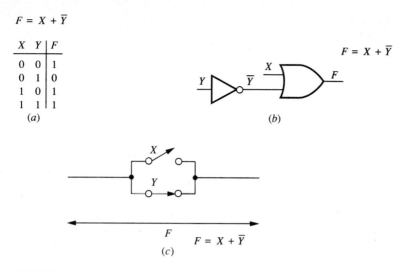

X	Y	F
0	0	1
0	1	0
1	0	1
1	1	1

(a)

(b)

$F = X + \bar{Y}$

(c)

FIGURE 1-10
Function $F = X + \bar{Y}$. (a) truth table, (b) logic circuit using logic symbols, (c) logic circuit using switches.

output F is shown on the output line. A logic circuit for $F = X + \bar{Y}$ using momentary or push-button switches is shown in Fig. 1-10c.

1-4-1 Digital Design Compared to Digital Analysis

When a logic designer starts with a functional description or specification of a circuit and ends with a logic diagram or circuit diagram that performs the required function(s) as we did in Fig. 1-10, this exercise is called synthesis or design. The reverse process of starting with a logic diagram or circuit diagram and obtaining its functional description or specification is called analysis. In this context we analyze logic diagrams and obtain their logic functions by knowing the functions for each of the symbols that make up the diagrams. In order to be able to design a logic circuit we first obtain the Boolean functions for the circuit then translate these equations into appropriate logic symbols. For simple functions this requires remembering which logic symbols are associated with certain mathematical expressions as shown in the mathematical expression/logic symbol summary in Fig. 1-11.

Figure 1-11 shows the mathematical expressions for OR and AND operations, which may contain more than two input variables, and their corresponding logic symbols and logic names. The distinctive shape symbol for the complement operation or the Inverter has only a single input compared to the distinctive shape symbols for the OR operation (OR element) and the AND operation (AND element) which may have more than two inputs, as indicated by the ellipsis points placed between the two input lines in the figure. A detailed discussion

Mathematical Expression	Mathematical Name	Logic Symbol		Logic Name
		Distinctive Shape	Rectangular Shape	
$A + B + C + \cdots$	OR Operation		≥ 1	OR Element
		Distinctive Shape	Rectangular Shape	
$A \cdot B \cdot C \cdot \cdots$	AND Operation		$\&$	AND Element
		Distinctive Shape	Rectangular Shape	
$\overline{A}$	Complement Operation		1	Inverter

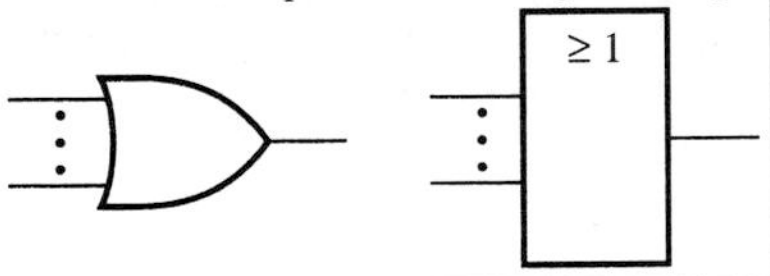

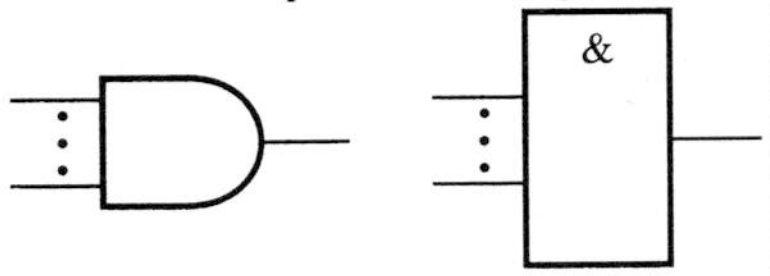

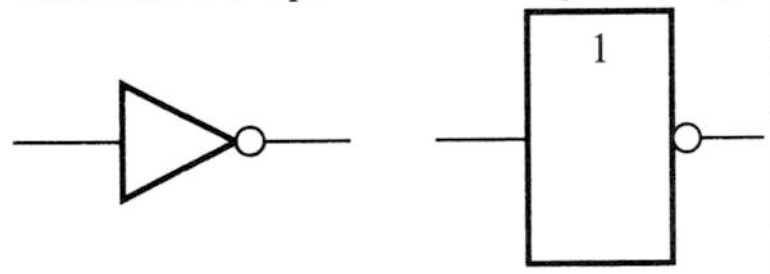

FIGURE 1-11
Mathematical expression/logic symbol summary.

of distinctive shape and rectangular shape logic symbols will be presented in the next section, Section 1-5.

Example 1-2

1. Analyze the switching circuit shown in Fig. E1-2a and obtain its Boolean function.
2. Use the Boolean function obtained for the switching circuit in Fig. E1-2a and show a circuit design using logic symbols for the expressions in the function.
3. Obtain the truth table for the function.

Solution

1. The functional notation in mathematical form for the circuit is

$$F(X, Y, Z) = (X \cdot Y + \overline{X}) \cdot Z$$

The function states that $F = 1$ (continuity exists across the entire circuit, $T = 1$) when X AND Y are pushed (continuity exists across the upper two switches X and Y, $T_x = 1$ and $T_y = 1$) OR X is not pushed (continuity exists across the lower switch X, $T_x = 1$) while at the same time Z is pushed (continuity exists across switch Z, $T_z = 1$). In other words, find the input conditions for the switches that will provide continuity across the entire circuit, and you have found the Boolean function for the circuit.

2. The logic diagram for the function $F(X, Y, Z)$ is shown in Fig. E1-2b. The logic diagram was generated by observing the operator order of precedence in the expression of the function, then substituting the corresponding logic symbol for

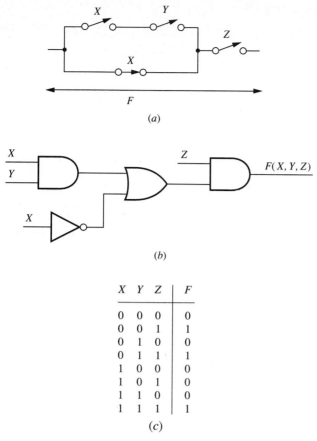

(a)

(b)

X	Y	Z	F
0	0	0	0
0	0	1	1
0	1	0	0
0	1	1	1
1	0	0	0
1	0	1	0
1	1	0	0
1	1	1	1

(c)

FIGURE E1-2

each mathematical expression. Observe that the expression inside the parentheses must be performed first.

3. The truth table is a tabular form that must contain all combinations of the independent variables X, Y, and Z. For n independent variables there are 2^n possible combinations as shown in Fig. E1-2c. The dependent variable F was obtained by substituting into the expression for F all possible row combinations of the independent variables $X\,Y\,Z$, 0 0 0 through 1 1 1. Each combination was evaluated separately using the truth tables for the OR, AND, and complement operators and then entered in the column under F for the appropriate row. An alternate and perhaps more organized way to obtain the dependent variable F would be to list separate columns for $X \cdot Y$, $\overline{X}$, $X \cdot Y + \overline{X}$, and $(X \cdot Y + \overline{X}) \cdot Z$ and then fill in these columns for each row of the truth table. The last column would represent the function F.

You should become adept at drawing logic symbols for Boolean functions. In a later chapter we will present a technique that allows you to draw the appropriate logic symbols for Boolean functions using a rather unique approach.

1-5 INTRODUCTION TO LOGIC SYMBOLS

The logic symbols presented in Fig. 1-11 are the symbols in *IEEE Standard Graphic Symbols for Logic Functions, ANSI/IEEE Std 91-1984*. This standard is based on the *International Electrotechnical Commission (IEC) Standard Publication 617-12, Graphical symbols for diagrams, Part 12: Binary logic elements, 1983*. These standards provide quite a few different concepts, especially the concept of dependency notation which serves as a means to obtain simplified symbols for complex circuits.

    Overview of IEEE Standard 91-1984, Explanation of Logic Symbols by Fred A. Mann, is presented in Appendix A. This overview provides an excellent summary of the graphic symbols for logic functions. Figure 1 in the *Overview* shows the general composition of a symbol, and this composition applies to both distinctive-shape and rectangular-shape symbols. Table I in the *Overview* lists general qualifying symbols, and Table II lists the qualifying symbols for inputs and outputs such as the negation symbol and the polarity symbol.

1-5-1 OR Elements

Either OR element symbol shown in Fig. 1-11 may be used to represent the OR operations in Boolean functions. The OR elements in Fig. 1-11 are shown with no qualifying symbols such as the negation symbol (represented by a small bubble or circle), or the polarity symbol (represented by a small wedge or right triangle) at their inputs or outputs. When we initially draw an OR element we should draw it in this pure form called the outline form. There are two outline forms for the OR element. These are the rectangular-shape form and the distinctive-shape form. The rectangular-shape AND element is the internationally agreed on logic symbol for the AND element, while the distinctive-shape AND element is the logic symbol more commonly used in publication and industry in the United States. The term "element" is used to represent all or part of any type of logic function within a single outline. The term "gate" is a more restrictive term used by manufacturers to describe devices that perform the basic logic functions such as the OR, AND, NOR, and NAND functions. The latter two functions will be discussed later in this chapter.

1-5-2 AND Elements

Either one of the AND element symbols shown in Fig. 1-11 may be used to represent AND operations in our Boolean equations. Like the OR element, there are two outline forms for the AND element. The first outline form shown in Fig. 1-11 represents the distinctive-shape AND element while the second outline form represents the internationally agreed on rectangular-shape AND element. Like the distinctive-shape OR element, the distinctive-shape AND element is used more commonly in the United States.

1-5-3   Qualifying Symbols

Without a method to add qualifying symbols to OR and AND elements drawn in pure outline form, the information contained in the Boolean functions could not always be represented graphically. The concept of symbol construction using a rectangular-shape logic symbol is illustrated in Figure 1 of the *Overview of IEEE Standard 91-1984*, in Appendix A. Double asterisks (**) indicate the preferred position and alternate position for the general qualifying symbols such as ≥ 1 and & when using rectangular-shape symbols for OR and AND elements respectively. What we want to stress about Figure 1 are the possible positions for qualifying symbols relating to inputs and outputs. Notice that input and output qualifying symbols may appear both outside and inside the outline form. The negation qualifying symbol is the small bubble or circle found on the output side (the right side) of the distinctive-shape and the rectangular-shape Inverter elements shown in Fig. 1-11. Negation qualifying symbols can also exist at the inputs and outputs of OR and AND elements as indicated in Figure 1 in the *Overview*, but we will discuss these applications in more detail later. Example 1-3 illustrates other qualifying symbols that can be used with logic elements.

Example 1-3. Read Sections 1 through 3 of the *Overview of IEEE Standard 91-1984* in Appendix A and draw the following qualifying symbols at an input.

(*a*) signal flow from right to left

(*b*) active-low input

(*c*) nonlogic connection such as a power supply connection

(*d*) analog signal on a digital symbol

Solution Each of the above qualifying symbols are shown at an input in Fig. E1-3. Observe that the qualifying symbols are used to designate the physical or logic characteristic of an input or output.

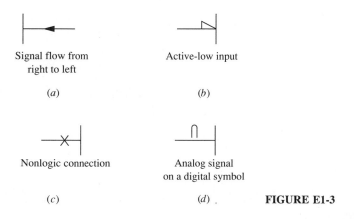

Signal flow from
right to left

(*a*)

Active-low input

(*b*)

Nonlogic connection

(*c*)

Analog signal
on a digital symbol

(*d*) .   **FIGURE E1-3**

When questions arise concerning standards for logic symbols, you can use Appendix A as a brief reference. *A Practical Introduction to the New Logic Symbols* by I. Kampel, and *Using Functional Logic Symbols and ANSI/IEEE Std 91-1984*, by F. A. Mann are listed in the reference section at the end of the chapter as additional references. However, *ANSI/IEEE Std 91-1984* and *IEC STANDARD Publication 617-12* can also be consulted for a more in-depth discussion of logic symbols.

1-5-4 Physical Integrated Circuit Devices

There are many manufacturers that provide physical devices that are capable of carrying out two-value Boolean functions. These devices can contain an integrated circuit (IC) made up of tens, hundreds, or thousands of transistors on a small silicon semiconductor crystal called a die or chip. Since the circuitry contains mainly transistors, diodes, and resistors which are all interconnected inside the chip, power consumption can be quite low and reliability quite high. The die is then welded to a frame as illustrated in Fig. 1-12. Its input and output leads are connected by thin gold wires to the packages leads or pins, the unit is encapsulated using glass, ceramic, or plastic, and then the unit is usually hermetically sealed.

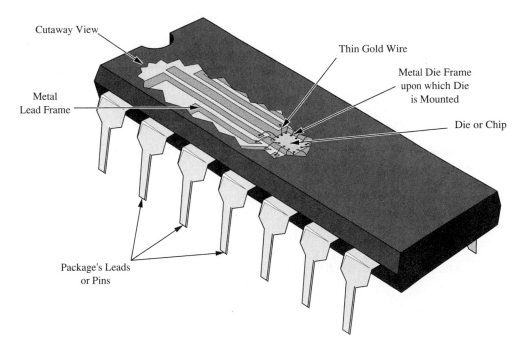

FIGURE 1-12
Cutaway view of an IC package showing the die, die frame, gold wire, lead frame, and package's leads or pins.

(a)   (b)   (c)

(d)   (e)

FIGURE 1-13
Packages for integrated circuits. (a) Dual-in-line package, (b) flat package, (c) surface mounting package, (d) plastic leaded chip carrier package, (e) pin grid array package.

Hermetic sealing guards against die contamination in many different environments including humidity.

Five different types of integrated circuit packages are shown in Fig. 1-13. The package shown in Fig. 1-13a is the very common dual-in-line package. The packages shown in Figs. 1-13b and c are the flat package (flat pack) and the surface mount (small outline) package. These packages are generally used in applications in which real estate on a printed circuit (pc) board is critical and/or a lower cost must be achieved for high volume application. The packages shown in Figs. 1-13d and e are the plastic leaded chip carrier (PLCC) package and the pin grid array (PGA) package which are used for very large IC designs especially when the pin count (i.e., the package inputs and outputs) for the designs become very large.

1-6   BOOLEAN ALGEBRA THEOREMS

A theorem, like a postulate, is a rule concerning a fundamental relationship between Boolean variables. Boolean algebra theorems are basic relationships that indicate that one expression is equivalent to another expression. Theorems can be proved algebraically using the Boolean postulates, that is, Huntington's postulates. They can also be proved using a combination of Boolean postulates with theorems that have already been proved. Theorems may also be proved graphically using a Venn diagram. The easiest and perhaps the most organized method of proving theorems results from using a truth table and the principle of perfect induction.

1-6-1 Principle of Duality

Once a theorem is proven, then by the principle of duality its dual is also proven. What this means is we do not really have to prove the dual of a proven theorem. To obtain the dual of a relationship simply interchange the identity elements (0 and 1) and the binary operators ($+$ and $\cdot$).

1-6-2 Proving Theorems Using the Boolean Postulates

Theorems are useful in that they provide a tool for mathematical manipulation similar to ordinary algebra. During the study of ordinary algebra, expressions were often presented, and you were asked to find a simpler expression. Theorems in Boolean algebra are what we use to simplify Boolean expressions. As we have shown, simpler expressions result in simpler circuits.

A few of the more useful theorems of Boolean algebra are listed in the table shown in Fig. 1-14. Our list is by no means exhaustive. Each theorem is presented

T1: $\overline{\overline{X}} = X$	Double Complementation or Double Negation Theorem
T2a: $X + X = X$ T2b: $X \cdot X = X$	Idempotency Theorem
T3a: $X + 1 = 1$ T3b: $X \cdot 0 = 0$	Identity Element Theorem
T4a: $X + X \cdot Y = X$ T4b: $X \cdot (X + Y) = X$	Absorption Theorem
T5a: $X + (Y + Z) = (X + Y) + Z$ T5b: $X \cdot (Y \cdot Z) = (X \cdot Y) \cdot Z$	Associative Theorem
T6a: $\overline{X + Y} = \overline{X} \cdot \overline{Y}$ T6b: $\overline{X \cdot Y} = \overline{X} + \overline{Y}$	DeMorgan's Theorem
T7a: $X \cdot Y + X \cdot \overline{Y} = X$ T7b: $(X + Y) \cdot (X + \overline{Y}) = X$	Adjacency Theorem
T8a: $X \cdot Y + \overline{X} \cdot Z + Y \cdot Z = X \cdot Y + \overline{X} \cdot Z$ T8b: $(X + Y) \cdot (\overline{X} + Z) \cdot (Y + Z) = (X + Y) \cdot (\overline{X} + Z)$	Consensus Theorem
T9a: $X + \overline{X} \cdot Y = X + Y$ T9b: $X \cdot (\overline{X} + Y) = X \cdot Y$	Simplification Theorem

FIGURE 1-14
Boolean algebra theorems.

with its dual, except for the first one which is the dual of itself. Theorems T1 through T3 are single variable theorems while Theorems T4 through T9 have two or more variables. Each of the theorems also has a name, and that name is listed across from the theorem in the table.

To illustrate how to prove theorems we will manipulate Boolean expressions into forms that can utilize the Boolean postulates. This is a purely algebraic approach and requires that you either have a ready reference or have throughly mastered the Boolean postulates, since we must utilize them in every step of a proof. Proving theorems by using the Boolean postulates is a deductive process which means we draw conclusions based on the stated premises.

Theorem T2a. $X + X = X$ Idempotency Theorem

Proof

$$X + X = (X + X) \cdot 1 \qquad \text{by P2b (Huntington's Postulate in Fig. 1-6)}$$
$$= (X + X) \cdot (X + \overline{X}) \qquad \text{by P5a}$$
$$= X + X \cdot \overline{X} \qquad \text{by P4a}$$
$$= X + 0 \qquad \text{by P5b}$$
$$= X \qquad \text{by P2a}$$

Observe in the second line of the proof that $(X + Y) \cdot (X + Z) = (X + X) \cdot (X + \overline{X})$ when $Y = X$ and $Z = \overline{X}$.

If we take the dual of the expression $X + X = X$ by interchanging the elements 0 and 1 (none exits in this case) and the binary operators $+$ and $\cdot$, we obtain the expression $X \cdot X = X$ which is also true by the principle of duality. In fact, every step in the proof of Theorem T2a ($X + X = X$, Idempotency Theorem) may be replaced by its dual thus providing a proof for the dual of Theorem T2a.

Theorem T3a. $X + 1 = 1$ Identity Element Theorem

Proof

$$X + 1 = X + X + \overline{X} \qquad \text{by P5a (Huntington's Postulate in Fig. 1-6)}$$
$$= X + \overline{X} \qquad \text{by T2a (Idempotency Theorem in Fig. 1-14)}$$
$$= 1 \qquad \text{by P5a}$$

The dual of $X + 1 = 1$ is $X \cdot 0 = 0$ is also true by the duality principle.

Theorem T4a. $X + X \cdot Y = X$ Absorption Theorem

Proof

$$X + X \cdot Y = X \cdot 1 + X \cdot Y \qquad \text{by P2b (Huntington's Postulate in Fig. 1-6)}$$

$$= X \cdot (1 + Y) \qquad \text{by P4b (AND is distributive over OR)}$$

$$= X \cdot 1 \qquad \text{by T3a (Identity Element Theorem in Fig. 1-14)}$$

$$= X \qquad \text{by P2b}$$

The dual of $X + X \cdot Y = X$ is $X \cdot (X + Y) = X$ by observing the order of precedence of the binary operators before they are interchanged. Theorem T4b, $X \cdot (X + Y) = X$, is also true.

Notice that the proofs of theorems T3a and T4a: (*a*) use the Boolean postulates, and (*b*) use a theorem that was just previously proved.

Example 1-4. One of most important theorems in Boolean algebra is Theorem T7a which is called the Adjacency Theorem. This theorem is utilized perhaps more than any other theorem in simplifying Boolean expressions. Prove the Adjacency Theorem, Theorem T7a, by the Boolean postulates.

Solution

Theorem T7a. $X \cdot Y + X \cdot \overline{Y} = X$ Adjacency Theorem

Proof

$$X \cdot Y + X \cdot \overline{Y} = X \cdot (Y + \overline{Y}) \qquad \text{by P4b (Huntington's Postulate in Fig. 1-6)}$$

$$= X \cdot 1 \qquad \text{by P5a}$$

$$= X \qquad \text{by P2b}$$

The dual of Theorem T7a, $(X + Y) \cdot (X + \overline{Y}) = X$, is also true by the principle of duality.

1-6-3   Proving Theorems Using Venn Diagrams

For a small number of variables we can also prove the Boolean algebra theorems using a graphical approach called a Venn diagram. The approach is closely connected with the algebra of sets which can be applied to two-valued Boolean algebra if we remember that the variables can only contain two values. A Venn diagram is constructed by first drawing a rectangle or square. The entire area inside the rectangle represents the value 1 (the shaded area) as shown in Fig. 1-15*a*. The expression for the area shaded in the Venn diagram is written below the diagram, and in this case the expression is 1. Figure 1-15*b* shows a Venn diagram with no shaded area, representing the expression 0, as written below the diagram. The procedure is simple. Below each Venn diagram we show an expression. The shaded area in the Venn diagram is a graphical representation of that expression.

An expression that contains a variable such as X is drawn as a circle inside the rectangle of the Venn diagram as shown in Fig. 1-15*c*. The entire area inside the circle represents the expression X when its value is 1 (shaded area).

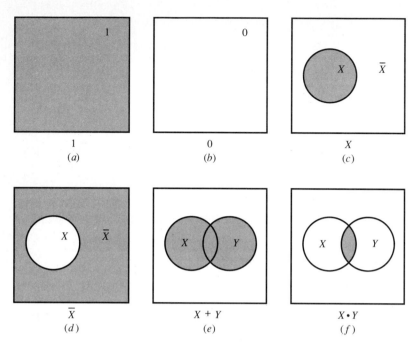

FIGURE 1-15
Venn diagram representations of expressions.

In Fig. 1-15d the expression below the diagram is $\overline{X}$ and the entire area outside the circle is shaded. The area outside the circle represents the expression $\overline{X}$ when its value is 1.

When multiple variables make up an expression, the circles in the rectangle of the Venn diagram must overlap as shown in Fig. 1-15e. For two variables there are 2^2 or 4 different areas to be considered, for three variables there are 2^3 or 8 different areas to be considered and so on for each additional variable. Venn diagrams drawn with circles are generally considered inadequate for more than three variables. The OR operator is used to obtain the combined area of two circles (in set theory this is called the union operator). The AND operator is used to obtain the overlapping area of two circles (in set theory this is called the intersection operator). The resulting OR operation (union) for two overlapping circles, and the resulting AND operation (intersection) for two overlapping circles are illustrated in Figs. 1-15e and f respectively. The expression written below each diagram is represented in the diagram by the shaded area.

Example 1-5. Prove DeMorgan's Theorem, $\overline{X + Y} = \overline{X} \cdot \overline{Y}$, Theorem T6a, utilizing a Venn diagram.

Solution The proof for theorem T6a is shown in Fig. E1-5. It is sometimes easier to draw multiple Venn diagrams (as we have done) to distinguish between various

Theorem T6a: $\overline{X + Y} = \overline{X} \cdot \overline{Y}$ (DeMorgan's Theorem)

Proof:

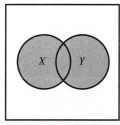

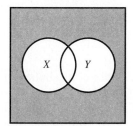

$X + Y$ Union of X, Y
(area in either X OR Y)

$\overline{X + Y}$ Complement of
(union of X, Y)
(area not in either X OR Y)

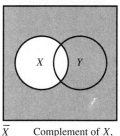

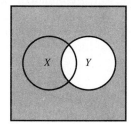

$\overline{X}$ Complement of X,
(area not in X)

$\overline{Y}$ Complement of Y,
(area not in Y)

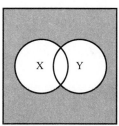

$\overline{X} \cdot \overline{Y}$ Intersection of $\overline{X}$, $\overline{Y}$
(area in $\overline{X}$ AND $\overline{Y}$)

$\therefore$ $\overline{X + Y} = \overline{X} \cdot \overline{Y}$ by graphical observation   **FIGURE E1-5**

areas than to shade the areas of interest on the same diagram. Since a theorem consists of an equality of two expressions, we have generated a separate Venn diagram for each expression as shown in Fig. E1-5.

The dual of Theorem T6a, $\overline{X \cdot Y} = \overline{X} + \overline{Y}$, is also true by the principle of duality.

Venn diagrams are generally used only to prove basic theorems. When a large number of Boolean variables are involved, the diagrams take too long to draw and are difficult to work with. Karnaugh maps (K maps) are rectangular versions of Venn diagrams and are easier to work with. Karnaugh maps are presented in Chapter 3.

1-6-4 Proving Theorems Using Perfect Induction

The format provided by a truth table is a tabular format in which a list is made of all possible combinations of binary values of the Boolean variables involved in the theorem. Recall from Definition 1-3, for n variables there are 2^n possible combinations of binary values. A column in the truth table is generated for each Boolean expression in a theorem using one operator at a time. The columns are filled in with 1s or 0s according to the expression listed in the column headings. Each row of the table is then evaluated using the appropriate operator for the expression. When two columns agree in every respect, we can say that the Boolean expressions for those columns are equal by the principle of perfect induction. The following examples illustrate the process of proving theorems using perfect induction.

Theorem T1. $\overline{\overline{X}} = X$ Double Complementation or Double Negation Theorem

Proof

X	$\overline{X}$	$\overline{\overline{X}}$
0	1	0
1	0	1

therefore, $\overline{\overline{X}} = X$ by perfect induction (column $\overline{\overline{X}}$ is identical to column X)

Theorem T5a. $X + (Y + Z) = (X + Y) + Z$ Associative Theorem

Proof

X	Y	Z	$Y + Z$	$X + (Y + Z)$	$X + Y$	$(X + Y) + Z$
0	0	0	0	0	0	0
0	0	1	1	1	0	1
0	1	0	1	1	1	1
0	1	1	1	1	1	1
1	0	0	0	1	1	1
1	0	1	1	1	1	1
1	1	0	1	1	1	1
1	1	1	1	1	1	1

therefore, $X + (Y + Z) = (X + Y) + Z$ by perfect induction (column $X + (Y + Z)$ is identical to column $(X + Y) + Z$)

Since we proved T5a, its dual T5b, $X \cdot (Y \cdot Z) = (X \cdot Y) \cdot Z$ is also true.

Theorem T8a. $X \cdot Y + \overline{X} \cdot Z + Y \cdot Z = X \cdot Y + \overline{X} \cdot Z$ Consensus Theorem

Proof

X	Y	Z	$X \cdot Y$	$\overline{X}$	$\overline{X} \cdot Z$	$Y \cdot Z$	$X \cdot Y + \overline{X} \cdot Z + Y \cdot Z$	$X \cdot Y + \overline{X} \cdot Z$
0	0	0	0	1	0	0	0	0
0	0	1	0	1	1	0	1	1
0	1	0	0	1	0	0	0	0
0	1	1	0	1	1	1	1	1
1	0	0	0	0	0	0	0	0
1	0	1	0	0	0	0	0	0
1	1	0	1	0	0	0	1	1
1	1	1	1	0	0	1	1	1

therefore,

    $X \cdot Y + \overline{X} \cdot Z + Y \cdot Z = X \cdot Y + \overline{X} \cdot Z$ by perfect induction (column $X \cdot Y + \overline{X} \cdot Z + Y \cdot Z$ is identical to column $X \cdot Y + \overline{X} \cdot Z$)

Theorem T8b, $(X + Y) \cdot (\overline{X} + Z) \cdot (Y + Z) = (X + Y) \cdot (\overline{X} + Z)$, is also true by the principle of duality.

Notice in the proof for Theorem T8a that we needed to know the operator table for three variables ORed together, or we simply OR two variables, remember the result, and then OR the result with the remaining variable. The OR operator table was generated when we proved Theorem T5a, the Associative Theorem. Constructing the AND operator table for more than two variables is left as an exercise for the student.

Example 1-6. Is the expression $X + \overline{X} \cdot Y$ equal to the expression $X + Y$?

Solution If $X + \overline{X} \cdot Y = X + Y$ we should be able to show that their truth table columns are identical and hence the equality holds by perfect induction.

X	Y	$\overline{X}$	$\overline{X} \cdot Y$	$X + \overline{X} \cdot Y$	$X + Y$
0	0	1	0	0	0
0	1	1	1	1	1
1	0	0	0	1	1
1	1	0	0	1	1

Since the columns for $X + \overline{X} \cdot Y$ and $X + Y$ are identical, the two expressions are equal by perfect induction. This proves the Simplification Theorem, Theorem T9a, and, by the principle of duality, its dual $X \cdot (\overline{X} + Y) = X \cdot Y$ is also true.

1-7 MINIMIZING BOOLEAN FUNCTIONS ALGEBRAICALLY

Expressions that contain fewer literals result in circuits that contain fewer logic symbols. Fewer logic symbols means fewer physical parts. Circuits with fewer parts are generally more reliable and cost less than circuits with more parts. This is the motivation for minimizing Boolean functions.

Now that we have proven a number of Boolean algebra theorems, we can use these theorems along with the Boolean postulates to manipulate Boolean functions into a simpler form. A simpler form is a form with fewer literals. To obtain fewer literals we minimize, reduce, or simplify Boolean functions or Boolean equations. In this context the terms all mean the same. As with ordinary algebra, it is often hard to know exactly when you have obtained an absolute minimum form of a function when you reduce an expression algebraically. Beginning in Chapter 3 we will present a more procedure-oriented method called the Karnaugh map method. This is considered the standard paper and pencil method to minimize Boolean functions. When a Karnaugh map is used it is easier to visually recognize patterns that result in minimum expressions. Algebraic manipulation, however, often requires our best ingenuity to obtain minimum expressions.

Consider the following Boolean function.

$$F(X, Y, Z) = \overline{X} \cdot \overline{Y} \cdot \overline{Z} + Y \cdot Z + Y \cdot \overline{Z}$$

Can we reduce this function? The answer is "yes" since $Y \cdot Z + Y \cdot \overline{Z}$ reduces to Y by the Adjacency Theorem, Theorem T7a. $\overline{X} \cdot \overline{Y} \cdot \overline{Z} + Y$ reduces to $\overline{X} \cdot \overline{Z} + Y$ by the Simplification Theorem, Theorem T9a. We know from postulate P3a that the expression $\overline{X} \cdot \overline{Y} \cdot \overline{Z} + Y$ is commutative as are the terms in $\overline{X} \cdot \overline{Y} \cdot \overline{Z}$ so we can write $\overline{X} \cdot \overline{Y} \cdot \overline{Z} + Y = Y + \overline{Y} \cdot (\overline{X} \cdot \overline{Z}) = Y + \overline{X} \cdot \overline{Z} = \overline{X} \cdot \overline{Z} + Y$. Side calculations or regroupings (such as we illustrate here) are often necessary. Arranging expressions and literals in the form that they were presented in Fig. 1-6 and Fig. 1-14 so that they are more easily recognized helps to insure the proper use of the postulates and theorems.

$$F = \overline{X} \cdot \overline{Y} \cdot \overline{Z} + Y \cdot Z + Y \cdot \overline{Z}$$

$$= \overline{X} \cdot \overline{Y} \cdot \overline{Z} + Y \qquad \text{by T7a}$$
(Adjacency Theorem in Fig. 1-14)

$$= \overline{X} \cdot \overline{Z} + Y \qquad \text{by T9a}$$
(Simplification Theorem in Fig. 1-14)

So, $F = \overline{X} \cdot \overline{Z} + Y$ is the same function in a reduced form, that is, a form with fewer literals. Does this represent a minimum form of the function? The answer is yes. This is a very simple function, and since it only has one form for each literal (each input variable is either complemented or uncomplemented), the function is in a minimum form, that is, a form with the smallest number of literals, and therefore cannot be further reduced. When input variables appear in both complemented and uncomplemented form, a function may or may not be reducible.

An example of a function that is not reducible is illustrated by the following expanded function. This function is not reducible since it is already in a minimum form. How do we know? The Adjacency Theorem, Theorem T7a, cannot be applied.

$$F(X, Y, Z) = \overline{X} \cdot \overline{Y} \cdot Z + X \cdot Y \cdot \overline{Z}$$

1-7-1   Minimizing a Function by First Expanding the Function

The following function may appear to be in a minimum form since none of the theorems we presented appear to directly provide a simpler expression. This function, however, is not in a minimum form.

$$F(X, Y, Z) = X \cdot \overline{Z} + X \cdot \overline{Y} \cdot Z + \overline{X} \cdot Z + \overline{X} \cdot Y \cdot \overline{Z}$$

A general technique that is recommended in this case (when there are three or more input variables and the theorems do not appear to provide a simpler expression), is to expand the function such that every product term contains one of the input variables in either its complemented or uncomplemented form (recall that product terms are expressions that contain literals that are only ANDed together). It seems contrary to normal reason to expand a function when trying to reduce it, but remember to be clever when using algebraic reduction. To expand a product term in a function containing a missing variable such as Y, the missing variable is supplied using Postulate P5a, $Y + \overline{Y} = 1$. In the first product term of our function, $X \cdot \overline{Z}$, we observe the input variable Y is missing so we do the following.

$$X \cdot \overline{Z} = X \cdot 1 \cdot \overline{Z} = X \cdot (Y + \overline{Y}) \cdot \overline{Z} = X \cdot Y \cdot \overline{Z} + X \cdot \overline{Y} \cdot \overline{Z}$$

The last two product terms, obtained by applying Postulate P4b, $X \cdot (Y + Z) = X \cdot Y + X \cdot Z$, are then substituted back into the function. Next, product term $\overline{X} \cdot Z$ is expanded in a similar manner to obtain $\overline{X} \cdot Y \cdot Z + \overline{X} \cdot \overline{Y} \cdot Z$. Our function now looks like this in expanded form:

$$F(X, Y, Z) = X \cdot Y \cdot \overline{Z} + X \cdot \overline{Y} \cdot \overline{Z} + X \cdot \overline{Y} \cdot Z + \overline{X} \cdot Y \cdot Z$$
$$+ \overline{X} \cdot \overline{Y} \cdot Z + \overline{X} \cdot Y \cdot \overline{Z}$$

This expanded form of a function is also known as the canonical or standard sum of products form (standard SOP form). This form gives us a unique starting point for reduction. Standard forms will be discussed in depth in the next section.

An expanded form will obviously reduce, but the question is, can we use the expanded form to reduce the function to a minimum form?

$$F(X, Y, Z) = X \cdot Y \cdot \overline{Z} + X \cdot \overline{Y} \cdot \overline{Z} + X \cdot \overline{Y} \cdot Z + \overline{X} \cdot Y \cdot Z$$

$$+ \overline{X} \cdot \overline{Y} \cdot Z + \overline{X} \cdot Y \cdot \overline{Z}$$

$$= Y \cdot \overline{Z} + X \cdot \overline{Y} \cdot \overline{Z} + X \cdot \overline{Y} \cdot Z + \overline{X} \cdot Y \cdot Z$$

$$+ \overline{X} \cdot \overline{Y} \cdot Z$$ by T7a

(i.e., $X \cdot Y \cdot \overline{Z} + \overline{X} \cdot Y \cdot \overline{Z} = Y \cdot \overline{Z}$ (Adjacency

Theorem

since $A \cdot B + A \cdot \overline{B} = A$ in Fig. 1-14)

for $A = Y \cdot \overline{Z}, B = X$)

$$= Y \cdot \overline{Z} + X \cdot \overline{Y} + \overline{X} \cdot Y \cdot Z + \overline{X} \cdot \overline{Y} \cdot Z$$ by T7a

(i.e., $X \cdot \overline{Y} \cdot \overline{Z} + X \cdot \overline{Y} \cdot Z = X \cdot \overline{Y}$)

$$= Y \cdot \overline{Z} + X \cdot \overline{Y} + \overline{X} \cdot Z$$ by T7a

(i.e., $\overline{X} \cdot Y \cdot Z + \overline{X} \cdot \overline{Y} \cdot Z = \overline{X} \cdot Z$)

So, $F(X, Y, Z) = Y \cdot \overline{Z} + X \cdot \overline{Y} + \overline{X} \cdot Z$, and is a minimum form.

By first expanding a function to its standard sum of products form we can utilize the Adjacency Theorem, Theorem T7a, repeatedly and thus reduce a function in an organized fashion using Boolean algebra. Before we expanded the original function, it was paired off a different way and thus did not appear to be easily reducible. Although algebraic reduction can always be used to obtain a minimum function, it is not easy to tell when you have reached a minimum. Two organized methods exist for selecting product terms to obtain a minimum function. These are the Karnaugh map method which we mentioned earlier and the Quine-McCluskey tabular method which will be presented in Chapter 4. Both of these methods are more popular than the pure algebraic reduction method.

One final example is in order, and that example is to minimize a function expressed in a product of sums form such as the following function.

$$F(A, B, C) = (A + B + \overline{C}) \cdot (\overline{B} + \overline{C})$$

A product of sums form gets its name from the fact that sum terms (recall that sum terms are expressions that contain literals that are only ORed together) are first evaluated and then ANDed together.

We can expand this function into a canonical or standard product of sums form (standard POS form) by providing the missing input variables in each of its sum terms. To expand a sum term in a function containing a missing variable such as A, the missing variable is supplied using Postulate P5b, $A \cdot \overline{A} = 0$. For example the sum term $(\overline{B} + \overline{C})$ is missing the A variable which we can supply as follows $(\overline{B} + \overline{C}) = (0 + \overline{B} + \overline{C}) = (A \cdot \overline{A} + \overline{B} + \overline{C}) = (A + \overline{B} + \overline{C}) \cdot (\overline{A} + \overline{B} + \overline{C})$. Notice that Postulate P4a, $X + Y \cdot Z = (X + Y) \cdot (X + Z)$ was also applied. Substituting this back into the function gives us the following canonical or standard product of sums form which we can reduce to a minimum form as follows.

$$F(A, B, C) = (A + B + \overline{C}) \cdot (A + \overline{B} + \overline{C}) \cdot (\overline{A} + \overline{B} + \overline{C})$$

$$= (A + B + \overline{C}) \cdot (A + \overline{B} + \overline{C}) \cdot (A + \overline{B} + C)$$

$$\cdot (\overline{A} + \overline{B} + \overline{C}) \qquad \text{by T2b}$$
$$\text{(Idem. Th.)}$$
$$(\text{i.e., } (A + \overline{B} + \overline{C}) = (A + \overline{B} + \overline{C})$$

$$\cdot (A + \overline{B} + \overline{C}))$$

$$= (A + \overline{C}) \cdot (A + \overline{B} + C) \cdot (\overline{A} + \overline{B} + \overline{C}) \qquad \text{by T7b}$$

$$(\text{i.e., } (A + B + \overline{C}) \cdot (A + \overline{B} + \overline{C}) = (A + \overline{C}) \qquad \text{(Adj. Th.)}$$

$$\text{since } (X + Y) \cdot (X + \overline{Y}) = X$$

$$\text{for } X = A + \overline{C}, \ Y = B)$$

$$= (A + \overline{C}) \cdot (\overline{B} + \overline{C}) \qquad \text{by T7b}$$

$$(\text{i.e., } (A + \overline{B} + \overline{C}) \cdot (\overline{A} + \overline{B} + \overline{C}) = (\overline{B} + \overline{C})) \qquad \text{(Adj. Th.)}$$

$$= \overline{C} \ + \ A \cdot \overline{B} \qquad \text{by P4a}$$

$$(\text{i.e., OR over AND distributive property}$$

$$\text{since } (X + Y) \cdot (X + Z) = X \ + \ Y \cdot Z$$

$$\text{for } X = \overline{C}, \ Y = A, \ Z = \overline{B})$$

Example 1-7. Reduce the following functions algebraically to a minimum form.

(a) $$F(X, Y) = Y \ + \ X \cdot \overline{Y}$$

(b) $$F(X, Y, Z) = \overline{Y} \cdot Z \ + \ \overline{X} \cdot Y \cdot Z$$

Solution

(a) $$F(X, Y) = Y \ + \ X \cdot \overline{Y}$$

$$= Y + X \qquad \text{by T9a}$$
$$\text{(Simp. Th.)}$$

(b) The function $F(X, Y, Z) = \overline{Y} \cdot Z \ + \ \overline{X} \cdot Y \cdot Z$ has three variables, and it is not obvious whether it will reduce or not. We will first expand the function into its standard sum of products form by supplying the missing input variables in the product terms. Then we can proceed to obtain a minimum form for the function.

$$F(X, Y, Z) = \overline{Y} \cdot Z \ + \ \overline{X} \cdot Y \cdot Z$$

$$= (X + \overline{X}) \cdot \overline{Y} \cdot Z \ + \ \overline{X} \cdot Y \cdot Z \qquad \text{by P2b and P5a}$$
$$\text{(Hunt. Post.)}$$

$$= X \cdot \overline{Y} \cdot Z \ + \ \overline{X} \cdot \overline{Y} \cdot Z \ + \ \overline{X} \cdot Y \cdot Z \qquad \text{by P4b}$$
$$(\cdot \text{ Dis. over } +)$$

$$= X \cdot \overline{Y} \cdot Z \ + \ \overline{X} \cdot \overline{Y} \cdot Z \ + \ \overline{X} \cdot \overline{Y} \cdot Z + \overline{X} \cdot Y \cdot Z \qquad \text{by T2a}$$
$$\text{(Idem. Th.)}$$
$$(\text{i.e., } \overline{X} \cdot \overline{Y} \cdot Z = \overline{X} \cdot \overline{Y} \cdot Z \ + \ \overline{X} \cdot \overline{Y} \cdot Z)$$

$$= \bar{Y} \cdot Z + \bar{X} \cdot \bar{Y} \cdot Z + \bar{X} \cdot Y \cdot Z \qquad \text{by T7a}$$
(Adj. Th.)

$$= \bar{Y} \cdot Z + \bar{X} \cdot Z \qquad \text{by T7a}$$

$$= (\bar{Y} + \bar{X}) \cdot Z \qquad \text{by P4b}$$
(• Dis. over +)

1-8 CANONICAL OR STANDARD FORMS FOR BOOLEAN FUNCTIONS

There are two unique representations for each Boolean function. One of these representations is called the canonical or standard sum of products form, and the other is called the canonical or standard product of sums form. The reason we present these forms is because functional comparisons can be made using these forms, and, in addition, many systematic minimization techniques use these forms. We shall also see that these two standard forms provide us with the most compact notation for specifying switching functions. The following discussion assumes that the students have had an introduction to the binary number system and understand simple conversions between the binary and the decimal number systems. If this is not the case, it would be better to cover the introductory material in Chapter 2 prior to presenting the material in this section.

1-8-1 Standard Sum of Products Form

First consider the problem. We are given a table of combinations such as the one shown in Fig. 1-16a which has three independent input binary variables X, Y, and Z, and one dependent output variable F. Our goal is to obtain an equation relating the output variable F to the input variables X, Y, and Z. This is generally referred to as finding the Boolean function to represent this truth table or any similar truth table. The particular unique function we want to find at this time is the standard sum of products form. A simple sketch illustrating the problem can be generated by utilizing a block diagram. Fig. 1-16b shows the input and output variables in the problem.

To obtain a standard sum of products form (standard SOP form) means to obtain a function that uses standard product terms. A standard product or standard

X	Y	Z	F
0	0	0	0
0	0	1	1
0	1	0	1
0	1	1	0
1	0	0	1
1	0	1	0
1	1	0	0
1	1	1	0

(a)

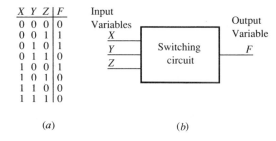

(b)

FIGURE 1-16
(a) Truth table for a problem,
(b) block diagram illustrating the independent input variables and the dependent output variable.

product term is the only product expression that contains all the independent input variables either in uncomplemented or complemented form. The variables in a standard product term represent a single row in a truth table such that on evaluating the expression for that row its value is 1. In order to satisfy this definition of a standard product term, an input variable is complemented when its value is 0 and uncomplemented when its value is 1. In Fig. 1-17 we show the standard product terms for the truth table shown in Fig. 1-16a. Observe that each standard product term contains all three of the independent binary variables X, Y, and Z in either their uncomplemented or complemented form. If we evaluate only one row in the truth table (any row will do) such as the row where $X = 0$, $Y = 1$, and $Z = 1$, the standard product term for that row, $\overline{X} \cdot Y \cdot Z$, evaluates to 1, and all other standard product terms for combination $XYZ = 011$ evaluate to 0. Since a minimum number of the product terms, namely 1, evaluates to 1 for any one of the 2^n possible combinations of n independent variables, standard product terms are called minterms and designated by a small m. The standard product term $\overline{X} \cdot Y \cdot Z$ is minterm 011 in binary or minterm 3 in decimal and is normally written m_3 as shown under the minterm designators column in Fig. 1-17.

It is particularly important that the number associated with the minterm is formed in the same order as the independent variables are listed in the truth table. Minterm 3 is a formulation of the binary values of the independent variables in the order $X = 0$, $Y = 1$, and $Z = 1$ as they are listed in the truth table in Fig. 1-17. If the independent variables were listed in the order $Y = 1$, $X = 0$, $Z = 1$ then the minterm for this row of the table would be $Y \cdot \overline{X} \cdot Z$, and its minterm designation would now be m_5.

Also shown in the table of combinations in Fig. 1-17 is a column where the characteristic numbers f_j are listed for switching function F. The subscripts associated with the characteristic numbers are the same as the numbers associated with the minterms in the table.

X	Y	Z	Standard product terms (minterms)	Minterm designators	Characteristic numbers for switching function F
0	0	0	$\overline{X} \cdot \overline{Y} \cdot \overline{Z}$	m_0	$f_0 = 0$
0	0	1	$\overline{X} \cdot \overline{Y} \cdot Z$	m_1	$f_1 = 1$
0	1	0	$\overline{X} \cdot Y \cdot \overline{Z}$	m_2	$f_2 = 1$
0	1	1	$\overline{X} \cdot Y \cdot Z$	m_3	$f_3 = 0$
1	0	0	$X \cdot \overline{Y} \cdot \overline{Z}$	m_4	$f_4 = 1$
1	0	1	$X \cdot \overline{Y} \cdot Z$	m_5	$f_5 = 0$
1	1	0	$X \cdot Y \cdot \overline{Z}$	m_6	$f_6 = 0$
1	1	1	$X \cdot Y \cdot Z$	m_7	$f_7 = 0$

FIGURE 1-17
Minterms, minterm designators, and characteristic numbers for switching function F for three independent binary variables.

Using the minterm designators and the characteristic numbers f_j for the switching function F, we can now write the general form for a standard sum of products for a function of three independent binary variables.

$$F(X, Y, Z) = f_0 \bullet m_0 \;+\; f_1 \bullet m_1 \;+\; f_2 \bullet m_2 \;+\; f_3 \bullet m_3 \;+\; f_4 \bullet m_4 \;+\; f_5 \bullet m_5$$
$$+\; f_6 \bullet m_6 \;+\; f_7 \bullet m_7 \tag{1-1}$$

Observe that the general form consists of each characteristic number ANDed with its corresponding minterm. These products in turn are all ORed (logically summed). Substituting the constant values for the characteristic numbers assigned to the function F in Fig. 1-17 provides us with the traditional or utility form of the standard sum of products.

$$F(X, Y, Z) = 0 \bullet m_0 \;+\; 1 \bullet m_1 \;+\; 1 \bullet m_2 \;+\; 0 \bullet m_3 \;+\; 1 \bullet m_4 \;+\; 0 \bullet m_5$$
$$+\; 0 \bullet m_6 \;+\; 0 \bullet m_7$$
$$F(X, Y, Z) = 0 \;+\; 1 \bullet m_1 \;+\; 1 \bullet m_2 \;+\; 0 \;+\; 1 \bullet m_4 \;+\; 0 \;+\; 0 \;+\; 0$$
$$F(X, Y, Z) = 1 \bullet m_1 \;+\; 1 \bullet m_2 \;+\; 1 \bullet m_4$$
$$F(X, Y, Z) = m_1 + m_2 + m_4$$

Since the expression on the right side of the last equation is formed from only standard product terms (minterms) for which the corresponding characteristic numbers of the function are 1, we know that the function will evaluate to 1 if and only if m_1 OR m_2 OR m_4 evaluates to 1; otherwise, the function will evaluate to 0. Since only one standard SOP form like this can be written for any particular switching function, the form is unique. Any other switching function that has this standard SOP form must represent the same function.

By writing the corresponding standard product term for each minterm designator, the function can be expressed in terms of its independent variables as follows.

$$F(X, Y, Z) = \overline{X} \bullet \overline{Y} \bullet Z \;+\; \overline{X} \bullet Y \bullet \overline{Z} \;+\; X \bullet \overline{Y} \bullet \overline{Z}$$

Any Boolean function written in a standard sum of products form can be written in a more compact form called the minterm compact form or minterm list form as shown in what follows.

$$F(X, Y, Z) = \Sigma\, m(1, 2, 4)$$

The numbers enclosed in parentheses are the decimal numbers of the respective minterms that make up the function. This form leads to the following generalized standard sum of products equation.

$$F = \Sigma\, m(\text{minterm numbers for the 1s of the function } F) \tag{1-2}$$

The standard sum of products form for any function written as a table of combinations (or truth table) can now be obtained in a straightforward manner.

Simply write the minterm compact form using the truth table to compile the list of the minterm numbers where the function is a 1. The minterm compact form may then be expanded to obtain the Boolean function in terms of the independent variables. The minterm compact form is actually just an alternate representation for the truth table of a function.

Example 1-8. Write the expression for the switching function $F(X, Y, Z) = \sum m(0, 2, 4, 7)$ in a standard SOP form in terms of its independent variables.

Solution First we expand the compact form to obtain the following minterm form.

$$F(X, Y, Z) = m_0 + m_2 + m_4 + m_7$$

Next we write the expression for each minterm in terms of the independent variables.

$$F(X, Y, Z) = \bar{X} \cdot \bar{Y} \cdot \bar{Z} \; + \; \bar{X} \cdot Y \cdot \bar{Z} \; + \; X \cdot \bar{Y} \cdot \bar{Z} \; + \; X \cdot Y \cdot Z$$

or

$$F(X, Y, Z) = \bar{X}\,\bar{Y}\bar{Z} + \bar{X}\,Y\bar{Z} + X\,\bar{Y}\bar{Z} + X\,Y\,Z.$$

The latter form is often used when writing Boolean functions since it requires less writing. This form uses implied AND operators in the product terms similar to ordinary algebra.

1-8-2   Standard Product of Sums Form

To obtain a standard product of sums form (standard POS form) means to obtain a function that uses standard sum terms. A standard sum or standard sum term is the only sum expression that contains all the independent variables either in uncomplemented or complemented form. The variables in a standard sum term represent a single row in a truth table such that on evaluating the expression for that row its value is 0. To satisfy this definition of a standard sum term, an input variable is complemented when its value is 1 and uncomplemented when its value is 0. In Fig. 1-18 we repeated the truth table that appeared in Fig. 1-17 with the addition of standard sum terms.

Observe that each standard sum term also contains all three of the independent binary variables X, Y, and Z in either their uncomplemented or complemented form. If we evaluate only one row in the truth table (again any row will do) such as the row where $X = 1$, $Y = 0$, and $Z = 1$, the standard sum term for that row, $\bar{X} + Y + \bar{Z}$, evaluates to 0. All other standard sum terms for combination $X\,Y\,Z = 101$ evaluate to 1. Since a maximum number of the sum terms, that is, all except one, evaluate to 1 for any one of the 2^n possible combinations of n independent variables, standard sum terms are called maxterms and designated by a capital M. The standard sum term $\bar{X} + Y + \bar{Z}$ is maxterm 101 in binary or maxterm 5 in decimal, and is normally written M_5 as shown under the maxterm designators column in Fig. 1-18.

X Y Z	Standard sum terms (maxterms)	Maxterm designators	Standard product terms (minterms)	Minterm designators	Characteristic numbers for switching function F
0 0 0	$X + Y + Z$	M_0	$\bar{X}\cdot\bar{Y}\cdot\bar{Z}$	m_0	$f_0 = 0$
0 0 1	$X + Y + \bar{Z}$	M_1	$\bar{X}\cdot\bar{Y}\cdot Z$	m_1	$f_1 = 1$
0 1 0	$X + \bar{Y} + Z$	M_2	$\bar{X}\cdot Y\cdot\bar{Z}$	m_2	$f_2 = 1$
0 1 1	$X + \bar{Y} + \bar{Z}$	M_3	$\bar{X}\cdot Y\cdot Z$	m_3	$f_3 = 0$
1 0 0	$\bar{X} + Y + Z$	M_4	$X\cdot\bar{Y}\cdot\bar{Z}$	m_4	$f_4 = 1$
1 0 1	$\bar{X} + Y + \bar{Z}$	M_5	$X\cdot\bar{Y}\cdot Z$	m_5	$f_5 = 0$
1 1 0	$\bar{X} + \bar{Y} + Z$	M_6	$X\cdot Y\cdot\bar{Z}$	m_6	$f_6 = 0$
1 1 1	$\bar{X} + \bar{Y} + \bar{Z}$	M_7	$X\cdot Y\cdot Z$	m_7	$f_7 = 0$

FIGURE 1-18
Maxterms, maxterm designators, minterms, minterm designators, and characteristic numbers for switching function F for three independent binary variables.

The order in which the independent variables are listed determines the binary number and the corresponding decimal number associated with the maxterms, minterms, and the characteristic numbers of the function.

Using the maxterm designators and the characteristic numbers f_j of the switching function F we can write the general form for a standard product of sums for a function of three independent binary variables as follows:

$$F(X, Y, Z) = (f_0 + M_0)\cdot(f_1 + M_1)\cdot(f_2 + M_2)\cdot(f_3 + M_3)\cdot(f_4 + M_4)$$
$$\cdot(f_5 + M_5)\cdot(f_6 + M_6)\cdot(f_7 + M_7) \tag{1-3}$$

Observe that the general form consists of each characteristic number ORed with its corresponding maxterm. These sums in turn are all ANDed. Substituting the constant values for the characteristic numbers assigned to the function F in Fig. 1-18 provides us with the traditional or utility form of the standard product of sums.

$$F(X, Y, Z) = (0 + M_0)\cdot(1 + M_1)\cdot(1 + M_2)\cdot(0 + M_3)\cdot(1 + M_4)$$
$$\cdot(0 + M_5)\cdot(0 + M_6)\cdot(0 + M_7)$$
$$F(X, Y, Z) = (0 + M_0)\cdot(1)\cdot(1)\cdot(0 + M_3)\cdot(1)\cdot(0 + M_5)\cdot(0 + M_6)$$
$$\cdot(0 + M_7)$$
$$F(X, Y, Z) = (0 + M_0)\cdot(0 + M_3)\cdot(0 + M_5)\cdot(0 + M_6)\cdot(0 + M_7)$$
$$F(X, Y, Z) = (M_0)\cdot(M_3)\cdot(M_5)\cdot(M_6)\cdot(M_7)$$

Since the expression on the right side of the last equation is formed from only standard sum terms (maxterms) for which the corresponding characteristic numbers of the function are 0, we know that the function will evaluate to 0 if, and only if, M_0 AND M_3 AND M_5 AND M_6 AND M_7 evaluate to 0, otherwise the function will evaluate to 1. This canonical or standard POS form is also unique, since only one form like this can be written for any function, and any function having the same standard POS form must represent the same function.

By writing the corresponding standard sum term for each maxterm designators, the function can be expressed in terms of its independent variables as follows.

$$F(X, Y, Z) = (X + Y + Z) \cdot (X + \overline{Y} + \overline{Z}) \cdot (\overline{X} + Y + \overline{Z}) \cdot (\overline{X} + \overline{Y} + Z)$$
$$\cdot (\overline{X} + \overline{Y} + \overline{Z})$$

Any Boolean function written in a standard product of sums form can also be written in a more compact form. The following maxterm compact form or maxterm list form provides a simpler representation for the function.

$$F(X, Y, Z) = \prod M(0, 3, 5, 6, 7)$$

The numbers enclosed in parentheses are the decimal numbers of the respective maxterms that make up the function. This form leads to the following generalized standard product of sums equation.

$$F = \prod M(\text{maxterm numbers for the 0s of the function } F) \qquad (1\text{-}4)$$

We can now obtain the standard product of sums form for any function written as a table of combinations. Simply write the maxterm compact form, using the truth table to compile the list of the maxterm numbers where the function is a 0. The maxterm compact form may then be expanded to obtain the Boolean function in terms of its independent variables. Like the minterm compact form, the maxterm compact form is an alternate representation for the truth table of a function.

The standard SOP form and the standard POS form of a function are just two different representations for the same function. For functions that are assigned a fixed output value of 0 or 1 for all the combinations specified by the input variables, that is, completely or fully specified functions, these forms are algebraically identical. What we mean by algebraically identical is that one form can be algebraically obtained from the other form. Functions that contain an output value other than 0 or 1 will be discussed in Section 3-6 in Chapter 3.

Example 1-9. Write the minterm and maxterm compact forms (the standard SOP and standard POS forms) for switching function F represented by the truth table shown in Fig. E1-9.

Solution Write the minterm compact form for function F by listing those numbers associated with the minterms in the table where the function is a 1.

$$F(A, B, C, D) = \sum m(0, 4, 6, 8, 9, 14)$$

Write the maxterm compact form for function F by listing those numbers associated with the maxterms in the table where the function is a 0.

$$F(A, B, C, D) = \prod M(1, 2, 3, 5, 7, 10, 11, 12, 13, 15)$$

A	B	C	D	F
0	0	0	0	1
0	0	0	1	0
0	0	1	0	0
0	0	1	1	0
0	1	0	0	1
0	1	0	1	0
0	1	1	0	1
0	1	1	1	0
1	0	0	0	1
1	0	0	1	1
1	0	1	0	0
1	0	1	1	0
1	1	0	0	0
1	1	0	1	0
1	1	1	0	1
1	1	1	1	0

FIGURE E1-9

1-8-3 Converting Between Standard Forms

To convert between the minterm compact form of a function and the maxterm compact form of the same function is a simple procedure. Start by creating a list of all the 2^n possible decimal numbers for the switching function where n is the number of independent variables. The range of decimal numbers is from 0 through $2^n - 1$. Compile a list of all the numbers where the function is a 1 and a separate list of all the numbers where the function is a 0. Next use the appropriate list of minterms or maxterms to obtain the required form. Given either the minterms or the maxterms, it should be simple to find the missing terms, knowing that the total number of minterms and maxterms equals 2^n.

Once the required compact form is obtained, the function can then be expanded and written in terms of the independent variables by writing the expression represented by each minterm or maxterm designator.

Example 1-10. Write the expression for the switching function $F(X, Y, Z) = \sum m(0, 2, 4, 7)$ in a standard POS form in terms of the independent variables.

Solution First, a list of all the possible decimal numbers for three ($n = 3$) independent variables is created as follows: (0,1,2,3,4,5,6,7). Since 0, 2, 4, 7 are the minterm numbers, the maxterm numbers are 1, 3, 5, 6. Using the maxterm compact form we obtain

$$F(X, Y, Z) = \prod M(1, 3, 5, 6)$$

Expanding this we obtain:

$$F(X, Y, Z) = M_1 \cdot M_3 \cdot M_5 \cdot M_6$$

By writing the corresponding standard sum term for each maxterm, the function may now be expressed in terms of the independent variables as follows.

Notice that parentheses must be used around each standard sum term to insure that the AND operators maintain the highest order of precedence.

$$F(X, Y, Z) = (X + Y + \bar{Z}) \cdot (X + \bar{Y} + \bar{Z}) \cdot (\bar{X} + Y + \bar{Z}) \cdot (\bar{X} + \bar{Y} + Z)$$

or

$$F(X, Y, Z) = (X + Y + \bar{Z})(X + \bar{Y} + \bar{Z})(\bar{X} + Y + \bar{Z})(\bar{X} + \bar{Y} + Z)$$

The second form uses implied AND operators as a short hand notation similar to ordinary algebra.

Remember the rule for writing a standard product term so that it evaluates to 1: An input variable is complemented when its value is 0 and uncomplemented when its value is 1. Compare this with the rule for writing a standard sum term so that it evaluates to 0: An input variable is complemented when its value is a 1 and uncomplemented when its value is 0.

For the input $XYZ = 000$ the minterm designator is m_0, and this represents the standard product term $\bar{X} \cdot \bar{Y} \cdot \bar{Z}$. For the same input $XYZ = 000$ its maxterm designator is M_0, representing the standard sum term $X + Y + Z$.

Minterm and maxterm definitions provide us with a relatively easy way to convert between the standard forms, as we shall now show. For any row in the table in Fig. 1-18, complementing the maxterm results in an expression equivalent to the minterm for that row. This can be represented by the following relationship:

$$\overline{M_j} = m_j \tag{1-5}$$

where j is the maxterm or minterm number for any particular row.

When complementing a standard product term, such as $\bar{X} \cdot \bar{Y} \cdot \bar{Z}$, or a standard sum term, such as $X + Y + Z$, the generalized forms for DeMorgan's theorem are useful. DeMorgan's theorem can be applied to three variables by applying it two times. Applying DeMorgan's theorem to n variables requires applying the theorem $n - 1$ times.

The generalized forms for DeMorgan's theorem can be written as follows with the variables $X1, X2, \ldots Xn$, each of which may represent a different expression.

T10a: $\overline{X1 + X2 + X3 + \cdots + Xn} = \overline{X1} \cdot \overline{X2} \cdot \overline{X3} \cdot \cdots \cdot \overline{Xn}$

T10b: $\overline{X1 \cdot X2 \cdot X3 \cdot \cdots \cdot Xn} = \overline{X1} + \overline{X2} + \overline{X3} + \cdots + \overline{Xn}$

These generalized forms are often used to complement an expression that contains two or more variables. These theorems state that a complement of a sum expression or a product expression is obtained by complementing each variable and interchanging the OR and AND operators. Using Theorem T10a or T10b, its dual, the relationship $\overline{M_j} = m_j$ can be easily confirmed for each of the rows in Fig. 1-18.

For example, for maxterm 0 representing the first row in the table, we can write

$$M_0 = X + Y + Z$$

complementing M_0, $\qquad \overline{M_0} = \overline{X + Y + Z}$

applying DeMorgan's theorem, T10a, $\qquad \overline{M_0} = \overline{X + Y + Z} = \overline{X} \cdot \overline{Y} \cdot \overline{Z}$

but, $\qquad m_0 = \overline{X} \cdot \overline{Y} \cdot \overline{Z}$

so, $\qquad \overline{M_0} = m_0$

In Fig. 1-19 we show the truth table for a function F, an Exclusive OR function, and its complement, $\overline{F}$. The following discussion will illustrate the validity of the following relationship.

$$\text{Standard POS form of } F = \overline{\text{Standard SOP form of } \overline{F}} \qquad (1\text{-}6)$$

This relationship states that the standard POS form of a function may be obtained by first obtaining the standard SOP form of the complement of the function, and then complementing.

The standard SOP form for the function $\overline{F}$ in Fig. 1-19 can be written from its truth table as follows.

$$\overline{F} = \sum m(\text{minterm numbers for the 1s of the function } \overline{F}) \qquad (1\text{-}7)$$

$$\overline{F} = \sum m(0, 3) = m_0 + m_3$$

Complementing both sides of this function, applying DeMorgan's theorem, and using the relationship $\overline{M_j} = m_j$ results in

$$F = \overline{m_0 + m_3} = \overline{m_0} \cdot \overline{m_3} = M_0 \cdot M_3$$

Notice this is the standard POS form for the function F in Fig. 1-19 as shown.

$$F = \prod M(\text{maxterm numbers for the 0s of the function } F)$$

$$F = \prod M(0, 3) = M_0 \cdot M_3$$

When obtaining the standard SOP form of the complement of a function such as F, it is not necessary to actually write the column for the complement of F as we did in Fig. 1-19. It is only necessary to remember to use the 0s of the function F to obtain the right side of the equation for the standard SOP form, and then equate the sum of the minterms to the complement of the function F. The procedure of forming a sum of minterms for the 0s of a function and then equating that result to the complement of the function is used quite often when

X	Y	F	$\overline{F}$
0	0	0	1
0	1	1	0
1	0	1	0
1	1	0	1

FIGURE 1-19
Exclusive OR function F and its complement.

working with Karnaugh maps in Chapter 3. Most engineers generally prefer to work with minterms rather than maxterms. This implies solving for the SOP form of a function rather than the POS form of a function. By solving for the standard SOP form of the complement of a function, one has effectively solved for the standard POS form of the function without using maxterms.

Now suppose we obtain the standard POS form for the function $\overline{F}$ in Fig. 1-19.

$$\overline{F} = \prod M(\text{maxterm numbers for the 0s of the function } \overline{F}) \qquad (1\text{-}8)$$

$$\overline{F} = \prod M(1, 2) = M_1 \cdot M_2$$

Complementing both sides of this function, applying DeMorgan's theorem, and using the relationship $\overline{M_j} = m_j$ results in

$$F = \overline{M_1 \cdot M_2} = \overline{M_1} + \overline{M_2} = m_1 + m_2$$

This equation is equivalent to the standard SOP form for the function F. This demonstrates the validity of the following relationship.

$$\text{Standard SOP form of } F = \overline{\text{Standard POS form of } \overline{F}} \qquad (1\text{-}9)$$

This relationship states that the standard SOP form of a function can be obtained by first obtaining the standard POS form of the complement of the function, and then complementing. This relationship is not used very often, but it does show how the standard forms are closely related.

Example 1-11

1. Given the following standard SOP forms of the function $\overline{F}$, obtain the standard POS forms of the function F.
   (a) $\overline{F} = m_3 + m_6$
   (b) $\overline{F} = m_0 + m_4 + m_7$
2. Given the following standard POS forms of the function $\overline{F}$, obtain the standard SOP forms of the function F.
   (a) $\overline{F} = M_2 \cdot M_5 \cdot M_7$
   (b) $\overline{F} = M_0 \cdot M_4 \cdot M_6 \cdot M_7$

Solution

1. (a) $\overline{F} = m_3 + m_6$
   $\quad F = \overline{m_3 + m_6} = \overline{m_3} \cdot \overline{m_6} = M_3 \cdot M_6$
   (b) $\overline{F} = m_0 + m_4 + m_7$
   $\quad F = \overline{m_0 + m_4 + m_7} = \overline{m_0} \cdot \overline{m_4} \cdot \overline{m_7} = M_0 \cdot M_4 \cdot M_7$
2. (a) $\overline{F} = M_2 \cdot M_5 \cdot M_7$
   $\quad F = \overline{M_2 \cdot M_5 \cdot M_7} = \overline{M_2} + \overline{M_5} + \overline{M_7} = m_2 + m_5 + m_7$
   (b) $\overline{F} = M_0 \cdot M_4 \cdot M_6 \cdot M_7$
   $\quad F = \overline{M_0 \cdot M_4 \cdot M_6 \cdot M_7} = \overline{M_0} + \overline{M_4} + \overline{M_6} + \overline{M_7}$
   $\quad\quad = m_0 + m_4 + m_6 + m_7$

When using standard forms we often find that, using various techniques, the expression on the right side of the equation can be reduced to a simpler expression containing fewer literals. Algebraic reduction is one such technique. The Karnaugh map method of reduction in Chapter 3 is the most popular method for a small number of variables (usually five variables or less). The Quine-McCluskey tabular method is still another method of reduction. The Quine-McCluskey method may also be used as a basis for computer reduction programs. We will cover the tabular method in Chapter 4. The list of minterms specified by the standard SOP form of a Boolean function is the starting point for the tabular method.

1-8-4   Comments on Functional Completeness

What device or set of devices does it take to implement any switching function? A device or set of devices that will implement any switching circuit is a functionally complete set. The following functions demonstrate any switching function can be expressed in either of the two basic forms that we have discussed.

Standard SOP forms

Equation E1. $F(X, Y) = \overline{X} \cdot Y \;+\; X \cdot \overline{Y}$

Equation E2. $F(X, Y, Z) = \overline{X} \cdot \overline{Y} \cdot \overline{Z} \;+\; \overline{X} \cdot Y \cdot \overline{Z} \;+\; X \cdot \overline{Y} \cdot \overline{Z} \;+\; X \cdot Y \cdot Z$

Standard POS forms

Equation E3. $F(X, Y) = (\overline{X} + \overline{Y}) \cdot (X + Y)$

Equation E4. $F(X, Y, Z) = (X + Y + \overline{Z}) \cdot (X + \overline{Y} + \overline{Z}) \cdot (\overline{X} + Y + \overline{Z}) \cdot (\overline{X} + \overline{Y} + Z)$

Notice the common operators in Equations E1 through E4. To implement these equations requires hardware devices called OR elements (or gates), AND elements (or gates), and Inverters. The OR elements and AND elements sometimes require more than two inputs as seen by Equations E2 and E4. As we know any switching function can be expressed in the form of a standard sum of products or in the form of a standard product of sums; therefore, any set of hardware devices that can perform all three operations, namely the OR, AND, and complement operations, form a set of functionally complete hardware devices.

A set of OR elements, AND elements, and Inverters is the most obvious functionally complete set of devices. In practice it is possible to realize any switching function with only OR elements and Inverters. Thus this set also represents a functionally complete set of devices. It is also possible to realize any switching function with only AND elements and Inverters; therefore, this set of devices is functionally complete.

Two other popular switching devices are the NOR element (or gate) which is a device that performs the OR operation followed by a complement operation;

and the NAND element (or gate) which is a device that performs the AND operation followed by a complement operation. A set of NOR elements is functionally complete and so is a set of NAND elements. Logic symbols for the NOR element are shown in Figs. 1-20*a*, and the corresponding truth table for a NOR function is shown in Fig. 1-20*b*. Logic symbols for the NAND element are

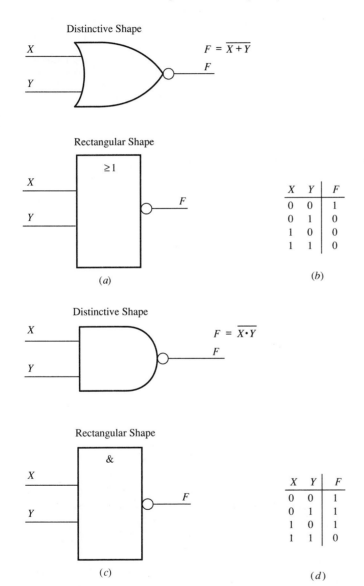

Distinctive Shape

$F = \overline{X+Y}$

Rectangular Shape

X	Y	F
0	0	1
0	1	0
1	0	0
1	1	0

(*a*)   (*b*)

Distinctive Shape

$F = \overline{X \cdot Y}$

Rectangular Shape

X	Y	F
0	0	1
0	1	1
1	0	1
1	1	0

(*c*)   (*d*)

FIGURE 1-20
(*a*) Logic symbols and Boolean equation for the NOR element, (*b*) truth table for the NOR function, (*c*) logic symbols and Boolean equation for the NAND element, and (*d*) truth table for the NAND function.

shown in Figs. 1-20c. The truth table for the NAND function is shown in Fig. 1-20d.

In Fig. 1-21 we list the five functionally complete sets of switching devices that we discussed above and show a basic circuit for each set that will perform the OR operation, the AND operation, and the complement operation. Three dots are

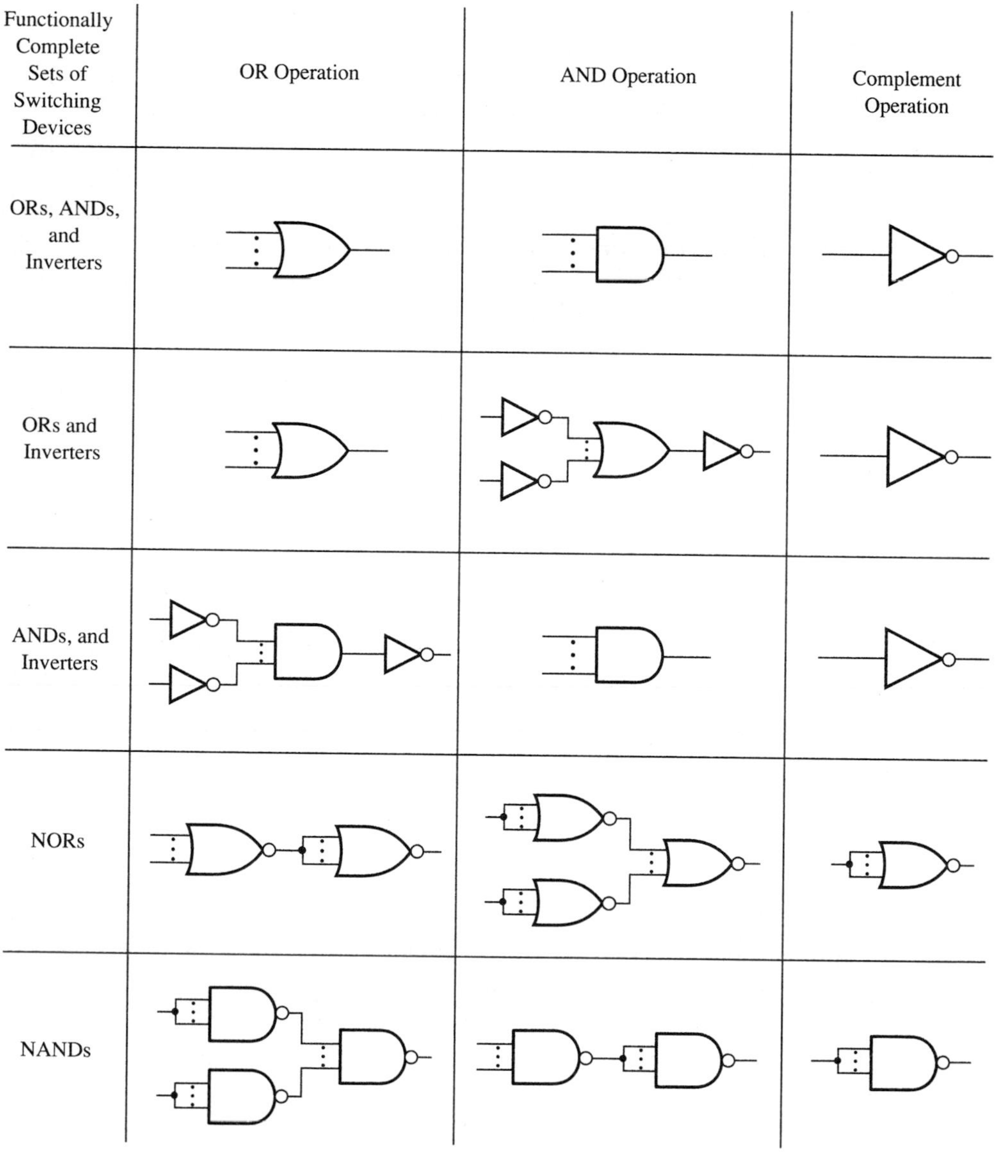

Functionally Complete Sets of Switching Devices	OR Operation	AND Operation	Complement Operation
ORs, ANDs, and Inverters			
ORs and Inverters			
ANDs, and Inverters			
NORs			
NANDs			

FIGURE 1-21
Basic circuits for functionally complete sets of switching devices.

shown between the device symbols with two inputs to indicate that multiple inputs are possible. The circuit for the AND operation using only ORs and Inverters is obtained by first writing the AND function ($A \cdot B$), and then using the postulates and theorems to manipulate the function into a form that uses only OR operations and complement operations, for example, $A \cdot B = \overline{\overline{A} \cdot \overline{B}} = \overline{\overline{A} + \overline{B}}$. The circuit can then be drawn using only ORs and Inverters in place of the OR operations and complement operations respectively. The same procedure also applies to the other elements used in the table, but these are left as an exercise for the student. The function performed by each circuit in Fig. 1-21 can be verified using perfect induction. The practice of using paralleled inputs as illustrated in Fig. 1-21 is logically correct; however, this technique is not the preferred method as discussed in Section 6-3 in Chapter 6.

Example 1-12. Verify by perfect induction that a set of NAND elements may be connected as shown in Fig. 1-21 to produce the OR operation, AND operation, and complement operation, thus representing a functionally complete set.

Solution The simplest case is the complement operation. In Fig. E1-12a we show a truth table with two independent variables X and Y and a column for the NAND operation $\overline{X \cdot Y}$. In Fig. E1-12b we show a subset of the truth table which has a column $X = Y$ which represents the NAND element with a single input, since both inputs (assuming it only has two—the simplest case) are effectively tied together with the column $\overline{X \cdot Y}$ representing its output. Creating a column $X = Y$ and filling it in, we can see that a two-input NAND element performs the Inverter

X	Y	NAND $\overline{X \cdot Y}$
0	0	1
0	1	1
1	0	1
1	1	0

(a)

$\Longrightarrow$

$X = Y$	NAND $\overline{X \cdot Y}$	Complemented input $\overline{X = Y}$
0	1	1
1	0	0

(b)

X	Y	NAND $\overline{X \cdot Y}$	Complemented NAND $\overline{\overline{X \cdot Y}}$	AND $X \cdot Y$
0	0	1	0	0
0	1	1	0	0
1	0	1	0	0
1	1	0	1	1

(c)

X	Y	Complemented inputs $\overline{X}$	$\overline{Y}$	NAND $\overline{\overline{X} \cdot \overline{Y}}$	OR $X + Y$
0	0	1	1	0	0
0	1	1	0	1	1
1	0	0	1	1	1
1	1	0	0	1	1

(d)

FIGURE E1-12

operation since, by perfect induction, column $\overline{X \cdot Y}$ and column $\overline{X = Y}$ are identical.

The next simplest case is the AND operation. To keep our example simple, we again use a truth table with two independent variables X and Y as shown in Fig. E1-12c and a column for the NAND operation $\overline{X \cdot Y}$. Since we want to verify that a complemented NAND results in the AND, we show a column for a complemented NAND $\overline{\overline{X \cdot Y}}$ and a column for the AND $X \cdot Y$. Both columns are equal; therefore, a complemented NAND performs the AND operation by perfect induction.

The last case is the OR operation which is shown in Fig. E1-12d. Again, we use only two variables for simplicity. The column headings indicate that the NAND operation of two complemented inputs $\overline{X}$ and $\overline{Y}$ performs the OR operation by perfect induction.

As we have illustrated, a set of NAND elements represents a functionally complete set.

A functionally complete set of hardware devices can be used to implement any switching circuit. This means that a complete personal computer or mainframe computer could be constructed from only NOR gates or from only NAND gates. While this idea is certainly feasible from a logic point of view, with the massive number of integrated circuits, (ranging from small-scale integration (SSI), through medium-scale integration (MSI), and large-scale integration (LSI), to very-large-scale integration (VLSI)) it is highly improbable that a designer would ever use this strategy in a modern design. Considerations such as power consumption, speed, physical size for the overall design, and of course competitive pricing, prohibit such an approach. The integrated circuits we mentioned above are classified in general according to their complexity. This broad classification is illustrated in Fig. 1-22. The classification is based on the number of gate-equivalent circuits in a single IC package. The term gate-equivalent circuits means the number of individual logic gates that would need to be connected to perform the same function.

Integrated Circuit Classification	Gate equivalent circuits
Small-Scale Integration (SSI)	Less than 12
Medium-Scale Integration (MSI)	12 to 99
Large-Scale Integration (LSI)	100 or more
Very Large-Scale Integration (VLSI)	Much more complex than LSI (sometimes taken as 1000 or more)

FIGURE 1-22
The gate equivalent circuit classification of integrated circuits.

1-9   SPECIFYING DESIGNS USING LOGIC DESCRIPTIONS

In a mathematical sense we tend to use the terms independent variables (or independent input variables) and dependent variables (or dependent output variables) when referring to mathematical functions. From an analog circuit or digital circuit perspective, it is more common to refer to these variables as signals, since signals are things that convey information, notice, or action.

When we talk about a logical description we are referring to a picture or an account of the digital nature of a circuit, not the actual circuit or logic diagram, but a logic description which may lead to the circuit. Since there is no definite rule comprising a logic description or specification, we will present some guidelines. A logic description often starts with an idea which when written down may be referred to as a language statement. The language statement may be considered a rough sketch or first cut at a logic description.

From a language statement in the form of either a mental conception or a written specification, five basic types of logic descriptions can evolve as illustrated in Fig. 1-23. These logic descriptions are (1) truth tables including minterm or maxterm compact forms, (2) logic statements of the type "if then else," (3) Karnaugh maps, (4) timing diagrams, and (5) state diagrams. The last two are more often utilized in sequential designs, while the first three are used in both combinational and sequential designs. Most designs probably originate from logic statements; however, all of our basic definitions for the functions OR, AND, complement, NOR, and NAND were provided using a truth table. Boolean functions and their corresponding logic diagrams may then be obtained using one of these five basic logic descriptions. We will only present logic descriptions for the truth table and logic statement at this time. Logic descriptions for Karnaugh maps are presented in Chapter 3. Logic descriptions for timing diagrams and state diagrams will be presented as appropriate when we discuss sequential logic design.

For now we will only consider the following two logic descriptions and the Boolean functions that result from these descriptions: the truth table and the logic statement.

1-9-1   The Language Statement Description

Consider a switching circuit that has three independent input signals. These signals (being two-valued) provide eight different combinations. Design the circuit to detect only those combinations that result in a binary number that represents a power of two.

The language statement for a problem is often open-ended and may not be exact enough to fully describe all the possibilities associated with the problem. One of the five logic descriptions illustrated in Fig. 1-23 usually fills this gap and provides an exact description of the problem.

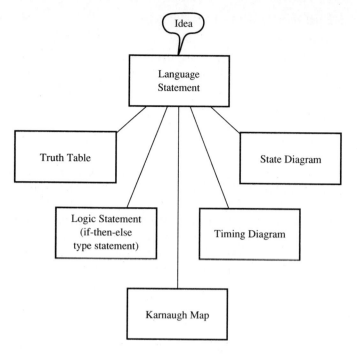

FIGURE 1-23
Evolution of a logic description from an idea.

1-9-2 The Truth Table Description

A truth table description of the problem is illustrated in Fig. 1-24 for input signals
A, B, and, *C* and output signal *F*. Observe that the output signal *F* is only 1 when
one of the input combinations results in a binary number that represents a power
of two—namely the combinations 001, 010, and 100. The truth table is perhaps
the most organized form of logic description since it is a tabular representation
of the language statement.

From the truth table description shown in Fig. 1-24 we can easily write
the standard sum of products form and the standard product of sums form of the
function as follows.

$$F(A,B,C) = \sum m(1,2,4)$$

$$= m_1 + m_2 + m_4$$

$$= \overline{A}\cdot\overline{B}\cdot C \;+\; \overline{A}\cdot B\cdot\overline{C} \;+\; A\cdot\overline{B}\cdot\overline{C}$$

$$F(A,B,C) = \prod M(0,3,5,6,7)$$

$$= M_0\cdot M_3\cdot M_5\cdot M_6\cdot M_7$$

$$= (A + B + C) \cdot (A + \overline{B} + \overline{C}) \cdot (\overline{A} + B + \overline{C}) \cdot (\overline{A} + \overline{B} + C)$$
$$\cdot (\overline{A} + \overline{B} + \overline{C})$$

Or using the 0s of the function F, i.e., the 1s of $\overline{F}$, we can write

$$\overline{F}(A,B,C) = \sum m(0,3,5,6,7)$$
$$= m_0 + m_3 + m_5 + m_6 + m_7$$
$$= \overline{A} \cdot \overline{B} \cdot \overline{C} + \overline{A} \cdot B \cdot C + A \cdot \overline{B} \cdot C + A \cdot B \cdot \overline{C} + A \cdot B \cdot C$$

1-9-3   The Logic Statement Description

A logic statement of the problem may be written as follows. If minterm 1 or minterm 2 or minterm 4 is true then F is true, else F is not true. What this logic description represents is quite similar to a Pascal computer programming if-then-else statement which can be written "If ((NOT A AND NOT B AND C) OR (NOT A AND B AND NOT C) OR (A AND NOT B AND NOT C)) then $F = 1$, else $F = 0$."

From the logic statement description when the logic expression after the "if" evaluates to 1, the statement after the "then" is carried out. For all other cases the statement after the "else" is carried out.

We can write a Boolean function by equating the logic expression after the "if" part of the statement to the variable identifier (F) after the "then" part of the statement.

$$F(A,B,C) = \overline{A} \cdot \overline{B} \cdot C + \overline{A} \cdot B \cdot \overline{C} + A \cdot \overline{B} \cdot \overline{C}$$

As we illustrated in Section 1-4, once we obtain a Boolean function or equation we can obtain a circuit implementation or diagram.

In practically all of our examples we have used variables or signals such as X, Y, Z, A, B, C, and F without specifying clearly what these letters represent. A letter, or in general a name, represented by a signal should provide some indication of the process being performed in order to convey the best possible meaning. In the early stage of a design, letters are often used for compactness

A	B	C	F
0	0	0	0
0	0	1	1
0	1	0	1
0	1	1	0
1	0	0	1
1	0	1	0
1	1	0	0
1	1	1	0

FIGURE 1-24
Truth table description.

and simplicity; however, to make a circuit implementation user-friendly, these letters are almost always changed to meaningful names that are associated with the process being performed. The names provided for signals are similar to variable identifiers assigned by a programmer. The more meaningful the identifier, the easier the program is to understand. Likewise the more meaningful the signal name, the easier the logic diagram is to understand. Using meaningful signal names (abbreviations or mnemonic phrases) that tend to indicate what is actually happening in a circuit is usually more helpful to the designer.

Example 1-13

1. Using meaningful signal names, write the Boolean function for the following design specification: If either the driver's door or the passenger's door of a two-door car is opened, or a switch on the dash is turned on, then the overhead light in the car will turn on; else, the overhead light will not turn on.
2. Show a block diagram and label the signals on the diagram.

Solution

1. First rewrite the logic statement in a computer programming if-then-else form, using abbreviated signal names rather than the statements they represent. If (DRIV_DOOR_OPEN OR PASS_DOOR_OPEN OR SWITCH_TURNED) then LIGHT_ON = 1 else LIGHT_ON = 0. Now we can write the function as

$$LIGHT\_ON = DRIV\_DOOR\_OPEN + PASS\_DOOR\_OPEN$$

$$+ SWITCH\_TURNED$$

2. Figure E1-13 shows the block diagram labeled with the input signals DRIV_DOOR_OPEN, PASS_DOOR_OPEN, and SWITCH_TURNED, and the output signal labeled LIGHT_ON.

Logic statements are written in whatever logic sequence best describes the function being performed on the signals. Canonical or standard forms therefore do not usually evolve from logic statement descriptions.

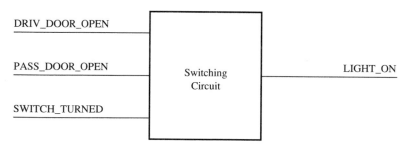

FIGURE E1-13

1-10   NUMBER OF DIFFERENT FUNCTIONS FOR n INDEPENDENT VARIABLES

For n independent input variables, how may different output functions are possible? For $n = 1$, four functions are possible, as shown in Table 1-1.

The functions $F0 = 0$ (all 0s case) and $F3 = 1$ (all 1s case) are trivial cases. The function $F1 = \overline{X}$ is the Inverter function, while the function $F2 = X$ is the Buffer function. You might also notice that $F0 = \overline{F3}$ and $F1 = \overline{F2}$, so half of the functions are complements of the other half.

For $n = 2$, 16 functions are possible as illustrated in Table 1-2.

Like the functions in Table 1-1, half of the functions in Table 1-2 are complements of the other half. The more useful functions are available as off-the-shelf packaged IC devices. The function $F1 = \overline{X + Y}$ is the NOR function. Its complement is function $F14$ which is the OR function. The function $F7 = \overline{X \cdot Y}$ is the NAND function, and its complement $F8$ is the AND function. The function $F6 = X \oplus Y$ is the Exclusive OR function, and its complement $F9$ is the Exclusive NOR function (Equivalence function). Function $F2 = \overline{X} \cdot Y$ and its complement $F13$, and $F4 = X \cdot \overline{Y}$ and its complement $F11$ are functions of two variables with one of the variables complemented. Function $F3 = \overline{X}$ is the Inverter function, and its complement $F12$ is the Buffer function. Function $F5 = \overline{Y}$ and its complement $F10$, as well as $F3$ and $F12$, are each single variable functions. And finally there are the trivial functions, $F0 = 0$ and its complement $F15$. Figure 1-25 represents a summary of the useful functions in Tables 1-1 and 1-2. Both distinctive-shape and rectangular-shape graphic symbols are presented in Fig. 1-25 as a ready reference.

Since there are 2^n rows for n independent variables in a truth table and each row can have a value of 1 or 0, the number of different functions may be represented as 2^{2^n}. For $n = 1$, $2^{2^1} = 2^2 = 4$, and for $n = 2$, $2^{2^2} = 2^4 = 16$. For $n = 3$, the number of different functions is 256, while for $n = 4$, the number of different functions is 65,536. This clearly shows that listing the different functions just for three, and especially for four independent input variables, would be quite a task, even when considering that half of the functions are complements of the other half.

Rather than trying to classify the myriad of possible functions for three or more independent variables, we will concentrate our efforts in Chapter 3 on minimizing Boolean functions using Karnaugh maps. After we present how this is accomplished, we will show in the following chapters how to implement minimized functions using OR, AND, NOR, and NAND graphic symbols.

TABLE 1-1
Single variable functions

X	$F0$	$F1$	$F2$	$F3$
0	0	1	0	1
1	0	0	1	1

TABLE 1-2
Two variable functions

X	*Y*	*F0*	*F1*	*F2*	*F3*	*F4*	*F5*	*F6*	*F7*	*F8*	*F9*	*F10*	*F11*	*F12*	*F13*	*F14*	*F15*
0	0	0	1	0	1	0	1	0	1	0	1	0	1	0	1	0	1
0	1	0	0	1	1	0	0	1	1	0	0	1	1	0	0	1	1
1	0	0	0	0	0	1	1	1	1	0	0	0	0	1	1	1	1
1	1	0	0	0	0	0	0	0	0	1	1	1	1	1	1	1	1

Truth Table	Boolean Function	Function Name	Distinctive-shape Graphic Symbol	Rectangular-shape Graphic Symbol
X Y F 0 0 0 0 1 0 1 0 0 1 1 1	$F = X\cdot Y$	AND	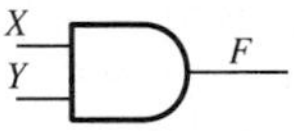	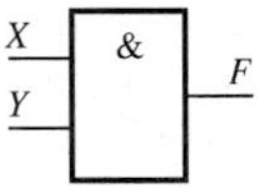
X Y F 0 0 0 0 1 1 1 0 1 1 1 1	$F = X + Y$	OR	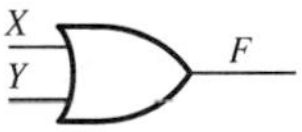	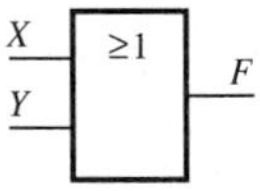
X F 0 0 1 1	$F = X$	Buffer	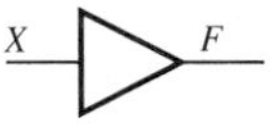	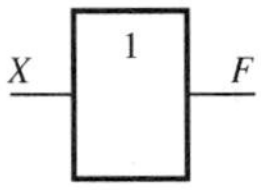
X Y F 0 0 1 0 1 1 1 0 1 1 1 0	$\overline{F} = X\cdot Y$ or $F = \overline{X\cdot Y}$	NAND	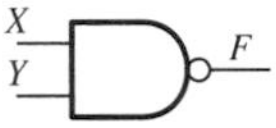	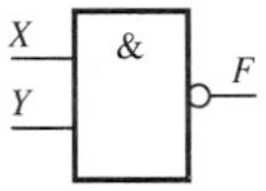
X Y F 0 0 1 0 1 0 1 0 0 1 1 0	$\overline{F} = X + Y$ or $F = \overline{X + Y}$	NOR	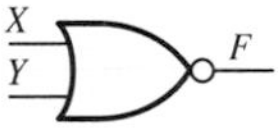	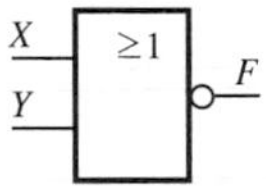
X F 0 1 1 0	$\overline{F} = X$ or $F = \overline{X}$	Inverter	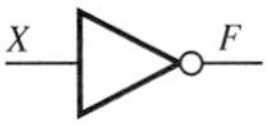	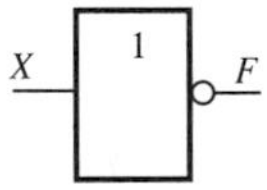
X Y F 0 0 0 0 1 1 1 0 1 1 1 0	$F = \overline{X}\cdot Y + X\cdot\overline{Y}$ or $F = X\oplus Y$	Exclusive OR	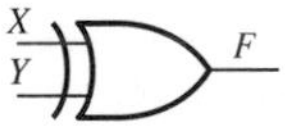	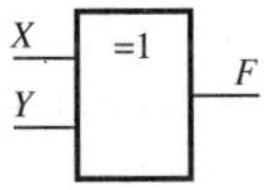
X Y F 0 0 1 0 1 0 1 0 0 1 1 1	$F = \overline{X}\cdot\overline{Y} + X\cdot Y$ or $F = \overline{X\oplus Y}$ or $F = X\odot Y$	Exclusive NOR or Equivalence	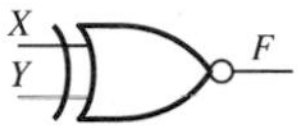	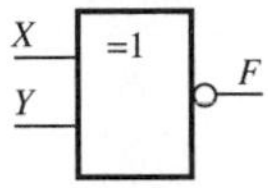

FIGURE 1-25
Summary of useful functions and their corresponding graphic symbols.

REFERENCES

1. Boole, G., *An Investigation of the Laws of Thought*, Dover Pub., New York, 1954 (First published in 1854).
2. Shannon, C. E., "A Symbolic Analysis of Relay and Switching Circuits," *Trans. AIEE*, Vol. 57, pp. 713–23, 1938.
3. Huntington, E. V., "Sets of Independent Postulates for the Algebra of Logic," *Trans. Amer. Math. Soc.*, Vol. 5, pp. 288–309, 1904.
4. Huntington, E. V., "New Sets of Independent Postulates for the Algebra of Logic, with Special Reference to Whitehead and Russell's Principia Mathematica" *Trans. Amer. Math. Soc.*, Vol. 35, pp. 274–304, 1933.
5. ANSI/IEEE Std 91-1984, *IEEE Standard Graphic Symbols for Logic Functions*, The Institute of Electrical and Electronic Engineers, Inc., New York, 1984.
6. IEC STANDARD Publication 617-12, *Graphic Symbols for Diagrams, Part 12: Binary logic elements*, International Electrotechnical Commission, Geneve, Swisse, 1983.
7. Kampel, I., *A Practical Introduction to the New Logic Symbols*, 2d ed., Butterworths, London, 1986.
8. Mann, F. A., *Using Functional Logic Symbols and ANSI/IEEE Std 91-1984*, Texas Instruments Incorporated, Dallas, Texas, 1987.
9. Dietmeyer, D. L., *Logic Design of Digital Systems*, 3d ed., Allyn and Bacon, Inc., Boston, Mass., 1988.
10. Mowle, F. J., *A Systematic Approach to Digital Logic design*, Addison-Wesley Pub. Co., Inc., Reading, Mass., 1976.
11. Mano, M. M., *Digital Design*, Prentice-Hall, Inc., Englewood Cliffs, New Jersey, 1984.
12. Torng, H. C., *Switching Circuits, Theory and Logic Design*, Addison-Wesley Pub. Co., Inc., Reading, Mass., 1972.
13. Edwards, H. E., *The Principles of Switching Circuits*, The M. I. T. Press., Cambridge, Mass., 1973.
14. Matisoo, J., "The superconducting computer," *Sci. Amer.*, Vol. 242, No. 5, pp. 50–65, May 1980.
15. Hill, J. H., Peterson, G. R., *Introduction to Switching Theory & Logic Design*, 3d ed., John Wiley & Sons, Inc., New York, 1981.

PROBLEMS

Section 1-2 A Mathematical Model

1-1. Draw a block diagram for the following switching circuit configurations. Provide symbols for the input and output variables.
   (*a*) 1 inputs and 2 outputs
   (*b*) 2 inputs and 1 output
   (*c*) 3 inputs and 2 outputs
   (*d*) 2 inputs and 4 outputs

1-2. All things being equal except the period of the clock, would a computer (computer 1) operating with a clock period of 83.33 ns be faster or slower than a computer (computer 2) operating with a clock period of 62.5 ns? Determine the frequency of operation of computer 1 and computer 2.

Section 1-3 The Algebra of Logic

1-3. Does Postulate P2a, $X + 0 = X$, hold true in ordinary algebra? Write the dual of P2a. Does the dual hold true in ordinary algebra?

1-4. Does the Postulate P3b, $X \cdot Y = Y \cdot X$, hold true in ordinary algebra? What mathematical property is expressed by P3b?

1-5. Write the rules for converting a postulate or an expression into a dual postulate or expression.

1-6. Postulate P4b, $X \cdot (Y + Z) = X \cdot Y + X \cdot Z$, tells us that logical multiplication is distributive with respect to logical addition. As we know, this principle also holds true in ordinary algebra. Write the dual of P4b. Is addition distributive with respect to multiplication in ordinary algebra?

1-7. Write Postulates P5a and P5b using each of the different types of complement operators.

1-8. Write the definition for the AND operator in truth table format using input variables A and B.

1-9. Write the truth table for the definition of the complement operator using input variable Z.

1-10. Write a complete set of equalities using only 1s and 0s that define the OR operator.

1-11. List the hierarchal order of the Boolean operators in Huntington's postulates including the complement operation. Evaluate the following expressions for $X = 1$, $Y = 0$, and $Z = 1$.
 (*a*) $X + X \cdot Y$
 (*b*) $\sim(\sim X + Y)$
 (*c*) $X' + Y + Z \cdot Y'$

Section 1-4   Digital Logic Functions

1-12. Draw a switching circuit using switches that will perform each of the following basic operations.
 (*a*) OR operation
 (*b*) AND operation
 (*c*) complement operation

1-13. Draw a switching circuit using logic symbols that will perform each of the following basic operations.
 (*a*) OR operation
 (*b*) AND operation
 (*c*) complement operation

1-14. Design a switching circuit using switches that will perform each of the following functions.
 (*a*) $F(X, Y, Z) = X + X \cdot Y \cdot Z$
 (*b*) $F(X, Y, Z) = X \cdot (X + Y + Z) + \overline{X}$
 (*c*) $F(X, Y, Z) = \overline{X} + X \cdot Y + Z$

1-15. Show a circuit implementation for the following functions using OR, AND, and Inverter logic symbols.
 (*a*) $F(X, Y, Z) = X + X \cdot Y \cdot Z$
 (*b*) $F(X, Y, Z) = X \cdot (X + Y + Z) + \overline{X}$
 (*c*) $F(X, Y, Z) = \overline{X} + X \cdot Y + Z$

1-16. Analyze the switching circuits shown in Fig. P1-16 and obtain the logic function for each circuit.

1-17. Analyze the switching circuits shown in Fig. P1-17 and obtain the logic function for each circuit.

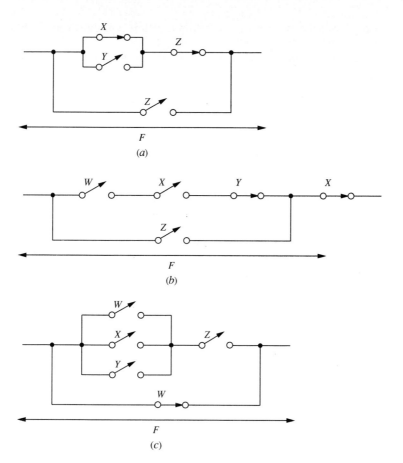

FIGURE P1-16

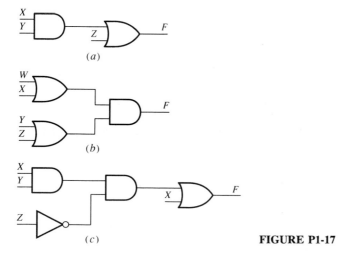

FIGURE P1-17

1-18. Draw a circuit for each switching circuit shown in Fig. P1-16 using logic symbols instead of switches.

1-19. Draw a truth table for each of the following functions and evaluate the output F for each row of the table for all possible combinations of the input variables.
(*a*) $F(X, Y) = X + X \cdot Y$
(*b*) $F(X, Y, Z) = X \cdot Y + X \cdot Z$
(*c*) $F(X, Y, Z) = X \cdot Y \cdot \overline{Z} + \overline{X} \cdot Y + Z$

Section 1-5   Introduction to Logic Symbols

1-20. Draw the logic symbols for two-, three-, and four-input AND elements.
(*a*) using distinctive-shape symbols
(*b*) using rectangular-shape symbols

1-21. Draw the logic symbols for two- three- and four-input OR elements.
(*a*) using distinctive-shape symbols
(*b*) using rectangular-shape symbols

1-22. Show the qualifying symbols for the following inputs or outputs.
(*a*) logic negation at an input
(*b*) bidirectional signal flow
(*c*) enable input
(*d*) output with more than usual output current capability

Section 1-6   Boolean Algebra Theorems

1-23. Prove the Idempotency Theorem, $X \cdot X = X$, using Boolean postulates.

1-24. Prove the Identity element Theorem, $X \cdot 0 = 0$, using Boolean postulates.

1-25. Prove the Absorption Theorem, $X \cdot (X + Y) = X$, using Boolean postulates.

1-26. Prove the DeMorgan's Theorem, $\overline{X \cdot Y} = \overline{X} + \overline{Y}$, using Boolean postulates. (Hint: Use proposition P5a and b to show that $W + \overline{W} = 1$, and $W \cdot \overline{W} = 0$ each hold true, for $W = X \cdot Y$, and $\overline{W} = \overline{X} + \overline{Y}$.)

1-27. Prove the Associative Theorem, $X + (Y + Z) = (X + Y) + Z$ using Venn diagrams.

1-28. Prove the Absorption Theorem, $X + X \cdot Y = X$, using Venn diagrams.

1-29. Prove the Adjacency Theorem, $(X + Y) \cdot (X + \overline{Y}) = X$, using Venn diagrams.

1-30. Prove the Absorption Theorem, $X \cdot (X + Y) = X$, by perfect induction.

1-31. Prove the Associative Theorem, $X \cdot (Y \cdot Z) = (X \cdot Y) \cdot Z$, by perfect induction.

1-32. Prove the Consensus Theorem, $(X + Y) \cdot (\overline{X} + Z) \cdot (Y + Z) = (X + Y) \cdot (\overline{X} + Z)$, by perfect induction.

1-33. Prove the Simplification Theorem, $X + \overline{X} \cdot Y = X + Y$, by perfect induction.

1-34. Determine if the relationship, $\overline{X} \cdot Y + X \cdot \overline{Y} = 1$ is true, using perfect induction.

1-35. Determine if the relationship, $Y \cdot Z + X \cdot Y + \overline{X} \cdot Y = Y$ is true, using perfect induction.

Section 1-7   Minimizing Boolean Functions Algebraically

1-36. Algebraically obtain a minimum form for each of the following Boolean functions. A minimum form is a form with the smallest number of literals.

(a) $F(A, B) = A \cdot B + A \cdot \bar{B}$

(b) $F(A, B, C) = A \cdot \bar{B} \cdot C + B \cdot C + B \cdot \bar{C}$

(c) $F(X, Y, Z) = \bar{Y} \cdot \bar{Z} + \bar{X} \cdot Y + \bar{X} \cdot Y \cdot Z + X \cdot Y \cdot \bar{Z}$

1-37. Simplify the following logic functions to the smallest number of literals.

(a) $F(X, Y, Z) = (\bar{Y} + \bar{Z}) \cdot (\bar{X} + Y) \cdot (\bar{X} + Y + Z) \cdot (X + Y + \bar{Z})$

(b) $F(A, B, C) = (A + B) \cdot (C + B) \cdot (A + \bar{B} + \bar{C})$

(c) $F(A, B, C) = (A + \bar{B} + C) \cdot (B + C) \cdot (B + \bar{C})$

1-38. Expand the following minimum functions to their canonical or standard sum of products form.

(a) $F(X, Y, Z) = Y + \bar{X} \cdot \bar{Z}$

(b) $F(A, B, C) = B + C$

(c) $F(M, R, S) = \bar{S} + \bar{M} \cdot R$

1-39. Expand the following minimum functions to their canonical or standard product of sums form.

(a) $F(X, Y, Z) = (Y) \cdot (\bar{X} + \bar{Z})$

(b) $F(A, B, C) = (B) \cdot (C)$

(c) $F(M, R, S) = (\bar{S}) \cdot (\bar{M} + R)$

1-40. Reduce the following Boolean functions to a minimum form.

(a) $F(W, X, Y, Z) = \bar{W} \cdot \bar{X} \cdot Y \cdot Z + X \cdot Y \cdot Z + W \cdot X \cdot Z$

(b) $F(A, B, C, D) = A \cdot \bar{B} \cdot \bar{D} + A \cdot B \cdot C \cdot \bar{D} + \bar{A} \cdot C \cdot \bar{D}$

(c) $F(W, X, Y, Z) = W \cdot \bar{X} \cdot \bar{Y} \cdot \bar{Z} + W \cdot X \cdot Y \cdot Z + \bar{W} \cdot \bar{X} \cdot Y \cdot \bar{Z}$

Section 1-8   Canonical or Standard Forms for Boolean Functions

1-41. Write each of the functions shown in the truth table in Fig. P1-41 in minterm compact form.

1-42. Write each of the following switching functions in terms of an expression using the independent variables in standard SOP form.

(a) $F(X, Y) = \Sigma\, m(1, 2, 3)$

(b) $F(X, Y, Z) = \Sigma\, m(0, 5, 6, 7)$

(c) $F(W, X, Y, Z) = \Sigma\, m(7, 10, 12, 14, 15)$

(d) $F(A, B, C, D) = \Sigma\, m(3, 6, 9, 11, 14)$

1-43. Write each of the functions shown in Fig. P1-41 in maxterm compact form.

1-44. Determine the maxterm compact forms for each of the minterm compact forms shown in problem 1-42.

1-45. Write an expression in standard POS form using the independent variables for each of the following maxterm compact forms.

X	Y	Z	F1	F2	F3	F4
0	0	0	0	0	1	0
0	0	1	0	1	1	0
0	1	0	0	0	1	1
0	1	1	1	1	0	1
1	0	0	1	1	0	1
1	0	1	1	0	1	1
1	1	0	0	1	0	1
1	1	1	1	0	0	0

FIGURE P1-41

(a) $F(X, Y) = \prod M(0, 2)$
(b) $F(X, Y, Z) = \prod M(3, 5, 7)$
(c) $F(W, X, Y, Z) = \prod M(2, 4, 6, 8, 10)$
(d) $F(A, B, C, D) = \prod M(1, 3, 5, 6, 8, 9, 13, 14)$

1-46. Make a table with columns for input variables *W, X, Y,* and *Z* and list in binary counting order all 16 possible rows.

(a) Make a column for minterms and write the standard product terms.
(b) Make a column for maxterms and write the standard sum terms.

1-47. Each of the following functions are written in a form that is equivalent to the complement of the standard POS form of *F*. Write each function in a standard POS form of *F*, and also in a standard SOP form of *F*. Leave each result in a compact or list form.

(a) $\overline{F}(P, Q, R, S) = \sum m(0, 5, 9, 13)$
(b) $\overline{F}(D, C, B, A) = \sum m(3, 6, 8, 14)$
(c) $\overline{F}(W, X, Y, Z) = \sum m(4, 5, 6, 7)$
(d) $\overline{F}(P, Q, R, S) = \sum m(12, 13, 14, 15)$

1-48. Each of the following functions are written in a form that is equivalent to the complement of the standard SOP form of *F*. Write each function in a standard SOP form of *F*, and also in a standard POS form of *F*. Leave each result in a compact or list form.

(a) $\overline{F}(J, K, L, M) = \prod M(7, 9, 12, 15)$
(b) $\overline{F}(B, J, M, Z) = \prod M(3, 6, 8, 14)$
(c) $\overline{F}(W, X, Y, Z) = \prod M(4, 7, 9, 13)$
(d) $\overline{F}(U, T, P, W) = \prod M(0, 3, 7, 9, 14)$

1-49. Draw a truth table representation for each of the following functions.

(a) $F(J, K, L, M) = \prod M(7, 9, 12, 15)$
(b) $\overline{F}(B, J, M, Z) = \prod M(3, 6, 8, 14)$
(c) $F(W, X, Y, Z) = \sum m(4, 7, 9, 13)$
(d) $\overline{F}(U, T, P, W) = \sum m(0, 3, 7, 9, 14)$

1-50. Write the expression for each of the following functions in terms of the independent variables.

(a) $F(V, W, X, Y, Z) = m_0 + m_6 + m_{23}$
(b) $F(U, V, W, X, Y, Z) = m_{19} + m_{22} + m_{30}$
(c) $F(U, V, W, X, Y, Z) = m_{25} + m_{35} + m_{47}$
(d) $F(T, U, V, W, X, Y, Z) = m_{55} + m_{67} + m_{93}$

1-51. Write the expression for each of the following functions in terms of the independent variables.

(a) $F(U, V, W, X, Y, Z) = M_0 \bullet M_6 \bullet M_{23}$
(b) $F(U, V, W, X, Y, Z) = M_5 \bullet M_9 \bullet M_{43}$
(c) $F(U, V, W, X, Y, Z) = M_7 \bullet M_{26} \bullet M_{56}$
(d) $F(U, V, W, X, Y, Z) = M_{19} \bullet M_{36} \bullet M_{62}$

1-52. Fig. P1-52 lists four functions that have been implemented using transistors and are available as off-the-shelf packaged IC devices that will perform these functions.

X	Y	F1	F7	F8	F14
0	0	1	1	0	0
0	1	0	1	0	1
1	0	0	1	0	1
1	1	0	0	1	1

FIGURE P1-52

OR Operation   AND Operation   Complement Operation

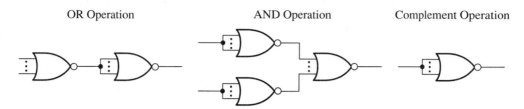

FIGURE P1-55

Write each function in a standard SOP form of F and reduce the function to a minimum equation. If necessary use double complementation and DeMorgan's theorem to write an equation in a familiar form, then give the name associated with the function.

1-53. Show how to manipulate the AND function $F = X \cdot Y$ using Boolean postulates and/or theorems so that it can be implemented with only OR elements and Inverters. Use parentheses to clearly indicate the order of the intended operations. Draw the circuit and verify that the circuit satisfies the AND function using perfect induction.

1-54. Using Boolean postulates and/or theorems manipulate the OR function $F = X + Y$ so that it can be implemented with only AND elements and Inverters. Use parentheses to clearly indicate the order of the intended operations. Draw the circuit and verify that the circuit satisfies the OR function using perfect induction.

1-55. Verify by perfect induction that a set of NOR elements may be connected as shown in Fig. P1-55 to produce the OR operation, AND operation, and complement operation, thus representing a functionally complete set.

1-56. Use Boolean postulates and/or theorems to manipulate the complement function $F = \overline{X}$ so that it can be implemented first with a NAND element and then with a NOR element.

1-57. Show how to manipulate the OR function $F = X + Y$ so that it can be implemented with only NAND elements. Use parentheses to clearly show the order of the intended operations.

1-58. Manipulate the AND function $F = X \cdot Y$ so that it can be implemented with only NANDs. Use parentheses to clearly indicate the order of the intended operations.

Section 1-9 Specifying Designs Using Logic Descriptions

1-59. Given the following language statement: Design a switching circuit that has three independent input signals. When only two inputs are 1 at the same time, the output is to be 1.

(*a*) Obtain the truth table logic description.

(*b*) Obtain the standard SOP form of the function.

(*c*) Obtain the complement of the standard POS form of the function.

1-60. For the following language statement: Design a switching circuit that has three independent inputs and two outputs. One of the outputs should be 1 when all the inputs are 1, and the other output should be 0 when two or less of the inputs are 0.

(*a*) Write the truth table logic description.

(*b*) Write the standard SOP form of each function.

(*c*) Write the complement of the standard POS form of each function.

1-61. Write a logic statement for the language statement given in problem 1-59. Write a Boolean function for the logic statement.

1-62. Write the Boolean function for the following logic statement: If task A is active or task B is not active and task C is active and task D in not active or task E is active then $F = 1$ else $F = 0$.

Section 1-10   Number of Different Functions for n Independent Variables

1-63. Write a relationship for the number of different Boolean functions possible for n independent variables. How many different functions are possible for the following number of independent variables? If a function and its complement is considered as only one function then how many functions are possible for the following number of independent variables?

(*a*) 2

(*b*) 3

(*c*) 4

(*d*) 5

CHAPTER

2

NUMBER SYSTEMS, NUMBER REPRESENTATIONS, AND CODES

2-1 INTRODUCTION AND INSTRUCTIONAL GOALS

In this chapter we cover the basic material required to understand number systems, number representations, and codes. If the student is already well versed in number systems, the following sections may not be necessary: Section 2-2, Number Systems; Section 2-3, The Binary Number System; and Section 2-4, Converting from Decimal to Other Number Systems. In Section 2-5, Number Representations, and Section 2-6, Arithmetic Operations, we discuss binary representations for negative numbers and show how computers use the indirect subtraction method to save hardware.

To illustrate how two-valued input variables are entered in a digital system, Section 2-7 discusses the keyboard as an input to a switching circuit. The ASCII character set is presented, and a small keyboard switch matrix is discussed to illustrate the importance of codes in studying digital design.

In Section 2-8, Short Introduction: Analog Interfaces to Switching Circuits, we present the functions of analog-to-digital conversion and digital-to-analog conversion. An application utilizing the Exclusive OR function is presented, and the conversion between binary and gray code is discussed. The section ends with a generalized interface example—a digital process controller, using an analog-to-digital converter (ADC) and a digital-to-analog coverter (DAC) with a digital processor. In some curricula the material in this chapter may be bypassed or, preferably, assigned as outside reading without loss of continuity.

After studying this chapter, you should be prepared to:

1. Convert mixed numbers, integer numbers, and fractional numbers to and from the decimal, binary, octal, and hexadecimal number systems.
2. Write binary expressions for positive and negative decimal numbers in sign magnitude, 1's complement, and 2's complement representations.
3. Perform binary subtraction indirectly using 1's and 2's complement numbers, and recognize when an answer is correct and when it is not because of overflow.
4. Determine how many bits are required to represent a specified number of keyboard characters.
5. Draw and label a block diagram for an analog-to-digital converter and a digital-to-analog converter, determine the resolution of an ADC, and determine the number of input bits required for a DAC for a given number of analog output steps.
6. Convert gray code numbers to equivalent binary numbers and binary numbers to equivalent gray code numbers.

2-2 NUMBER SYSTEMS

Number systems may be symbolic, such as the Roman number system, or they may be weighted-positional, such as the Arabic number system we normally use. First we will briefly discuss the Roman number system, and then we will concentrate on number systems that are associated with digital design. The binary number system, with only two values, is the ideal number system for two-valued devices; consequently, we will concentrate on the binary number system. The octal, hexadecimal, and decimal number systems will also be discussed; but from the standpoint of digital design, these number systems are simply a convenience for humans.

2-2-1 The Roman Number System

Modern number systems are weighted-positional number systems as opposed to symbolic number systems such as the Roman, where a symbol always represents the same weight. Consider the Roman number system, in which the following fixed values are assigned to a set of symbols: I = 1, V = 5, X = 10, L = 50, C = 100, D = 500, and M = 1000. If a symbol is repeated or lies to the right of another symbol, then the value of the symbol is additive. The value of the combination XX is 10 + 10, or 20, and the value of the combination CXX is 100 + 10 + 10, or 120. If a symbol is repeated or lies to the left of a higher-valued symbol, then the value of the symbol is subtractive. The value of the combination XXC is $-10 - 10 + 100$, or 80. The value of the combination of XLVIII may be written as $-10 + 50 + 5 + 3$, or 48.

The movie industry often uses Roman numerals to show the release dates of films. The dials of some clocks and watches as well as the cornerstones of many older buildings utilize Roman numerals. At one time it was also considered proper to list the copyright dates of published materials using Roman numerals. Consider a book containing the following copyright dates: MCML, MCMLII, and MCMLX. The first of these combinations of symbols represent the year 1950. What are the other two years?

2-2-2   Weighted-Positional Number Systems

In studying number systems, it is important to keep in mind that we are discussing arithmetic operators rather than the binary operators OR and AND. It is traditional in computer engineering to use the symbols $+$ and $\bullet$ as the OR and AND binary operators, respectively; however, it is traditional in mathematics, engineering, physics, etc. to use the symbols $+$ and $\times$ as the plus and multiplication arithmetic operators. It is usually clear from the context of the discussion which symbols are being used.

Now consider a weighted-positional number system using Arabic numerals. The value of a particular digit in a weighted-positional number system is dependent on its position in the sequence of digits relative to a radix point. If the radix point is not shown in the sequence of digits, it is implied to be located just to the right of the last digit. For example, the decimal number 269. is often written as 269 with an implied radix point. The symbol used for the radix point in some countries is a comma rather than a period. For example, the number 4,5 (4 is the integer part and ,5 is the fractional part) in these countries would be equivalent to the number 4.5 (4 is the integer part and .5 is the fractional part) in the United States. The decimal number system is the universal number system. The decimal number 269 can be expanded into a polynomial function as shown below.

$$200 = 2 \times 10^2$$
$$60 = 6 \times 10^1$$
$$\underline{9 = 9 \times 10^0}$$
$$269 = 2 \times 10^2 + 6 \times 10^1 + 9 \times 10^0 = (N)_{10}$$

In fact, a number in any weighted-positional number system can be represented using a sequence of digits representing the integer part of the number, followed by a radix point, and another sequence of digits representing the fractional part of the number:

$$(N)_r = d_j \ldots d_3 d_2 d_1 d_0 . d_{-1} d_{-2} d_{-3} \ldots d_{-k}$$

Such a number has $j + 1$ integer digits, (d_j being the most significant digit, MSD, and d_0 being the least significant digit, LSD). This number also has k fractional digits (d_{-1} being the MSD, and d_{-k} being the LSD). N represents the number, and r represents the base or radix of the number. In the decimal number system the base is 10, and powers of 10 are used to express the number as a polynomial function like the one above.

Numbers in other weighted-positional number systems can be easily converted to equivalent decimal numbers using the following polynomial function. Each digit in the number system to be converted is multiplied by its corresponding radix raised to the power indicated by the position of the digit in the number. These products are next added to obtain the base 10 or decimal number.

$$(N)_r = d_j \ldots d_3 d_2 d_1 d_0 . d_{-1} d_{-2} d_{-3} \ldots d_{-k}$$

$$= d_j r^j + \cdots + d_3 r^3 + d_2 r^2 + d_1 r^1 + d_0 r^0 + d_{-1} r^{-1} + d_{-2} r^{-2}$$

$$+ d_{-3} r^{-3} + \cdots + d_{-k} r^{-k}$$

$$= (M)_{10}$$

A summary of the weighted-positional number systems that we will discuss is shown in Fig. 2-1. The table in Fig. 2-1 begins with base 10 numbers (decimal numbers) and lists base 2 (binary numbers), base 8 (octal numbers), and finally base 16 (hexadecimal numbers). Binary numbers can contain only two values, represented by the numbers 0 and 1. Octal numbers can contain only the following eight values: 0, 1, 2, 3, 4, 5, 6, and 7. Hexadecimal numbers (hex numbers) can

Base 10 $(N)_{10}$ Decimal	Base 2 $(N)_2$ Binary	Base 8 $(N)_8$ Octal	Base 16 $(N)_{16}$ Hexadecimal
0	0	0	0
1	1	1	1
2	10	2	2
3	11	3	3
4	100	4	4
5	101	5	5
6	110	6	6
7	111	7	7
8	1000	10	8
9	1001	11	9
10	1010	12	A
11	1011	13	B
12	1100	14	C
13	1101	15	D
14	1110	16	E
15	1111	17	F
16	10000	20	10
21	10101	25	15
25	11001	31	19
85	1010101	125	55
170	10101010	252	AA

FIGURE 2-1
Weighted-positional number systems.

contain the same values as the decimal numbers 0 through 9 and, in addition, A = 10, B = 11, C = 12, D = 13, E = 14, and F = 15. All together, there are 16 values (0 through F).

2-3 THE BINARY NUMBER SYSTEM

Binary numbers are associated with the set BN$\{0, 1\}$ as compared to decimal numbers, which are associated with the set DN$\{0, 1, 2, 3, 4, 5, 6, 7, 8, 9\}$. The numbers 0 and 1 used in this context are not necessarily the logic 0 value and the logic 1 value we used for the Boolean variables, but they can be. The characteristics of the binary number system parallel the characteristics of the decimal number system in practically every respect. Digits in a binary number are customarily called *bits* (*bi*nary dig*its*). Eight bits are referred to as a byte, and four bits are referred to as half a byte or a nibble.

The major difference between the binary number system and the decimal number systems is the fact that binary numbers have only two values. This is a very important distinction since we learned in Chapter 1 that devices used in the construction of digital circuits and systems operate with only two values. The binary number system is the natural number system used to describe two-valued digital circuits and systems.

The binary number system is a weighted-positional number system. Each bit has a value relative to its position in the number. Using a polynomial function, we can express a binary number in powers of 2 and obtain its decimal equivalent as follows.

$$(N)_2 = (1111.111)_2$$

$$= 1 \times 2^3 + 1 \times 2^2 + 1 \times 2^1 + 1 \times 2^0 + 1 \times 2^{-1} + 1 \times 2^{-2} + 1 \times 2^{-3}$$

$$= 1 \times 8 + 1 \times 4 + 1 \times 2 + 1 \times 1 + 1 \times \frac{1}{2} + 1 \times \frac{1}{4} + 1 \times \frac{1}{8}$$

$$= 8 + 4 + 2 + 1 + \frac{1}{2} + \frac{1}{4} + \frac{1}{8}$$

$$= 15 + \frac{7}{8}$$

$$= (15.875)_{10}, 15.875 \text{ decimal, or } 15.875$$

If we number the bits in a binary number to correspond to the exponents of its base or radix, we have a way of identifying each bit in the number. For example, consider the following 12-bit binary number which has 9 bits for its integer part and 3 bits for its fractional part.

Bit positions	8	7	6	5	4	3	2	1	0		−1	−2	−3
Positional weights	256	128	64	32	16	8	4	2	1		$\frac{1}{2}$	$\frac{1}{4}$	$\frac{1}{8}$
Binary number	1	0	1	1	0	0	1	0	1	.	1	0	1

If bit position n (often shortened to read bit n) represents a bit in a binary number with a positional weight of w, then bit $(n - 1)$ has a positional weight of $w/2$, and bit $(n + 1)$ has a positional weight of $2w$. Bit positions provide a convenient way of identifying each binary digit associated with a binary number. The most significant digit (MSD) for the integer part of the binary number $(101100101.101)_2$ is bit 8, while the least significant digit (LSD) is bit -3. In working with binary numbers, the MSD is also the most significant bit (MSB), and the LSD is also the least significant bit (LSB).

By identifying the positional weight associated with each bit in a binary number, the conversion to decimal is quite easy for binary numbers with a small number of bits; however, the conversion of octal and hexadecimal numbers to decimal is a little more difficult, as illustrated in the following example.

Example 2-1. Convert the binary number 100111, the octal number 271, and the hexadecimal number 1D2 to their equivalent decimal numbers.

Solution

$$(100111)_2 = 32 + 4 + 2 + 1 \quad \text{(using positional weights}$$
$$\text{for bits 5, 2, 1, and 0)}$$

$$= 39 \text{ decimal}$$

$$(271)_8 = 2 \times 8^2 + 7 \times 8^1 + 1 \times 8^0$$

$$= 2 \times 64 + 7 \times 8 + 1 \times 1 \quad \text{(observe the positional weights}$$
$$\text{for digit positions 2, 1, and 0)}$$

$$= 128 + 56 + 1$$

$$= 185 \text{ decimal}$$

$$(1D2)_{16} = 1 \times 16^2 + 13 \times 16^1 + 2 \times 16^0$$

$$= 1 \times 256 + 13 \times 16 + 2 \times 1 \quad \text{(observe the positional weights}$$
$$\text{for digit positions 2, 1, and 0)}$$

$$= 256 + 208 + 2$$

$$= 466 \text{ decimal}$$

Observe that the hexadecimal digit D was entered in the polynomial function as 13, the equivalent decimal number. The equivalent decimal numbers of the hexadecimal digits A through F are listed in Fig. 2-1.

2-3-1   Using Octal and Hexadecimal Numbers

Binary numbers that contain more than 5 or 6 bits are not only hard to remember but hard to accurately transcribe from one sheet of paper to another. For this reason, octal numbers and hexadecimal numbers are mainly used as a shorthand for binary numbers. Both octal numbers and hexadecimal numbers provide a convenient method of expressing binary numbers using fewer digits. A by-product of using the decimal, octal, or hexadecimal number system instead of the binary number system is a significant reduction in the amount of input data (data entered from a keyboard) and output data (data dumped by a printer), especially during computer program development, troubleshooting, and debugging.

Octal and hexadecimal numbers are particularly easy to convert to and from binary numbers, since their bases are powers of 2. To convert from binary to octal, write each group of three bits to the right and to the left of the radix point as illustrated below using the corresponding equivalent octal digit shown in Fig. 2-1. Leading and trailing zeroes are shown in parentheses.

Binary number	(0)10	101	110	011	.	101	111	1(00)
Octal number	2	5	6	3	.	5	7	4

so, $(10101110011.1011111)_2 = (2563.574)_8$

To convert from binary to hexadecimal, write each group of four bits to the right and to the left of the radix point as illustrated below using the corresponding equivalent hexadecimal digit shown in Fig. 2-1. It is not necessary to use Fig. 2-1 if one uses the positional weights for each group of bits and converts the group first to its equivalent decimal digit and then to its equivalent hexadecimal digit. For example, in the following illustration, group $(1011)_2 = 8 + 2 + 1 = (11)_{10} = (B)_{16}$. It is important to keep in mind that the decimal digits 10 through 15 represent the hexadecimal digits A through F.

Binary number	(0)101	0111	0011	.	1011	111(0)
Hexadecimal number	5	7	3	.	B	E

so, $(10101110011.1011111)_2 = (573.BE)_{16}$

To convert from octal or hexadecimal back to binary, just reverse the procedure and write the three bits or four bits that represent each octal or hexadecimal digit, respectively.

Notice in the illustrations above that leading zeroes (shown in parentheses) in the integer part of the binary number are of no consequence; however, one must add a sufficient number of trailing zeroes (shown in parentheses) at the end of the fractional part of the binary number to complete a group of three binary bits or a group of four binary bits before converting those bits to their respective octal or hexadecimal digits. When converting from octal or hexadecimal, all three or four bits, respectively, must be supplied for the binary bits corresponding to each octal or hexadecimal digit, except leading zeroes for the most significant group of bits in the integer part and trailing zeroes for the least significant group of bits in the fractional part.

Example 2-2

1. Convert the binary number 10001110 to equivalent octal and hexadecimal numbers.
2. Convert the hexadecimal number C3.F4 to an equivalent binary number.
3. Convert the octal number 1370.14 to an equivalent binary number.

Solution

1. Group three bits at a time starting at the radix point. For an integer, the position of the radix point is located on the right side of the least significant bit; therefore, we obtain

$$(10\ 001\ 110)_2 = (2\ 1\ 6)_8$$

Grouping four bits at a time, starting at the radix point, we obtain

$$(1000\ 1110)_2 = (8\ E)_{16}$$

2. Writing the four bits for each digit in the hexadecimal number, we obtain

$$(C3.F4)_{16} = (1100\ 0011.1111\ 0100)_2$$

3. Writing the three bits for each digit in the octal number, we obtain

$$(1370.14)_8 = (001\ 011\ 111\ 000.001\ 100)_2$$

Practicing engineers and computer scientists often rely on special hand-held calculators or computer utility programs to obtain number conversions for decimal, binary, octal, and hexadecimal numbers, and to perform arithmetic and Boolean operations for these number systems.

Keep in mind that the actual value of a number in any number system except binary is converted to a two-valued variable for a computer or switching circuit that utilizes that number as its input. This is true because switching circuits only need their input values to be two-valued variables, and they in turn generate only two-valued variables at their outputs. Decimal, octal, and hexadecimal numbers are simply equivalent numbers provided for human convenience, since we humans do not relish working with long strings of 1s and 0s.

With hand-held calculators and computer programs readily available, the only reason for a discussion of number systems is to present the concepts. When the tools are available, they are almost always used to convert between different number systems to save time and, we hope, human errors. The concepts tell us how the results may be arrived at without using the tools, and they also provide us with an insight on how the tools may have been written.

2-4 CONVERTING FROM DECIMAL TO OTHER NUMBER SYSTEMS

How do we convert from the decimal number system to any other number system in general? First let us work with the integer part of the number, i.e., the part to the left of the radix point.

2-4-1   Repeated Radix Division Technique for Integers

To show the technique, the following small integer number will be used.

$$N_i = d_3 r^3 + d_2 r^2 + d_1 r^1 + d_0 r^0$$

or

$$N_i = d_3 r^3 + d_2 r^2 + d_1 r + d_0$$

Dividing by the radix r results in

$$\frac{N_i}{r} = \frac{d_3 r^3}{r} + \frac{d_2 r^2}{r} + \frac{d_1 r}{r} + \frac{d_0}{r}$$

When the division is carried out, the quotient is

$$d_3 r^2 + d_2 r + d_1, \text{with a remainder of } d_0$$

Continued radix division of each remaining quotient allows us to obtain a set of remainders. These remainders are then used to represent the integer part of the number in the required base by writing each remainder digit in its correct position.

$$(N_i)_r = d_3 d_2 d_1 d_0$$

From our discussion, the least significant digit (d_0) is found first, and the most significant digit (d_3) is found last.

In Example 2-3, repeated radix division is used to convert a decimal number to equivalent binary, octal, and hexadecimal numbers. The advantage of the repeated radix division technique is that it can be programmed. Once a 0 quotient is reached, it is not necessary to continue the division process since only insignificant leading 0s will result.

Example 2-3. Use the repeated radix division technique to convert the decimal number 651 to equivalent binary, octal, and hexadecimal numbers.

Solution The solutions are illustrated in Figs. E2-3a, b, and c. Different organizations for carrying out the process are possible, but the concept of obtaining remainders remains the same. The last remainder is the MSD and the first remainder is the LSD. Notice that 11 was converted to B in the first step of the conversion in Fig. E2-3c.

Observe that $(1\ 010\ 001\ 011)_2 = (1213)_8$, and $(10\ 1000\ 1011)_2 = (28B)_{16}$. Calculations like this can be made to check your results.

2-4-2   Repeated Radix Multiplication Technique for Fractions

Numbers do not always appear as integers; consequently, one needs to understand the concept of converting fractions or the fractional part of a mixed number from

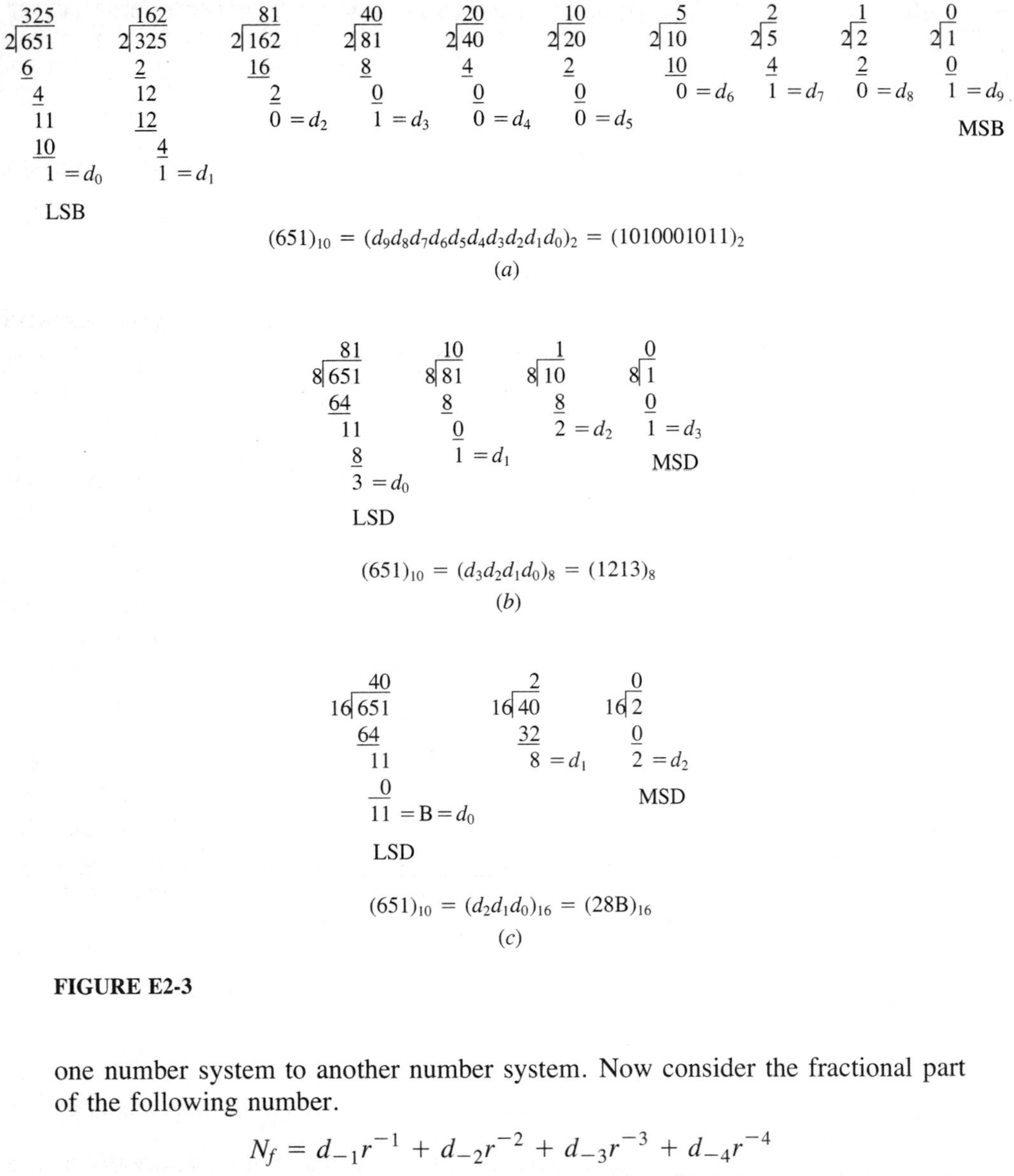

```
  325        162        81        40        20        10         5        2        1        0
2)651      2)325      2)162     2)81      2)40      2)20      2)10     2)5      2)2      2)1
  6          2         16         8         4         2         10        4        2        0
  4         12          2         0         0         0          0 = d6   1 = d7   0 = d8   1 = d9
 11         12          0 = d2    1 = d3    0 = d4    0 = d5                                MSB
 10          4
  1 = d0     1 = d1
 LSB
```

$$(651)_{10} = (d_9d_8d_7d_6d_5d_4d_3d_2d_1d_0)_2 = (1010001011)_2$$

(a)

```
   81          10          1          0
8)651       8)81        8)10       8)1
  64           8           8          0
  11           0           8          0
   8           1 = d1      2 = d2     1 = d3
   3 = d0                             MSD
  LSD
```

$$(651)_{10} = (d_3d_2d_1d_0)_8 = (1213)_8$$

(b)

```
   40            2            0
16)651       16)40        16)2
  64           32            0
  11            8 = d1       2 = d2
   0                        MSD
  11 = B = d0
  LSD
```

$$(651)_{10} = (d_2d_1d_0)_{16} = (28B)_{16}$$

(c)

FIGURE E2-3

one number system to another number system. Now consider the fractional part of the following number.

$$N_f = d_{-1}r^{-1} + d_{-2}r^{-2} + d_{-3}r^{-3} + d_{-4}r^{-4}$$

Multiplying by the radix r results in

$$rN_f = rd_{-1}r^{-1} + rd_{-2}r^{-2} + rd_{-3}r^{-3} + rd_{-4}r^{-4}$$

When the multiplication is carried out, the product that results is the coefficient d_{-1}, plus the remaining fraction $d_{-2}r^{-1} + d_{-3}r^{-2} + d_{-4}r^{-3}$.

Continued radix multiplication of each remaining fraction allows us to obtain a set of coefficients. These coefficients are then used to represent the fractional part of the number in the required base by writing each coefficient digit in its correct position.

$$(N_f)_r = .d_{-1}d_{-2}d_{-3}d_{-4}$$

Notice that the most significant digit (d_{-1}) is found first, followed by each less significant digit. This points to the fact that this process may be continued until you are satisfied with the number of digits obtained; that is, either the exact value is reached or the required accuracy is reached for a good approximation.

The technique of repeated radix multiplication is easy to use, and, like repeated radix division, the process can also be programmed.

> **Example 2-4.** Use the repeated radix multiplication technique to convert the decimal fractional number .719 to equivalent binary, octal, and hexadecimal numbers.
>
> *Solution* The solutions are illustrated in Figs. E2-4a, b, and c. Alternate organizations are also possible for repeated radix multiplication. The principle is to obtain the integer coefficients, and these are used to obtain the fractional part of the number. The first integer coefficient is the MSD and the last integer coefficient you elect to obtain is the LSD. More accuracy may be obtained by continuing the process as long as the remaining fractional part is not zero. Observe that $(.1011)_2 = 1/2 + 1/8 + 1/16 = (.6875)_{10}$ not $(.719)_{10}$ as we might expect. However, $(.5601)_8 = 5/8 + 6/64 + 1/4096 = (.718994)_{10}$, which illustrates that a larger number of bits (12 bits when the octal number is converted to binary) provides greater accuracy.

2-4-3   Add-the-Weights Method

Another method for converting integer decimal numbers to binary is known as the add-the-weights method. This method involves listing the positional weights of a binary number beginning with a weight of 1 in the first bit position (bit 0), 2 in the second bit position, 4 in the third bit position, and so on until the last positional weight 2^n in the $(n + 1)$th bit position (bit n) is either equal to or greater than the value of the decimal integer to be converted. The conversion process begins by placing a 1 under the largest weight equal to or less than the decimal integer. Then, by inspection, a 1 is placed under each additional weight added to the first weight that does not cause the sum to exceed the value of the decimal integer to be converted. A 0 is placed under each positional weight that is not used, that is, a positional weight that will cause the sum to exceed the value of the decimal integer. When the sum of the positional weights agrees with the decimal integer, the binary value of the number may be written as shown by the following example for the decimal number 42.

64	32	16	8	4	2	1
	1	0	1	0	1	0

$$
\begin{array}{cccc}
.719 & .438 & .876 & .752 \\
\underline{\times 2} & \underline{\times 2} & \underline{\times 2} & \underline{\times 2} \\
1.438 & .876 & 1.752 & 1.504
\end{array}
$$

$$
\begin{array}{cccc}
d_{-1} = 1 & d_{-2} = 0 & d_{-3} = 1 & d_{-4} = 1 \\
\text{MSB} & & & \text{LSB}
\end{array}
$$

$(.719)_{10} = (.d_{-1}d_{-2}d_{-3}d_{-4})_2 = (.1011)_2$ (4-bit approximation)

(*a*)

$$
\begin{array}{cccc}
.719 & .752 & .016 & .128 \\
\underline{\times 8} & \underline{\times 8} & \underline{\times 8} & \underline{\times 8} \\
5.752 & 6.016 & .128 & 1.024
\end{array}
$$

$$
\begin{array}{cccc}
d_{-1} = 5 & d_{-2} = 6 & d_{-3} = 0 & d_{-4} = 1 \\
\text{MSD} & & & \text{LSD}
\end{array}
$$

$(.719)_{10} = (.d_{-1}d_{-2}d_{-3}d_{-4})_8 = (.5601)_8$ (12-bit approximation
when octal is
converted to binary)

(*b*)

$$
\begin{array}{cccc}
.719 & .504 & .064 & .024 \\
\underline{\times 16} & \underline{\times 16} & \underline{\times 16} & \underline{\times 16} \\
4\ 314 & 3\ 024 & 384 & 144 \\
\underline{7\ 19} & \underline{5\ 04} & \underline{0\ 64} & \underline{0\ 24} \\
11.504 & 8.064 & 1.024 & 0.384
\end{array}
$$

$$
\begin{array}{cccc}
d_{-1} = 11 = \text{B} & d_{-2} = 8 & d_{-3} = 1 & d_{-4} = 0 \\
\text{MSD} & & & \text{LSD}
\end{array}
$$

$(.719)_{10} = (.d_{-1}d_{-2}d_{-3}d_{-4})_{16} = (.B810)_2$ (16-bit approximation
when hexadecimal is
converted to binary)

(*c*)

FIGURE E2-4

Since $42 = 32 + 8 + 2$, 42 decimal is equivalent to 101010 binary. The add-the-weights method also works for converting fractional decimal numbers to binary but is harder to use since some decimal fractions do not have an exact binary equivalent, or an exact binary equivalent may take too many bits to represent.

2-5   NUMBER REPRESENTATIONS

Until now we have considered only positive numbers. When we first learned the decimal number system we were taught that positive numbers are represented with a plus sign ($+$) preceding the most significant digit of the number while negative numbers are represented with a minus sign ($-$) preceding the most significant digit of the number. The absence of a sign preceding a decimal number has an implied plus sign. This representation is appropriately called sign magnitude (SM) representation or sign magnitude notation.

Sign magnitude representation is natural for humans, but not for computers. Computer circuits can use binary numbers expressed in a sign magnitude representation if they are designed to contain the following three functional elements: an Adder, a Subtractor, and a Magnitude comparator. These three functional elements are required to handle binary addition and subtraction using the same rules you are familiar with for decimal numbers. The Magnitude comparator is required to determine the larger of the two magnitudes before subtraction is performed. Multiplication by repeated addition and division by repeated subtraction is also possible.

Computers are generally not designed to interpret a plus sign ($+$) or a minus sign ($-$). To keep the coding simple, a 0 is used in place of a plus sign and a 1 is used in place of a minus sign (symbols already existing in the binary system).

Example 2-5. Convert the following positive and negative numbers to a sign magnitude representation suitable for a computer designed to use binary numbers with 8 integer bits.

(*a*) $(+1001101)_2$

(*b*) $(-10001)_2$

(*c*) $(+1001.01)_2$

(*d*) $(-111.001)_2$

Solution

(*a*) $(+1001101)_2 = (01001101)_2$

(*b*) $(-10001)_2 = (10010001)_2$

(*c*) $(+1001.01)_2 = (00001001.01)_2$

(*d*) $(-111.001)_2 = (10000111.001)_2$

In this example, a 0 is substituted for a plus sign and a 1 is substituted for a minus sign for the most significant integer bit, i.e., bit 7. This process is referred to as encoding or supplying the code for the plus symbol (whose code is 0) and the minus symbol (whose code is 1). The most significant digit or most significant bit of a binary number expressed in sign magnitude representation is called the sign bit. When the sign bit is 0 the number is positive, and when the sign bit is 1 the number is negative. Notice that zero filling is necessary for the sign magnitude representations (b) through (d) since the computer requires binary numbers that contain 8 integer bits.

Positive and negative numbers can also be expressed by two other number representations. These are the diminished radix complement (DRC) or ($r - 1$)'s complement representation, and the radix complement (RC) or *r*'s complement representation. For binary numbers the base or radix $= 2$; therefore, the DRC or $(r - 1)$'s complement representation is the 1's complement representation while the RC or *r*'s complement representation is the 2's complement representation. When applied to the base 10 or decimal number system, the $(r - 1)$'s complement representation is the 9's complement representation while the *r*'s complement representation is the 10's complement representation. Table 2-1 shows the comparison of a few positive and negative 8-bit numbers using three different binary number representations.

The 1's complement and 2's complement representations are not instinctive representations for humans, but both are very simple for computers. Computer circuits today are more often designed to utilize binary numbers expressed in a 1's or 2's complement representation than sign magnitude representation. This is true because both addition and subtraction of positive and negative binary numbers expressed in either a 1's or 2's complement representation can be handled by using only an Adder and a Complementor.

Positive binary numbers in a sign magnitude representation, 1's complement representation, and 2's complement representation are expressed the same as shown in Table 2-1 using 8 bits for each representation; however, the corresponding negative numbers in each representation are expressed differently. The first bit on the left (the most significant bit) is the sign bit in all three of the binary representations. Positive numbers have a sign bit $= 0$ while negative numbers have a sign bit $= 1$.

The following observations can be made by analyzing Table 2-1. The largest positive 8-bit number in all three representations is 127 instead of 255 since the sign bit is present. Only the 2's complement representation has a single binary number expression (all zeroes) for both $+0$ and -0. The sign magnitude representation and the 1's complement representation have two different binary number expressions for $+0$ and -0 that must be accounted for in circuits designed to

TABLE 2-1
Comparison table for binary number representations using 8 bits

Decimal number	Sign magnitude representation	1's complement representation	2's complement representation
+127	01111111	01111111	01111111
+31	00011111	00011111	00011111
+3	00000011	00000011	00000011
+0	00000000	00000000	00000000
−0	10000000	11111111	00000000
−3	10000011	11111100	11111101
−31	10011111	11100000	11100001
−127	11111111	10000000	10000001
−128	—	—	10000000

handle these representations. The sign magnitude representation for any negative number is simply the same arrangement of bits, 0 through 6, for the corresponding positive number, while the sign bit changes from 0 to 1. Notice that the negative numbers for all three binary representations are different. For an 8-bit negative number, only the number -128 exists in the 2's complement representation. The largest N-bit negative number in 2's complement representation is the number's straight unsigned binary value. The number -128 in Table 2-1 in the 2's complement representation is therefore 10000000.

To see how negative numbers are obtained in Table 2-1 in a 1's complement representation, consider the binary expression for the decimal number $+127$, which is 01111111. If we logically complement every bit in the positive number 01111111 from bit 7 down to bit 0, we obtain the negative number 10000000 (simply apply the logical complement operator to every bit; see Chapter 1). As one can see from Table 2-1, 10000000 is the expression for -127 in 1's complement representation. Therefore,

$$+127 = 01111111$$

$$\text{1's complement of } 01111111 = 10000000$$

But

$$10000000 = -127$$

The above calculation demonstrates that the 1's complement of $N = -N$.

The terms complement, logical complement, true complement, and 1's complement are all used somewhat interchangeably. Each of these terms means to perform a bitwise logical complement of a binary number N. The result of complementing each bit of a binary number is the conversion of a number to its negative value if it was positive, or to its positive value if it was negative, that is, the binary number is negated (its magnitude remains the same but its sign is changed). Each of the binary numbers shown in Table 2-1 in the column labeled 1's complement representation is shown in both its positive form and negative form. To convert from one form to the other simply requires complementing one form to obtain the other form as the 1's complement representation column indicates.

It should be obvious that double complementation of a binary number results in the original binary number. For example, $+5 = 00000101$; the complement of $00000101 = 11111010$, or -5; the complement of $11111010 = 00000101$, or $+5$.

Definition 2-1. The $(r - 1)$'s complement of a number N is defined by the relationship, $r^i - r^{-f} - N$, where N is the number to be negated, r is the radix, i is the number of integer digits, and f is the number of fractional digits.

Definition 2-1 has been written for the general case of a mixed number for any weighted positional number system, and it applies to integer numbers and fractional numbers as well. Keep in mind that when numbers are expressed in a diminished radix complement or 1's complement representation, the first, or most significant, bit is the sign bit.

Example 2-6. Use Definition 2-1 to obtain the $(r - 1)$'s complement for each of the following binary numbers expressed in a 1's complement representation.

(a) 01011, a positive integer (decimal equivalent, +11)

(b) 1.001, a negative fraction (decimal equivalent, −0.75)

(c) 11100010110.1001101, a negative mixed number (decimal equivalent, −233.390625)

Solution

(a) $N = 01011, r = 2, i = 5, f = 0$

$$r^i - r^{-f} - N = 2^5 - 2^{-0} - N$$

$$= 100000 - 1 - 01011$$

$$= 11111 - 01011$$

$$= 10100, \text{ a negative integer (decimal equivalent, } -11)$$

To obtain the 1's complement of an integer binary number N, simply change each 0 to a 1 and each 1 to a 0 in the number N.

(b) $N = 1.001, r = 2, i = 1, f = 3$

$$r^i - r^{-f} - N = 2^1 - 2^{-3} - 1.001$$

$$= 10 - .001 - 1.001$$

$$= 1.111 - 1.001$$

$$= 0.110, \text{ a positive fraction (decimal equivalent, } +0.75)$$

Again, all that is required is to obtain the 1's complement of a fractional binary number N is to change each 0 to a 1 and each 1 to a 0 in the number N.

(c) Using the results of (a) and (b), the

1's complement of 11100010110.1001101

is simply 00011101001.0110010

(decimal equivalent, +233.390625)

It helps to write the original number and then write its complement immediately under it (as we show here), especially for large numbers, so that mistakes will be reduced to a minimum. To obtain the decimal equivalent of a negative number (a number with a sign bit of 1) in a 1's complement representation, complement the number to find its positive expression, evaluate the positive number to obtain its decimal equivalent, then place a minus sign in front of the number.

It may not be quite as obvious to see how negative numbers are obtained in Table 2-1 in a 2's complement representation until you realize that you first obtain the 1's complement and then add 1 to the least significant bit as indicated by the following equation.

$$\text{2's complement of } N = (\text{1's complement of } N) + 1_{\text{LSB}}$$

This equation will provide the 2's complement of any binary number N, including 0, if you remember that the 2's complement of N must contain the same number of bits beginning with the most significant bit as N. In Table 2-1, under the 2's complement representation column, the number for $+31$ is 00011111; complementing each bit of this number results in 11100000, and then adding 1 to the least significant bit results in 11100001, which is -31.

A quick inspection method that can be used to write the 2's complement of a binary number may be stated as follows: Scan the number from its least significant bit, writing down each bit up to and including the first 1, then complement each bit after the first 1.

Example 2-7

1. Each of the following binary numbers is expressed in a 2's complement representation. Obtain the equivalent decimal number for each binary number. Next write the 2's complement of each of the binary numbers.
   (*a*) 010001
   (*b*) 111110
   (*c*) 0.1101
   (*d*) 011011.011
2. Express the following decimal numbers as binary numbers in both a 1's and 2's complement representations using 8 bits (7 bits plus a sign bit).
   (*a*) -68
   (*b*) $+7$
   (*c*) -10

Solution

1. First obtain the equivalent decimal numbers.
   (*a*) 010001 is a positive number $= +17$
   (*b*) 111110 is a negative number, therefore the 2's complement of 111110 $=$ 000010 $= +2$, so 111110 must be -2.
   (*c*) 0.1101 is a positive number $= +0.8125$
   (*d*) 011011.011 is a positive number $= +27.375$
       The 2's complements of each binary number are written as follows:
   (*a*) 2's complement of 010001 $=$ 101111
   (*b*) 2's complement of 111110 $=$ 000010
   (*c*) 2's complement of 0.1101 $=$ 1.0011
   (*d*) 2's complement of 011011.011 $=$ 100100.101
   These expressions represent the decimal numbers -17, $+2$, -0.8125, and -27.375, respectively.

2. First write the positive expression for each positive decimal number in 1's and 2's complement representations as follows. If the expression desired is for a negative number, take the 1's or 2's complement of the expression for the positive number to obtain the expression for the negative number.

	Decimal	1's complement representation	2's complement representation
(a)	68	01000100	01000100
	−68	10111011	10111100
(b)	+7	00000111	00000111
(c)	10	00001010	00001010
	−10	11110101	11110110

Notice that the expressions for the positive decimal numbers in both the 1's and 2's complement representations are the same. The expressions for the negative decimal numbers are different, and are obtained using the definitions for the 1's and 2's complements of N, respectively.

Using the inspection method, it is obvious that if we take the 2's complement of the expression 10111100, or −68, we obtain 01000100, or +68, which is the value that we started with. Double 2's complementation, like double complementation, results in the original binary number.

2-6   ARITHMETIC OPERATIONS

Since computers use only binary numbers to perform internal operations, we will discuss arithmetic operations with the binary number system only. The operations of addition, subtraction, multiplication, and division can be carried out in exactly the same manner with binary numbers as they are with decimal numbers with the exception that there are two values to work with (0 and 1) in the binary number system compared to ten values in the decimal number system (0 through 9).

Examples of decimal addition and subtraction and corresponding binary addition and subtraction are illustrated below for positive binary numbers using 8 bits (7 bits plus a sign bit in a 2's complement representation).

	Decimal	Binary	Notation
Addition			
Augend	14	00001110	$N1$
Addend	7	00000111	$N2$
Sum	21	00010101	$N1 + N2$
Direct Subtraction			
Minuend	14	00001110	$N1$
Subtrahend	7	00000111	$N2$
Difference	7	00000111	$N1 - N2$

The following example shows decimal subtraction and the corresponding binary subtraction (by addition of the 2's complement of the subtrahend). This indirect subtraction method is the method used by a number of computers since it requires only an Adder circuit and a Complementor circuit.

$$N1 - N2 = N1 + (-N2)$$

$$N1 - N2 = N1 + (2\text{'s complement of } N2)$$

since $-N2 = 2$'s complement of $N2$.

	Decimal	Binary	Notation
Subtraction by addition of 2's complement of subtrahend (indirect subtraction)			
Minuend	14	00001110	$N1$
Subtrahend	7	00000111	$N2$
		2's C. of $00000111 = \underline{11111001}$	$-N2$
Difference	7	Sum 00000111	$N1 + (-N2)$

Notice that the carry that occurs out of the sign bit position is simply ignored when binary numbers are added in a 2's complement representation. Subtraction by addition can also be performed for binary numbers expressed in a 1's complement representation. For this case, $-N2 = 1$'s complement of $N2$; however, the carry that occurs out of the sign bit position must be kept and added back to the least significant bit of the sum. This is called end-around carry, and it applies only for binary numbers expressed in a 1's complement representation.

Since multiplication can always be performed by repeated addition, and division can be performed by repeated subtraction, the addition process can be used to perform all four arithmetic operations (addition, subtraction, multiplication, and division) when a 1's or 2's complement number representation is used.

The number of bits a binary number is allowed to occupy is fixed by the design of a particular computer and is called the *data word size*, or simply *word size*. If the word size is n bits and the result of an addition operation exceeds n bits or the word size of the computer being used, then overflow has occurred. Overflow occurs when the sign bit contains the wrong value (it contains a 1 when two positive numbers are added or a 0 when two negative numbers are added). Properly designed circuitry flags the user that overflow has occurred. Actions that can be taken to eliminate overflow are to use an extended precision addition mode if available, or to rescale the data. The following two examples indicate an overflow condition for a 2's complement representation using an 8-bit word size.

127	01111111		−96	10100000
+64	01000000		−84	10101100
+191	10111111		−180	01001100

(Any number larger than +127 will not fit in this 8-bit data word, thus creating overflow; sign bit is 1, but should be 0.)

(Any number smaller than −128 will not fit in this 8-bit data word, thus creating overflow; sign bit is 0, but should be 1.)

2-7 THE KEYBOARD AS AN INPUT TO A SWITCHING CIRCUIT

When you type on the keyboard of a computer or enter commands on the keypad of a microwave oven, a TV driver (remote control unit), or a calculator, how do the numbers and letters designated on the switches get mapped or transformed to binary, or two-valued, numbers? We have seen that input variables must be two-valued for a switching circuit to operate. Where do these two-valued inputs come from?

This book uses predominantly Boolean variables and the algebra of logic to teach you how to design switching circuits. This assumes, of course, that the variables you will be working with are only two-valued. In this section we will briefly cover the concept associated with keyboard codes. These topics will give you a better understanding of the origin of two-valued variables in the real world.

2-7-1 The ASCII Character Set

When you want to enter a character at a computer keyboard, you simply press the push button with the appropriate character. Each push button on a keyboard or a keypad is assigned a sequence of binary digits called a code. The code is therefore a table of binary numbers with the required number of bits to effectively transform, or map, each keyboard or keypad switch closure with its own special binary number in the table (in the case of the shift and control keys, two, or sometimes three, keys all pressed at the same time also have a special code). Practically every personal computer on the market today uses the same keyboard code. The code we are referring to is the ASCII (pronounced as´-key) character set code shown in Fig. 2-2. ASCII is the abbreviation for "American Standard Code for Information Interchange." As you can see, we show the code for each ASCII character not only as a binary number but also as a decimal, octal, and hexadecimal number (for easy reference). Since the ASCII code is only a 7-bit code, octal and hexadecimal numbers in the table are written assuming the code has an eighth bit with a value of 0.

Since the ASCII character set contains 128 different characters, 7 bits, or binary digits, are required to represent all 128 characters. For Y characters the smallest number of two-valued variables X that are required to represent all the characters can be found using the following relationship:

$$2^X \geq Y$$

It is possible that a keyboard using the ASCII character set could contain 128 separate push buttons or keys. However, in practice a keyboard this large would be unwieldy. Therefore, a smaller keyboard is used, with special keys (such as the shift key and the control key) in conjunction with the normal keys generating the additional codes. For example, the capital letter M is generated by pressing the key for M while holding down the shift key.

Dec	Binary	Octal	Hex	ASCII	Dec	Binary	Octal	Hex	ASCII
0	0000000	000	00	NULL	32	0100000	040	20	space
1	0000001	001	01	SOH	33	0100001	041	21	!
2	0000010	002	02	STX	34	0100010	042	22	"
3	0000011	003	03	ETX	35	0100011	043	23	#
4	0000100	004	04	EOT	36	0100100	044	24	$
5	0000101	005	05	ENQ	37	0100101	045	25	%
6	0000110	006	06	ACK	38	0100110	046	26	&
7	0000111	007	07	BELL	39	0100111	047	27	'
8	0001000	010	08	BS	40	0101000	050	28	(
9	0001001	011	09	HT	41	0101001	051	29	)
10	0001010	012	0A	LF	42	0101010	052	2A	*
11	0001011	013	0B	VT	43	0101011	053	2B	+
12	0001100	014	0C	FF	44	0101100	054	2C	,
13	0001101	015	0D	CR	45	0101101	055	2D	-
14	0001110	016	0E	SO	46	0101110	056	2E	.
15	0001111	017	0F	SI	47	0101111	057	2F	/
16	0010000	020	10	DLE	48	0110000	060	30	0
17	0010001	021	11	DC1	49	0110001	061	31	1
18	0010010	022	12	DC2	50	0110010	062	32	2
19	0010011	023	13	DC3	51	0110011	063	33	3
20	0010100	024	14	DC4	52	0110100	064	34	4
21	0010101	025	15	NAK	53	0110101	065	35	5
22	0010110	026	16	SYNC	54	0110110	066	36	6
23	0010111	027	17	ETB	55	0110111	067	37	7
24	0011000	030	18	CAN	56	0111000	070	38	8
25	0011001	031	19	EM	57	0111001	071	39	9
26	0011010	032	1A	SUB	58	0111010	072	3A	:
27	0011011	033	1B	ESC	59	0111011	073	3B	;
28	0011100	034	1C	FS	60	0111100	074	3C	<
29	0011101	035	1D	GS	61	0111101	075	3D	=
30	0011110	036	1E	RS	62	0111110	076	3E	>
31	0011111	037	1F	US	63	0111111	077	3F	?

(a)

FIGURE 2-2
ASCII character set code.

Dec	Binary	Octal	Hex	ASCII	Dec	Binary	Octal	Hex	ASCII
64	1000000	100	40	@	96	1100000	140	60	‘
65	1000001	101	41	A	97	1100001	141	61	a
66	1000010	102	42	B	98	1100010	142	62	b
67	1000011	103	43	C	99	1100011	143	63	c
68	1000100	104	44	D	100	1100100	144	64	d
69	1000101	105	45	E	101	1100101	145	65	e
70	1000110	106	46	F	102	1100110	146	66	f
71	1000111	107	47	G	103	1100111	147	67	g
72	1001000	110	48	H	104	1101000	150	68	h
73	1001001	111	49	I	105	1101001	151	69	i
74	1001010	112	4A	J	106	1101010	152	6A	j
75	1001011	113	4B	K	107	1101011	153	6B	k
76	1001100	114	4C	L	108	1101100	154	6C	l
77	1001101	115	4D	M	109	1101101	155	6D	m
78	1001110	116	4E	N	110	1101110	156	6E	n
79	1001111	117	4F	O	111	1101111	157	6F	o
80	1010000	120	50	P	112	1110000	160	70	p
81	1010001	121	51	Q	113	1110001	161	71	q
82	1010010	122	52	R	114	1110010	162	72	r
83	1010011	123	53	S	115	1110011	163	73	s
84	1010100	124	54	T	116	1110100	164	74	t
85	1010101	125	55	U	117	1110101	165	75	u
86	1010110	126	56	V	118	1110110	166	76	v
87	1010111	127	57	W	119	1110111	167	77	w
88	1011000	130	58	X	120	1111000	170	78	x
89	1011001	131	59	Y	121	1111001	171	79	y
90	1011010	132	5A	Z	122	1111010	172	7A	z
91	1011011	133	5B	[	123	1111011	173	7B	{
92	1011100	134	5C	\	124	1111100	174	7C	|
93	1011101	135	5D	]	125	1111101	175	7D	}
94	1011110	136	5E	^	126	1111110	176	7E	~
95	1011111	137	5F	_	127	1111111	177	7F	DEL

(b)

FIGURE 2-2
(*continued*)

Example 2-8. What is the minimum number of binary variables that can be used to represent the 12 pushbuttons of a standard telephone key pad?

Solution

$$2^X \geq 12$$

$X = 4$; therefore a minimum of four binary variables is required.

2-7-2 A Small Keyboard Switch Matrix

In Fig. 2-3 we show a switch matrix for a small telephone-type keypad and its connection to a single-chip microcomputer. The microcomputer (for example, an

Control characters:

NUL	Null	DC1	Device control 1
SOH	Start of heading	DC2	Device control 2
STX	Start of text	DC3	Device control 3
ETX	End of text	DC4	Device control 4 (stop)
EOT	End of transmission	NAK	Negative acknowledge
ENQ	Enquiry	SYN	Synchronous idle
ACK	Acknowledge	ETB	End of transmission block
BELL	Bell (audible signal)	CAN	Cancel
BS	Backspace	EM	End of medium
HT	Horizontal tabulation	SUB	Substitute
LF	Line feed	ESC	Escape
VT	Vertical tabulation	FS	File separator
FF	Form feed	GS	Group separator
CR	Carriage return	RS	Record separator
SO	Shift out	US	Unit separator
SI	Shift in	DEL	Delete
DLE	Data link escape		

Graphic characters:

space	Space (normally nonprinting)	_	Underline
'	Apostrophe (closing single quotation mark; acute accent)	`	Grave accent (opening single quotation mark)
<	Less than	{	Opening brace
\	Reverse slant	|	Vertical line
[	Opening bracket	~	Overline (tilde)
^	Circumflex		

(*c*)

FIGURE 2-2
(*continued*)

Intel 8748, 49, or 51) has an internal memory programmed to provide binary output data on output lines OP3(MSB) through OP0(LSB) on the output port. The keypad is scanned one row at a time using the following four binary codes: 0111(row 3), 1011(row 2), 1101(row 1), and 1110(row 0). As each row is scanned, the program reads the input binary data from input lines IP2 through IP0 on the input port to monitor a key closure. A key closure occurs when one of the input bits returns a value of 0. The program can then issue its key code at the data port via bits D7 through D0. Only 4 out of 8 bits at the data port are required to provide the key code for a 12-pushbutton switch matrix.

Figure 2-4 illustrates the binary number for each pushbutton on the keypad switch matrix shown in Fig. 2-3. It is interesting to note that computers often have keyboards with a special microcomputer similar to the one illustrated in Fig. 2-3 dedicated to reading the keyboard switch matrix and issuing the unique key code for each key being pressed. This provides more time for the main computer to work on its computational tasks, thus speeding up the computer's processing capability.

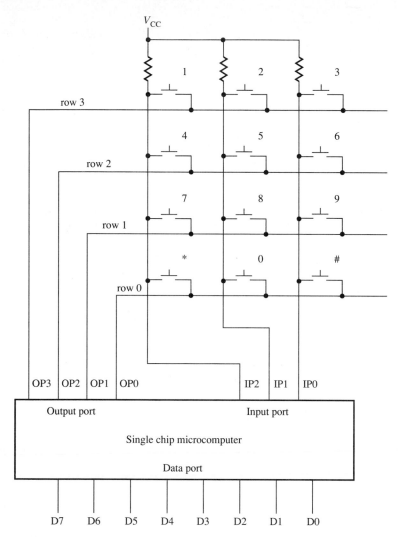

FIGURE 2-3
A switch matrix for a telephone keypad interfaced with a microcomputer.

Many different types and varieties of codes are used in digital circuits. Codes for calculators, for example, range from binary-coded decimal for keyboard entry to seven-segment output codes for driving light-emitting diode or liquid crystal displays. When a particular code is needed, a code converter circuit can be designed that will meet the requirement. Code converter design is handled just like any other digital design and will be discussed as appropriate. The topic of codes is quite broad and is closely related to the field of communications. Codes used for transmission of digital data over landlines or satellite links require error-free transmission. Error detection and correction codes have also been devised

	OP3	OP2	OP1	OP0	IP2	IP1	IP0	
row 3	0	1	1	1	1	1	1	none pressed
	0	1	1	1	0	1	1	key 1 pressed
	0	1	1	1	1	0	1	key 2 pressed
	0	1	1	1	1	1	0	key 3 pressed
row 2	1	0	1	1	1	1	1	none pressed
	1	0	1	1	0	1	1	key 4 pressed
	1	0	1	1	1	0	1	key 5 pressed
	1	0	1	1	1	1	0	key 6 pressed
row 1	1	1	0	1	1	1	1	none pressed
	1	1	0	1	0	1	1	key 7 pressed
	1	1	0	1	1	0	1	key 8 pressed
	1	1	0	1	1	1	0	key 9 pressed
row 0	1	1	1	0	1	1	1	none pressed
	1	1	1	0	0	1	1	key * pressed
	1	1	1	0	1	0	1	key 0 pressed
	1	1	1	0	1	1	0	key # pressed

FIGURE 2-4
Binary number for each push button of the keypad switch matrix shown in Fig. 2-3.

for these applications, and these codes are used in memory systems in computers subject to continuous use. Table 2-2 lists a few of the common binary codes for the decimal digits. Some of these codes will be used at a later time.

2-8   SHORT INTRODUCTION: ANALOG INTERFACES TO SWITCHING CIRCUITS

The next question you may ask is: How are analog voltages processed by a digital computer or switching circuit? Special circuits have been developed that can

TABLE 2-2
Common binary codes

Decimal	BCD 8421	XS3	2421	2-out-of-5	$84-2-1$	Biquinary(2-out-of-7) 5043210
0	0000	0011	0000	00011	0000	0100001
1	0001	0100	0001	00101	0111	0100010
2	0010	0101	0010	00110	0110	0100100
3	0011	0110	0011	01001	0101	0101000
4	0100	0111	0100	01010	0100	0110000
5	0101	1000	1011	01100	1011	1000001
6	0110	1001	1100	10001	1010	1000010
7	0111	1010	1101	10010	1001	1000100
8	1000	1011	1110	10100	1000	1001000
9	1001	1100	1111	11000	1111	1010000

sample the value of an analog voltage over a particular time period and convert the analog value to a binary number. Circuits also exist for the inverse process of converting a binary number to an analog voltage. In this section we will discuss both analog-to-digital and digital-to-analog conversion from a fundamental input-output point of view.

2-8-1 The Function of Analog-to-Digital Conversion

When a circuit is designed to accept an analog voltage at its input and convert it to a binary number at its output the circuit is called an analog-to-digital converter, or an ADC. A block diagram of an ADC is shown in Fig. 2-5.

A block diagram is used because we are only interested in the relationship of the variables at the input and the output of the device rather than the internal circuit structure. An analog-to-digital converter quantizes or encodes specific values of its analog input as equivalent binary codes at its output. The binary codes that are produced at the output have an uncertainty or quantizing error of $(\pm\frac{1}{2})$LSB. The number of bits in the binary output of an ADC establishes its resolution. If 14 bits are used on the output, the least significant bit of the converter has a resolution of $1/(2^{14})$ or $1/16{,}384$ times the analog input voltage full scale range (FSR); that is, an analog voltage range from 0 to 5 volts (FSR of 5 volts) has a LSB output resolution of 0.305 millivolts for a 14-bit binary output. Resolution can be expressed as

$$\text{Resolution of LSB} = \frac{V_{FSR}}{2^n}$$

where LSB is the least significant bit, V_{FSR} is the input voltage full-scale range, and n is the number of bits on the output.

Conversion speed, settling time, and accuracy are just a few of the important key specifications associated with analog-to-digital converters. Each of these quantities is associated with circuit specification instead of the function performed by the ADC (which is all we are concerned with for this brief introduction).

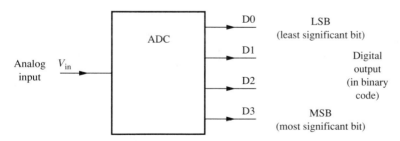

FIGURE 2-5
Block diagram of a low resolution analog-to-digital converter.

Example 2-9. Determine the resolution for an ADC with an analog voltage input range of 0 to 5 volts and an output with 8 bits.

Solution The resolution of the LSB at the output of the ADC is

$$\text{Resolution of LSB} = \frac{V_{\text{FSR}}}{2^n} = \frac{5-0}{2^8} = \frac{5}{256} = 20 \text{ mV}$$

2-8-2   The Function of Digital-to-Analog Conversion

Digital outputs are available for driving digital circuits, but sometimes those outputs must drive analog devices. The process of a circuit that converts a binary number at its input to an analog voltage at its output is called a digital-to-analog converter or a DAC. Figure 2-6 shows a block diagram of a simple 4-bit DAC. A DAC with an input resolution of n bits has 2^n analog output steps.

The 4-bit resolution DAC shown in Fig. 2-6 can only achieve 2^4, or 16, different values at its output. If the number of input bits for a DAC were 8 bits, then the output voltage could be any one of 2^8, or 256, different values, allowing greater relative accuracy (the maximum deviation of the output voltage relative to a straight line from zero to full scale).

In many cases analog-to-digital conversions that involve rotary motion are accomplished using optical or magnetic means (as in auto ignition systems that do not contain points). In these cases the pulses generated are proportional to the engine's angular speed in revolutions/minute. Consider a plate with transparent slits. The plate is attached to the distributor shaft. A fixed light source is mounted on one side of the plate and a fixed photocell is mounted on the other side. As the plate rotates, pulses from the photocell can be picked up and used to determine angular speed. By using a stationary pick-up coil, the same result can be achieved by magnetic detection. In this case a plate containing a magnetic is attached to the distributor shaft. Pulses from the moving plate are picked up by the stationary coil.

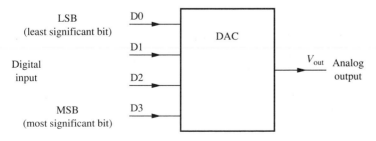

FIGURE 2-6
Block diagram of a simple 4-bit digital-to-analog converter.

A shaft-angle encoder is a device for measuring the angular position of a rotating shaft as illustrated in Fig. 2-7a. To keep our example simple we show only two binary output variables X and Y; however, encoders with many output variables are available. The resolution for 2 output bits is only $\frac{1}{4}$ of a full revolution, or 90 degrees. The purpose of our simple device is to convert the angular position of a shaft into four separate quadrants identified by different binary numbers. The binary numbers can then be used by a switching circuit to measure the position of the shaft.

Encoders of this type are designed using a special code called a gray code. The gray code, or unit distance code as it is sometime called, allows only one

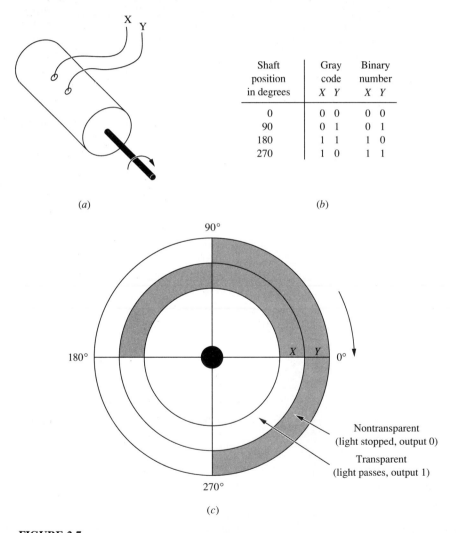

Shaft position in degrees	Gray code X Y	Binary number X Y
0	0 0	0 0
90	0 1	0 1
180	1 1	1 0
270	1 0	1 1

(a)   (b)

(c)

FIGURE 2-7
(a) A shaft-angle encoder, (b) input shaft position in degrees, output 2-bit gray code and equivalent binary number, (c) photographically encoded disk.

bit to change between successive quadrants. When more than one bit is allowed to change between successive quadrants any misalignment in the encoder can produce a momentary false output code. The output 2-bit gray code, its equivalent binary number, and the corresponding input shaft position in degrees for our simple shaft-angle encoder is shown in Fig. 2-7*b*.

   An example of a photographically encoded disk that could be mounted to the shaft and used with two light sources and two photocells to indicate the angular position of the shaft is illustrated in Fig. 2-7*c*. When light shines through the transparent area, the bit associated with that output is a binary 1; otherwise it is a binary 0. An alternate design could be a set of two brushes rubbing against a disk that has conducting and insulated areas such that a conducting area produces a binary 1 while an insulated area produces a binary 0. These examples serve to illustrate the principle involved in the encoding process.

2-8-3   The Exclusive OR Function (Modulo-2 Addition)

The Exclusive OR function or operation is used to convert a binary number to its gray code equivalent or to convert a gray code number to its binary equivalent. The Exclusive OR operator is a plus sign enclosed in a circle, and it performs an operation on two bits as indicated by the following equalities. This operation is also known as Modulo-2 arithmetic addition and can be used in the design of adder circuits.

$$0 \oplus 0 = 0$$

$$0 \oplus 1 = 1$$

$$1 \oplus 0 = 1$$

$$1 \oplus 1 = 0$$

Beginning with a binary number represented by the following bits

$$B_n \; B_{n-1} \; B_{n-2} \quad \ldots \quad B_1 \; B_0$$

the equivalent gray code number is represented as

$$B_n \quad B_n \oplus B_{n-1} \quad B_{n-1} \oplus B_{n-2} \quad \ldots \quad B_1 \oplus B_0$$

   This relationship allows us to generate Table 2-3, which converts 4-bit binary to gray code.

   Notice as you scan the gray code numbers in Table 2-3 from 0000 to 1000 that only one bit changes as you move from one line to an adjacent line either up or down in the list. This is true even from the bottom line 1000 back to the top line 0000; that is, only one bit changes when wrap-around occurs. Codes with this property are called cyclic codes. This is the property that makes the gray code a particularly useful code for some applications.

   Gray code is also called reflective binary code. Beginning with the 2-bit gray code shown in Fig. 2-7*b*, 3-bit gray code can be obtained by writing the

TABLE 2-3
Binary to gray code
conversion table

Binary number	Gray code
0000	0000
0001	0001
0010	0011
0011	0010
0100	0110
0101	0111
0110	0101
0111	0100
1000	1100
1001	1101
1010	1111
1011	1110
1100	1010
1101	1011
1110	1001
1111	1000

reflection of the 2-bit code (just like a mirror would reflect each bit) below the existing 2-bit code. By assigning a third bit of 0 (most significant bit) to the original 2-bit code and a third bit of 1 (most significant bit) to the reflected 2-bit code, we obtain 3-bit gray code. Using the same procedure and reflecting the 3-bit gray code, we can write 4-bit gray code as shown in Table 2-3.

Since there is only a unit distance between each line in the gray code, it is also called a unit distance code (UDC). The unit distance property is utilized throughout Chapter 3 for Karnaugh maps when assigning numbers to adjacent squares in the maps.

Beginning with a gray code number represented by the bits

$$G_n \quad G_{n-1} \quad G_{n-2} \quad \cdots \quad G_1 \quad G_0$$

the equivalent binary number is represented as

$$G_n \quad G_n \oplus G_{n-1} \quad G_n \oplus G_{n-1} \oplus G_{n-2} \quad \cdots$$
$$G_n \oplus G_{n-1} \oplus G_{n-2} \oplus \cdots \oplus G_1 \oplus G_0$$

Example 2-10. Obtain the equivalent gray code number for the binary number 10011. Also obtain the equivalent binary number for the gray code number 01101.

Solution

Binary number	10011
Gray code number	11010

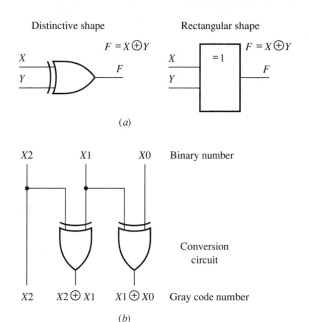

Distinctive shape

$F = X \oplus Y$

X
Y
F

Rectangular shape

X =1 $F = X \oplus Y$
Y F

(*a*)

$X2$ $X1$ $X0$ Binary number

Conversion
circuit

$X2$ $X2 \oplus X1$ $X1 \oplus X0$ Gray code number

(*b*)

FIGURE 2-8
(*a*) Logic symbols for Exclusive
OR element or gate, (*b*) binary to
gray code converter using
Exclusive OR elements.

Gray code number   01101

Binary number   01001

A simple check is to use the result and reverse the process. If the original number is not restored, a mistake has been made.

The logic symbol for an off-the-shelf IC device that performs the Exclusive OR or Modulo-2 addition function is shown in Fig. 2-8*a*. This device is called an Exclusive OR element or Exclusive OR gate, and it performs the function $X \oplus Y$ at its output when supplied with two binary variables X and Y at its inputs. The conversion operation for a simple 3-bit binary number to a gray code number can be implemented utilizing two Exclusive OR elements as shown in Fig. 2-8*b*. The conversion operation for a gray code number to binary number using Exclusive OR elements (or gates) is left as an exercise for the student.

2-8-4   Digital Process Control

Without analog-to-digital converters and corresponding digital-to-analog converters, analog measurements could not be made and analog processes could not be controlled using digital computers. Simulators using digital computers to control their simulation routines, such as aircraft simulators, including the space shuttle simulator, would not be possible. The data accumulated by aircrafts in the area of wind speed, altitude, compass heading, fuel consumption, and so on, are all analog waveforms that must somehow be converted to binary numbers. Conversions

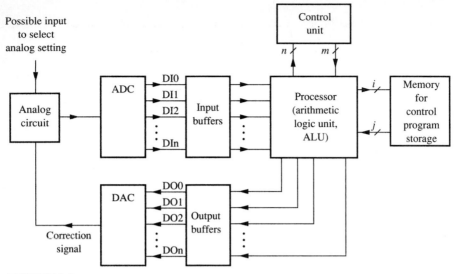

FIGURE 2-9
Simple block diagram of a digital process controller.

from analog to digital and also from digital to analog may be accomplished using off-the-shelf IC devices that perform these functions.

Many different control applications in industry utilize digital computers to control analog processes. The automotive industry is constantly introducing innovative computer-controlled devices for their cars. These include monitoring and controlling the engine's temperature, the interior temperature, and the antiskid breaking systems. A simple block diagram of a digital process controller is illustrated in Fig. 2-9. This is a typical example of an analog process being controlled by a digital processor (perhaps by a single chip microcomputer) via an appropriate ADC and a DAC. This system is running in a closed loop mode between the analog domain and the digital domain.

Brief as this introduction may be, it should provide you with a basic understanding of how analog inputs can be supplied to switching circuits. Throughout the rest of our discussions in this text we will concentrate primarily on digital switching circuits and leave the analog domain for future studies.

REFERENCES

1. Peatman, J. B., *Digital Hardware Design,* McGraw-Hill Book Company, New York, 1980.
2. Johnson, E. L., and Karim, M. A., *Digital Design, A Pragmatic Approach,* Prindle, Weber & Schmidt Publishers, Boston, Mass., 1987.
3. McCluskey, E. J., *Logic Design Principles,* Prentice-Hall, Inc., Englewood Cliffs, New Jersey, 1986.
4. Mano, M. M., *Digital Design,* Prentice-Hall, Inc., Englewood Cliffs, New Jersey, 1984.
5. Hill, J. H., and Peterson, G. R., *Introduction to Switching Theory & Logic Design,* 2d ed., John Wiley & Sons, Inc., New York, 1974.

6. Torng, H. C., *Switching Circuits, Theory and Logic Design,* Addison-Wesley, Reading, Mass., 1972.
7. Dietmeyer, D. L., *Logic Design of Digital Systems,* 2d ed., Allyn and Bacon, Boston, Mass., 1978.
8. Edwards, H. E., *The Principles of Switching Circuits,* The M. I. T. Press., Cambridge, Mass., 1973.
9. Mowle, F. J., *A Systematic Approach to Digital Logic Design,* Addison-Wesley, Reading, Mass., 1976.

PROBLEMS

Section 2-2   Number Systems

2-1. Convert the following Roman numerals to decimal.
   (*a*) IMICIX
   (*b*) MCMLVI
   (*c*) CCMLVXII
   (*d*) MMLLCCXXII

2-2. Write the following decimal numbers as polynomial functions.
   (*a*) 23
   (*b*) 4087
   (*c*) 39.28

2-3. Write the numbers $(36)_8$ and $(E5.3)_{16}$ as polynomial functions.

Section 2-3   The Binary Number System

2-4. Write the value for each of the following bits for the binary number 10001001.0101.
   (*a*) bit 2
   (*b*) bit -2
   (*c*) bit 8
   (*d*) bit -4
   (*e*) bit 4

2-5. Convert the following binary numbers to decimal numbers using polynomial functions.
   (*a*) 11011
   (*b*) 110110
   (*c*) 1101101
   (*d*) 11011.110

2-6. Convert the following nibbles or bytes to equivalent decimal numbers.
   (*a*) 1110
   (*b*) 1011
   (*c*) 10101010
   (*d*) 00110110

2-7. Convert the following numbers to their decimal equivalents using polynomial functions.
   (*a*) $(27431)_8$
   (*b*) $(476620)_8$
   (*c*) $(1234.567)_8$
   (*d*) $(11011110)_8$

(e) (FFFCC)$_{16}$

(f) (123430)$_{16}$

(g) (E2B4.5E7)$_{16}$

(h) (11011110)$_{16}$

2-8. Convert the following binary numbers to octal and hex numbers.

(a) 10001010

(b) 11110000

(c) 100000011.111

(d) 11001100.1

2-9. Make a table with a column for powers of two, and list the numbers 2^0 through 2^{15}. Make a column for decimal and list the equivalent decimal numbers. Make a wide column for binary and list the equivalent binary numbers. Make columns for octal and hexadecimal numbers and list the equivalent numbers.

2-10. Convert the following octal and hex numbers to binary numbers.

(a) (3451)$_8$

(b) (65473)$_8$

(c) (563451)$_8$

(d) (7657.1100)$_8$

(e) (BDE)$_{16}$

(f) (13F5)$_{16}$

(g) (563.4512)$_{16}$

(h) (1.1)$_{16}$

2-11. Make a table with a column for hexadecimal numbers and list the numbers 00 through 20. Make a column for binary and list the equivalent binary numbers.

Section 2-4   Converting from Decimal to Other Number Systems

2-12. Use repeated radix division to convert the following decimal numbers to binary, octal, and hexadecimal.

(a) 23

(b) 47

(c) 268

2-13. Use repeated radix multiplication to convert the following decimal numbers to binary, octal, and hexadecimal.

(a) .75

(b) .475

(c) .96

2-14. Convert the decimal number 12.87 to numbers in the following bases.

(a) base-2

(b) base-8

(c) base-16

2-15. Convert the decimal number 234.56 to numbers in the following bases.

(a) binary

(b) octal

(c) hexadecimal

2-16. Use the add-the-weights method to convert the following decimal integers to binary.

(a) 17

(b) 37

(c) 56

(d) 72

(e) 475

Section 2-5   Number Representations

2-17. Write the following decimal numbers in a sign magnitude representation using the number of bits specified in each case.

(a) 10, using 6 bits

(b) 29, using 6 bits

(c) 51, using 8 bits

(d) 75, using 8 bits

(e) 327, using 12 bits

2-18. Write the following binary numbers in sign magnitude representation. What is the minimum number of bits required to represent each number in sign magnitude representation?

(a) +1001010

(b) −11110000

(c) −11001100.1

(d) +100000011.111

2-19. Obtain the 1's complement of the following binary numbers.

(a) 1010110

(b) 01011010011

(c) 010010.0101

(d) 101110101.1001

2-20. Write the following decimal numbers in a 1's complement representation using 8 bits.

(a) +6

(b) −25

(c) +125

(d) −126

2-21. Obtain the decimal values for the following binary numbers expressed in a 1's complement representation. (In each case the most significant bit is the sign bit.)

(a) 0010110

(b) 00011010011

(c) 110010.0101

(d) 100000101.1000

2-22. Obtain the 2's complement of the following binary numbers. (In each case the most significant bit is the sign bit.)

(a) 1010110

(b) 01011010011

(c) 010010.0101

(d) 101110101.1001

2-23. Write the following decimal numbers in a 2's complement representation using 8 bits.

(a) +9

(b) −36

   (c) +85
   (d) −114

2-24. Obtain the decimal values for the following binary numbers expressed in a 2's complement representation (in each case the most significant bit is the sign bit).
   (a) 0110110
   (b) 01011010111
   (c) 110001.0111
   (d) 101100101.1010

2-25. Write the following decimal numbers in 1's and 2's complement representations using 8 bits.
   (a) +7
   (b) −37
   (c) +91
   (d) −113

2-26. Write the following decimal numbers using 16 bits in sign magnitude representation, 1's complement representation, and 2's complement representation.
   (a) +38
   (b) −192
   (c) +389
   (d) −4751

Section 2-6   Arithmetic Operations

2-27. Carry out the following arithmetic operations in both decimal and 2's complement representation. Use 7 bits plus a sign bit for the numbers expressed in 2's complement representation.
   (a) 6 − 3
   (b) 95 + 27
   (c) 101 − 46
   (d) 39 − 17

2-28. Carry out the following arithmetic operations in both decimal and 1's complement representation. Use 7 bits plus a sign bit for the numbers expressed in 1's complement representation.
   (a) 17 − 12
   (b) 42 + 25
   (c) 123 − 76
   (d) 27 − 15

2-29. Which of the following arithmetic operations are correct? The numbers are all expressed in a 2's complement representation using 5 bits plus a sign bit. Explain your answers.
   (a) $N1 = 001000$
                $N2 = 011111$
        $N1 + N2 = 100111$
   (b) $N1 = 010100$
                   $N2 = 101011$
                  $-N2 = 010101$
        $N1 + (-N2) = 101001$

(c) $N1 = 100011$
           $N2 = 101101$
$N1 + N2 = 010000$
(d) $N1 = 010000$
               $N2 = 000011$
              $-N2 = 111101$
$N1 + (-N2) = 001101$

2-30. Perform $N1 + N2$, and $N1 + (-N2)$ for the following 8-bit numbers expressed in a 2's complement representation. Verify your answers by using decimal addition and subtraction and explain any anomalies.
(a) $N1 = 00110010$, $N2 = 11111101$
(b) $N1 = 10001110$, $N2 = 00001101$
(c) $N1 = 11111010$, $N2 = 10010101$
(d) $N1 = 00010010$, $N2 = 10011101$

2-31. Repeat 2-30 assuming that the values for the numbers $N1$ and $N2$ are expressed in a 1's complement representation.

Section 2-7   The Keyboard as an Input to a Switching Circuit

2-32. Determine the minimum number of bits required to represent all the characters on a keyboard that has the following number of keys.
(a) 9
(b) 16
(c) 22
(d) 36
(e) 104

2-33. If the following number of bits are used to represent all the characters on a keyboard, determine how many characters can be represented on each keyboard.
(a) 4
(b) 5
(c) 6
(d) 7
(e) 8

2-34. BCD is an abbreviation for Binary Coded Decimal. In this code, straight binary is used to represent the decimal numbers 0 through 9 with a minimum number of bits (four). XS3 is an abbreviation for Excess 3. In this code, straight binary + 0011 is used to represent the decimal numbers 0 through 9 with a minimum number of bits (again four). Make a table for the decimal numbers 0 through 9, the BCD representation for each decimal number, and the XS3 representation for each decimal number. Verify your results with Table 2-2.

Section 2-8   Short Introduction: Analog Interfaces to Switching Circuits

2-35. Determine the resolution of the least significant bit for the following ADCs.
(a) Analog full scale voltage range 5 volts, 4 output bits

    (*b*) Analog full scale voltage range 7.5 volts, 10 output bits

    (*c*) Analog full scale voltage range 20 volts, 6 output bits

    (*d*) Analog full scale voltage range 9 volts, 12 output bits

2-36. Given the requirements listed below, find the minimum number of output bits for each ADC.

    (*a*) An output resolution no larger than 0.25 volts and a full scale voltage range of 5 volts.

    (*b*) An output resolution no larger than 70 millivolts and a full scale voltage range of 10 volts.

    (*c*) An output resolution no larger than 8 millivolts and a full scale voltage range of 12 volts.

    (*d*) An output resolution no larger than 3/4 millivolts and a full scale voltage range of 12 volts.

2-37. Find the minimum number of input bits for each DAC given the following requirements.

    (*a*) The analog output voltage will have no less than 32 output steps.

    (*b*) The analog output voltage will have no less than 128 output steps.

    (*c*) The analog output voltage will have no less than 425 output steps.

    (*d*) The analog output voltage will have no less than 1024 output steps.

2-38. What is the required resolution in degrees of each shaft-angle encoder with the following number of output bits.

    (*a*) 3 output bits

    (*b*) 4 output bits

    (*c*) 5 output bits

    (*d*) 7 output bits

2-39. Write the expressions for converting the binary numbers represented by the following bits to equivalent gray code numbers.

    (*a*) $X3\ X2\ X1\ X0$

    (*b*) $X4\ X3\ X2\ X1\ X0$

    (*c*) $X5\ X4\ X3\ X2\ X1\ X0$

2-40. Write the expressions for converting the gray code numbers represented by the following bits to equivalent binary numbers.

    (*a*) $X3\ X2\ X1\ X0$

    (*b*) $X4\ X3\ X2\ X1\ X0$

    (*c*) $X5\ X4\ X3\ X2\ X1\ X0$

2-41. Convert the following binary numbers to equivalent gray code numbers and check each result by converting the resulting gray code numbers back to equivalent binary numbers.

    (*a*) 01110

    (*b*) 10110

    (*c*) 01100

    (*d*) 11101

2-42. Draw a circuit using Exclusive OR elements that will perform the code conversion on the binary numbers in Prob. 2-41 and produce the required gray code output. Label each input and output bit.

2-43. Convert the following gray code numbers to equivalent binary numbers and check each result by converting the resulting binary numbers back to equivalent gray code numbers.

(*a*) 011101
(*b*) 101111
(*c*) 011001
(*d*) 111010

2-44. Draw a circuit using Exclusive OR elements that will perform the code conversion on the gray code numbers in Prob. 2-43 and produce the required binary output. Label each input and output bit.

CHAPTER
3

MINIMIZING FUNCTIONS USING MAPS

3-1   INTRODUCTION AND INSTRUCTIONAL GOALS

In this chapter we will concentrate on the presentation of the Karnaugh map logic description and its unique graphical role in the reduction of Boolean functions. We have already shown in Chapter 1 how to obtain a Boolean function in a minimum form using Boolean algebra reduction. Karnaugh maps are considered easier to use than Boolean propositions and Boolean theorems for minimizing functions because they provide a graphical viewpoint.

Section 3-2, Drawing, Filling, and Reading a Karnaugh Map, begins with two variable maps. Section 3-3, Three-Variable Karnaugh Maps, shows how to read product terms and provides the steps to follow in order to obtain a reduced expression for a function. In Section 3-4, Four- and Five-Variable Karnaugh Maps, larger sizes of maps are introduced. When there are multiple output functions to be minimized, Section 3-5 Multiple Function Minimization discusses one approach that can be used. Section 3-6 Don't Care Output Conditions introduces functions that contain output conditions that can be treated as either a 0 or a 1 and how to use these "don't cares" to obtain a minimum function. In Section 3-7, Alternate Functional Forms, Boolean theorems are used to write functions in alternate forms so two-level implementations can be drawn using combinations of AND, OR, NAND, and NOR gates.

Some of the new terms we will discuss include *p*-squares, *p*-subcubes, adjacent squares, adjacent *p*-squares, overlay adjacency, prime implicants, essential prime implicants, redundant prime implicants, don't care output conditions, AND/OR forms, and functional forms.

This chapter should prepare you to

1. Construct two-, three-, and four-variable Karnaugh maps from Venn diagrams.
2. Draw, fill, and read simplified expressions from Karnaugh maps consisting of two, three, four, and five variables.
3. Use Karnaugh maps to obtain minimum SOP expressions for the 1s (or 0s) of functions.
4. Plot and reduce partially reduced functions using Karnaugh maps.
5. Perform multiple function minimization using Karnaugh maps.
6. Effectively use don't care outputs in Karnaugh maps to obtain minimum functions.
7. Write alternate functional forms for the minimum SOP and the minimum POS forms of a function.
8. Draw two-level circuit implementations for functions represented in a minimum form.
9. Design code converters.

3-2 DRAWING, FILLING, AND READING A KARNAUGH MAP

The Karnaugh map is a pictorial or graphical method that relies heavily on the ability of our minds to perceive patterns. As it turns out, our minds can handle this task remarkably well. For a small number of input variables, usually five or less, a Karnaugh map can be drawn and labeled, the map filled in or plotted, and, finally, the Boolean function entered into the map can be simplified to a minimum form without too much difficulty.

3-2-1 Two-Variable Karnaugh Maps

First we have to draw the map. One of the easier ways to visualize the formation of a Karnaugh map is by rearranging a Venn diagram made up with rounded rectangles rather than traditional circles. In Fig. 3-1 we show a two-variable Venn diagram and corresponding two-variable Karnaugh map representations. Like the two-variable Venn diagram, the two-variable Karnaugh map has four distinguishable areas. These areas are identified on the Venn diagram and also on the Karnaugh map by using the standard product terms for two variables, i.e., the minterms. To assist in plotting the map, an alternate two variable map is shown using minterm designators. The latter map may be used as a ready reference to assist in filling in a two-variable map for a function when the function is given in one of the standard compact forms.

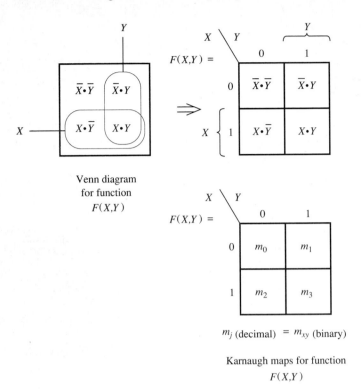

Venn diagram
for function
$F(X,Y)$

m_j (decimal) $= m_{xy}$ (binary)

Karnaugh maps for function
$F(X,Y)$

FIGURE 3-1
Two-variable Venn diagram and corresponding two-variable Karnaugh maps.

To continue the simplification process, we must next show how to fill in a map. One easy way to fill in a map is to utilize a description of the Boolean function either in truth table format or in one of the canonical or standard compact forms. The minterm compact form is perhaps the one most often used. Each minterm in the minterm compact form represents a 1 for the function, and those minterms that are not present represent a 0 for the function. The following two-variable function contains three minterms. The function can be entered in a two-variable Karnaugh map by placing a 1 in each square representing the minterms in the list.

$$F(X, Y) = \sum m(1, 2, 3)$$

Squares that are not identified by a minterm in the function can be filled in with 0s. The present function contains only one 0 which is represented by the square m_0 on the Karnaugh map shown in Fig. 3-2*a*.

You should observe that the two-variable Karnaugh map is arranged so that the areas on the map vary by a change in only one literal. Grouping adjacent areas that either contain all 1s or all 0s allows simplification by utilizing the Adjacency Theorem T7*a*

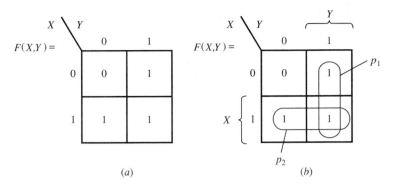

FIGURE 3-2
(*a*) Filling in a Karnaugh map, (*b*) grouping adjacent 1s in a Karnaugh map.

$$A \cdot B \ + \ A \cdot \overline{B} = A$$

which is a result of Postulate P5*b* ($\cdot$ is distributive over $+$) and Postulate P5*a*

$$B + \overline{B} = 1$$

The map in Fig. 3-2*a* contains two groups of adjacent 1s. It is standard practice to enclose adjacent 1s (or 0s) by circling them as shown in the map in Fig. 3-2*b*. When adjacent 1s (0s) are circled, they are said to be covered. To allow easy identification of each group, each group of 1s (0s) is identified by a product term designator, p_x (or r_x) where $x = 1, 2, \ldots$. Product term p_1 shown in Fig. 3-2*b* is formed from minterms m_1 and m_3 as follows:

$$p_1 = m_1 + m_3 = \overline{X} \cdot Y \ + \ X \cdot Y$$

$$= (\overline{X} + X) \cdot Y$$

$$= Y$$

The area on the map where the 1s for product term p_1 are contained can be seen to include all of the area for variable Y. Remember to think in terms of the area being covered by adjacent 1s or adjacent 0s, just as we do for Venn diagrams; only now, the area is not shaded.

Numbers associated with product terms have no significance other than to differentiate one product term from another. We could have just as easily used letters such as the Greek alphabet to make the differentiation.

Product term p_2 shown in Fig. 3-2*b* is formed from minterms m_2 and m_3 as follows.

$$p_2 = m_2 + m_3 = X \cdot \overline{Y} \ + \ X \cdot Y$$

$$= X \cdot (\overline{Y} + Y)$$

$$= X$$

The area on the map where the 1s for product term p_2 are contained can be seen to include all of the area for variable X.

By observing the 1s in a Karnaugh map and circling adjacent groups of powers of two, i.e., pairs, quadruples, octets, etc., we can apply Boolean reduction by inspection of the map. Obtaining the product terms from a Karnaugh map is referred to as "reading the map." To form the function represented in the map, simply read the map for the circled 1s, OR the product terms together, then set the expression that results equal to the function name. It should be obvious that if each individual 1 in a map is circled, the standard SOP form of the function will be obtained. By grouping product terms as shown in Fig. 3-2b and reading the map for these circled groups of 1s, we obtain the function in a minimum sum of products form as follows.

$$F(X, Y) = p_1 + p_2 = Y + X$$

In a similar fashion we can observe the 0s in a Karnaugh map, circle adjacent groups of powers of two, and read the map for the product terms for the 0s. To form the function represented in the map, the product terms for the circled 0s are ORed together, and the resulting expression is set equal to the complement of the function name. Referring to the two-variable map shown in Fig. 3-3, we notice one group of adjacent pairs indicated on the map by product term r_1 and another group indicated by product term r_2. The first group is important to recognize since it occurs quite often and is called "end around adjacency." By grouping adjacent 0s as shown in the two-variable map in Fig. 3-3 and reading the map for these circled groups, we obtain the complement of the function in a minimum sum of products form as follows.

$$\overline{F}(X, Y) = r_1 + r_2$$

$$= m_0 + m_2 + m_3 + m_2$$

$$= \overline{X} \bullet \overline{Y} + X \bullet \overline{Y} + X \bullet Y + X \bullet \overline{Y}$$

$$= \overline{Y} + X$$

Notice by simple inspection that the squares where the 0s for product term r_1 are contained include all of the area for variable $\overline{Y}$, and the squares where the 0s for product term r_2 are contained include all of the area for variable X.

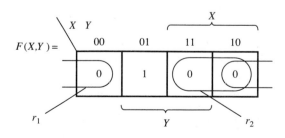

FIGURE 3-3
Grouping adjacent 0s in a Karnaugh map.

Complementing both sides of the complemented function results in the following minimum product of sums form for the function F.

Since,

$$\text{Standard POS form of } F = \overline{\text{Standard SOP form of } \overline{F}}$$

it follows that

$$\text{Minimum POS form of } F = \overline{\text{Minimum SOP form of } \overline{F}}$$

so

$$F(X, Y) = Y \cdot \overline{X}$$

Obtaining the two variable Karnaugh map used in Fig. 3-3 from a corresponding two-variable Venn diagram is left as an exercise for the student.

In 1952, E. W. Veitch published a paper entitled "A Chart Method for Simplifying Truth Functions," in which he first proposed the idea that "squares that are adjacent, either vertically or horizontally, differ in the value of only one of their literals." A year later M. Karnaugh published his paper entitled "The Map Method for Synthesis of Combinational Logic Circuits." In his paper Karnaugh used maps to demonstrate Veitch's idea, and today these maps are generally referred to as Karnaugh maps in his honor. The Karnaugh map is an important reduction tool for Boolean functions since it allow us to quickly group adjacent 1s or 0s pictorially or graphically and also to quickly read the product terms represented by each group by simply evaluating the area.

3-3 THREE-VARIABLE KARNAUGH MAPS

A three-variable Venn diagram and corresponding three-variable Karnaugh map representations are shown in Fig. 3-4. The three-variable Karnaugh map has eight distinguishable areas, each represented by the standard product terms shown on the Venn diagram. As an easy reference, an alternate three-variable Karnaugh map is shown using just the minterm designators. The latter map using the minterm designators may be used as a ready reference for filling in a three-variable map for a function when the function is given in one of the standard compact forms.

Consider the following three-variable function represented in minterm compact form. The Karnaugh map representation for this function is shown in Fig. 3-5.

$$F(X, Y, Z) = \sum m(0, 1, 3, 5)$$

Definition 3-1. To differentiate between squares that contain 1s and those that contain 0s, the squares that contain 1s will be called *p*-squares and those that contain 0s will be called *r*-squares.

Both *p*-squares and *r*-squares are squares that contain the standard product terms which we call minterms. The function represented in the Karnaugh map

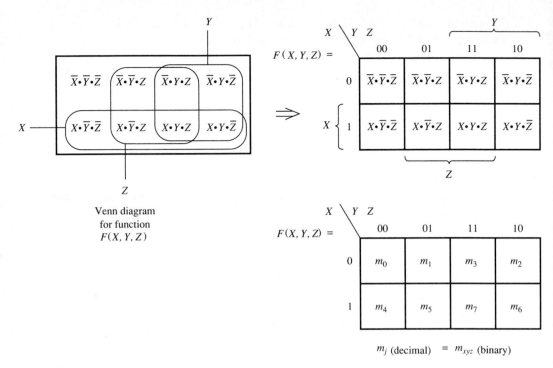

FIGURE 3-4
Three-variable Venn diagram and corresponding three-variable Karnaugh maps.

in Fig. 3-5 has four p-squares represented by the 1s in the squares for minterms 0, 1, 3, and 5. The function also has four r-squares represented by the 0s in the squares for minterms 2, 4, 6, and 7. If a designer only wants to obtain a sum of products for a function (which is often the case), the r-squares need not be filled in.

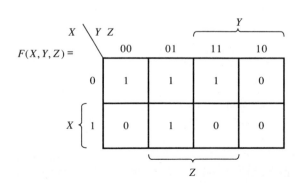

FIGURE 3-5
Filling in p-squares (squares filled in by 1s) and r-squares (squares filled in by 0s).

Subcube containing 2^i adjacent squares	Subcube size i	Number of literals k in a product term ($n =$ number of input variables) $k = n - i$ where $n \geq i$
1	0-cube	n
2	1-cube	$n - 1$
4	2-cube	$n - 2$
8	3-cube	$n - 3$
16	4-cube	$n - 4$
32	5-cube	$n - 5$

FIGURE 3-6

Relationship between adjacent squares, subcube size, and number of literals in a product term.

Definition 3-2. A subcube on a map is defined as the area of the map that covers or contains a set of all adjacent squares that have the same value (either all 1s or all 0s).

It may not be obvious, but to satisfy this definition each subcube represents a product term that contains 2^i adjacent squares where subcube size i is any positive integer including 0. The number of adjacent squares in a subcube or product term must therefore be a number contained in the following set of numbers { 1, 2, 4, 8, 16, 32 . . . }. For n input variables, each product term represented by a subcube of size i can only contain k literals where $k = n - i$ for $n \geq i$ as illustrated in the table in Fig. 3-6. It is common in technical literature to represent subcube size i as an i-cube; therefore, subcube size 3 is a 3-cube. If a subcube contains 4 adjacent squares, for example, the subcube size must be a 2-cube and the number of literals represented by this subcube must be $k = n - 2$. If a map had 4 input variables ($k = n - 2 = 4 - 2 = 2$), then two literal would be required to represent the product term. If a map had 2 input variables ($k = n - 2 = 2 - 2 = 0$), then no literals would be required since the function would be fixed at a 1 or a 0 value (all subcubes 1 or all subcubes 0).

Definition 3-3. A subcube that is formed entirely of p-squares is a p-subcube, and a subcube that is formed entirely of r-squares is a r-subcube.

3-3-1 A Three-Variable Cube Representation of a Function

The representation of a function by a three-variable cube is presented to provide historical background and is not recommended as a tool for minimizing Boolean functions. In Fig. 3-7b we show a three-variable cube (TVC) representation of the function shown in the three-variable Karnaugh map in Fig. 3-7a. The 1s of the functions are represented on the three-variable cube by the vertices marked with a large dot. These correspond to minterms 0,2,3,5,6, and 7. The 0s of the function corresponding to minterms 1 and 4 are represented on the TVC by the vertices not marked with a large dot. Product term p_1 represents the p-subcube for the p-squares represented by minterms 0 and 2 circled on the Karnaugh map.

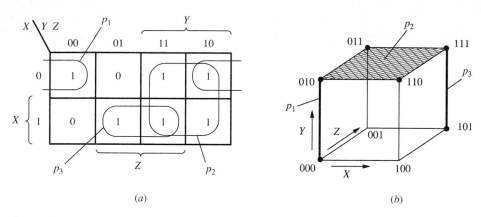

FIGURE 3-7
(*a*) Identifying *p*-subcubes on a Karnaugh map, (*b*) identifying corresponding *p*-subcubes on a three-variable cube.

The same *p*-subcube on the three-variable cube is represented by the group of adjacent vertices for minterms 0 and 2 connected by a darker line. The adjacency of minterms 0 and 2 on the Karnaugh map is referred to as end around adjacency. The adjacency of minterms 0 and 2 is more apparent when we observe the cube, since the vertices that represent these minterms lie next to one another along the *Y* axis. Product term p_2 represents the *p*-subcube covering the adjacent minterms 2,3,6, and 7 on the Karnaugh map. The same *p*-subcube on the three-variable cube is represented by the group of adjacent vertices shown grouped together by the cross-hatched plane. Product term p_3 represents the *p*-subcube for minterms 5 and 7 on the Karnaugh map. The *p*-subcube for the corresponding vertices is shown on the three-variable cube by a darker line that connects the adjacent vertices for minterms 5 and 7.

In practice, Karnaugh maps are more useful than multivariable cubes for the representation and minimization of Boolean functions. The idea of a subcube no doubt originated from the cube representation of a function. In a physical sense, a subcube is subdivision of a cube, hence the name subcube. The term cube is used to describe a specific grouping of adjacent 1s or 0s where a 0-cube represents a single square, a 1-cube represents two adjacent squares, a 2-cube represents four adjacent squares, et cetera, as illustrated in the table in Fig. 3-6.

3-3-2 Reading the Product Terms

Next we show how to read the product terms that represent the subcubes (either *p*-subcubes or *r*-subcubes) in a Karnaugh map. As we have indicated in Fig. 3-6, the number of literals in a product term may be determined by the following relationship:

$$\text{Number of literals} = \text{number of input variables} - \text{subcube size}$$

or,
$$k = n - i$$

This relationship tells us how many literals a product term must contain, but it does not tell us which literals to use. The following rules help determine which literals to use:

1. The literals used to make up a product term are those input variables with fixed values over the area of the subcube.

2. A literal is negated only if its value over the area of the subcube is fixed at 0.

As a check on the formation of a product term, it should be noted that each product term must always evaluate to 1 for its respective input combinations, that is, the input combinations read along the edges of the map.

For the three-variable Karnaugh map in Fig. 3-7a we show three product terms, p_1, p_2, and p_3. Product term p_1 is a 1-cube, while p_2 is a 2-cube and p_3 is a 1-cube. Product term p_1 contains 2 literals ($k = n - i = 3 - 1 = 2$), p_2 contains 1 literal ($k = n - i = 3 - 2 = 1$), and p_3 contains 2 literals. Since the area of the map represented by p_1 has its X value fixed at 0, its Y value not fixed, and its Z value fixed at 0, its two literals consist of $\overline{X}$ and $\overline{Z}$. Therefore, $p_1 = \overline{X} \cdot \overline{Z}$. Substituting the values of the fixed variables over the area of the subcube representing p_1, we see that $p_1 = \overline{X} \cdot \overline{Z} = \overline{0} \cdot \overline{0} = 1$. This tells us that the product has been properly formed for fixed value $X = 0$ and $Z = 0$ over the area of the subcube.

The area of the map represented by p_2 has its Y value fixed at 1, so product term $p_2 = Y$. Product term $p_3 = X \cdot Z$ since variable X has its value fixed at 1, and variable Z has its value fixed at 1. Both p_2 and p_3 evaluate to 1 for their respective input combinations over the area of their subcubes.

Example 3-1. Determine the number of literals that will appear in each of the product terms represented in the map in Fig. E3-1. Obtain each product term and verify that it evaluates to 1 for its respective input combinations. Use the product terms to write the complement of the function in a sum of products form.

Solution For a three-variable map $n = 3$, product terms r_1, r_2, and r_3 consist of 2 literals, 2 literals, and 1 literal respectively. Product term $r_1 = \overline{Y} \cdot Z$, product term $r_2 = Y \cdot \overline{Z}$, and product term $r_3 = X$.

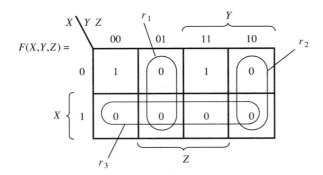

FIGURE E3-1

For r_1 the input combinations read along the edges of the map are $Y = 0$, $Z = 1$ so $\overline{Y} \cdot Z = \overline{0} \cdot 1 = 1$. The input combinations for r_2 are $Y = 1$, $Z = 0$ so $Y \cdot \overline{Z} = 1 \cdot \overline{0} = 1$. Product term r_3 has an input of $X = 1$ resulting in $r_3 = X = 1$.

$$\overline{F}(X, Y, X) = r_1 + r_2 + r_3$$

$$= \overline{Y} \cdot Z + Y \cdot \overline{Z} + X$$

3-3-3 Steps to Follow to Obtain a Reduced Expression

The algorithmic steps to follow to obtain a reduced sum of products (SOP) expression for a function from reading a Karnaugh map are listed below. When using p-subcubes, the SOP expression obtained is equated to the function name; however, the SOP expression that is obtained when using r-subcubes is equated to the complement of the function name. All p-subcubes are obtained by using the 1s for the function in the Karnaugh map. All r-subcubes are obtained by using the 0s for the function in the Karnaugh map.

1. Start by circling all p-subcubes (r-subcubes) for isolated 0-cubes.
2. Next circle all p-subcubes (r-subcubes) for 1-cubes which are not completely contained in a larger p-subcube (r-subcube).
3. Continue in this fashion by circling all p-subcubes (r-subcubes) for (2, 3, 4, . . .)-cubes which are not completely contained in a larger p-subcube (r-subcube).
4. Write the product terms or implicants for the p-subcubes (r-subcubes) and OR them together. Equate the resulting SOP expression to the function name (complement of the function name) as illustrated by the following relationships.

$$F = p_1 + p_2 + p_3 + \cdots$$

$$(\overline{F} = r_1 + r_2 + r_3 + \cdots)$$

When a product term of a function (or complement of a function) assumes the value of 1, so does the function. In other words, when a product term assumes the value of 1, this implies that the function also assumes the value of 1. For this reason it is common practice to call a product term of a function an implicant.

Definition 3-4. A prime implicant of a function is a product term that represents a subcube that is not completely contained in a larger subcube.

Definition 3-5. An essential prime implicant of a function is a prime implicant that provides the only covering for a minterm, and must be used in the set of product terms that is used to express the function in a minimum form.

To obtain a minimum SOP or POS expression, the trick is to recognize and group only the essential prime implicants (if any exist). In other words, choose a set of *p*-subcubes (*r*-subcubes) that includes every *p*-square (*r*-square) at least once. In general, it is desirable to make the selected subcubes as large and as few in number as possible to cover all the *p*-squares (*r*-squares). Keep in mind that there is no guarantee that the result will be a minimum SOP or POS expression especially when there are no essential prime implicants. The reduction procedure outlined above, however, will yield a minimum solution in the majority of cases.

> **Definition 3-6.** A necessary prime implicant of a function is a prime implicant that is required when a particular set of product terms is used to express a function in a minimum form.

The Karnaugh map shown in Fig. 3-8*a* has two essential prime implicants considering only the 1s of the function *F1*. The 1s of function *F2* shown in Fig.

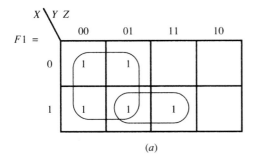

(a)

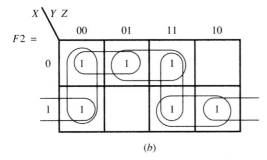

(b)

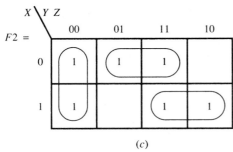

(c)

FIGURE 3-8
(a) Karnaugh map with essential prime implicants, (b) Karnaugh map with no essential prime implicants, (c) Selecting a minimum set of prime implicants to express the function *F2* (in (b)) in a minimum form.

3-8*b* form what is referred to as a cyclic map. This map has six prime implicants, none of which is essential because two different sets of prime implicants can be chosen to express the function in a minimum form. One combination that can be used to represent a minimum SOP expression for function *F2* is shown in Fig. 3-8*c*. Each of the prime implicants used to cover the cyclic map in Fig. 3-8*c* are necessary prime implicants, even though none of them are essential prime implicants. Since there is an alternate set of prime implicants that can be chosen to obtain a minimum SOP expression in Fig. 3-8*c*, this illustrates that reduced functions are not necessarily unique. If you will recall, there are two canonical or standard forms and these forms are unique; however, reduced forms for functions are very often not unique. In the case of cyclic functions, to find a minimum function can require finding all possible solutions and then comparing them to determine the minimum of all the solutions.

Definition 3-7. A redundant prime implicant of a function is a prime implicant that represents a subcube that is completely covered by other essential or necessary prime implicants, and is not required in the set of product terms that is used to express the function in a minimum form.

If minterms 0 and 1 in Fig. 3-8*c* were circled, forming a 1-cube, then that 1-cube would be a redundant prime implicant and should not be used in the set of product terms for the function. The 1-cube formed from minterms 1 and 3 and the 1-cube formed from minterms 0 and 4 completely cover the 1-cube formed from minterms 0 and 1.

Example 3-2

1. For the map shown in Fig. E3-2, write all the essential prime implicants for the 1s of the function in the map. Write a minimum SOP expression for the function.

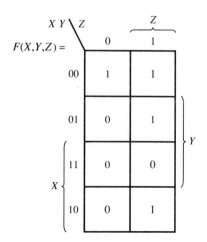

FIGURE E3-2

2. Write all the essential prime implicants for the 0s of the function in the map. Write a minimum SOP expression for the complement of the function.

Solution

1. For the 1s, the essential prime implicants are

$$\overline{X} \cdot \overline{Y}$$

$$\overline{X} \cdot Z$$

$$\overline{Y} \cdot Z$$

Since all the 1s in the function are covered by the essential prime implicants, these implicants when ORed together form a minimum SOP expression for the function F as follows.

$$F = \overline{X} \cdot \overline{Y} \ + \ \overline{X} \cdot Z \ + \ \overline{Y} \cdot Z$$

2. For the 0s, the essential prime implicants are

$$Y \cdot \overline{Z}$$

$$X \cdot Y$$

$$X \cdot \overline{Z}$$

Every 0 in the function is covered by the essential prime implicants. ORing these essential prime implicants together results in the following expression for the complement of the function F.

$$\overline{F} = Y \cdot \overline{Z} \ + \ X \cdot Y \ + \ X \cdot \overline{Z}$$

The three-variable Karnaugh map used in Fig. E3-2 can be easily obtained by simply rearranging a three-variable Venn diagram. This exercise is left for the student.

3-3-4 Plotting and Reducing Partially Reduced Functions

Sometimes a function or complement of a function is not expressed in either a standard form or a minimum form such as the following partially reduced function.

$$F(X, Y, Z) = X \cdot (\overline{Z} \ + \ \overline{Y} \cdot Z) \ + \ \overline{X} \cdot (Z \ + \ Y \cdot \overline{Z})$$

To plot this function on a Karnaugh map, Boolean algebra can be used to transform the expression on the right side of the function into a sum of products expression. In this case simply apply the Distributive Postulate P4*b*, $A \cdot (B + C) = A \cdot B \ + \ A \cdot C$, twice to change the factored form of the function into a form where all the product terms are summed, that is, ORed together.

$$F(X, Y, Z) = X \cdot \overline{Z} \ + \ X \cdot \overline{Y} \cdot Z \ + \ \overline{X} \cdot Z \ + \ \overline{X} \cdot Y \cdot \overline{Z}$$

Using the procedure presented in Chapter 1, the function can now be expanded such that every product term contains one of the input variables in either its complemented or uncomplemented form. The minterm compact form can then be written as follows.

Expanding the function we obtain

$$F(X, Y, Z) = X \cdot Y \cdot \bar{Z} + X \cdot \bar{Y} \cdot \bar{Z} + X \cdot \bar{Y} \cdot Z + \bar{X} \cdot Y \cdot Z + \bar{X} \cdot \bar{Y} \cdot Z$$
$$+ \bar{X} \cdot Y \cdot \bar{Z}$$
$$= m_6 + m_4 + m_5 + m_3 + m_1 + m_2$$
$$= \sum m(1, 2, 3, 4, 5, 6)$$

Using the minterm compact form, the function can now be plotted on a map as shown in Fig. 3-9. Once the minterm compact form is obtained, plotting the function is easy; however, Boolean algebra must be applied to obtain the compact form.

There is an alternate way to plot a sum of products expression on a Karnaugh map. This method does not require the function to be expanded into its minterm compact form. Recall that each product term of a function (complement of a functions) must always evaluate to 1 for the combination of inputs representing its subcube. The alternate way to plot a sum of products expression is to obtain the combination of inputs that causes each product term to evaluate to 1 and use that information to plot each subcube.

For the sum of products form of the function

$$F(X, Y, Z) = X \cdot \bar{Z} + X \cdot \bar{Y} \cdot Z + \bar{X} \cdot Z + \bar{X} \cdot Y \cdot \bar{Z}$$

we see by inspection

$$X \cdot \bar{Z} = 1 \qquad \text{when } X = 1, \text{ and } Z = 0,$$
$$X \cdot \bar{Y} \cdot Z = 1 \qquad \text{when } X = 1, Y = 0, \text{ and } Z = 1,$$
$$\bar{X} \cdot Z = 1 \qquad \text{when } X = 0, \text{ and } Z = 1, \text{ and finally}$$
$$\bar{X} \cdot Y \cdot \bar{Z} = 1 \qquad \text{when } X = 0, Y = 1, \text{ and } Z = 0.$$

To plot the function F on a three-variable map, a 1 is entered in each square representing the area where the input variables are fixed. The same procedure

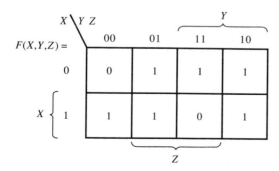

FIGURE 3-9
Plotting a function from its minterm compact form.

can be used to plot the 0s of a function F after a SOP expression for $\overline{F}$ is obtained.

The 1s (or 0s) for a function are usually plotted on one map as shown in Fig. 3-9. To clearly illustrate the procedure, the product terms have been labeled p_1 through p_4 and are shown plotted on a separate map in Fig. 3-10. To plot the p-subcube for product term p_1, simply place a 1 in each square that represents the areas for $X = 1$ and $Z = 0$ (the minterms 4 and 6 on the map in Fig. 3-10a). The p-subcube for product term p_2 is plotted by placing a 1 in the square that represents the area for $X = 1$, $Y = 0$, and $Z = 1$ (minterm 5 on the map in Fig. 10b). The p-subcubes for product terms p_3 and p_4 are plotted in a similar manner as shown on the maps in Figs. 10c and d. It is easy to see that if we had plotted the 1s for each p-subcube on the same map, that map would look just like the map in Fig. 3-9.

The 1s of the function in Fig. 3-9 represent a cyclic map. To obtain a minimum SOP expression requires choosing a minimum set of prime implicants to cover all the 1s in the map as we did for the cyclic map shown in Fig. 3-8c.

Example 3-3

  1. Plot the Karnaugh map for the following functions by plotting the subcubes for the individual product terms.

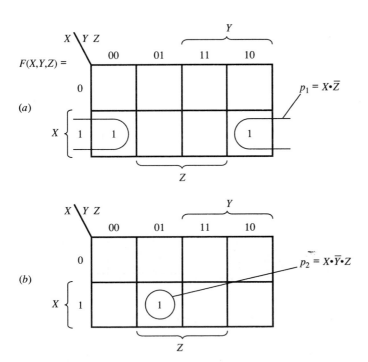

FIGURE 3-10
Plotting the p-subcubes for a function in SOP form (a) through (d).

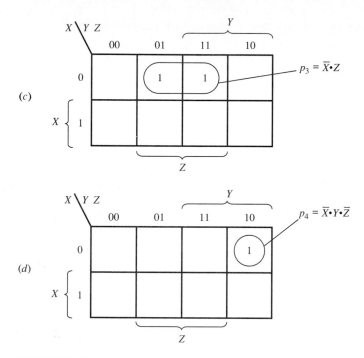

(c)

(d)

FIGURE 3-10
(*continued*)

     (a) $F1(X,Y,Z) = \overline{Y} \cdot \overline{Z} + Y \cdot \overline{Z} + \overline{X} \cdot \overline{Y} \cdot \overline{Z} + X \cdot Y \cdot Z + X \cdot Y$
     (b) $\overline{F2}(X,Y,Z) = Y \cdot (Z + X \cdot \overline{Z}) + X \cdot \overline{Y} \cdot \overline{Z}$

2. Obtain a minimum SOP expression for $F1$ and $\overline{F2}$.

Solution

1. First we convert the function into a SOP form if it is not already in a SOP form. The function $F1$ was given in a SOP form. The Karnaugh map for $F1$ is plotted

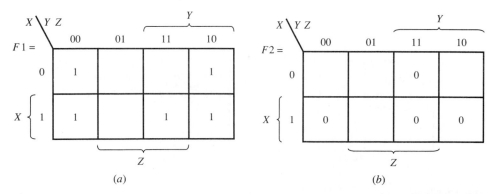

(a)

(b)

FIGURE E3-3

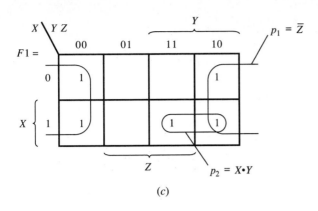

(c)

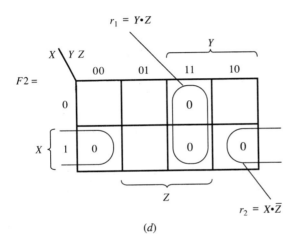

(d)

FIGURE E3-3
(*continued*)

in Fig. E3-3*a*. The blank squares represent *r*-squares (0s), and since these will not be used they are not filled in.

We must convert the function $\overline{F2}$ into a SOP form.

$$\overline{F2}(X, Y, Z) = Y \cdot Z + X \cdot Y \cdot \overline{Z} + X \cdot \overline{Y} \cdot \overline{Z}$$

The Karnaugh map for $\overline{F2}$ is plotted in Fig. E3-3*b*. The blank squares represent *p*-squares (1s), and since these will not be used they are not filled in.

2. The two essential prime implicants shown in Fig. E3-3*c* result in the following minimum SOP form for *F*1.

$$F1 = p_1 + p_2 = \overline{Z} + X \cdot Y$$

The two essential prime implicants shown in Fig. E3-3*d* result in the following minimum SOP form for $\overline{F2}$.

$$\overline{F2} = r_1 + r_2 = Y \cdot Z + X \cdot \overline{Z}$$

3-4   FOUR- AND FIVE-VARIABLE KARNAUGH MAPS

Many different functions can be represented using four variables ($2^{2^n} = 2^{2^4} = 65,536$) and five variables ($2^{2^n} = 2^{2^5} = 4,294,967,296$); however, Karnaugh maps for these functions require only 2^n squares. A four-variable Karnaugh map thus contains 16 squares, and a five-variable Karnaugh map contains 32 squares.

3-4-1   Using Four-Variable Maps

Using a four-variable Venn diagram, we can construct four-variable Karnaugh maps as shown in Fig. 3-11. To assist in plotting four-variable maps, an alternate map is shown filled in with minterm designators. Like the two-variable map and the three-variable map, the four-variable map is arranged so only one literal can change between any two adjacent squares, thus allowing the Adjacency Theorem T7a, ($A \cdot B + A \cdot \bar{B} = A$), to be used to simplify functions plotted in the map. Using the adjacency theorem, each square along the top row (left column) of the map is adjacent to a corresponding square along the bottom row (right column) of the map. End around adjacency exists on the four-variable Karnaugh map in Fig. 3-11 for the following sets of squares along the top and bottom rows of the map $\{m_0,m_8\}$, $\{m_1,m_9\}$, $\{m_3,m_{11}\}$, and $\{m_2,m_{10}\}$. End around adjacency also exists along the left and right columns of the map for the following sets of squares $\{m_0,m_2\}$, $\{m_4,m_6\}$, $\{m_{12},m_{14}\}$, and $\{m_8,m_{10}\}$.

It is also possible to construct a four-variable cube representation for a four variable function by drawing a cube inside of a cube. The vertices (which represent minterms) of the outer cube can be accessed just like the three-variable cube illustrated in Fig. 3-7b, allowing the value of the fourth variable to be fixed at 0. Corresponding vertices of the intercube can be accessed by the same three variables as the outer cube, allowing the value of the fourth variable to be fixed at 1. It may be easier to identify adjacent sets of minterms on a four-variable cube since they lie along the same axis, lie in the same plane, or belong to the same inner or outer cube, but because a four-variable cube is harder to draw than an equivalent four-variable map representation, the cube representation is not used as a design tool. The construction of a four-variable cube is left as an exercise for the student.

Consider the following switching function:

$$F(W,X,Y,Z) = \sum m(3,4,6,7,9,11,12,14,15)$$

This function is shown plotted in Fig. 3-12 using the minterms listed in the minterm compact form. Those squares not included in the compact form represent the 0s of the function and are not filled in. The p-subcubes circled for the function each represent an essential prime implicant. Using the essential prime implicants, the following minimum sum of products expression is obtained for the function.

$$F(W,X,Y,Z) = p_1 + p_2 + p_3$$
$$= X \cdot \bar{Z} + Y \cdot Z + W \cdot \bar{X} \cdot Z$$

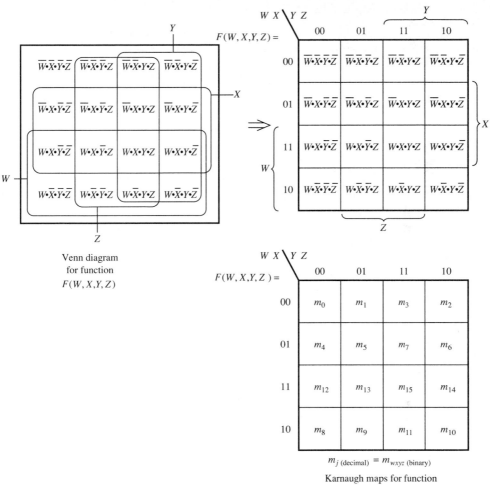

FIGURE 3-11
Four-variable Venn diagram and corresponding four-variable Karnaugh maps.

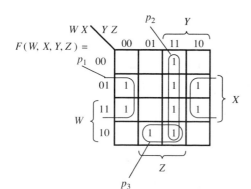

FIGURE 3-12
Four-variable map with *p*-subcubes circled
to obtain a minimum SOP expression.

3-4-2   Generating a Check List

As a check on our work, we can calculate the number of literals that each product term should contain by noting the size of its subcube. Product terms p_1 and p_2 are both 2-cubes; therefore, they each contain $(k = n - i = 4 - 2 = 2)$ 2 literals. Product term p_3 is a 1-cube. It contains $(k = n - i = 4 - 1 = 3)$ 3 literals. If the number of literals obtained for a product term does not agree with the number calculated for the same product term, there is an error to be corrected.

Another check that should be made is to verify that each product term evaluates to 1 for its respective input combinations as read along the edges of the map. The input combinations for p_1 $(X = 1, Z = 0)$ show that $p_1 = X \cdot \bar{Z} = 1$. The input combinations for p_2 $(Y = 1, Z = 1)$ show that $p_2 = Y \cdot Z = 1$. The input combinations for p_3 $(W = 1, X = 0, Z = 1)$ show that $p_3 = W \cdot \bar{X} \cdot Z = 1$. If any of the product terms fail to evaluate to 1 for their respective input combinations, there is an error to be corrected.

In Fig. 3-13 we show the same function plotted on a four-variable Karnaugh map using the 0s of the function instead of the 1s. The r-subcubes circled, when ORed together, represent a minimum sum of products expression for the complement of the function. Prime implicants r_1 and r_3 are essential, while prime implicant r_2 is one of two possible prime implicants that can be used to cover the 0 for m_2. The following minimum sum of products expression can be written for the complement of the function.

$$\bar{F}(W, X, Y, Z) = r_1 + r_2 + r_3$$

$$= \bar{X} \cdot \bar{Z} + \bar{W} \cdot \bar{Y} \cdot Z + X \cdot \bar{Y} \cdot Z$$

By filling in a checklist table like the one in Table 3-1, we can verify at a glance that the number of literals in the table agrees with the number of literals found for each product term of $\bar{F}$. Also each product term evaluates to 1 for its respective input combinations.

Example 3-4. Obtain a minimum SOP expression for the following function. Also obtain a minimum POS expression for the same function.

$$F(A, B, C, D) = \prod M(0, 1, 4, 5, 9)$$

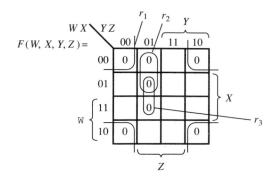

FIGURE 3-13
Four-variable map with r-subcubes circled to obtain a minimum SOP expression.

TABLE 3-1
Checklist table for the product terms of $\overline{F}$

Product term	Subcube size i	Number of literals $k = n - i$	Input combinations	Result
r_1	2-cube	$k = 4 - 2 = 2$	$X = 0, Z = 0$	$\overline{X} \cdot \overline{Z} = 1$
r_2	1-cube	$k = 4 - 1 = 3$	$W = 0, Y = 0,$ $Z = 1$	$\overline{W} \cdot \overline{Y} \cdot Z = 1$
r_3	1-cube	$k = 4 - 1 = 3$	$X = 1, Y = 0,$ $Z = 1$	$X \cdot \overline{Y} \cdot Z = 1$

Solution The function is shown plotted in Fig. E3-4.
Reading the map we can write the following minimum SOP expression for F.

$$F = C + A \cdot B + A \cdot \overline{D}$$

Reading the map we can also write the following minimum SOP expression for $\overline{F}$.

$$\overline{F} = \overline{A} \cdot \overline{C} + \overline{B} \cdot \overline{C} \cdot D$$

A minimum POS expression for F can be written by using the relationship

minimum POS form of $F = \overline{\text{minimum SOP form of } \overline{F}}$

and applying DeMorgan's theorem:

$$F = (A + C) \cdot (B + C + \overline{D})$$

3-4-3   Using Five-Variable Maps

Karnaugh maps larger than five variables are seldom used. Sometimes however, it is convenient to draw a five-variable map without resorting to a computer program to perform the reduction. One version of a five-variable map is shown in Fig.

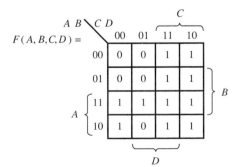

FIGURE E3-4

3-14. In this version, squares on the top map are adjacent to the squares on the bottom map when the two maps are superimposed. This version of a five-variable map uses the property of overlay adjacency. If desired, the concept of using two four-variable maps to form a five-variable map can also be extended to using two five-variable maps to represent a six-variable map. When using two four-variable maps to represent a five-variable map, the bottom map is labeled in the usual manner, and the top map is labeled in a similar fashion beginning with m_{16} and proceeding through m_{31} as shown in Fig. 3-14. Examples for several minterms are also illustrated in Fig. 3-14. To keep the diagram simple, only the minterm designators are used on the map. Like all the previous maps, the five-variable map is arranged so only one literal can change between any two adjacent squares including the additional property due to overlay adjacency.

Another common version of a five-variable Karnaugh map uses foldout adjacency. First a four-variable map is drawn, and then a second four-variable map is drawn to the immediate right of the first map. Squares on the right map that match up with squares on the left map when the map is folded back on itself are adjacent. The construction of a five-variable map using foldout adjacency is left as an exercise for the reader.

Example 3-5. Obtain a minimum SOP expression for the function F:

$F(V, W, X, Y, Z) =$

$$\sum m(3, 4, 7, 9, 11, 12, 15, 16, 18, 19, 20, 23, 24, 26, 27, 28, 31)$$

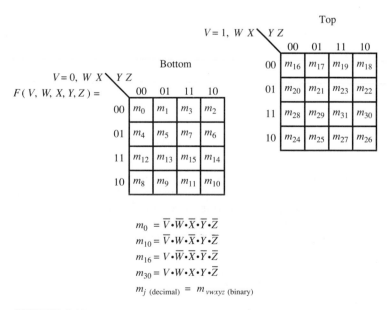

$m_0 = \overline{V} \cdot \overline{W} \cdot \overline{X} \cdot \overline{Y} \cdot \overline{Z}$
$m_{10} = \overline{V} \cdot W \cdot \overline{X} \cdot Y \cdot \overline{Z}$
$m_{16} = V \cdot \overline{W} \cdot \overline{X} \cdot \overline{Y} \cdot \overline{Z}$
$m_{30} = V \cdot W \cdot X \cdot Y \cdot \overline{Z}$

$m_{j \ (\text{decimal})} = m_{vwxyz \ (\text{binary})}$

FIGURE 3-14
Five-variable Karnaugh map using overlay adjacency.

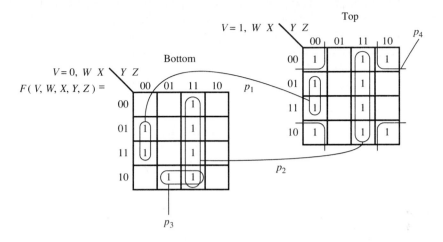

FIGURE E3-5

Solution The function is plotted in the five-variable map shown in Fig. E3-5. From the map we can write the following minimum SOP expression for F.

$$F = p_1 + p_2 + p_3 + p_4$$
$$= X \cdot \overline{Y} \cdot \overline{Z} \; + \; Y \cdot Z \; + \; \overline{V} \cdot W \cdot \overline{X} \cdot Z \; + \; V \cdot \overline{X} \cdot \overline{Z}$$

Product terms p_1 and p_2 exhibit the property of overlay adjacency. Variable V is therefore absent from these product terms, since its value is not fixed.

3-5 MULTIPLE FUNCTION MINIMIZATION

A problem we have not addressed up to this point is multiple output function minimization. Switching circuits are often required to have more than one output. Take the case of the two functions listed below.

$$F1 = \overline{Z} \; + \; \overline{X} \cdot \overline{Y}$$
$$F2 = \overline{X} \cdot \overline{Y} \cdot Z \; + \; Y \cdot \overline{Z}$$

These minimum functions are obtained by reading the maps in Fig. 3-15a independently. One simplified method of measuring the complexity of a function is to count the number of literals in the function. The total literal count of a set of independent functions is, therefore, the sum of the literals in each individual function. The method of counting literals is easy to use but it ignores the number of product terms ORed together in SOP expressions or the number of sum terms ANDed together in POS expressions. For a more accurate measure of circuit complexity that uses second level gating, that is, it includes both the sum terms and product terms of a function, see Rhyne in the list of references at the end of the chapter. To keep the following discussion simple, we will use the less accurate literal count method as a measure of function complexity. The literal

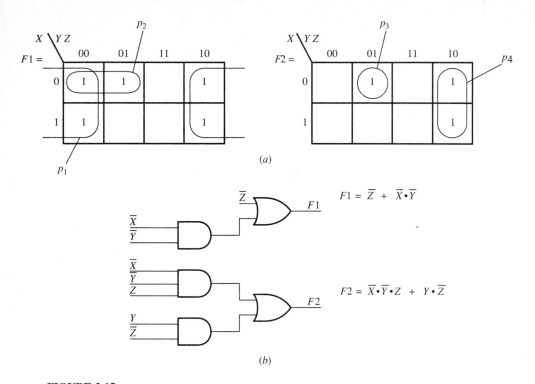

(a)

(b)

FIGURE 3-15
(a) $F1$ and $F2$ obtained using independent minimizations, (b) Independent circuit implementations for $F1$ and $F2$.

count of $F1$ is 3 literals while the literal count of $F2$ is 5 literals. The total literal count of the set of functions $F1$ and $F2$ taken independently is 8 literals. In Fig. 3-15b we show circuit implementations for $F1$ and $F2$.

Multiple function minimization means to obtain a combined set of reduced output expressions such that the expressions are no longer independent of one another. This means that the expressions will have shared product terms, that is, product terms that are the same. A product term is only generated once. Afterward it is used as needed by other expressions in the combined set. A shared product term is generated by covering a prime implicant representing a subcube on one of the maps with the same subcube on a different map. Several attempts are often required to obtain a combined set of expressions with a minimum number of literals. The goal is to obtain a combined set of minimum SOP (or POS) expressions with a complexity in terms of total literal count less than the total literal count of an independent set of minimum SOP (or POS) expressions. The concept of multiple function minimization is easy to understand but usually difficult to execute.

In Fig. 3-16a we show the same maps for functions $F1$ and $F2$ with a slightly different covering of the 1s on the maps. On the map for $F2$, the covering of minterm 1 is represented by prime implicant p_2. Minterm 1 on map $F1$ is

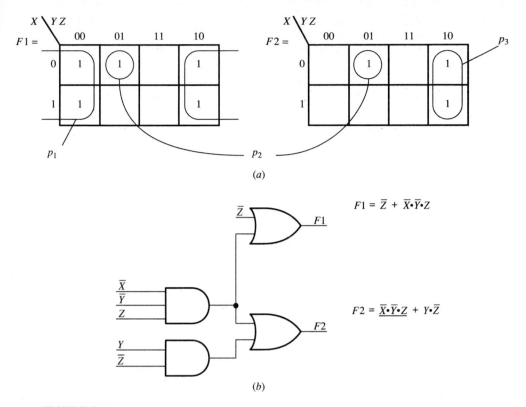

FIGURE 3-16
(*a*) *F*1 and *F*2 obtained using multiple function minimization, (*b*) combined circuit implementation for *F*1 and *F*2.

not a prime implicant, but if it is also covered as a 0-cube, i.e., the same subcube as map *F*2, then prime implicant p_2 can be shared between the two output expressions. This can be seen by obtaining the combined set of functions in SOP form for these coverings.

$$F1 = \overline{Z} + \overline{X} \cdot \overline{Y} \cdot Z$$
$$F2 = \overline{X} \cdot \overline{Y} \cdot Z + Y \cdot \overline{Z}$$

We show the prime implicant for p_2 underlined in the second function as a reminder that it will only be generated once for the first function and then shared as many times as it is needed. Underlined terms are often called *no cost* or *0 cost terms* since the literals in the underlined terms are not counted. Counting the number of literals for the combined functions *F*1 and *F*2, we obtain 4 literals for *F*1, and 2 literals for *F*2, for a total of 6 literals. This set of combined minimum SOP expressions therefore provides fewer literals than the total number of literals required by the independent minimum SOP expressions. The combined SOP expressions therefore require less hardware to implement as shown in Fig. 3-16*b*. Observe that the output of the three-input AND gate is only generated once, and is shared when it is needed again, thus providing a circuit implementation with fewer literals and fewer components.

The problem of minimization of multiple functions is more appropriate for a computer program. This is true since it is a difficult task to manually search for the minimum number of literals that can result from covering corresponding subcubes on several maps. As the number of output functions or the number of input variables is increased, the search becomes more difficult and takes longer to perform.

Example 3-6

1. Using Karnaugh maps, circle the p-subcubes to obtain an independent minimum SOP expression for each of the following functions. Indicate the literal count for each p-subcube ($k = n - i$), and calculate the complexity in terms of total literal count without writing the independent minimum SOP expressions.

$$F1(A, B, C, D) = \sum m(4, 6, 7, 8, 9, 10, 11, 12, 13, 14, 15)$$

$$F2(A, B, C, D) = \sum m(4, 6, 7, 8, 9, 10, 11, 12)$$

2. Using Karnaugh maps, circle the p-subcubes to obtain a combined set of minimum SOP expressions for the same functions. Indicate the literal count for each p-subcube and calculate the complexity without writing the combined minimum SOP expressions.
3. Write the set of functions that has the overall simplest SOP complexity.

Solution

1. The independent maps in Fig. E3-6a indicate a complexity of 13 literals for the two functions considered independently.

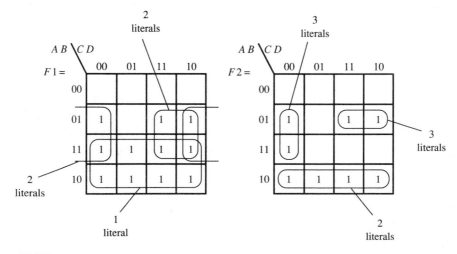

FIGURE E3-6a

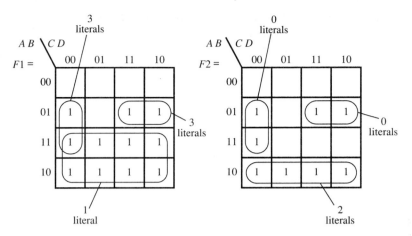

FIGURE E3-6*b*

2. The combined maps in Fig. E3-6*b* indicate a complexity of 9 literals by allowing a sharing of product terms.

3. The combined set of functions in this case has the simplest SOP complexity and are listed below. The shared terms (0 cost terms) are shown underlined.

$$F1 = B \cdot \overline{C} \cdot \overline{D} \ + \ \overline{A} \cdot B \cdot C \ + \ A$$

$$F2 = \underline{B \cdot \overline{C} \cdot \overline{D}} \ + \ \underline{\overline{A} \cdot B \cdot C} \ + \ A \cdot \overline{B}$$

3-6   DON'T CARE OUTPUT CONDITIONS

All the output variables for the functions we have covered thus far have been assigned a fixed value of 0 or 1 for all the combinations specified by the input variables. Functions of this type are called completely or fully specified functions. There are times, however, when certain input variable values may never occur simultaneously as in the case of any of the binary codes listed in Table 2-2 in Chapter 2. The Excess 3 code (XS3 code), for example, has 4 bits to represent 10 coded values out of a possible 16 combinations, while the 2-out-of-5 code has 5 bits to represent 10 coded values out of a possible 32 combinations. Six of the combinations for the XS3 code and twenty-two of the combinations for the 2-out-of-5 code (those not representing the decimal numbers 0 through 9) should not occur, but if they do and you don't care, the output conditions that occur from these combinations are classified as don't care output conditions.

A don't care output condition is usually represented by one of the following symbols in a truth table or Karnaugh map: "*X*," "*d*," or "−." When don't care outputs (often referred to as simply don't cares) occur in a Karnaugh map, they may be used to enlarge certain subcubes (either *p*-subcubes when grouping the 1s or *r*-subcubes when grouping the 0s), providing prime implicants that contain a smaller number of literals to simplify the function. Don't cares in a map are never used to form additional subcubes all by themselves.

X	Y	Z	F1	F2
0	0	0	1	1
0	0	1	1	0
0	1	0	1	X
0	1	1	0	0
1	0	0	0	1
1	0	1	0	X
1	1	0	X	X
1	1	1	0	X

FIGURE 3-17
Truth table with don't care output conditions indicated by Xs.

Observe the truth table shown in Fig.3-17. It has one don't care in the output column for $F1$ and four don't cares in the output column for $F2$ as indicated by the Xs in these columns.

Functions that include don't cares are said to be incompletely specified. To accommodate the don't cares in minterm compact and maxterm compact forms, the functions for these forms may be modified to include the list of don't cares as follows:

$$F1(X, Y, Z) = \sum m(0, 1, 2) + d(6)$$

$$F1(X, Y, Z) = \prod M(3, 4, 5, 7) \cdot d(6)$$

$$F2(X, Y, Z) = \sum m(0, 4) + d(2, 5, 6, 7)$$

$$F2(X, Y, Z) = \prod M(1, 3) \cdot d(2, 5, 6, 7)$$

The Karnaugh map descriptions for these functions are shown in Fig.3-18. In the map for $F1$ the don't care at minterm 6 allows two alternatives for writing a prime implicant to include minterm 2. p_3 is one prime implicant and p_1 is the second prime implicant. In the map for $F1$ the don't care at minterm 6 allows r_2 to increase in size from a 1-cube when only 0s are used to a 2-cube requiring 1 less literal.

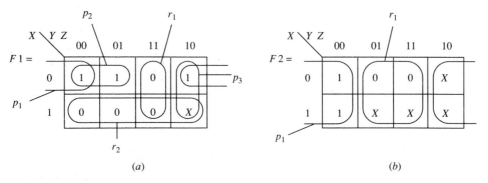

(a) (b)

FIGURE 3-18
Karnaugh maps with don't care output conditions indicated by Xs.

When a don't care is used both as a 1 and as a 0 to obtain a minimum SOP expression and a minimum POS expression, the two expressions cannot be algebraically identical since the value of the square represented by the don't care is not fixed. However, when a don't care is used only as a 1 or only as a 0 to obtain a minimum SOP expression and a minimum POS expression, then the two expressions will be algebraically identical, that is, one expression can be algebraically obtained from the other expression. Reading the map for function $F1$ in Fig. 3-18a, two different minimum SOP expressions can be written for $F1$ as indicated below.

$$F1 = p_1 + p_2 = \overline{X} \cdot \overline{Z} + \overline{X} \cdot \overline{Y}$$

or,
$$F1 = p_2 + p_3 = \overline{X} \cdot \overline{Y} + Y \cdot \overline{Z}$$

Only a single minimum POS expression can be written for $F1$, however.

$$\overline{F1} = r_1 + r_2 = Y \cdot Z + X$$
$$F1 = (\overline{Y} + \overline{Z}) \cdot \overline{X}$$

The first minimum SOP expression is algebraically identical to the minimum POS expression, since the value of the don't care at minterm 6 is fixed at 0 by the selection of prime implicants for both expressions. The second minimum SOP expression, however, is not algebraically identical to the minimum POS expression since the don't care at minterm 6 is used both as a 1 and as a 0 by the selection of the prime implicants for both expressions.

It can be observed that both minimum SOP expressions for function $F1$ have a literal count of 4 literals, while the minimum POS expression for function $F1$ has a literal count of only 3 literals. When working with incompletely specified functions, one form of the function may be simpler than the other form as we have demonstrated here (this is also true for completely specified functions). To obtain the simplest minimum SOP or POS expression for a function (the expression with the least number of literals) requires examining both a minimum SOP expression as well as a minimum POS expression (or minimum SOP expression for the complement of a function).

Now observe the map for $F2$ in Fig. 3-18b. Both p_1 and r_1 can increase in size from a 1-cube, when only 1s or 0s are used, to a 2-cube, taking advantage of the appropriate don't care outputs. The following minimum SOP expressions are obtained for $F2$ and $\overline{F2}$ by reading the map for $F2$ in Fig. 3-18b.

$$F2 = p_1 = \overline{Z}$$
$$\overline{F2} = r_1 = Z$$

Both the minimum SOP and the minimum POS expressions in this case are algebraically identical, as we would expect, since each don't care has been restricted to only one value by our choice of prime implicants.

Example 3-7. Obtain a set of minimum SOP expressions for a switching circuit that will accept XS3 code at its input and provide BCD code at its output. Show a block diagram for a XS3 to BCD code converter with the inputs and outputs labeled. Show a circuit diagram using AND and OR element circuit symbols. In the circuit diagram, assume the input variables are available in uncomplemented as well as complemented form. Show a circuit to generate the complemented input variables if only the uncomplemented input variables are available.

Solution The problem is stated in terms of the truth table providing four inputs and four outputs shown below. The input combinations are the XS3 encoded combinations listed as $I3$ through $I0$, while the output combinations are the corresponding BCD encoded combinations listed as $F3$ through $F0$. Outputs for the six input combinations (1101, 1110, 1111, 0000, 0001, and 0010) that do not represent the decimal digits 0 through 9 are don't cares and are represented as such in the table under each output column using the symbol X as shown below.

Decimal digits	XS3				BCD			
	I3	*I2*	*I1*	*I0*	*F3*	*F2*	*F1*	*F0*
0	0	0	1	1	0	0	0	0
1	0	1	0	0	0	0	0	1
2	0	1	0	1	0	0	1	0
3	0	1	1	0	0	0	1	1
4	0	1	1	1	0	1	0	0
5	1	0	0	0	0	1	0	1
6	1	0	0	1	0	1	1	0
7	1	0	1	0	0	1	1	1
8	1	0	1	1	1	0	0	0
9	1	1	0	0	1	0	0	1
	1	1	0	1	X	X	X	X
	1	1	1	0	X	X	X	X
	1	1	1	1	X	X	X	X
	0	0	0	0	X	X	X	X
	0	0	0	1	X	X	X	X
	0	0	1	0	X	X	X	X

By observing the truth table, one can readily write the function for $F0$. It is just the $I0$ column complemented, $F0 = \overline{I0}$. Using the truth table, the Karnaugh maps shown in Figs. E3-7a through c are drawn and filled in.

Using a minimum set of prime implicants to cover the 1s for each function independently, we obtain the following minimum SOP expressions. Don't cares are only used to enlarge p-subcubes and are not used to form prime implicants by themselves.

$$F3 = p_1 + p_2 = I3 \cdot I2 + I3 \cdot I1 \cdot I0$$

$$F2 = p_3 + p_4 + p_5 = \overline{I2} \cdot \overline{I0} + \overline{I2} \cdot \overline{I1} + I2 \cdot I1 \cdot I0$$

$$= \overline{I2} \cdot (\overline{I0} + \overline{I1}) + I2 \cdot I1 \cdot I0$$

$$F1 = p_6 + p_7 = \overline{I1} \cdot I0 + I1 \cdot \overline{I0}$$

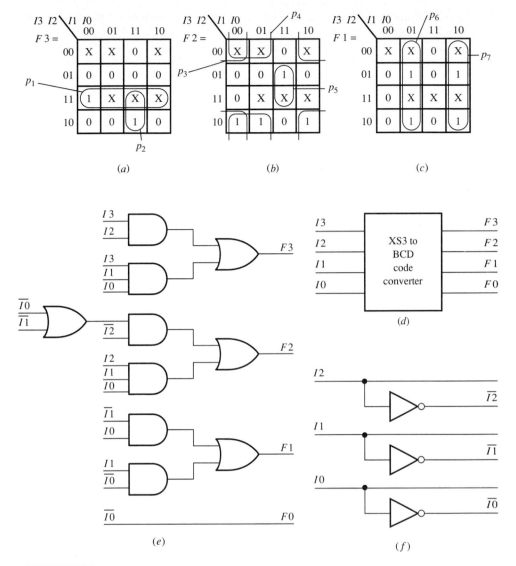

FIGURE E3-7

A block diagram illustrating the input and output variables for the XS3 to BCD code converter is shown in Fig. E3-7d. The circuit for the code converter using AND and OR elements is illustrated in Fig. E3-7e. The circuit used to obtain the complemented variables from the uncomplemented variables is shown in Fig. E3-7f.

The individual circuit for the function $F3$ was implemented using the following procedure. The AND operators in the function have the highest precedence, and so the symbols for the two AND elements were drawn first. This is called the first level of the implementation. The OR operator in the function *F3* has the second order of precedence, so the symbol for the OR element was drawn next. This is called the second level of the implementation. The outputs for the first level of the implementation are then connected to the second level inputs. Finally, the inputs (*I3*, *I2*, *I3*, *I1*, and *I0*) for the first level are labeled and the output (*F3*) for the second level is labeled. Since two levels of logic are used for implementations of *F3* and *F1*, these circuits are called 2-level circuit implementations. The implementation for *F2* is a 3-level circuit implementation to allow the use of 2-input OR gates, the more common off-the-shelf devices (3-input OR gates are not generally available as off-the-shelf devices).

Obtaining a set of minimum POS expressions for a XS3 to BCD code converter is left as an exercise for the student.

3-7   ADDITIONAL FUNCTIONAL FORMS

As we have discussed, there are two possible minimum forms for a Boolean function. These are the minimum SOP form obtained from the 1s of the function, and the minimum POS form obtained from the 0s of the function. Both of these forms can be implemented using the following elements or gates: AND, OR, NAND, and NOR. Up to now we have only shown how to use AND and OR gates to implement functions. To illustrate how functions can be implemented using combinations other than ANDs and ORs, consider the incompletely specified function represented by the Karnaugh map shown in Fig. 3-19.

The minimum SOP expression and minimum POS expression for *F* can be written as follows. Since product terms p_1 and r_1 use the same don't care at minterm 5 (the don't care value is not fixed), the two minimum expressions are not algebraically identical.

$$F = p_1 + p_2 = \overline{A} \cdot \overline{C} + A \cdot B \quad \text{(function for minimum SOP expression)}$$

$$\overline{F} = r_1 + r_2 = C + A \cdot B$$

$$F = \overline{C} \cdot (\overline{A} + \overline{B}) \qquad \text{(function for minimum POS expression)}$$

Beginning with the minimum SOP expression in AND/OR form, we can derive additional functional forms using the theorems of Boolean algebra, that is, using algebraic transformations. The name associated with each of the following

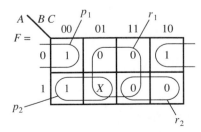

FIGURE 3-19
An incompletely specified function.

functional forms is obtained by observing the order of precedence of the operators in the function. The name of each form also represents the gate types for the first level and the second level of the circuit implementation for the function.

Functional forms

$$F = \overline{A} \cdot \overline{C} + A \cdot \overline{B} \qquad \text{AND/OR form}$$

$$= \overline{\overline{\overline{A} \cdot \overline{C}} + \overline{A \cdot \overline{B}}} = \overline{(\overline{A} \cdot \overline{C}) \cdot (A \cdot \overline{B})} \qquad \text{NAND/NAND form}$$

$$= \overline{(A + C) \cdot (\overline{A} + B)} \qquad \text{OR/NAND form}$$

$$= \overline{(A + C)} + \overline{(\overline{A} + B)} \qquad \text{NOR/OR form}$$

It is instructive to notice the cyclic nature of the derived forms. The AND/OR form is the SOP expression for the function. Applying the double complementation theorem followed by DeMorgan's theorem to this form results in the NAND/NAND form of the function. Applying DeMorgan's theorem again we obtain the OR/NAND form. Continuing in this manner, we finally obtain the form called the NOR/OR form. Applying DeMorgan's theorem once more we circle back to the original AND/OR form. A 2-level circuit implementation for each of these SOP derived forms is shown in Fig. 3-20.

Beginning with the minimum POS expression in OR/AND form, we can also derive the following additional functional forms using algebraic transformations.

Functional forms

$$F = \overline{C} \cdot (\overline{A} + \overline{B}) \qquad \text{OR/AND form}$$

$$= \overline{\overline{\overline{C} \cdot (\overline{A} + \overline{B})}} = \overline{C + \overline{(\overline{A} + \overline{B})}} \qquad \text{NOR/NOR form}$$

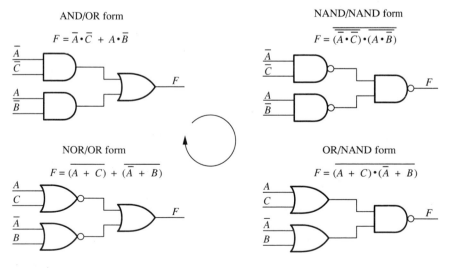

FIGURE 3-20
SOP derived forms.

$$= \overline{C + (A \cdot B)} \qquad \text{AND/NOR form}$$

$$= \overline{C} \cdot \overline{(A \cdot B)} \qquad \text{NAND/AND form}$$

Applying DeMorgan's theorem to the last form in the group (the NAND/AND form) results in the OR/AND form, the first form in the group, illustrating the cyclic nature of the process. Fig. 3-21 illustrates a 2-level circuit implementation for each of these POS derived forms.

Of the many possible 2-level circuit forms for a function, the AND/OR form and the OR/AND form are the ones predominantly used. Both of these forms only involve simple AND, OR, and complement operations on single variables. Forms other than the AND/OR form and the OR/AND form should not be considered as a serious design tool for gate level circuit design, since the algebraic transformations required to obtain these forms are generally considered too error prone. An easy to apply top-down design approach will be introduced in Chapter 5 as the tool we recommend for obtaining gate level designs utilizing AND, OR, NAND, and NOR gates.

Example 3-8. Show a NAND/NAND form of circuit implementation for the minimum functions obtained in Example 3-7 for the XS3 to BCD code converter. Assume that the input variables are available in complemented form as well as uncomplemented form.

Solution Applying the double complementation theorem followed by DeMorgan's theorem to the minimum SOP expressions for $F3$, $F2$, and $F1$ obtained in Example 3-7 results in the following NAND/NAND forms.

$$F3 = I3 \cdot I2 + I3 \cdot I1 \cdot I0 = \overline{\overline{I3 \cdot I2} + \overline{I3 \cdot I1 \cdot I0}}$$

$$= \overline{(I3 \cdot I2)} \cdot \overline{(I3 \cdot I1 \cdot I0)}$$

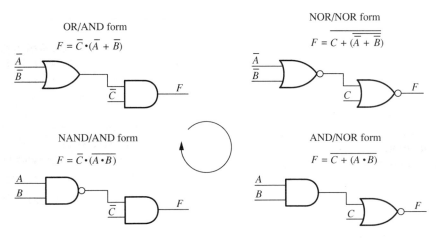

OR/AND form

$$F = \overline{C} \cdot (\overline{A} + \overline{B})$$

NOR/NOR form

$$F = \overline{C + (\overline{A} + \overline{B})}$$

NAND/AND form

$$F = \overline{C} \cdot \overline{(A \cdot B)}$$

AND/NOR form

$$F = \overline{C + (A \cdot B)}$$

FIGURE 3-21
POS derived forms.

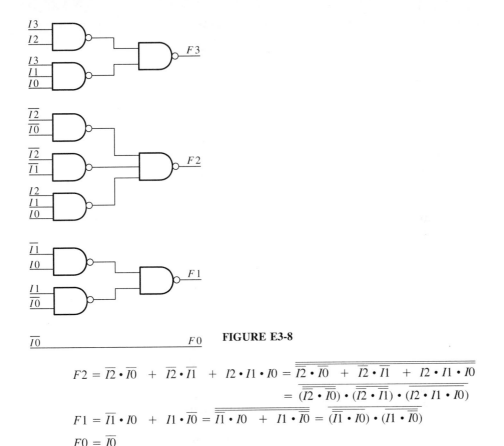

FIGURE E3-8

$$F2 = \overline{I2} \cdot \overline{I0} \;+\; \overline{I2} \cdot \overline{I1} \;+\; I2 \cdot I1 \cdot I0 = \overline{\overline{I2} \cdot \overline{I0}} \;+\; \overline{\overline{I2} \cdot \overline{I1}} \;+\; \overline{I2 \cdot I1 \cdot I0}$$

$$= \overline{(\overline{I2} \cdot \overline{I0}) \cdot (\overline{I2} \cdot \overline{I1}) \cdot (I2 \cdot I1 \cdot I0)}$$

$$F1 = \overline{I1} \cdot I0 \;+\; I1 \cdot \overline{I0} = \overline{\overline{I1} \cdot I0} \;+\; \overline{I1 \cdot \overline{I0}} = \overline{(\overline{I1} \cdot I0) \cdot (I1 \cdot \overline{I0})}$$

$$F0 = \overline{I0}$$

The circuit implementation for the code converter using only NAND elements or gates is shown in Fig. E3-8. The larger number of complementations required to obtain NAND/NAND forms from AND/OR forms requires extra care to ensure that no complements are accidently dropped.

3-8 COMMENTS ON USEFULNESS OF KARNAUGH MAPS

With the availability of programmable logic devices (PLDs) which will be discussed later, some will argue that Karnaugh maps are not necessary since simpler functions are often not needed. In some cases this may be true but there are still cases where functions with fewer literals are necessary. The following statement was made by Karnaugh concerning his map method. "The map method, in its present form, is likely to be useful in two ways: as a pedagogic device, for the introduction of ideas about logic circuits and their synthesis, and also as a desk-top aid to the working engineer." New large scale integrated devices and programmable devices have reduced the need for Karnaugh maps; however, on occasions, these new devices represent an overkill. In these cases the use of small-scale integrated devices or medium-scale integrated devices provide a more economical solution, with Karnaugh maps still the most convenient reduction tool.

REFERENCES

1. Bartee, T. C., *Digital Computer Fundamentals,* 5th ed., McGraw-Hill Book Company, New York, 1981.
2. Veitch, E. W., "A Chart Method for Simplifying Truth Functions." *Proc. of the ACM,* pp. 127–133, May, 1952.
3. Karnaugh, M., "The Map Method for Synthesis of Combinational Logic Circuits", *Trans. AIEE, Comm. and Electronics,* Vol. 72, Part I, pp. 593–599, Nov., 1953.
4. Hill, J. H., and Peterson, G. R., *Introduction to Switching Theory & Logic Design,* 3d ed., John Wiley & Sons, New York, 1981.
5. Roth, C. H., Jr., *Fundamentals of Logic Design,* 3d ed., West Publishing Co., St. Paul, Minnesota, 1985.
6. Edwards, H. E., *The Principles of Switching Circuits,* The M.I.T. Press., Cambridge, Mass., 1973.
7. Mano, M. M., *Digital Design,* Prentice-Hall, Englewood Cliffs, New Jersey, 1984.
8. Rhyne, V. T., Jr., *Fundamentals of Digital System Design,* Prentice-Hall, Englewood Cliffs, New Jersey, 1973.
9. Dietmeyer, D. L., *Logic Design of Digital Systems,* 3d ed., Allyn and Bacon, Boston, Mass., 1988.

PROBLEMS

Section 3-2   Drawing, Filling, and Reading a Karnaugh Map

3-1. Show how to obtain the two-variable Karnaugh map in Fig. 3-3 from a two-variable Venn diagram. Fill in corresponding areas on the Venn Diagram and the Karnaugh map with the same product terms.

3-2. Enter each of the following functions on a Karnaugh map and find a minimum SOP form for each function. Identify each product term by circling adjacent groups of 1s.

(a) $F(X,Y) = \sum m(0,1)$
(b) $F(X,Y) = \sum m(1,3)$
(c) $F(X,Y) = \sum m(2,3)$
(d) $F(X,Y) = \prod M(1,3)$

3-3. Enter each of the following functions on a Karnaugh map and find a minimum SOP form for each function. Identify each product term by circling adjacent groups of 1s.

(a) $F(X,Y) = \sum m(0,1,2)$
(b) $F(X,Y) = \sum m(0,2,3)$
(c) $F(X,Y) = \prod M(2)$

3-4. Enter each of the following functions on a Karnaugh map and find a minimum POS form for each function. Identify each product term by circling adjacent groups of 0s.

(a) $F(X,Y) = \sum m(0,1)$

(b) $F(X,Y) = \sum m(1,3)$
(c) $F(X,Y) = \sum m(2,3)$
(d) $F(X,Y) = \prod M(1,3)$

3-5. Enter each of the following functions on a Karnaugh map and find a minimum POS form for each function. Identify each product term by circling adjacent groups of 0s.

(a) $F(X,Y) = \sum m(3)$
(b) $F(X,Y) = \sum m(1)$
(c) $F(X,Y) = \prod M(0,1,3)$

Section 3-3   Three-Variable Karnaugh Maps

3-6. Draw a Karnaugh map for each of the following three-variable functions and fill in all the p-squares and the r-squares on each map. Show groups of adjacent pairs and adjacent quadruples of p-squares by circling them. Write each function in a minimum form using the p-squares.

(a) $F(X,Y,Z) = \sum m(0,2,4,6)$
(b) $F(X,Y,Z) = \sum m(0,2,3,6,7)$
(c) $F(X,Y,Z) = \sum m(0,1,4,5,6,7)$

3-7. Draw a Karnaugh map for each of the following three-variable functions and fill in all the p-squares and the r-squares on each map. Show groups of adjacent pairs and adjacent quadruples of r-squares by circling them. Write each function in a minimum form using the r-squares.

(a) $F(X,Y,Z) = \sum m(0,2,4,6)$
(b) $F(X,Y,Z) = \sum m(0,2,3,6,7)$
(c) $F(X,Y,Z) = \sum m(0,1,4,5,6,7)$

3-8. Obtain a three-variable cube representation for each of the following functions. Represent each p-square on the TVC using a large dot. Group adjacent pairs of p-squares on the TVC using darker lines. Group adjacent quadruples by cross-hatched planes such that a minimum number of product terms cover all the p-squares of the function. Write each function in a minimum form using the p-squares. Check your work using a three-variable Karnaugh map.

(a) $F(X,Y,Z) = \sum m(0,2,4,6)$
(b) $F(X,Y,Z) = \sum m(0,2,3,6,7)$
(c) $F(X,Y,Z) = \sum m(0,1,4,5,6,7)$

3-9. Obtain the product terms for p_1, p_2, and p_3 circled in Fig. P3-9, and verify that each product term evaluates to 1 for its respective input combinations. Use the p-subcubes to write the function in a sum of products form.

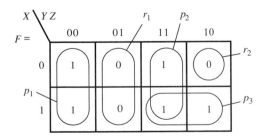

FIGURE P3-9

3-10. Obtain the product terms for r_1 and r_2 shown circled in Fig. P3-9 and verify that each product term evaluates to 1 for its respective input combinations. Use the r-subcubes and write the complement of the function in a sum of products form.

3-11. Write all the essential prime implicants for the 1s of the function in the map shown in Fig. P3-11. Use these implicants to obtain a minimum SOP expression for the function.

3-12. Write all the essential prime implicants for the 0s of the function in the map shown in Fig. P3-11. Use these implicants to obtain a minimum SOP expression for the complement of the function.

3-13. Starting with a three-variable Venn diagram, show how to obtain the three-variable Karnaugh map shown in Fig. P3-11.

3-14. Obtain minimum SOP expressions for the following Boolean functions using Karnaugh maps.
(*a*) $F(A,B,C) = \sum m(0,2,3,4,6,7)$
(*b*) $F(A,B,C) = \sum m(0,1,2,3,7)$
(*c*) $F(A,B,C) = \sum m(2,3,4,5)$

3-15. Obtain minimum POS expressions for the following Boolean functions using Karnaugh maps.
(*a*) $F(A,B,C) = \sum m(0,2,3,4,6,7)$
(*b*) $F(A,B,C) = \sum m(0,1,2,3,7)$
(*c*) $F(A,B,C) = \sum m(2,3,4,5)$

3-16. Obtain minimum SOP expressions and minimum POS expressions for the following Boolean functions using Karnaugh maps.
(*a*) $F(A,B,C) = \prod M(2,3)$
(*b*) $F(A,B,C) = \sum m(4,6)$
(*c*) $F(A,B,C) = \prod M(1,2,3,5,7)$

3-17. Using Karnaugh maps, obtain minimum SOP expressions for the following Boolean functions.
(*a*) $F(X,Y,Z) = \bar{X} \cdot \bar{Y} \cdot Z + X \cdot \bar{Y} \cdot Z + X \cdot Y$
(*b*) $F(X,Y,Z) = \bar{X} \cdot (\bar{Y} \cdot \bar{Z} + Y \cdot \bar{Z}) + X \cdot (\bar{Y} \cdot Z + Y \cdot Z)$
(*c*) $\bar{F}(X,Y,Z) = \bar{X} + \bar{X} \cdot Z + \bar{X} \cdot Y + X \cdot Y \cdot Z$

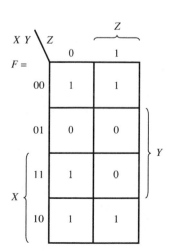

FIGURE P3-11

3-18. Using Karnaugh maps, obtain minimum POS expressions for the following Boolean functions.

(a) $F(X,Y,Z) = \bar{X} \cdot \bar{Y} \cdot Z + X \cdot \bar{Y} \cdot Z + X \cdot Y$

(b) $F(X,Y,Z) = \bar{X} \cdot (\bar{Y} \cdot \bar{Z} + Y \cdot \bar{Z}) + X \cdot (\bar{Y} \cdot Z + Y \cdot Z)$

(c) $\bar{F}(X,Y,Z) = \bar{X} + \bar{X} \cdot Z + \bar{X} \cdot Y + X \cdot Y \cdot Z$

3-19. Using Karnaugh maps, obtain minimum SOP expressions for the following Boolean functions. Hint: convert each POS form into a SOP form.

(a) $F(X,Y,Z) = X \cdot \bar{Y} \cdot Z + X \cdot \bar{Y} \cdot \bar{Z} + X \cdot Y \cdot \bar{Z}$

(b) $F(X,Y,Z) = X \cdot (Y \cdot Z + Y \cdot \bar{Z}) + X \cdot (\bar{Y} \cdot Z + Y \cdot Z)$

(c) $\bar{F}(X,Y,Z) = \bar{X} + \bar{X} \cdot Z + \bar{X} \cdot Y + \bar{X} \cdot Y \cdot Z$

(d) $F(X,Y,Z) = (X + Y) \cdot (X + \bar{Y} + Z)$

(e) $\bar{F}(X,Y,Z) = (X + Z) \cdot (\bar{Y} + Z) \cdot (\bar{X} + Y) \cdot (X + \bar{Y} + Z)$

3-20. Using Karnaugh maps, obtain minimum POS expressions for the following Boolean functions. Hint: convert each POS form into a SOP form.

(a) $F(X,Y,Z) = X \cdot \bar{Y} \cdot Z + X \cdot \bar{Y} \cdot \bar{Z} + X \cdot Y \cdot \bar{Z}$

(b) $F(X,Y,Z) = X \cdot (Y \cdot Z + Y \cdot \bar{Z}) + X \cdot (\bar{Y} \cdot Z + Y \cdot Z)$

(c) $\bar{F}(X,Y,Z) = \bar{X} + \bar{X} \cdot Z + \bar{X} \cdot Y + \bar{X} \cdot Y \cdot Z$

(d) $F(X,Y,Z) = (X + Y) \cdot (X + \bar{Y} + Z)$

(e) $\bar{F}(X,Y,Z) = (X + Z) \cdot (\bar{Y} + Z) \cdot (\bar{X} + Y) \cdot (X + \bar{Y} + Z)$

Section 3-4 Four- and Five-Variable Karnaugh Maps

3-21. Obtain an equivalent Karnaugh map for each of the Venn diagrams shown in Fig. P3-21. Label the minterm designators in the appropriate squares of each map. The Karnaugh maps are alternate versions that are used by various designers.

3-22. Obtain minimum SOP expressions for the following Boolean functions using Karnaugh maps.

(a) $F(W,X,Y,Z) = \sum m(2,3,6,7,8,9,12,13)$

(b) $F(W,X,Y,Z) = \sum m(0,3,4,5,6,7,11,12,13,14,15)$

(c) $F(W,X,Y,Z) = \prod M(0,2,5,7,8,10)$

3-23. Obtain minimum SOP expressions for the 0s of the following Boolean functions using Karnaugh maps.

(a) $F(W,X,Y,Z) = \sum m(2,3,6,7,8,9,12,13)$

(b) $F(W,X,Y,Z) = \sum m(0,3,4,5,6,7,11,12,13,14,15)$

(c) $F(W,X,Y,Z) = \prod M(0,2,5,7,8,10)$

3-24. From the discussion of the four-variable cube in Section 3-4, draw a four-variable cube representation for a four-variable function $F(W,X,Y,Z)$. Plot the function $F(W,X,Y,Z) = \sum m(0,4,5,6,7,8,10,12,14)$ on the four-variable cube using a large dot to represent each p-square. Write a minimum expression for the function. Check your work using a four-variable Karnaugh map.

3-25. Using Karnaugh maps, obtain minimum SOP expressions for F for the following Boolean functions.

(a) $F(W,X,Y,Z) = \bar{W} \cdot Y \cdot \bar{Z} + X \cdot Y \cdot \bar{Z} + W \cdot X + W \cdot Z + X \cdot \bar{Y} \cdot \bar{Z}$

(b) $F(W,X,Y,Z) = Y \cdot (\bar{Z} + W \cdot \bar{X} \cdot Z) + \bar{Y} \cdot \bar{Z} + \bar{W} \cdot X$

(c) $\bar{F}(W,X,Y,Z) = \bar{W} \cdot Z + W \cdot \bar{X} \cdot \bar{Z} + X \cdot Z + \bar{W} \cdot \bar{Z}$

3-26. Using Karnaugh maps, obtain minimum SOP expressions for $\bar{F}$ for the following Boolean functions.

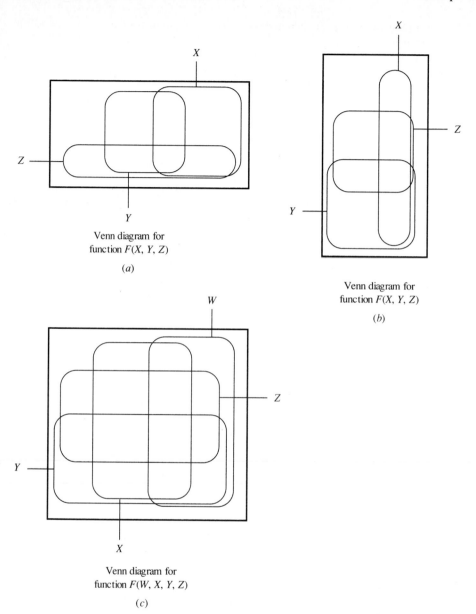

Venn diagram for
function $F(X, Y, Z)$

(a)

Venn diagram for
function $F(X, Y, Z)$

(b)

Venn diagram for
function $F(W, X, Y, Z)$

(c)

FIGURE P3-21

(a) $F(W,X,Y,Z) = \overline{W} \cdot Y \cdot \overline{Z} + X \cdot Y \cdot \overline{Z} + W \cdot X + W \cdot Z + X \cdot \overline{Y} \cdot \overline{Z}$
(b) $\underline{F}(W,X,Y,Z) = Y \cdot (\overline{Z} + W \cdot \overline{X} \cdot Z) + \overline{Y} \cdot \overline{Z} + \overline{W} \cdot X$
(c) $\overline{F}(W,X,Y,Z) = \overline{W} \cdot Z + W \cdot \overline{X} \cdot \overline{Z} + X \cdot Z + \overline{W} \cdot \overline{Z}$

3-27. Obtain minimum SOP expressions for the following Boolean functions using Karnaugh maps.

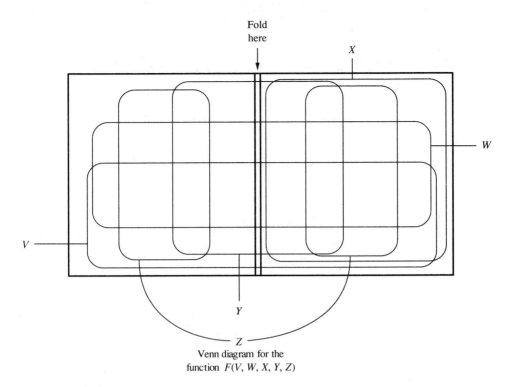

Fold
here

X

W

V

Y

Z

Venn diagram for the
function $F(V, W, X, Y, Z)$

FIGURE P3-29

(a) $F(V,W,X,Y,Z) =$
$\sum m(0,1,4,5,6,8,9,12,13,14,15,16,17,18,20,21,24,25,26,28,29,31)$
(b) $F(V,W,X,Y,Z) = \sum m(3,5,8,9,10,11,13,17,19,24,25,26,27)$
(c) $F(V,W,X,Y,Z) = \prod M(2,3,6,7,12,13,14,15,18,19,22,23,25,27,28,29)$

3-28. Obtain minimum SOP expressions for the 0s of the following Boolean functions using Karnaugh maps.
(a) $F(A,B,C,D,E) =$
$\sum m(0,1,4,5,6,8,9,12,13,14,15,16,17,18,20,21,24,25,26,28,29,31)$
(b) $F(A,B,C,D,E) = \sum m(3,5,8,9,10,11,13,17,19,24,25,26,27)$
(c) $F(A,B,C,D,E) = \prod M(2,3,6,7,12,13,14,15,18,19,22,23,25,27,28,29)$

3-29. For the five-variable Venn diagram shown in Fig. P3-29, obtain a five-variable Karnaugh map. Notice that adjacent areas on the Venn diagram are opposite the middle of the diagram. This Karnaugh map is referred to as a foldout adjacency map. Label the minterm designators on the Karnaugh map. Show adjacent areas for 1-cubes, 2-cubes, and 3-cubes across the fold of the five-variable Karnaugh map.

Section 3-5 Multiple Function Minimization

3-30. Use multiple function minimization as much as possible to obtain the simplest set of SOP expressions for the *p*-subcubes of the following functions. Specify the combined complexity in total literal count.

(a) $F1(X,Y,Z) = \sum m(0,2,4,6,7)$
   $F2(X,Y,Z) = \sum m(2,6,7)$
(b) $F1(X,Y,Z) = \sum m(1,3,4,7)$
   $F2(X,Y,Z) = \sum m(3,4,6,7)$
(c) $F1(A,B,C) = \sum m(1,2,4,6)$
   $F2(A,B,C) = \sum m(0,1,2,6,7)$
   $F3(A,B,C) = \sum m(1,2,6)$

3-31. Use multiple function minimization as much as possible to obtain the simplest set of SOP expressions for the *r*-subcubes of the following functions. Specify the combined complexity in total literal count.
(a) $F1(X,Y,Z) = \sum m(0,2,4,6,7)$
   $F2(X,Y,Z) = \sum m(2,6,7)$
(b) $F1(X,Y,Z) = \sum m(1,3,4,7)$
   $F2(X,Y,Z) = \sum m(3,4,6,7)$
(c) $F1(A,B,C) = \sum m(1,2,4,6)$
   $F2(A,B,C) = \sum m(0,1,2,6,7)$
   $F3(A,B,C) = \sum m(1,2,6)$

3-32. Obtain the simplest combined set of SOP expressions for the *r*-subcubes of the following functions. Find the total literal count of each combined set.
(a) $F1(A,B,C,D) = \sum m(4,5,6,7,9,14)$
   $F2(A,B,C,D) = \sum m(0,1,2,3,9,14)$
(b) $F1(W,X,Y,Z) = \sum m(0,2,7,8,10,15)$
   $F2(W,X,Y,Z) = \sum m(7,9,11,13,15)$

3-33. Obtain the simplest combined set of SOP expressions for the *p*-subcubes of the following functions. Find the total literal count of each combined set.
(a) $F1(A,B,C,D) = \sum m(4,5,6,7,9,14)$
   $F2(A,B,C,D) = \sum m(0,1,2,3,9,14)$
(b) $F1(W,X,Y,Z) = \sum m(0,2,7,8,10,15)$
   $F2(W,X,Y,Z) = \sum m(7,9,11,13,15)$

Section 3-6   Don't Care Output Conditions

3-34. Obtain minimum SOP expressions and minimum POS expressions for the following Boolean functions using Karnaugh maps.
(a) $F(A,B,C) = \prod M(2,3) \cdot d(4,6)$
(b) $F(A,B,C) = \sum m(4,6) + d(0,1,2)$
(c) $F(A,B,C) = \prod M(1,2,3,5,6) \cdot d(4)$

3-35. The minimum SOP expression of a certain function is algebraically identical to the minimum POS expression of the function. Explain how this can occur when don't cares are present in the function.

3-36. The minimum SOP expression of a certain function is not algebraically identical to the minimum POS expression of the function. Explain how this can occur when don't cares are present in the function.

3-37. Obtain minimum SOP expressions for the following Boolean functions using Karnaugh maps.
(a) $F(W,X,Y,Z) = \sum m(2,3,6,7,8,9,12,13) + d(0,4,15)$
(b) $F(W,X,Y,Z) = \sum m(0,3,4,5,6,7,11,12,13,14,15) + d(2,8,9)$
(c) $F(W,X,Y,Z) = \sum m(0,2,5,7,8,10,13) + d(1,9,11)$

3-38. Show that the following minimum expressions are algebraically identical.
(a) $F1 = \overline{X} \cdot Y + X \cdot \overline{Y}$ minimum SOP expression
   $F1 = (X + Y) \cdot (\overline{X} + \overline{Y})$ minimum POS expression
(b) $F2 = A \cdot \overline{C} + B \cdot C$ minimum SOP expression
   $F2 = (A + C) \cdot (B + \overline{C})$ minimum POS expression

3-39. Obtain minimum POS expressions for the following Boolean functions using Karnaugh maps.
(a) $F(W,X,Y,Z) = \sum m(2,3,6,7,8,9,12,13) + d(4,10,14)$
(b) $F(W,X,Y,Z) = \sum m(0,4,5,6,7,11,12,13,14,15) + d(2,3)$
(c) $F(W,X,Y,Z) = \sum m(0,2,5,7,8,10) + d(12,13)$

3-40. Obtain minimum SOP expressions for the following Boolean functions using Karnaugh maps.
(a) $F(W,X,Y,Z) = \sum m(2,3,6,7,8,9,12,13) + d(4,10,14)$
(b) $F(W,X,Y,Z) = \sum m(0,4,5,6,7,11,12,13,14,15) + d(2,3)$
(c) $F(W,X,Y,Z) = \sum m(0,2,5,7,8,10) + d(12,13)$

3-41. Obtain a set of minimum POS expressions for the functions $F3$, $F2$, $F1$, and $F0$ in Example 3-7. Show a circuit diagram using AND and OR element circuit symbols. In the circuit diagram, assume that the input variables are available in uncomplemented as well as complemented form.

3-42. Obtain minimum SOP expressions and minimum POS expressions for the following Boolean functions using Karnaugh maps.
(a) $F(A,B,C,D,E) =$
   $\sum m(1,3,5,6,9,13,14,19,22,30) + d(0,2,8,10,12,15,18,24,26)$
(b) $F(A,B,C,D,E) = \prod M(4,5,6,7,14,15,22,23) \cdot d(2,8,30,31)$

3-43. Obtain a set of minimum SOP expressions for a switching circuit that will accept BCD code at its input and provide 2421 code at its output. Show a circuit diagram using AND and OR element circuit symbols. In the circuit diagram assume the input variables are available in uncomplemented as well as complemented form.

Section 3-7 Additional Functional Forms

3-44. Obtain a minimum SOP equation for Prob. 3-37a. Show a 2-level circuit implementation for the equation using only NAND gates.

3-45. Draw a minimum NOR/NOR implementation for the Boolean function in Prob. 3-37c.

3-46. Show an AND/NOR implementation with a minimum number of gates for the Boolean function in Prob. 3-40c.

3-47. Draw a minimum 2-level circuit implementation for the function in Prob. 3-39b in OR/NAND form.

3-48. Show 2-level circuit implementations with a minimum number of gates in each of the following forms for the function in Prob. 3-38a.
(a) AND/OR
(b) NAND/NAND
(c) NOR/NOR
(d) NAND/AND

3-49. Draw a minimum 2-level circuit implementation in NAND/NAND form for the following Boolean function.
$F = (A + C) \cdot (B + \overline{C})$

3-50. Draw a minimum 2-level circuit implementation in NOR/NOR form for the following Boolean function.

$$F = A \cdot \overline{C} + B \cdot C$$

3-51. Show a NOR/NOR form of circuit implementation for the BCD to 2421 code converter in Prob. 3-43 using minimum functions. Assume that the input variables are available in complemented form as well as uncomplemented form.

CHAPTER 4

ADDITIONAL MINIMIZATION TECHNIQUES

4-1 INTRODUCTION AND INSTRUCTIONAL GOALS

In this chapter we present two more methods for minimizing Boolean functions. The first method is the tabular minimization method, and the second method is the map-entered variable method. Tabular minimization, also called the Quine-McCluskey method, provides insight into computer reduction techniques. The map-entered variable method extends the usefulness of the Karnaugh map method presented in Chapter 3.

In Section 4-2, Tabular Minimization, the purpose of tabular minimization is discussed. Background information concerning the method is presented in this section along with a brief overview of the two steps required in the process. To keep the introduction simple, Section 4-3, Minimizing Two-Variable Functions, starts by introducing the process using simple two-variable functions. The abbreviated binary method is utilized first since it is easy to recognize that adjacent product terms combine using binary numbers. In Section 4-4, Minimizing Three- and Four-Variable Functions, the discussion continues with larger functions. Section 4-5, The Abbreviated Decimal Method, presents tabular minimization using decimal numbers in order to help cut down the writing and speed up the comparison process for hand calculations. Tabular minimization may be omitted without loss of continuity.

In Section 4-6, Karnaugh Map Descriptions with Map-Entered Variables, an extension of the Karnaugh map method is presented, adding additional power to this proven design tool. The concept of adding variables to an existing design is presented to illustrate the flexibility of map-entered variables. This chapter is concluded with a procedure that illustrates how standard Karnaugh maps can be reduced in size.

This chapter should prepare you to

1. Minimize Boolean functions using tabular minimization.
2. Given a set of prime implicants, obtain or verify that a minimum expression has been obtained using a prime implicant table.
3. Use the abbreviated decimal method of tabular minimization.
4. Minimize Boolean functions using map-entered variables.
5. Use the concept of adding variables to an existing design.
6. Reduce the map size of standard Karnaugh maps.

4-2   TABULAR MINIMIZATION

Our purpose for presenting the tabular minimization for reducing Boolean functions is to provide the theory for reducing Boolean functions. For humans, the method is tedious and error-prone, especially for a large number of variables. The concept of tabular minimization, however, is very important for the following reasons: (1) tabular minimization can be used as a foundation for future studies, and (2) tabular minimization can also be programmed, allowing the computer to be used as a tool for reducing Boolean functions.

The concept of tabular minimization was originally formulated by Quine in 1952. McCluskey improved upon it in 1956. For this reason, the tabular minimization process of minimizing Boolean functions is often referred to as the Quine-McCluskey method. The process consists of two separate steps. Tabular minimization usually starts with a listing of the specified minterms for the 1s (or 0s) of a function and any don't cares (the unspecified minterms). Using a rather ingenious organizational scheme, all the prime implicants of a function are first obtained by utilizing the Adjacency Theorem, $X \cdot Y + X \cdot \overline{Y} = X$. This step is usually referred to as step 1 or the determination of prime implicants. If the prime implicants obtained in step 1 represent a complete covering of the function without any duplicate coverings, then these prime implements are all essential prime implicants and are ORed together to form a minimum SOP expression for the function (complement of the function).

A step 2 is necessary in many cases when there are no essential prime implicants or the prime implicants obtained in step 1 do not represent a minimum covering of the function. Step 2 involves creating a prime implicant table. This table provides a means of identifying the smallest number of prime implicants that can be used to cover the 1s (or 0s) of the function.

Don't cares must always be used in step 1 to find all the prime implicants; however, don't cares must never be used in step 2 since their inclusion at this point could prevent us from obtaining the smallest number of prime implicants to cover only the specified outputs of the function (the 1s or the 0s).

As a paper and pencil tool, tabular minimization becomes more difficult to use as the number of variables is increased. When Boolean functions with more than five variables must be reduced most designers turn to computer programs, which are often based on tabular minimization, to carry out the task. We will begin our discussion with the simplest of functions and increase in complexity from two variables to five variables, illustrating all the important concepts of tabular minimization.

4-3 MINIMIZING TWO-VARIABLE FUNCTIONS

To keep the explanation simple, we begin our discussion with the following two-variable function.

$$F(X, Y) = \sum m(0, 2, 3)$$

The first step in the tabular minimization process is to list the minterms for the 1s and any don't cares for the function as shown in Fig. 4-1. The table shown in Fig. 4-1 represents the structure for step 1 in the tabular procedure. Step 1 provides a method of obtaining all the prime implicants for the 1s of a function (or the 0s if we elect to use the minterms for the 0s of a function).

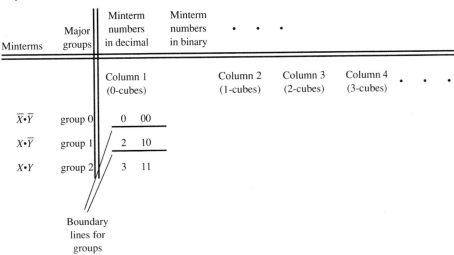

FIGURE 4-1
Structure for step 1 in the tabular process.

Major groups	2 variables	3 variables	4 variables
Group 0	00	000	0000
Group 1	01 10	001 010 100	0001 0010 0100 1000
Group 2	11	011 101 110	0011 0101 0110 1001 1010 1100
Group 3		111	0111 1011 1101 1110
Group 4			1111

FIGURE 4-2
Major groups of minterms for 2, 3, and 4 variables.

The minterms for the 0s of the function are used when it is desirable to obtain a minimum SOP expression for the complement of the function, which can then be written as a minimum POS expression.

When using pencil and paper, the only things that we need to write down from the structure shown in Fig. 4-1 are the decimal minterm numbers, the binary minterm numbers, the boundary lines separating the major groups, the column numbers, and a check mark beside each subcube that is combined with another subcube. We do not show any check marks in the table in Fig. 4-1 since we have not started to combine adjacent product terms. Check marks only appear as we start to form larger subcubes.

The first task is to correctly fill in column 1 by organizing the minterms for the 1s (0s) such that they are placed in the correct major group for ease of visual comparison. Group 0 can contain only one minterm—the minterm that contains no 1s when it is written in its binary representation as shown in Fig. 4-2. Group 1 can contain minterms with only one 1, group 2 can contain minterms with only two 1s, and so on as shown in Fig. 4-2 for two, three, and four variables. It is important that a line be drawn to separate the minterms in each major group.

4-3-1   Abbreviated Binary Method

To obtain a tabular minimization for the 1s of the function, we list the minterms of the 1s of the function in their proper major groups as shown in Fig. 4-3 in a stripped down version of the table. Group 0 contains minterm 0 written first in decimal and then in binary as 0 00. You may recall that $m_0 = \overline{X} \cdot \overline{Y}$. We could use this minterm product in column 1 and fill in the table as 0 $\overline{X} \cdot \overline{Y}$, but this

Step 1

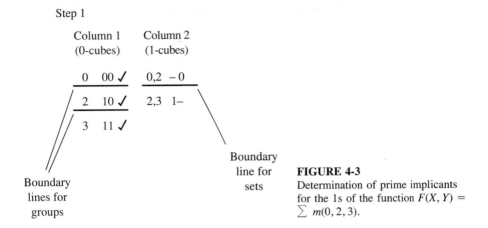

FIGURE 4-3
Determination of prime implicants
for the 1s of the function $F(X, Y) =$
$\sum m(0, 2, 3)$.

requires more writing and therefore the abbreviated binary notation 0 00 is used instead.

Minterm 2 represented in binary is 10, so it is listed in group 1, the next group. First a line is drawn under 0 00 to separate group 0 from group 1, and 2 10 is listed in column 1. The entry 2 10 represents $2\ X \bullet \overline{Y}$ in group 1 of column 1. Since this is the last entry in column 1 for group 1, another line is drawn under 2 10 to separate group 1 from the next group, group 2.

Group 2 represents the minterms of the 1s of the function with only two 1s present when the minterms are written in binary notation. Minterm 3 is listed in column 1 as 3 11 rather than $3\ X \bullet Y$ to simplify the representation. Decimal minterm numbers are used to provide identification, which is especially useful when minterms are combined to form larger subcubes.

By placing the minterms for the 1s (or 0s) of the function into the groups as indicated above, only the minterms in a specified group, group j, and the minterms in the group immediately following it, group $j + 1$, can be combined to form a larger subcube. A combination can only take place between two minterms when they differ by a single bit. It turns out that two minterms can differ only by a single bit when the minterm number is larger in the following group, group $j + 1$, not smaller. When two minterms are combined, a larger subcube is formed in the same manner that a larger subcube is formed by two adjacent 1s (0s) on a Karnaugh map.

When we use a binary representation, how do we spot minterms that will combine to form a larger subcube? We do this by looking for minterms in the table that only differ from each other by one bit. The bit that is different is replaced by a dash. The dash means that the variable in that position is left out when the combined expression is converted from binary back into a product expression. Using Boolean algebra, the Adjacency Theorem is used to combine minterm 0 and minterm 2 as follows: $\overline{X} \bullet \overline{Y} + X \bullet \overline{Y} = \overline{Y}$. Converting minterm 0 and minterm 2 to the special two-part decimal binary representation illustrated in the table in Fig. 4-3 results in 0 00 and 2 10, respectively. Since 00 and 10 differ from each other by only one bit, the most significant bit, they may be combined to form a 1-cube, which is written in column 2 as 0,2 −0. A line is drawn under 0,2 −0 in column 2 since no more combinations between group 0 and group 1 entries in column 1 are possible. The 0,2 represents the two minterms 0 and 2 that were combined. The − in the result of the combination −0 indicates that variable X is missing from the 1-cube representation, while the 0 indicates that the product term now representing the 1-cube represents $\overline{Y}$. Observe that this is the same result obtained by using straight Boolean algebra. All other combinations are handled in a similar fashion. Using the encoding scheme represented by the tabular method speeds up the table entry process as well as the visual process for manual calculations.

Combining minterm 2 (2 10) and minterm 3 (3 11) results in the 1-cube 2,3 1−, which represents X. Notice that each time a larger subcube is formed it is written in the next column to the right of the present column. We put check marks beside the terms 0 00 and 2 10, and 2 10 and 3 11 to indicate that they were used to form larger subcubes. Each set of combined product terms must be separated by a line to clearly indicate the boundary of the set. Step 1 of the tabular method is completed when product terms from groups j and $j + 1$ (separated by the boundary lines for the groups) and product terms from sets k and $k + 1$ (separated by the boundary lines for the sets) will no longer combine to form larger subgroups. The result of step 1 is a list of all the prime implicants of a function (for the 1s or the 0s). This list is obtained by collecting all the terms in any of the columns that have missing check marks beside them. No check marks are beside the product terms in column 2, indicating that 0,2 −0 and 2,3 1− (separated by the boundary line for the sets) will not combine since −0 and 1− do not differ by only one bit. As we have indicated, −0 represents $\overline{Y}$ and 1− represents X, so the two prime implicants for the 1s of the function $F(X,Y) = \sum m(0,2,3)$ are $\overline{Y}$ and X. ORing these prime implicants together and equating them to the function name results in the reduced function $F(X,Y) = \overline{Y} + X$. Since the expression $\overline{Y} + X$ cannot be further reduced algebraically, a minimum SOP expression for the function has been obtained using tabular minimization.

4-3-2   Using a Prime Implicant Table

To obtain the smallest number of prime implicants of a function, a prime implicant table may need to be constructed. As we indicated earlier, construction of the prime implicant table is step 2 of tabular minimization. A prime implicant table can also be used to verify that a given SOP expression is a minimum SOP expression. The prime implicant table for our simple function is shown in Fig. 4-4.

In the prime implicant table shown in Fig. 4-4, the prime implicants are listed along the left side of the table while the minterms for the 1s of the function are listed along the top of the table. The decimal minterm numbers listed with the prime implicants are now used as an aid to fill in the table. Minterms 0 and 2 have check marks under them in the table on row 0,2 −0 to represent the fact

Step 2

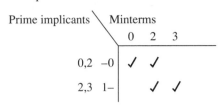

FIGURE 4-4
Prime implicant table for the 1s of the function
$F(X, Y) = \sum m(0, 2, 3)$.

that they are covered by prime implicant 0,2 −0. Likewise, minterms 2 and 3
have check marks under them in the table on row 2,3 1− to represent the fact
that they are covered by prime implicant 2,3 1−. A choice must now be made to
select the smallest number of prime implicants to cover all the minterms. When
a minterm is covered by only one prime implicant, that prime implicant is an
essential prime implicant.

   Prime implicant 0,2 −0 is an essential prime implicant since it is the only
prime implicant that covers minterm 0. The minterms that prime implicant 0,2 −0
covers are marked off by a horizontal line along that row. We add a vertical line
through each minterm which appears on more than one row as shown in Fig. 4-5.
An alternate way to indicate the minterms that are covered by a prime implicant is
to provide a check mark along the bottom of the prime implicant table as shown in
Fig. 4-5. The only minterm left to be covered by the function is minterm 3, which
must be covered by prime implicant 2,3 1−. Since minterm 3 is only covered
by prime implicant 2,3 1− this prime implicant is also an essential prime implicant.

   Since both prime implicants 0,2 −0 ($\bar{Y}$) and 2,2 1− (X) are essential, we
have confirmed with a prime implicant table that the SOP expression $\bar{Y} + X$ does
indeed represent a minimum SOP expression.

   Example 4-1. Obtain a minimum SOP expression for the 0s of the following
   function using tabular minimization.

$$F(A, B) = \prod M(1, 3) \cdot d(2)$$

   Solution First we set up the table for step 1 to obtain the prime implicants of the
   0s of the function as shown in Fig. E4-1a. Notice that minterm 2 (the don't care

Step 2

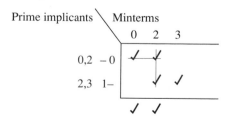

FIGURE 4-5
Prime implicant table for the 1s of the function
$F(X, Y) = \sum m(0, 2, 3)$ showing two different
methods for indicating minterms that are
covered by a prime implicant.

Step 1: Determination of prime implicants   **Step 1: Determination of prime implicants**

Column 1 (0-cubes)	Column 2 (1-cubes)
1 01	
2 10	
3 11	

Column 1 (0-cubes)	Column 2 (1-cubes)
1 01 √	1,3 −1
2 10 √	2,3 1−
3 11 √	

(a)   *(b)*

Step 2: Prime implicant table

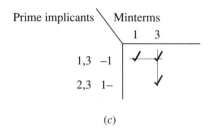

(c)

FIGURE E4-1

minterm) must be included to insure we obtain the largest subcubes for the prime implicants of the function.

Combining adjacent product terms, we obtain the table as shown in Fig. E4-1*b*. The two terms that will not combine are shown unchecked in the table; these terms are the prime implicants of the function.

A prime implicant table is shown in Fig. E4-1*c*. Notice that only the specified 0s of the function (1 and 3) are included in the list of minterms labeled across the top of the table. Don't cares are never used in the prime implicant table. Prime implicant 1,3 −1 representing *B* is an essential prime implicant, and it also covers all the minterms of the function. Prime implicant 2,3 1− is redundant.

The minimum SOP expression for the 0s of the function is therefore $\overline{F}(A,B) = B$. This result can also be easily confirmed using a Karnaugh map.

4-4   MINIMIZING THREE- AND FOUR-VARIABLE FUNCTIONS

Now consider the following three-variable function.

$$F(X,Y,Z) = \sum m(0,2,3,5,6,7)$$

Tabular minimization for this function is carried out in the same manner as for our simpler examples, only now the process takes longer due to twice as many combinations as a result of the additional variable. The minimization can

Step 1

Column 1 (0-cubes)	Column 2 (1-cubes)	Column 3 (2-cubes)
0 000 √	0,2 0-0	no combinations
2 010 √	2,3 01- √	2,3,6,7 -1-
	2,6 -10 √	2,6,3,7 -1-
3 011 √		
5 101 √	3,7 -11 √	
6 110 √	5,7 1-1	
	6,7 11- √	
7 111 √		

FIGURE 4-6
Determination of prime implicants for the 1s
of the function $F(X, Y, Z) = \sum m(0,2,3,5,6,7)$.

be carried out using the 1s of the function or the 0s of the function. For this example we will use the 1s of the function.

Step 1 involves determining the prime implicants. The table for this determination is shown in Fig. 4-6. First the minterms must be listed in the major groups (group 0 through group 3) as shown in the figure, and separated by boundary lines for the groups. Notice that all the minterms in column 1 are checked off, indicating that every minterm in group j combined with at least one other minterm in group $j + 1$. It is only necessary to put a single check mark beside terms that combine with more than one term. In column 2, the two decimal numbers separated by a comma indicate the minterms in column 1 that were combined. For example, 3,7 -11 is a combination of 3 011 in group 2 and 7 111 in group 3. In other words, 3 011 is in some group j, 7 111 is in the next highest group $j + 1$, and they differ by only one bit; therefore, they can be combined.

A boundary line is drawn in column 2 to separate the set of combined minterms in group 0 and group 1 of column 1, and another boundary line is drawn to separate the set of combined minterms in group 1 and group 2 of column 1. When checking for product terms that will combine in column 2, only product terms that are separated by a boundary line have to be compared. Only product terms in adjacent sets will combine; that is, product terms in set k will only combine with products terms in set $k + 1$. The product terms that are shown checked in column 2 will combine. No product terms in the first two adjacent sets in column 2 will combine. This is indicated in the table in Fig. 4-6 by writing "no combinations" followed by a boundary line. Product term 2,3 01- will combine with product term 6,7 11-, and 2,6 -10 will combine with 3,7 -11, since a single boundary line separates each of the combinable product terms and each of the combinable product terms differs by only one bit. The resulting product terms are 2,3,6,7 $-1-$ and 2,6,3,7 $-1-$, which are listed in the next column, column 3. Two product terms in column 2 and two product terms in column 3 will not combine, thus providing four prime implements for the 1s of the function.

The prime implicant table for our three-variable function is shown in Fig. 4-7. An asterisk has been placed beside the first, second, and fourth prime implicants in the prime implicant table to indicate that these provide the smallest

Step 2

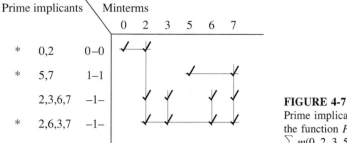

FIGURE 4-7
Prime implicant table for the 1s of
the function $F(X, Y, Z) =$
$\sum m(0, 2, 3, 5, 6, 7)$.

number of prime implicants covering all the minterms of the function. We choose
the first and second prime implicants because they are essential. Prime implicant
$0,2\ 0-0$ $(\overline{X} \cdot \overline{Z})$ is essential because m_0 must be covered by this implicant. Prime
implicant $5,7\ 1-1$ $(X \cdot Z)$ is also essential because m_5 must be covered by this
implicant. Prime implicant $2,6,3,7$ $-1-$ (Y) is not essential, but it covers all the
rest of the minterms not already covered, thus using the smallest number of prime
implicants to do so. Prime implicant $2,3,6,7$ $-1-$ is a duplicate prime implicant
since it covers the same minterms as prime implicant $2,6,3,7$ $-1-$. A minimum
SOP expression for the function can now be written as follows.

$$F(X, Y, Y) = \overline{X} \cdot \overline{Z} + X \cdot Z + Y$$

This function is shown plotted on a three-variable Karnaugh map in Fig.
3-7a in Chapter 3. The subcubes circled on the Karnaugh map represent the same
implicants chosen for a minimum SOP expression from the prime implicant table.

The following four-variable function will now be minimized using tabular
minimization. Notice that the function has don't cares that must be used in step
1, but must not be used in step 2.

$$F(W, X, Y, Z) = \sum m(2, 3, 6, 8, 10, 11, 12, 14, 15) + d(1, 7, 13)$$

Since the function is incompletely specified, we do not know in advance
whether a minimum SOP expression for the 1s or the 0s will be simpler. We
will obtain the minimum SOP expression for the 0s because with fewer minterms
involved in the tabular process, there are fewer combinations to consider, making
our discussion a little easier. Rewriting the function in terms of its 0s, we list the
maxterm compact form.

$$F(W, X, Y, Z) = \prod M(0, 4, 5, 9) \cdot d(1, 7, 13)$$

The table used to determine the prime implicants is shown in Fig. 4-8. The
minterms for the don't care outputs are included in the table to allow larger
subcubes to be formed where possible. Manual minimization of functions using
the tabular method now becomes more tedious. This is true especially when a
large number of product terms must be compared and comparisons are carried

Step 1

Column 1 (0-cubes)	Column 2 (1-cubes)	Column 3 (2-cubes)
0 0000 √	0,1 000– √ 0,4 0-00 √	0,1,4,5 0-0– 0,4,1,5 0-0–
1 0001 √ 4 0100 √	1,5 0-01 √ 1,9 –001 √ 4,5 010– √	1,5,9,13 – –01 1,9,5,13 – –01
5 0101 √ 9 1001 √	5,7 01-1 5,13 –101 √ 9,13 1-01 √	
7 0111 √ 13 1101 √		

FIGURE 4-8

Determination of prime implicants for the 0s of the function $F(W, X, Y, Z) = \Pi M(0, 4, 5, 9) \cdot d(1, 7, 13)$, using the abbreviated binary method.

out too fast, since terms that can combine may be overlooked. When minterms in adjacent groups and product terms in adjacent sets will no longer combine, then all possible product terms have been combined.

For this example, we end up with five prime implicants as indicated by the absence of a check mark beside the uncombined product terms. Observe that only three of the five product terms are unique. Applying the Idempotency Theorem, $X + X = X$, two of the prime implicants can be eliminated by inspection. These are duplicate (redundant) prime implicants. Using the encoded abbreviated notation, the following SOP expression for the prime implicants for the 0s of the function can be written as follows.

$$5,7 \quad 01-1 \quad + \quad 0,1,4,5 \quad 0-0- \quad + \quad 0,4,1,5 \quad 0-0-$$
$$+ \quad 1,5,9,13 \quad - -01 \quad + \quad 1,9,5,13 \quad - -01$$

Decoding each of these product terms results in the following expression.

$$\overline{W} \cdot X \cdot Z + \overline{W} \cdot \overline{Y} + \overline{W} \cdot \overline{Y} + \overline{Y} \cdot Z + \overline{Y} \cdot Z$$

Eliminating the duplicate prime implicants provides the following expression containing only unique prime implicants.

$$\overline{W} \cdot X \cdot Z + \overline{W} \cdot \overline{Y} + \overline{Y} \cdot Z$$

In the prime implicant table shown in Fig. 4-9 we only need to list the three unique prime implicants that were obtained in step 1, that is, those that are different. When don't care outputs are included in the function, remember not to include them in the minterm list across the top of the table.

After filling in the prime implicant table, look for columns that contain a single minterm. Columns with single minterms occur at m_0, m_4, and m_9. Prime implicants 0,1,4,5 0–0– ($\overline{W} \cdot \overline{Y}$) and 1,5,9,13 – –01 ($\overline{Y} \cdot Z$) that cover these columns are essential prime implicants, and an asterisk is placed beside these implicants to show that they are being used to cover the function. The rows containing m_0, m_4, and m_9 are crossed off in the prime implicant table since

Step 2

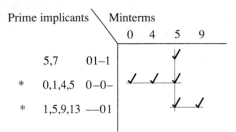

Prime implicants \ Minterms

		0	4	5	9
5,7	01–1			✓	
* 0,1,4,5	0–0–	✓	✓	✓	
* 1,5,9,13	––01			✓	✓

FIGURE 4-9
Prime implicant table for the 0s of the function
$F(W, X, Y, Z) = \Pi M(0, 4, 5, 9) \cdot d(1, 7, 13)$
using the abbreviated binary method.

they are now covered. Minterm 5 is included in at least one of these prime implicants, so a vertical line through the column for m_5 removes this minterm from further consideration. Remember to choose the smallest number of prime implicants (those containing the smallest number of literals) that cover the required minterms; otherwise a minimum expression will not be obtained. Prime implicant 5,7 $01-1$ ($\overline{W} \cdot X \cdot Z$) is a redundant prime implicant and is not used to cover the function. A minimum SOP expression for the 0s of the function can be written as follows.

$$\overline{F}(W, X, Y, Z) = \overline{W} \cdot \overline{Y} + \overline{Y} \cdot Z$$

It is interesting to note that if prime implicant selection is difficult or complicated to make for a Karnaugh map, a prime implicant table can be constructed to obtain the smallest number of prime implicants that cover the required minterms.

Example 4-2. Obtain a minimum SOP expression for the 1s of the following function using the abbreviated binary method of tabular minimization.

$$F(A, B, C, D) = \prod M(1, 3, 4, 5, 6, 7) \cdot d(9, 15)$$

Solution Since we are asked to obtain the expression for the 1s of the function, the function is written in minterm compact form as follows.

$$F(A, B, C, D) = \sum m(0, 2, 8, 10, 11, 12, 13, 14) + d(9, 15)$$

In Fig. E4-2a we show step 1 for determining the prime implicants. To reduce the amount of comparisons that must be made in column 3, we identified the duplicate implicants and drew a line through them indicating that they can be ignored. The duplicate prime implicants in column 4 were also lined out. The only two unique prime implicants remaining are 0,2,8,10 $-0-0$ ($\overline{B} \cdot \overline{D}$) and 8,9,10,11,12,13,14,15 $1---$ (A).

Step 2 for the abbreviated binary method is shown in Fig. E4-2b. The prime implicant table is drawn using the two unique prime implicants obtained in step 1. Only the minterms for the 1s are listed across the top of the table. Since both minterms are essential, they must both appear in the minimum SOP expression for the 1s of the function written below.

$$F(A, B, C, D) = \overline{B} \cdot \overline{D} + A$$

Step 1: Determination of prime implicants

Column 1 (0-cubes)	Column 2 (1-cubes)	Column 3 (2-cubes)	Column 4 (3-cubes)
0 0000 ✓	0,2 00-0 ✓	0,2,8,10 -0-0	no combinations
	0,8 -000 ✓	0,8,2,10 ~~-0-0~~	
2 0010 ✓			8,9,10,11,12,13,14,15 1---
8 1000 ✓	2,10 -010 ✓	8,9,10,11 10-- ✓	~~8,9,12,13,10,11,14,15 1---~~
	8,9 100- ✓	8,9,12,13 1-0- ✓	~~8,10,12,14,9,11,13,15 1---~~
9 1001 ✓	8,10 10-0 ✓	~~8,10,9,11 10--~~	
10 1010 ✓	8,12 1-00 ✓	8,10,12,14 1--0 ✓	
12 1100 ✓		~~8,12,9,13 1-0-~~	
	9,11 10-1 ✓	~~8,12,10,14 1--0~~	
11 1011 ✓	9,13 1-01 ✓		
13 1101 ✓	10,11 101- ✓	9,11,13,15 1--1 ✓	
14 1110 ✓	10,14 1-10 ✓	~~9,13,11,15 1--1~~	
	12,13 110- ✓	10,11,14,15 1-1- ✓	
15 1111 ✓	12,14 11-0 ✓	~~10,14,11,15 1-1-~~	
		12,13,14,15 11-- ✓	
	11,15 1-11 ✓	~~12,14,13,15 11--~~	
	13,15 11-1 ✓		
	14,15 111- ✓		

(*a*)

Step 2: Prime implicant table

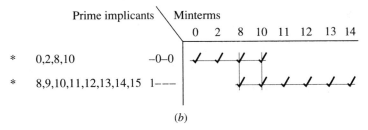

Prime implicants	Minterms	0	2	8	10	11	12	13	14
* 0,2,8,10	-0-0	✓	✓	✓	✓				
* 8,9,10,11,12,13,14,15	1---			✓	✓	✓	✓	✓	✓

(*b*)

FIGURE E4-2

4-5 THE ABBREVIATED DECIMAL METHOD

The abbreviated binary method is not the only method of tabular minimization. Other tabular minimization methods have been developed to help reduce the amount of writing that must be done when filling out the tables in step 1 and step 2. One method that reduces the amount of writing and helps make faster comparisons is called the abbreviated decimal method. First the 1s or 0s are written in binary in column 1 in their proper groups. Each binary representation is then followed by the decimal equivalent number. In other words, reverse the order of the decimal and binary representation for the minterms in column 1. For discussion purposes, consider the 0s of the following function.

Step 1

Column 1 (0-cubes)	Column 2 (1-cubes)	Column 3 (2-cubes)
0000 0 √	0,1 (1) √	0,1,4,5 (1,4)
	0,4 (4) √	~~0,4,1,5 (4,1)~~
0001 1 √		
0100 4 √	1,5 (4) √	1,5,9,13 (4,8)
	1,9 (8) √	~~1,9,5,13 (8,4)~~
0101 5 √	4,5 (1) √	
1001 9 √		
	5,7 (2)	
0111 7 √	5,13 (8) √	
1101 13 √	9,13 (4) √	

FIGURE 4-10

Determination of prime implicants for the 0s of the function $F(W, X, Y, Z) = \Pi M(0, 4, 5, 9) \cdot d(1, 7, 13)$ using the abbreviated decimal method.

$$F(W, X, Y, Z) = \prod M(0, 4, 5, 9) \cdot d(1, 7, 13)$$

Step 1 for the abbreviated decimal method is shown in Fig. 4-10. Minterms in column 1 are compared for adjacency by subtracting the decimal numbers representing each minterm in group j from the decimal numbers representing each minterm in group $j + 1$. Only when the decimal number in group j is smaller than the decimal number in group $j + 1$ and the difference is a power of two can the minterms be combined (adjacency exists). These minterms are then listed in column 2 and checked off of column 1. To illustrate the abbreviated decimal method, compare the solution shown in Fig. 4-10 with the same solution using the abbreviated binary method shown in Fig. 4-8.

The decimal numbers representing the product terms in column 2 can be obtained rather routinely as shown in step 1 in Fig. 4-10. Differences representing positive powers of two are listed in parentheses. All other differences are ignored. For example, $1 - 0 = 1$, so 0,1 (1) is recorded as an entry in column 2 and a check mark is placed beside minterm 0 and 1 in column 1. As before, one must remember to draw the boundary lines for each set. After all minterm combinations for group 0 and group 1 in column 1 have been listed in column 2, the boundary line for the first set must be drawn immediately after 0,4 (4) as shown in Fig. 4-10.

The decimal numbers representing the product terms in any of the other columns are obtained by comparing the numbers in parentheses in set k with the numbers in parentheses in set $k + 1$. When a match occurs, subtract the first minterm in each group outside the parentheses. When the difference represents only a positive power of two, that difference is added to the list of differences in parentheses, and the result is placed in the next column. For example, consider 0,4 (4), which combines with 1,5 (4) to provide 0,4,1,5 (4,1). Since the number in parentheses (4) provides a match and the difference between the first minterm of each of the outer groups of numbers (0,4 and 1,5) is a positive power of two (difference equals 1), adjacency exists. The 1 is added to the 4 in the list of differences shown in parentheses, and the result appears in the next col-

Step 2

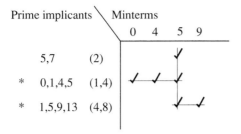

FIGURE 4-11

Prime implicant table for the 0s of the function $F(W, X, Y, Z) = \Pi M(0, 4, 5, 9) \cdot d(1, 7, 13)$ using the abbreviated decimal method.

umn, column 3. A check is placed beside each product term that combines with another product term to signify that these product terms are not prime implicants.

The abbreviated decimal method, as you have observed, is a method to shorten the amount of writing and detailed visual comparisons that must be made when using tabular minimization. By comparing the product terms in Fig. 4-10 and the corresponding product terms in Fig. 4-8, it is easy to see that the list of differences for each of the decimal product terms consists of only the missing variable(s) in the corresponding binary product terms.

Changes to the prime implicant table using the abbreviated decimal method are very minor as shown in Fig. 4-11.

Once the required prime implicants (the essential or necessary prime implicants) are chosen for a minimum SOP expression, it is necessary to convert the decimal number representations for the prime implicants back into a binary representation so that an expression can be determined. To convert the essential prime implicant 1,5,9,13 (4,8) to an expression, first write the binary representation for the largest minterm in the group (13) as 1101. Then cross out the bit that represents 4 (0100) and the bit that represents 8 (1000). The result is $--01$, the representation for the expression $\overline{Y} \cdot Z$. In a similar manner, the expression for 0,1,4,5 (1,4) is $\overline{W} \cdot \overline{Y}$. This is obtained by writing 5 (the largest minterm in the group) as 0101 for the 4 bits representing the inputs of the function, then crossing out the bit that represents 1 (0001) and the bit that represents 4 (0100) for a result of $0-0-$ or $\overline{W} \cdot \overline{Y}$. A minimum SOP expression for the 0s of the function may then be written as follows.

$$\overline{F}(W, X, Y, Z) = \overline{W} \cdot \overline{Y} + \overline{Y} \cdot Z$$

Example 4-3. Use the abbreviated decimal method to obtain a minimum SOP expression for the 1s of the following function.

$$F(V, W, X, Y, Z) = \sum m(3, 4, 7, 9, 11, 12, 15, 16, 18, 19, 20, 23, 24, 26, 27, 28, 31)$$

Solution We start by making a list of the minterms for the 1s of the function, using five bits for the binary representation so we can place these numbers in the correct major groups in step 1.

$$3 = 00011$$

$$4 = 00100$$

$$7 = 00111$$

$$9 = 01001$$

$$11 = 01011$$

$$12 = 01100$$

$$15 = 01111$$

$$16 = 10000$$

$$18 = 10010$$

$$19 = 10011$$

$$20 = 10100$$

$$23 = 10111$$

$$24 = 11000$$

$$26 = 11010$$

$$27 = 11011$$

$$28 = 11100$$

$$31 = 11111$$

Step 1 is shown in Fig. E4-3a. Drawing lines through duplicate implicants in column 3 prior to combining product terms in set k with product terms in set k + 1 helps reduce the work required for hand calculation. Duplicate prime implicants in column 4 are also removed from consideration in the prime implicant table by lining them out.

The prime implicant table is generated for the unchecked nonduplicate prime implicants as shown in Fig. E4-3b. The selected prime implicants for a minimum SOP expression for the 1s of the function are shown with an asterisk beside them in the prime implicant table, and these implicants result in the following minimum SOP function.

$$F(V, W, X, Y, Z) = \overline{V} \cdot W \cdot \overline{X} \cdot Z \ + \ X \cdot \overline{Y} \cdot \overline{Z} \ + \ V \cdot \overline{X} \cdot \overline{Z} \ + \ Y \cdot Z$$

By using the abbreviated decimal method, combinations can be spotted a little more quickly. This speeds up the process of tabular minimization by hand. Computer minimization is still the desired end result, especially for functions with six or more variables. Procedures have also been developed for the tabular method for minimizing multiple outputs (see Bartee 61, McCluskey and Schorr 62, and McCluskey 86 in the references). Generally computers utilize these procedures since multiple output problems typically contain too many variables to easily solve manually by the tabular method or by Karnaugh maps.

Step 1: Determination of prime implicants

Column 1 (0-cubes)	Column 2 (1-cubes)	Column 3 (2-cubes)	Column 4 (3-cubes)
00100 4 √	4,12 (8) √	4,12,20,28 (8,16)	no combinations
10000 16 √	4,20 (16) √	4,20,12,28 (16,8)	
―――――	16,18 (2) √	16,18,24,26 (2,8)	3,7,11,15,19,23,27,31 (4,8,16)
00011 3 √	16,20 (4) √	16,20,24,28 (4,8)	3,7,19,23,11,15,27,31 (4,16,8)
01001 9 √	16,24 (8) √	16,24,18,26 (8,2)	3,11,19,27,7,15,23,31 (8,16,4)
01100 12 √	―――――	16,24,20,28 (8,4)	
10010 18 √	3,7 (4) √	―――――	
10100 20 √	3,11 (8) √	3,7,11,15 (4,8) √	
11000 24 √	3,19 (16) √	3,7,19,23 (4,16) √	
―――――	9,11 (2)	3,11,7,15 (8,4)	
00111 7 √	12,28 (16) √	3,11,19,27 (8,16) √	
01011 11 √	18,19 (1) √	3,19,7,23 (16,4)	
10011 19 √	18,26 (8) √	3,19,11,27 (16,8)	
11010 26 √	20,28 (8) √	18,19,26,27 (1,8)	
11100 28 √	24,26 (2) √	18,26,19,27 (8,1)	
―――――	24,28 (4) √	―――――	
01111 15 √	―――――	7,15,23,31 (8,16) √	
10111 23 √	7,15 (8) √	7,23,15,31 (16,8)	
11011 27 √	7,23 (16) √	11,15,27,31 (4,16) √	
―――――	11,15 (4) √	11,27,15,31 (16,4)	
11111 31 √	11,27 (16) √	19,23,27,31 (4,8) √	
	19,23 (4) √	19,27,23,31 (8,4)	
	19,27 (8) √		
	26,27 (1) √		
	―――――		
	15,31 (16) √		
	23,31 (8) √		
	27,31 (4) √		

(a)

FIGURE E4-3

4-6 KARNAUGH MAP DESCRIPTIONS WITH MAP-ENTERED VARIABLES

The power inherent in the graphical presentation of a Karnaugh map should stand out clearly after drudging through the many comparisons that are required when using tabular minimization. A technique is available to allow Karnaugh maps to be made into a considerably more powerful tool by allowing the function values to contain not only the constants (1s, 0s, and don't cares) but also external variables. This technique for extending the usefulness of Karnaugh maps is attributed to T. E. Osborne who utilized the concept of infrequently-used variables in U.S. Patent No. 3566160, filed in June 1966.

> **Definition 4-1.** An infrequently-used variable is the name given to a variable or expression that is entered into a cell of a Karnaugh map. Another name for an infrequently-used variable is a map-entered variable.

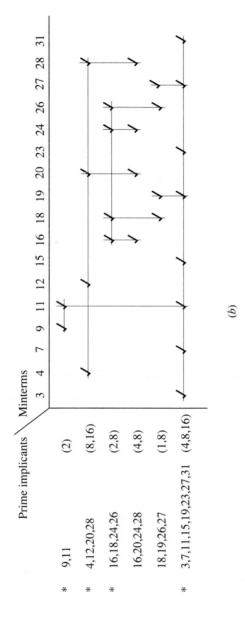

FIGURE E4-3
(*continued*)

4-6-1 Two-Variable Maps Using Map-Entered Variables

Consider the Karnaugh map shown in Fig. 4-12. This two-variable map contains the map variables W and X, as well as the map-entered variables Y and $\overline{Z}$ that are entered into the cells of the map. Fig. 4-12 specifies a four-variable function using only a two-variable map. Notice that a map-entered variable may consist of either a single variable or a Boolean expression. The object of using such a map is to allow us to specify functions of more than n variables using only an n-variable map. Since a map utilizing map-entered variables is compressed, reading the map for the product terms for a minimum SOP expression becomes more involved. Just being able to represent a larger function on a smaller size map, however, extends the usefulness of the Karnaugh map as a tool.

      The map variables used on the perimeter of a map are called the dominant variables. The variables or expressions that are entered in the cells of the map are the infrequently-used variables and are referred to as map-entered variables.

      The function represented by the Karnaugh map shown in Fig. 4-12 can be written in general terms using the minterm designators and the characteristic numbers for the function as follows:

$$F(W, X, Y, Z) = m_0 \cdot f_0 \ + \ m_1 \cdot f_1 \ + \ m_2 \cdot f_2 \ + \ m_3 \cdot f_3$$

where m_0 through m_3 are the minterms of the dominant variables (W,X) on the perimeter of the map, and f_0 through f_3 are the characteristic numbers representing the values of the function entered in the cells of the map at each respective minterm location.

$$m_0 = \overline{W} \cdot \overline{X}, \ f_0 = Y$$

$$m_1 = \overline{W} \cdot X, \ f_1 = 1$$

$$m_2 = W \cdot \overline{X}, \ f_2 = \overline{Z}$$

$$m_3 = W \cdot X, \ f_3 = - \quad \text{(For obvious reasons } - \text{ is used to represent a don't care instead of an } X.)$$

Therefore,

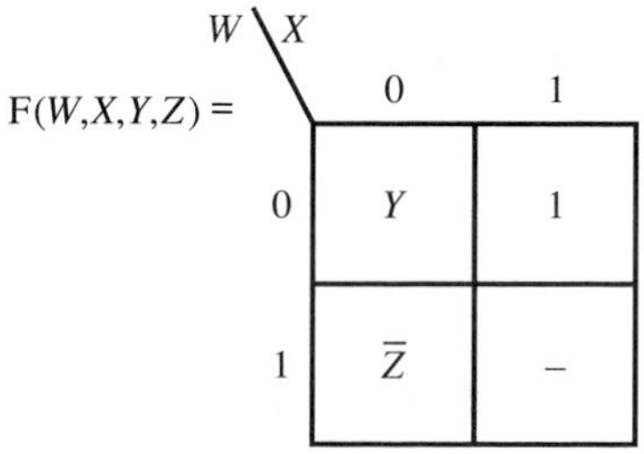

$$F(W,X,Y,Z) =$$

FIGURE 4-12
Two-variable map with map-entered variables Y and $\overline{Z}$.

$$F(W,X,Y,Z) = \overline{W} \bullet \overline{X} \bullet Y + \overline{W} \bullet X \bullet 1 + W \bullet \overline{X} \bullet \overline{Z} + W \bullet X \bullet -$$

represents the Boolean function expressed by the map shown in Fig. 4-12.

The function can be written in minterm compact form as follows:

$$F(W,X,Y,X) = \sum m(0 \bullet Y, 1, 2 \bullet \overline{Z}) + d(3)$$

where $m = m(W,X)$

This form of the function is a sort of pseudo-canonical or pseudo-standard form in terms of the minterms using the dominant input variables W and X. The infrequently-used variables are Y and $\overline{Z}$.

Like any Boolean function, this function can be reduced to a minimum form in several ways. The function will first be reduced to a minimum SOP form algebraically. Later we will present the rules for reducing the function using its Karnaugh map representation. Parentheses are used to identify product terms that can be combined.

$$F(W,X,Y,Z) = \overline{W} \bullet \overline{X} \bullet Y + \overline{W} \bullet X \bullet 1 + W \bullet X \bullet Z + W \bullet X \bullet -$$

$$F(W,X,Y,Z) = (\overline{W} \bullet \overline{X} \bullet Y + \overline{W} \bullet X \bullet 1) + (W \bullet \overline{X} \bullet \overline{Z} + W \bullet X \bullet 1)$$
$$\text{letting “}- = 1\text{”}$$

$$F(W,X,Y,Z) = \overline{W} \bullet Y + \overline{W} \bullet X + W \bullet \overline{Z} + W \bullet X$$
$$\text{using the Simplification Theorem}$$

$$F(W,X,Y,Z) = \overline{W} \bullet Y + W \bullet \overline{Z} + (\overline{W} \bullet X + W \bullet X)$$

$$F(W,X,Y,Z) = \overline{W} \bullet Y + W \bullet \overline{Z} + X$$
$$\text{using the Adjacency Theorem}$$

The purpose of Karnaugh maps is to provide a tool for obtaining reduced functions. The steps to follow to obtain a minimum SOP expression for the 1s of a function for a Karnaugh map using map-entered variables are listed below.

Step 1. Choose a set of p-subcubes that results in a minimum expression for the function by covering each map-entered variable in turn, treating other map-entered variables as 0s and all 1s as don't cares.

Step 2. Choose a set of p-subcubes that results in a minimum expression for the function by covering the 1s in the map that are not double-covered in step 1, treating all map-entered variables as 0s and all double-covered 1s as don't cares.

Definition 4-2. A 1 in a map using map-entered variables is double-covered if it is covered twice, once by a subcube containing a map-entered variable, and a second time by a subcube containing the complement of the same map-entered variable.

In Fig. 4-13a we show the p-subcubes that result by applying step 1 to the map in Fig. 4-12. The p-subcube for p_1 covers Y, treating $\overline{Z}$ as 0. The p-subcube for p_2 covers $\overline{Z}$, treating Y as 0. The following procedure is used

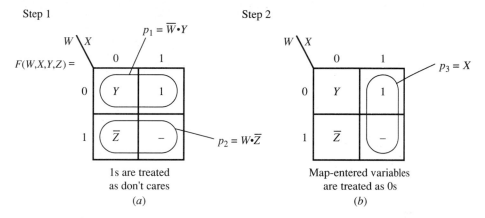

FIGURE 4-13
Selecting a minimum covering of 1s for a map using map-entered variables, (a) step 1, and (b) step 2.

to obtain the product terms for map-entered variables. The process is simply to obtain each product term in the normal manner and then to AND the map-entered variable with the result. The product term for p_1 is $\overline{W}$ ANDed with the map-entered variable Y; therefore, $p_1 = \overline{W} \cdot Y$. The product term p_2 is W ANDed with the map-entered variable $\overline{Z}$, so $p_2 = W \cdot \overline{Z}$.

    The p-subcubes that result from applying step 2 to the map in Fig. 4-12 are shown in Fig. 4-13b. Each product term is obtained in the normal manner. For product term p_3 we obtain X. Time and effort are saved by using only one map when circling the product terms, but until the concept is understood, it may be wise to apply each step using a separate map.

    An example of a double-covered 1 on a two-variable map using map-entered variables is illustrated in Fig. 4-14a. The 1 in minterm position 3 is covered by the p-subcube p_1 that covers map-entered variable Z and also by the p-subcube p_2 that covers map-entered variable $\overline{Z}$. Since the 1 is a double-covered 1 when step 1 is applied, covering it in step 2 by p-subcube p_3 results in a product term that is nonessential. The fact that the product term for p_3 is nonessential is easily verified by observing the equivalent map shown in Fig. 4-14b. A double-covered 1 should therefore only be used as a don't care in step 2 to help cover other 1s that are not double-covered that may be present in the map. When the p-subcubes are chosen in step 1, if a 1 can be double-covered, a check should be made to see if such a covering leads to a simpler minimum SOP expression.

    Other expressions such as two or more variables ORed together or two or more variables ANDed together may also be plotted in the cells of a map. For an OR expression, a set of p-subcubes is chosen that results in a minimum expression for the function by covering each map-entered variable in the OR expression in turn, treating other map-entered variables as 0s and all 1s as don't cares. For an AND expression, a set of p-subcubes is chosen that results in a minimum

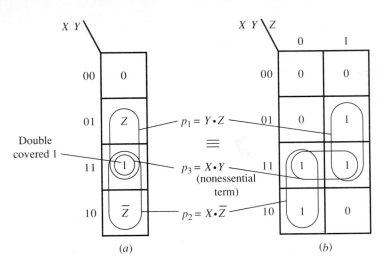

FIGURE 4-14
(*a*) Double-covered 1 in a map using map-entered variables, (*b*) same function in a larger map showing the double-covered 1 as a nonessential product term.

expression for the function by simply covering the AND expression while treating other map-entered variables as 0s and all 1s as don't cares.

Example 4-4. Plot the following function using a two-variable map utilizing map-entered variables. Obtain a reduced SOP expression for the four-variable function from the two-variable map.

$$F(A,B,C,D) = \sum m(0 \cdot (C + \overline{D}), 1, 2 \cdot C)$$

where $m = m(A, B)$.

Solution The function is shown plotted in Fig. E4-4. The p-subcubes that result from applying step 1 and step 2 are shown circled on the same map. The p-subcube for p_1 covers C, treating $\overline{D}$ as zero; therefore, in minterm position 0, $C + \overline{D} = C + 0 = C$, and a 1-cube is formed from the Cs in m_0 and m_2. The p-subcube for p_2 also forms a 1-cube using m_0 and m_1 to cover map-entered variable $\overline{D}$. In this case in minterm position 0, $C + \overline{D} = 0 + \overline{D} = \overline{D}$. The p-subcube for p_3 only covers one p-square; therefore, it represents a 0-cube. The concept of a 0-cube, 1-cube, etc., can still be used to describe the separate subcubes, but the relationship for the number of literals ($k = n - i$) in an expression representing a subcube no longer exists.

A minimum SOP expression for the 1s of the function is given below.

$$F(A,B,C,D) = p_1 + p_2 + p_3$$
$$= \overline{B} \cdot C + \overline{A} \cdot \overline{D} + \overline{A} \cdot B$$

An interesting exercise is to plot the original function on a four-variable Karnaugh map to confirm that the function given above is a minimum SOP expression. This is left as an exercise for the student.

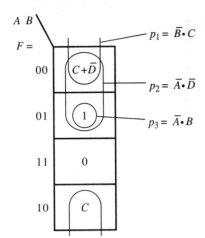

A B

$F =$

$p_1 = \bar{B} \cdot C$

$p_2 = \bar{A} \cdot \bar{D}$

$p_3 = \bar{A} \cdot B$

FIGURE E4-4

4-6-2   Three-Variable Maps Using Map-Entered Variables

In Fig. 4-15 we show a three-variable Karnaugh map with a total of six variables. The three dominant variables A, B, and C are used as map variables, while the other three infrequently-used variables D, E, and $\bar{G}$ form literals or expressions that are entered into the map.

The map was plotted using the following function.

$$F(A, B, C, D, E, G) = \sum m(0, 1 \cdot D, 4, 6, 7 \cdot (E \cdot \bar{G}))$$

where $m = m(A, B, C)$.

In Fig. 4-16a the p-subcubes for step 1 are shown, and in Fig. 4-16b we show the p-subcubes for step 2. The p-subcube covering m_0 and m_1 results in product term p_1, covering the map-entered variable D, while the p-subcube covering m_6 and m_7 results in product term p_2, covering the expression $E \cdot \bar{G}$. ORing the product terms for step 1 and step 2, we obtain the following minimum SOP expression for the function.

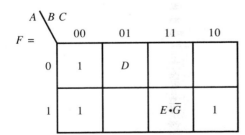

FIGURE 4-15
A three-variable Karnaugh map containing six variables (three dominant variables $A\ B\ C$ and three map-entered variables $D\ E\ \bar{G}$) for the function $F(A, B, C, D, E, G) = \sum m(0, 1 \cdot D, 4, 6, 7 \cdot (E \cdot \bar{G}))$.

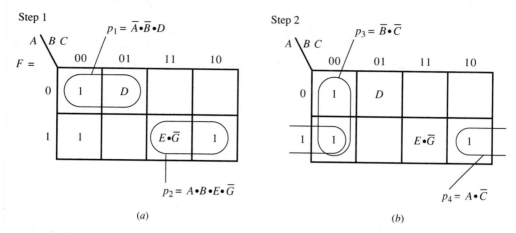

FIGURE 4-16
Solving a three-variable Karnaugh map using map-entered variables (*a*) step 1, and (*b*) step 2.

$$F(A, B, C, D, E, G) = p_1 + p_2 + p_3 + p_4$$

$$= \overline{A} \cdot \overline{B} \cdot D + A \cdot B \cdot E \cdot \overline{G} + \overline{B} \cdot \overline{C} + A \cdot \overline{C}$$

4-7 ADDING VARIABLES TO A DESIGN

It is interesting to note that Osborne used the map-entered variable technique mainly to obtain reduced functions in sequential designs (we will present sequential design at a later time).

To illustrate the real power of the concept of infrequently-used variables, consider the following scenario. The four-variable map shown in Fig. 4-17*a* was obtained when a design was needed to provide an output from four address lines $A3$, $A2$, $A1$, and $A0$. The output was designed to be a 1 only when any three or more adjacent address bits on these particular address lines were 1s; therefore, the output is 1 only for the three conditions $A3 \cdot A2 \cdot A1 \cdot A0$, $A3 \cdot A2 \cdot A1 \cdot \overline{A0}$, and $\overline{A3} \cdot A2 \cdot A1 \cdot A0$.

Later a decision was made to provide an additional variable Y to qualify the all 1s condition (the output would be 1 when all the address bits were 1 only if Y is 1). Without redrawing and filling in a five-variable map, the infrequently used Y variable (used only once in this case) can simply be added to the existing map. To provide the additional control input, Y is added at m_{15} as shown in Fig. 4-17*b*. Applying step 1 requires only one 1-cube to cover the Y; however, applying step 2 requires two 0-cubes, one for each of the 1s for a minimum covering of the function as shown in Fig. 4-17*b*. A minimum SOP expression for the 1s of the function $F2$ may be written as follows.

$$F2(A3, A2, A1, A0, Y) = p_1 + p_2 + p_3$$

$$= A2 \cdot A1 \cdot A0 \cdot Y + \overline{A3} \cdot A2 \cdot A1 \cdot A0$$

$$+ A3 \cdot A2 \cdot A1 \cdot \overline{A0}$$

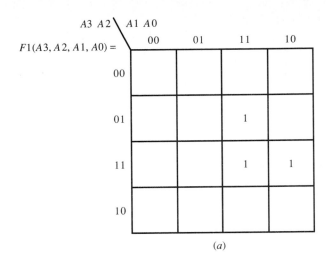

$F1(A3, A2, A1, A0) =$

(a)

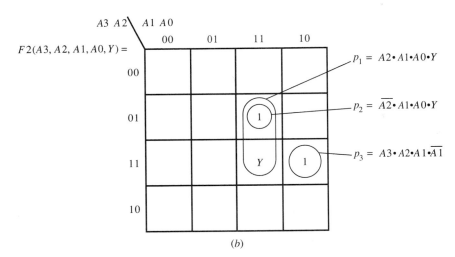

$F2(A3, A2, A1, A0, Y) =$

$P_1 = A2 \cdot A1 \cdot A0 \cdot Y$

$P_2 = \overline{A2} \cdot A1 \cdot A0 \cdot Y$

$P_3 = A3 \cdot A2 \cdot A1 \cdot \overline{A1}$

(b)

FIGURE 4-17
(a) Original design, (b) modified design using a map-entered variable.

The design flexibility provided by allowing variables to be added to an existing Karnaugh map greatly extends the usefulness of the Karnaugh map concept.

Example 4-5. Plot the following function and obtain a minimum SOP expression for the 1s of the function

$$F(U, V, W, X, Y, Z) = \sum m(0, 1, 2 \cdot V, 5 \cdot X, 8, 10 \cdot V) \; + \; d(9, 13)$$

where $m = m(U, W, Y, Z)$.

Solution A four-variable map using map-entered variables for the function is plotted in Fig. E4-5. Dashes are used for the don't care outputs. The variable X in the map

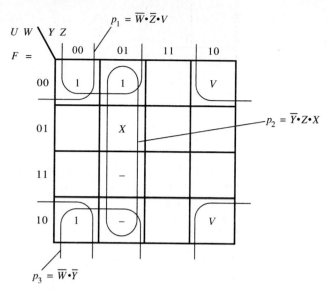

FIGURE E4-5

is not a don't care, but one of the inputs. Reading the map, we obtain the following minimum SOP expression for the 1s of the function. Remember to think in terms of the two-step procedure when covering the 1s for the function for a map-entered variable.

$$F(U, V, W, X, Y, Z) = p_1 + p_2 + p_3$$
$$= \overline{W} \cdot \overline{Z} \cdot V + \overline{Y} \cdot Z \cdot X + \overline{W} \cdot \overline{Y}$$

4-8   REDUCING THE MAP SIZE

Standard Karnaugh maps (maps without map-entered variables) can be reduced in size by entering variables in the cells of a reduced map. The mathematics of the procedure will now be presented. Consider the following three-variable truth table written using the characteristic numbers for the function F.

A	B	C	F
0	0	0	f_0
0	0	1	f_1
0	1	0	f_2
0	1	1	f_3
1	0	0	f_4
1	0	1	f_5
1	1	0	f_6
1	1	1	f_7

A Karnaugh map for this function is shown in Fig. 4-18.

$A\ B\ \backslash C$

$F\ =$

	0	1
00	f_0	f_1
01	f_2	f_3
11	f_6	f_7
10	f_4	f_5

FIGURE 4-18

Three-variable Karnaugh map using the characteristic numbers of the function.

The function can be expressed in SOP form in the following manner.

$$F = m_0 \cdot f_0 \ + \ m_1 \cdot f_1 \ + \ m_2 \cdot f_2 \ + \ m_3 \cdot f_3 \ + \ m_4 \cdot f_4 \ + \ m_5 \cdot f_5$$
$$+ \ m_6 \cdot f_6 \ + \ m_7 \cdot f_7$$

but

$$F = \overline{A} \cdot \overline{B} \cdot \overline{C} \cdot f_0 \ + \ \overline{A} \cdot \overline{B} \cdot C \cdot f_1 \ + \ \overline{A} \cdot B \cdot \overline{C} \cdot f_2 \ + \ \overline{A} \cdot B \cdot C \cdot f_3$$
$$+ \ A \cdot \overline{B} \cdot \overline{C} \cdot f_4 \ + \ A \cdot \overline{B} \cdot C \cdot f_5 \ + \ A \cdot B \cdot \overline{C} \cdot f_6 \ + \ A \cdot B \cdot C \cdot f_7$$

and

$$F = \overline{A} \cdot \overline{B} \cdot (\overline{C} \cdot f_0 \ + \ C \cdot f_1) \ + \ \overline{A} \cdot B \cdot (\overline{C} \cdot f_2 \ + \ C \cdot f_3)$$
$$+ \ A \cdot \overline{B} \cdot (\overline{C} \cdot f_4 \ + \ C \cdot f_5) \ + \ A \cdot B \cdot (\overline{C} \cdot f_6 \ + \ C \cdot f_7)$$

The last expression is simply a reorganization of the three-variable truth table.

A	B	C	F	F
0	0	0	f_0	
				$\overline{C} \cdot f_0 \ + \ C \cdot f_1$
0	0	1	f_1	
0	1	0	f_2	
				$\overline{C} \cdot f_2 \ + \ C \cdot f_3$
0	1	1	f_3	
1	0	0	f_4	
				$\overline{C} \cdot f_4 \ + \ C \cdot f_5$
1	0	1	f_5	
1	1	0	f_6	
				$\overline{C} \cdot f_6 \ + \ C \cdot f_7$
1	1	1	f_7	

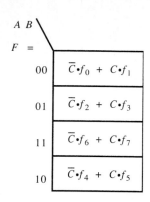

FIGURE 4-19
Reduced Karnaugh map for a three-variable function.

The function can also be plotted on a reduced Karnaugh map using variables A and B as the dominant or map variables and variable C as the map-entered variable as shown in Fig. 4-19. The characteristic numbers, f_x, can be 1s, 0s, or don't cares ($-$s). This development shows the general procedure for reducing the size of a Karnaugh map by one variable. An n-variable map can be reduced to n different $(n - 1)$-variable maps that all represent the same function, each with a different map-entered variable. Since each reduced map represents the same function, one should choose the dominant variables so that the reduction process is easy to accomplish.

Example 4-6. Plot two different one-variable Karnaugh maps for the two-variable Karnaugh map shown in Fig. E4-6a. Obtain a minimum SOP expression for each of the maps.

Solution Figure E4-6b shows the original map and the two possible reduced one-variable maps. The rules for obtaining a minimum SOP expression are different for the normal map and the reduced maps. The results, however, are the same (see figure).

Figure 4-20 illustrates the reduction of a four-variable map. The minimum SOP expression for the normal map and the reduced map is obtained in each case to show that the two maps do indeed represent the same function. The reduced map shown in Fig. 4-20 is one of four possible three-variable maps for the function.

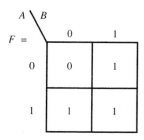

FIGURE E4-6a

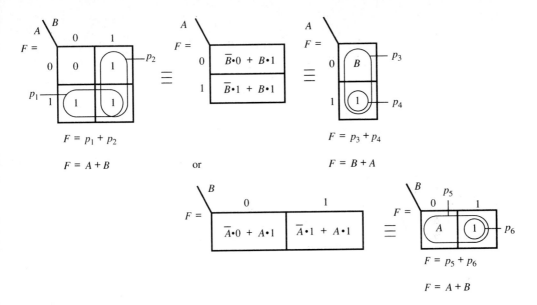

FIGURE E4-6*b*

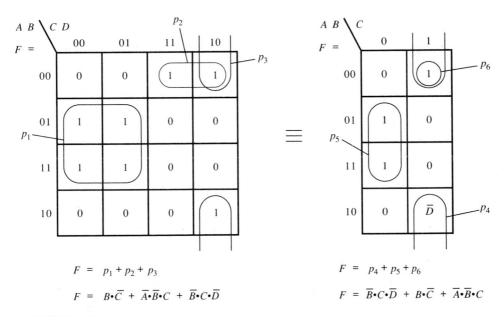

$$F = p_1 + p_2 + p_3$$

$$F = B{\cdot}\overline{C} + \overline{A}{\cdot}\overline{B}{\cdot}C + \overline{B}{\cdot}C{\cdot}\overline{D}$$

$$F = p_4 + p_5 + p_6$$

$$F = \overline{B}{\cdot}C{\cdot}\overline{D} + B{\cdot}\overline{C} + \overline{A}{\cdot}\overline{B}{\cdot}C$$

FIGURE 4-20
Three-variable map and one of four possible reduced maps of the function.

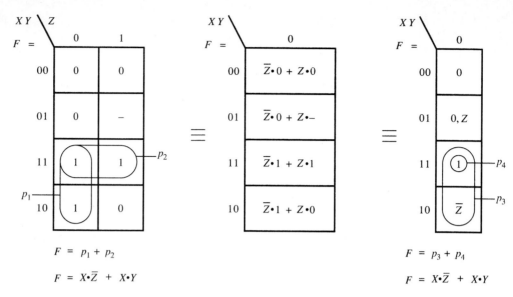

$$F = p_1 + p_2$$

$$F = X \cdot \bar{Z} + X \cdot Y$$

$$F = p_3 + p_4$$

$$F = X \cdot \bar{Z} + X \cdot Y$$

FIGURE 4-21
Effect of a don't care output in a reduced map.

When a normal Karnaugh map has don't care entries, obtaining a reduced map of the function is a little more complicated, as illustrated in Fig. 4-21. The don't care outputs in the normal map results in the map-entered variable entry of $\bar{Z} \cdot 0 + Z \cdot -$ in minterm position 1 of the reduced map. The expression $\bar{Z} \cdot 0 + Z \cdot -$ evaluates to 0 for a don't care value of 0 ($- = 0$) and Z for a don't care value of 1 ($- = 1$), and is represented as 0,Z since either value can be chosen. Compare this to a don't care that just as easily could be represented in a map as 0,1 since it can assume either a 0 or a 1. The values 0 and Z (0,Z) are both entered in the cell for minterm 1 on the reduced Karnaugh map. For this case, either 0 or Z can be used to obtain a minimum function.

Obtaining a reduced map from a normal map and being able to add infrequently used variables to an existing design are two ways in which the power and versatility of the Karnaugh map technique can be utilized as a more useful design tool.

Example 4-7. Plot the following truth table on a three-variable Karnaugh map and obtain a minimum SOP expression for the function. The dashes in the truth table represent don't care outputs.

A	B	C	D	F
0	0	0	0	0
0	0	0	1	0
0	0	1	0	0
0	0	1	1	0
0	1	0	0	1

(table is continued on following page)

A	B	C	D	F
0	1	0	1	0
0	1	1	0	1
0	1	1	1	0
1	0	0	0	0
1	0	0	1	0
1	0	1	0	–
1	0	1	1	–
1	1	0	0	1
1	1	0	1	0
1	1	1	0	–
1	1	1	1	1

Solution There are four possible reduced three-variable maps that can be plotted for a four-variable function. The simplest reduced map is obtained by choosing variables $A\ B\ C$ to represent the dominant variables and variable D to represent the map-entered variable as shown by sectioning the truth table illustrated in Fig. E4-7a. Each section of the truth table is plotted in the reduced map shown in Fig. E4-7b, and a minimum SOP expression for the function is obtained.

With practice, an n-variable map (for $n \le 5$) can be reduced to an $(n-1)$-variable map by inspection. In Fig. E4-7c, the four-variable truth table function is first plotted on a four-variable map and then reduced to a three-variable map by inspection. In the reduced map, variables $A\ C\ D$ are used as the dominant variables, and variable B is used as the map-entered variable. Notice that the minimum SOP expression for each of the maps is the same, as we would expect.

A	B	C	D	F	F
0	0	0	0	0	
0	0	0	1	0	0
0	0	1	0	0	
0	0	1	1	0	0
0	1	0	0	1	
0	1	0	1	0	$\overline{D}$
0	1	1	0	1	
0	1	1	1	0	$\overline{D}$
1	0	0	0	0	
1	0	0	1	0	0
1	0	1	0	–	
1	0	1	1	–	$\overline{D}\cdot- + D\cdot-$
1	1	0	0	1	
1	1	0	1	0	$\overline{D}$
1	1	1	0	–	
1	1	1	1	1	$\overline{D}\cdot- + D\cdot 1$

(a)

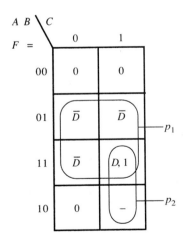

$$F = p_1 + p_2$$

$$F = B\cdot\overline{D} + A\cdot C$$

(b)

FIGURE E4-7

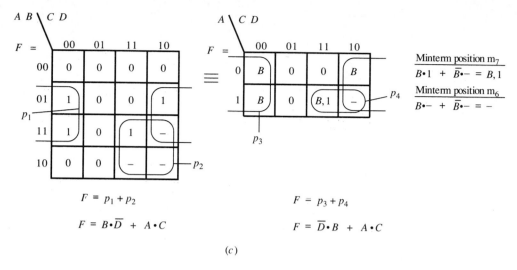

$$F = p_1 + p_2$$

$$F = B \cdot \overline{D} + A \cdot C$$

$$F = p_3 + p_4$$

$$F = \overline{D} \cdot B + A \cdot C$$

(c)

FIGURE E4-7 *(continued)*

REFERENCES

1. Bartee, T. C., Computer Design of Multiple Output Logical Networks, *IRE Trans. Electron. Computers,* vol. EC-10, no. 1, pp. 21–30, March 1961.
2. Kohavi, Z., *Switching and Finite Automata Theory,* 2d ed., McGraw-Hill, New York, 1978.
3. Hill, J. H., and Peterson, G. R., *Introduction to Switching Theory & Logic Design,* 3d ed., John Wiley & Sons, New York, 1981.
4. McCluskey, E. J., and H. Schorr, Essential Multiple-Output Prime Implicants, in Mathematical Theory of Automata, *Proc. Polytech. Inst. Brooklyn Symp.,* vol. 12, pp. 437–457, April 1962.
5. Roth, Jr., C. H., *Fundamentals of Logic Design,* 3d ed., West Publishing, St. Paul, Minnesota, 1985.
6. Mano, M. M., *Digital Design,* Prentice-Hall, Englewood Cliffs, New Jersey, 1984.
7. McCluskey, E. J., *Logic Design Principles,* Prentice-Hall, Englewood Cliffs, New Jersey, 1986.
8. Rhyne, Jr., V. T., *Fundamentals of Digital Systems Design,* Prentice-Hall, Englewood Cliffs, New Jersey, 1973.
9. Chirlian, P. M., *Analysis and Design of Digital Circuits and Computer Systems,* Matrix Publishers, Champaign, Illinois, 1976.
10. Fletcher, W. I., *An Engineering Approach to Digital Design,* Prentice-Hall, Englewood Cliffs, New Jersey, 1980.
11. Clare, C. R., *Designing Logic Systems Using State Machines,* McGraw-Hill, New York, 1973.

PROBLEMS

Section 4-3   Minimizing Two-Variable Functions

4-1. Obtain minimum SOP expressions for the 1s of the following functions using the abbreviated binary method of tabular minimization.
  (a) $F(X,Y) = \sum m(0,1)$
  (b) $F(X,Y) = \sum m(2,3)$
  (c) $F(X,Y) = \sum m(0,2)$
  (d) $F(X,Y) = \sum m(0,3)$

4-2. Obtain minimum SOP expressions for the 0s of the following functions using the abbreviated binary method of tabular minimization.

(a) $F(X,Y) = \sum m(0,1)$
(b) $F(X,Y) = \sum m(2,3)$
(c) $F(X,Y) = \sum m(0,2)$
(d) $F(X,Y) = \sum m(0,3)$

4-3. Find a minimum SOP form for each function using tabular minimization.

(a) $F(X,Y) = \sum m(0,1,2)$
(b) $F(X,Y) = \sum m(0,2,3)$
(c) $F(X,Y) = \sum m(0,1,3)$

4-4. Find a minimum POS form for each function using tabular minimization.

(a) $F(X,Y) = \sum m(3)$
(b) $F(X,Y) = \sum m(1)$
(c) $F(X,Y) = \sum m(2)$

Section 4-4 Minimizing Three- and Four-Variable Functions

4-5. Use tabular minimization to obtain a minimum SOP form for each of the following functions.

(a) $F(X,Y,Z) = \sum m(0,3,4,6)$
(b) $F(X,Y,Z) = \sum m(0,2,3,6,7)$
(c) $F(X,Y,Z) = \sum m(0,2,4,5,6,7)$

4-6. Obtain minimum SOP expressions for the 1s of the following functions using the abbreviated binary method.

(a) $F(A,B,C) = \sum m(0,2,3,4,6,7) + d(5)$
(b) $F(A,B,C) = \sum m(0,1,2,3,7) + d(6)$
(c) $F(A,B,C) = \prod M(0,7) \cdot d(1)$

4-7. Obtain minimum SOP expressions for the 1s of the following functions using tabular minimization.

(a) $F(W,X,Y,Z) = \sum m(2,3,6,7,8,9,12,13)$
(b) $F(W,X,Y,Z) = \sum m(0,3,4,5,6,7,11,12,13,14,15)$
(c) $F(W,X,Y,Z) = \sum m(0,2,5,7,8,10) + d(13,15)$

Section 4-5 The Abbreviated Decimal Method

4-8. Obtain minimum SOP expressions for the 0s of the following functions using the abbreviated decimal method.

(a) $F(W,X,Y,Z) = \sum m(2,3,6,7,8,9,12,13) + d(0,4,15)$
(b) $F(W,X,Y,Z) = \prod M(0,3,4,5,6,7,11,12,13,14,15) \cdot d(2,8,9)$
(c) $F(W,X,Y,Z) = \sum m(0,1,2,3,4,5,6,7,9,11,13,15) + d(10,14)$

4-9. Obtain minimum SOP expressions for the 1s of the following functions using the abbreviated decimal method.

(a) $F(V,W,X,Y,Z)$
$= \prod M(0,1,4,5,6,8,9,12,13,14,15,16,17,18,20,21,24,25,26,28,29,31)$
(b) $F(V,W,X,Y,Z) = \sum m(3,5,8,9,11,13,19,25,26) + d(10,17,24,27)$

4-10. Reduce the following functions to minimum SOP expressions using the tabular method.

(a) $F(X,Y,Z) = \bar{X} \cdot \bar{Y} \cdot Z + X \cdot \bar{Y} \cdot Z + X \cdot Y$
(b) $F(A,B,C) = \bar{A} \cdot (\bar{B} \cdot \bar{C} + \bar{B} \cdot C) + A \cdot (\bar{B} \cdot C + B \cdot C)$
(c) $\bar{F}(X,Y,Z) = \bar{X} + \bar{X} \cdot Z + \bar{X} \cdot Y + X \cdot Y \cdot Z$

Section 4-6 Karnaugh Map Descriptions with Map-Entered Variables

4-11. Reduce the following functions to minimum SOP expressions using map-entered variables.

(a) $F(X,Y,Z) = \sum m(0, 1 \cdot Z, 2)$ where $m = m(X, Y)$

(b) $F(X,Y,Z) = \sum m(1, 2 \cdot Z, 3 \cdot \overline{Z}) + d(0)$ where $m = m(X, Y)$

(c) $F(X,Y,Z) = \sum m(0 \cdot Z, 3) + d(1)$ where $m = m(X, Y)$

4-12. Reduce the following functions to minimum SOP expressions using map-entered variables.

(a) $F(A,B,C,D) = \sum m(0, 1 \cdot (C + D), 3 \cdot C)$ where $m = m(A, B)$

(b) $F(A,B,C,D) = \sum m(0 \cdot C, 2 \cdot C \cdot D, 3)$ where $m = m(A, B)$

(c) $F(W,X,Y,Z) = \sum m(0 \cdot Y \cdot \overline{Z}, 1, 2 \cdot \overline{Z}) + d(3)$ where $m = m(W, X)$

4-13. Obtain minimum SOP expressions for the following functions using map-entered variables.

(a) $F(X,Y,Z) = \overline{X} \cdot \overline{Y} \cdot Z + X \cdot \overline{Y} \cdot Z + X \cdot Y$ where $m = m(X, Y)$

(b) $F(X,Y,Z) - \overline{X} \cdot \overline{Y} \cdot \overline{Z} + \overline{X} \cdot Y + X \cdot Y \cdot Z$ where $m - m(X, Y)$

(c) $F(X,Y,Z) = \overline{X} + X \cdot Y \cdot Z$ where $m = m(X, Y)$

(d) $F(X,Z,Y) = \overline{X} \cdot \overline{Z} + X \cdot \overline{Z} \cdot \overline{Y}$ where $m = m(X, Z)$

4-14. Reduce the following functions to minimum SOP expressions using map-entered variables.

(a) $F(A,B,C,D) = \sum m(0, 2 \cdot D, 4 \cdot \overline{D}, 6)$ where $m = m(A, B, C)$

(b) $F(A,B,C,D) = \sum m(2, 4, 5, 7 \cdot D) + d(6)$ where $m = m(A, B, C)$

(c) $F(A,B,D,C) = \sum m(1, 2 \cdot C, 5, 6 \cdot C) + d(7)$ where $m = m(A, B, D)$

Section 4-7 Adding Variables to a Design

4-15. Reduce the following functions to minimum SOP expressions using map-entered variables.

(a) $F(U,V,W,X,Y,Z) = \sum m(0 \cdot X, 1, 3 \cdot Y, 6 \cdot \overline{Z})$ where $m = m(U, V, W)$

(b) $F(A,B,C,D,E) = \sum m(2, 3 \cdot D \cdot E, 7 \cdot E) + d(4, 6)$ where $m = m(A, B, C)$

(c) $F(T,U,V,W,X,Y,Z) = \sum m(0 \cdot W, 1 \cdot X, 3 \cdot Y, 4, 6 \cdot \overline{Z}) + d(5, 7)$ where $m = m(T, U, V)$

4-16. Obtain minimum SOP expressions for the following functions using map-entered variables.

(a) $F(V,W,X,Y,Z) = \sum m(1, 6 \cdot Z, 7 \cdot Z, 9, 10 \cdot \overline{Z}, 11 \cdot \overline{Z}, 15) + d(14)$ where $m = m(V, W, X, Y)$

(b) $F(A,B,C,D,E,G) = \sum m(2 \cdot (E + G), 4, 5, 9 \cdot \overline{G}, 10 \cdot (E + G), 12, 13) + d(6, 14)$ where $m = m(A, B, C, D)$

A	B	C	F1	F2	F3	F4
0	0	0	0	0	0	0
0	0	1	1	$\overline{D}$	1	1
0	1	0	0	0	0	0
0	1	1	1	1	$\overline{D} \cdot \overline{E}$	1
1	0	0	–	–	–	–
1	0	1	–	–	–	–
1	1	0	0	0	0	0
1	1	1	1	1	1	$D + \overline{E}$

FIGURE P4-17

4-17. The function *F1* specified by the truth table in Fig. P4-17 was modified three times using the expression at the minterm location shown in each case. Find a minimum SOP expression for each of the functions *F2* through *F4*. A dash in the table represents a don't care output.

Section 4-8 Reducing the Map Size

4-18. Plot the following three-variable functions on two-variable Karnaugh maps and obtain minimum SOP expressions for each function. Show a circuit implementation using only AND gates and OR gates. The dash in the table represents a don't care output.

A	*B*	*C*	(*a*) *F1*	(*b*) *F2*	(*c*) *F3*
0	0	0	1	1	1
0	0	1	0	1	0
0	1	0	1	0	1
0	1	1	1	1	0
1	0	0	1	1	1
1	0	1	0	0	0
1	1	0	1	0	0
1	1	1	0	0	-

4-19. For each of the Karnaugh maps shown in Fig. P4-19, show a reduced Karnaugh map one size smaller. Write a minimum SOP expression for each reduced map and each normal map.

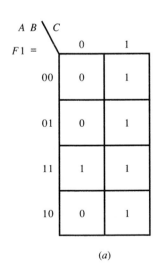

(*a*)

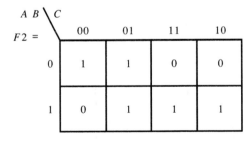

(*b*)

FIGURE P4-19

A B \ C D

F 3 =

	00	01	11	10
00	1	0	0	1
01	0	0	1	1
11	0	0	1	1
10	1	0	1	1

(c)

FIGURE P4-19
(*continued*)

4-20. Plot the following four-variable functions for a BCD to XS3 code converter on three-variable Karnaugh maps and obtain minimum SOP expressions for each function. The Xs in the table represent don't care outputs.

BCD				XS3			
I3	I2	I1	I0	F3	F2	F1	F0
0	0	0	0	0	0	1	1
0	0	0	1	0	1	0	0
0	0	1	0	0	1	0	1
0	0	1	1	0	1	1	0
0	1	0	0	0	1	1	1
0	1	0	1	1	0	0	0
0	1	1	0	1	0	0	1
0	1	1	1	1	0	1	0
1	0	0	0	1	0	1	1
1	0	0	1	1	1	0	0
1	0	1	0	X	X	X	X
1	0	1	1	X	X	X	X
1	1	0	0	X	X	X	X
1	1	0	1	X	X	X	X
1	1	1	0	X	X	X	X
1	1	1	1	X	X	X	X

PART
II

COMBINATIONAL
LOGIC DESIGN

CHAPTER
5

TOP-DOWN DESIGN PROCESS FOR GATE-LEVEL COMBINATIONAL LOGIC DESIGN

5-1 INTRODUCTION AND INSTRUCTIONAL GOALS

In Section 5-2, Gate-Level Combinational Logic Design, we begin this chapter by listing a set of design steps for combinational logic design. Next, a unique top-down design process is introduced to provide the novice and the experienced designer with a well-organized design approach for gate-level design. This section discusses the first two steps in the top-down design process: (1) using logic descriptions to obtain Boolean functions or their complements in SOP form, and writing the functions in design form (available form), and (2) obtaining signal lists (choosing the logic convention for each available signal to match its signal source or signal destination). This presentation covers the concepts of available signals, available forms of Boolean functions using available signals, and signal lists for Boolean functions. Signal matching and mismatching are a practical side of digital design that is often left uncovered, and these ideas are also presented in Section 5-2.

Section 5-3, Logic Conventions and Polarity Indication opens with a discussion of the positive logic convention. Practical consideration is given to the logic levels and noise margins associated with digital design. The negative logic convention and direct polarity indication are then presented. Additional information is provided concerning the functional interpretation of the logic symbols used in drawing logic diagrams. The qualifying symbols, that is, the negation symbol and the polarity symbol are presented, and the specific type of diagrams these symbols are associated with is discussed.

Using Design Documentation to Obtain Functional Logic Diagrams is the topic heading of Section 5-4. In this section the logic layout part, the negation indication part, and the polarity indication part of the design documentation procedure are presented. Step 3 of the top-down design process involves obtaining design documentation, while Step 4 involves obtaining functional logic diagrams using design documentation. Both of these design steps are thoroughly discussed in Section 5-4 where the design for such practical circuits as the half adder, the full adder, and the carry-lookahead concept for binary adder circuits is found.

An abundance of practical design examples for gate level combinational designs are presented in Section 5-5, Applying the Top-down Design Process to Other Realistic Problems. Each example is preceded by a discussion concerning a similar design, and each design presented utilizes Steps 1 through 4 of the top-down design process. The designs presented in this section include gate-level designs for the following circuit types:

Binary subtractors (half subtractors and full subtractors)

Binary comparators (half comparators and full comparators)

Code converters

Encoders (including priority encoders)

Decoders

Demultiplexers

Multiplexers or Data Selectors

Exclusive OR and Exclusive NOR

Binary Multipliers

In the last section, Section 5-6, Reviewing the Top-Down Design Process, we revisit the realistic problem of signal matching and mismatching. The gate-level design of a binary to seven-segment hexadecimal character generator is presented, and the design choice involves selecting either a common cathode or common anode seven-segment display device. The choice of one of the seven-segment displays results in a signal mismatching problem that is easily solved by the top-down design process. To illustrate the versatility of the top-down design procedure, a design is carried out for a problem containing both positive and negative logic signals in both complemented and uncomplemented forms.

This chapter should prepare you to

1. Write Boolean functions in design form or available form.
2. Obtain signal lists for designs such that available signals match their sources or destinations.
3. Write voltage truth tables for AND and OR elements that use either negation symbols or polarity symbols.
4. Write the truth table design descriptions for functions represented by the following names: Binary adder, Binary subtractor, Binary comparator, Code converter, Encoder, Decoder, Demultiplexer, Multiplexer (Data Selector), Exclusive OR, Exclusive NOR, and Binary Multiplier.
5. Use the top-down design process to design gate-level combinational logic circuits, and draw functional logic diagrams prepared using the positive logic convention.
6. Use the top-down design process to design gate-level combinational logic circuits, and draw functional logic diagrams prepared using direct polarity indication.
7. Utilize block diagrams of basic cascadable circuits to construct circuits of the same type using a larger number of bits.
8. Design gate-level circuits that contain any combination of positive and negative logic signals, including complemented and uncomplemented signals.

5-2   GATE-LEVEL COMBINATIONAL LOGIC DESIGN

First we present a set of generalized combinational logic design steps followed by the set of design steps we intend to follow called the top-down design process.

5-2-1   Generalized Combinational Design Steps

The following general design steps are presented in many books on digital logic and also in some manufacturer's handbooks as part of their logic tutorial.

1. Obtain the Boolean functions from the logic descriptions.
2. Assume a single logic convention (usually positive logic) for all signal names in the functions.
3. Obtain realizable logic circuit diagrams from the Boolean functions using the logic symbols chosen from data books. (There is another part to this step that is seldom discussed: the identification and correction of any signal mismatching problems.)

5-2-2 The Top-Down Design Process

The design steps that we propose to follow for gate-level logic are listed below.

1. Use logic descriptions to obtain Boolean functions or their complements in SOP form, then write the functions in design form (available form).
2. Obtain the signal lists, (choose the logic convention for each available signal to match its signal source or signal destination).
3. Obtain the design documentation for the circuits (document the layout steps for the functional logic diagrams).
4. Obtain the functional logic diagrams.
5. Obtain the realizable logic diagrams.
6. Obtain the detailed logic diagrams.

The first step is the top level in the design process, and is performed first, with each succeeding step performed in turn, as illustrated in flow chart form in Fig. 5-1.

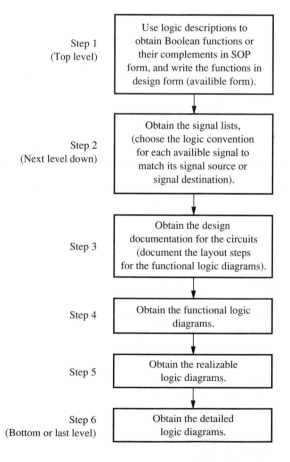

Step 1 (Top level) — Use logic descriptions to obtain Boolean functions or their complements in SOP form, and write the functions in design form (availible form).

Step 2 (Next level down) — Obtain the signal lists, (choose the logic convention for each availible signal to match its signal source or signal destination).

Step 3 — Obtain the design documentation for the circuits (document the layout steps for the functional logic diagrams).

Step 4 — Obtain the functional logic diagrams.

Step 5 — Obtain the realizable logic diagrams.

Step 6 (Bottom or last level) — Obtain the detailed logic diagrams.

FIGURE 5-1
Top-down design process.

In Step 1 (top level), an available signal name may be provided in an uncomplemented form such as X, or in a complemented form such as $\sim Y$. A tilde is used for the complemented form so that it is clear in a Boolean function when a signal is indeed available in complemented form. For example, if $F = A \cdot \overline{B}$, it would not be clear whether the signal $\overline{B}$ represents the available signal B or the available signal $\overline{B}$ unless we add the comment that the signal B is available in its complemented form. The expression $\sim B$ provides that information when the function is written using the tilde, as in $F = A \cdot \sim B$. The form $F = A \cdot \overline{B}$ will be referred to as the normal form of the Boolean function, while the form $F = A \cdot \sim B$ will be referred to as the design form or available form.

Example 5-1. For the Boolean function

$$F = W \cdot X + \overline{Y} \cdot Z + W \cdot \overline{Y} + X \cdot Y \cdot \overline{Z}$$

write the function in design form for the following available signals.

(a) F, W, X, $\sim Y$, Z

(b) F, $\sim W$, X, $\sim Y$, Z

(c) F, W, $\sim X$, Y, $\sim Z$

(d) $\sim F$, W, $\sim X$, $\sim Y$, Z

Solution

(a) $F = W \cdot X + \sim Y \cdot Z + W \cdot \sim Y + X \cdot \overline{\sim Y} \cdot \overline{Z}$

(b) $F = \overline{\sim W} \cdot X + \sim Y \cdot Z + \overline{\sim W} \cdot \sim Y + X \cdot \overline{\sim Y} \cdot \overline{Z}$

(c) $F = W \cdot \overline{\sim X} + \overline{Y} \cdot \sim Z + W \cdot \overline{Y} + \overline{\sim X} \cdot Y \cdot \sim Z$

(d) $\overline{\sim F} = W \cdot \overline{\sim X} + \sim Y \cdot Z + W \cdot \sim Y + \overline{\sim X} \cdot \overline{\sim Y} \cdot Z$

In Step 2 (next level down), the signal list is simply a list of the available signal names and their corresponding logic conventions. The positive and negative logic conventions will be discussed in the following section. This step involves choosing the logic convention for each available signal in the Boolean functions such that each input signal matches its signal source, and each output signal matches its signal destination as illustrated in Fig. 5-2.

Referring to Fig. 5-2, signals A and $\sim B$ in circuit 3 should be positive logic signals to match the outputs of circuit 1. Signals C and $\sim D$ should be negative logic signals to match the outputs of circuit 2. Signal F should be a positive logic signal, since it is used to drive circuit 4 which is a positive logic circuit. The signal list for circuit 3 can be written as follows.

Signal list: $F, A, \sim B, C[\text{NL}], \sim D[\text{NL}]$

Where the notation [NL] is appended after the signal names C and $\sim D$ it is an abbreviation for [negative logic]. The signal names without any appended notation are positive logic signals. We could also append [PL] to each positive logic signal name, but it is not necessary and it requires extra writing. From

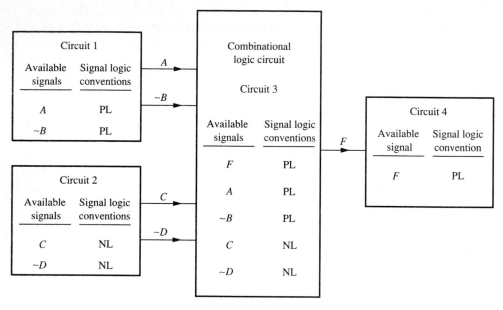

FIGURE 5-2
Logic convention matching technique.

a glance at the signal list, you can tell that available signals F, A, and $\sim B$ are positive logic signals while available signals C and $\sim D$ are negative logic signals.

> **Example 5-2.** For the following signal list, draw a block diagram showing the source and destination of each signal and its corresponding logic convention. Signal F is an output in the following list and the rest of the signals are inputs.
>
> Signal list: $F[NL], X, Y, \sim Z$

Solution The block diagram in Fig. E5-2 shows the signal list for circuit 2. Circuit 2 is being driven by circuit 1; therefore, each signal in circuit 2 has the same logic

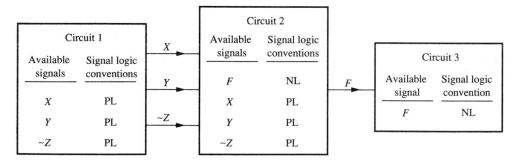

FIGURE E5-2

convention as the corresponding signal in circuit 1. The output signal from circuit 2 is driving circuit 3; it also has the same logic convention as specified in circuit 3.

It may seem strange to have signals with two logic conventions in the same problem. Often a designer will ignore negative logic signals for the time being, and use all the signals as though they were positive logic signals. After the Boolean functions are obtained and positive logic circuits drawn, the designer would then add or delete Inverters as necessary (along signal lines bridging positive and negative logic regions) to fix each signal mismatching problem. A signal mismatching problem occurs when a signal line, that provides a 1-state that is high, is expected to drive another signal line that requires a 1-state, that is low, or, when a signal line that provides a 0-state, that is low, is expected to drive another signal line that requires a 0-state that is high. The concept of high and low will now be discussed in relationship to the logic states of 1 and 0.

5-3 LOGIC CONVENTIONS AND POLARITY INDICATION

As we have seen thus far, each signal name in a logic equation can take on only a true (1) or a false (0) value. Digital logic circuits, however, deal primarily with voltages and currents, and in most cases with the former. Real off-the-shelf IC devices require input signals that consist of two non-overlapping voltage ranges as shown in Fig. 5-3. When logic symbols are used to represent real devices, it is necessary that a relationship between the logic states of 1 and 0 and the logic voltages that represent these states be established. This relationship is established by using either a positive logic convention, a negative logic convention, or direct polarity.

5-3-1 The Positive Logic Convention

When the proper input signals are applied to real devices, these devices respond with two non-overlapping voltage ranges at their outputs. The higher of the two ranges is represented by the letter H, and the lower of the two ranges is represented by the letter L. When a high level (H) represents an external logic state of 1 and a low level (L) represents an external logic state of 0 as illustrated in Fig. 5-3, the positive logic convention is being used. The area between the two distinct ranges is the transition region. A signal passes through the transition region only when it changes state from low to high or from high to low.

When discussing signals and states in a circuit, it is customary to refer to the logic diagram of the circuit. The term "external logic state" means a logic state on an input or output signal line outside a symbol outline in the circuit diagram, as opposed to an "internal logic state" which means a logic state on an input or output signal line inside a symbol outline as illustrated in Fig. 5-4. If an external

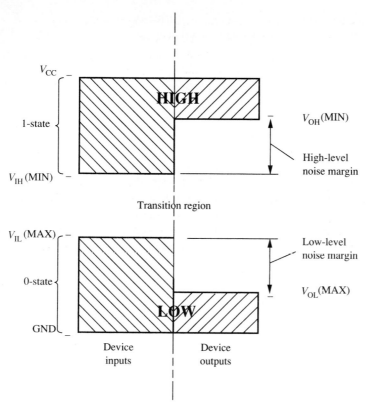

FIGURE 5-3
The positive logic convention relating an external 1-state to a high-level voltage range, and an external 0-state to a low-level voltage range.

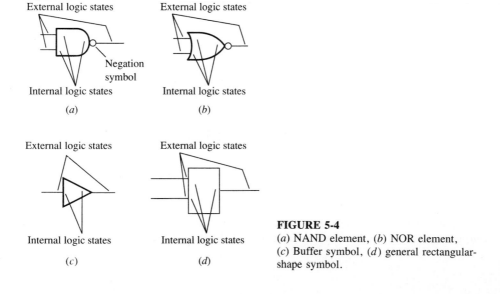

FIGURE 5-4
(*a*) NAND element, (*b*) NOR element,
(*c*) Buffer symbol, (*d*) general rectangular-shape symbol.

qualifying symbol such as the negation symbol ($\circ$) is present on an input signal line, then the external logic state is on the input line prior to the negation symbol; however, if an external qualifying symbol is present on an output signal line, then the external logic state is on the output line beyond the qualifying symbol.

Besides showing the positive logic convention, Fig. 5-3 also shows real input and output device characteristics. Devices with the characteristics shown in Fig. 5-3 that operate with a single positive supply voltage have a low level (L) closer to ground (GND), and a high level (H) closer to the power supply voltage (V_{CC}) of typically +5 volts. The non-overlapping nature of the high and low voltage levels is distinctly shown in Fig. 5-3 for both device outputs and device inputs. Device inputs are designed to accept a larger range for both high-level and low-level voltages than the device outputs are designed to supply. This is intentional so device outputs can drive other device inputs with plenty of noise margin. The technology used in the design of devices determines the noise margin that can be achieved in the design. A list of device technology abbreviations for digital logic devices is provided in Table 5-1. If two devices are interfaced (connected) that use different technologies, a designer must check the respective input and output characteristics of each device to determine if they will work together, and if not how additional circuitry can be added or other devices can be substituted to insure that they will work together.

With today's technology, Complementary Metal-Oxide Semiconductor (CMOS) has a much better noise margin than its competitor, bipolar Transistor-Transistor Logic (TTL). For CMOS and TTL digital circuit types see Appendix D. Noise margin is a DC voltage specification that indicates the immunity of a circuit to adverse operating conditions. High-level noise margins can be obtained by consulting a manufacturer's data books and recording the minimum highs for a device from the data sheet. These numbers are then used in the following relationship to calculate the high-level noise margin.

$$V_{NH} = V_{OH}(MIN) - V_{IH}(MIN)$$

TABLE 5-1
Device technology abbreviations

STD TTL	Standard Transistor-Transistor Logic
LS	Low Power Schottky TTL
S	Schottky TTL
ALS	Advanced Low Power Schottky TTL
AS	Advanced Schottky TTL
F or FAST	Fairchild Advanced Schottky TTL
AC	Advanced Complementary Metal-Oxide Semiconductor
ACT	Advanced CMOS TTL (compatible)
HC	High Speed CMOS
HCT	High Speed CMOS TTL (compatible)
BCT	Bipolar CMOS Technology

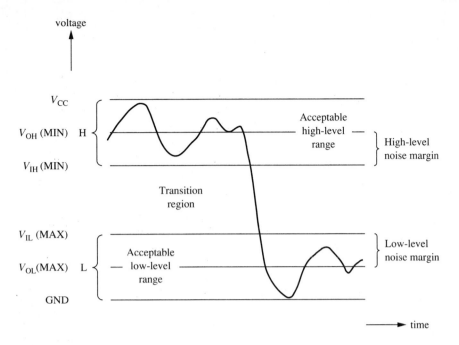

FIGURE 5-5
Switching waveform illustrating the importance of noise margin. Noise margin provides a guard band
to insure a noisy high or noisy low signal stays in it acceptable range.

Similarly, low-level noise margins can be obtained by calculating the difference in voltage between the maximum lows using the following relationship. High-level and low-level noise margins are illustrated graphically in Fig. 5-3.

$$V_{NL} = V_{IL}(MAX) - V_{OL}(MAX)$$

When comparing two devices, the device with the greater separation between its minimum highs and its maximum lows has better noise margin. The expression "noise margin" implies that an output signal that is feeding a corresponding input can have an AC noise voltage riding on top of it. The peak value of the AC noise voltage cannot be in excess of the high-level noise margin when the signal is high, or of the low-level noise margin when the signal is low, as shown in Fig. 5-5.

5-3-2 The Negative Logic Convention

When a high level (H) is used to represent an external logic state of 0 and a low level (L) is used to represent an external logic state of 1 as shown in Fig. 5-6, a negative logic convention is being used. One of the first off-the-shelf IC device families, called Resistor-Transistor Logic (RTL), made use of the negative

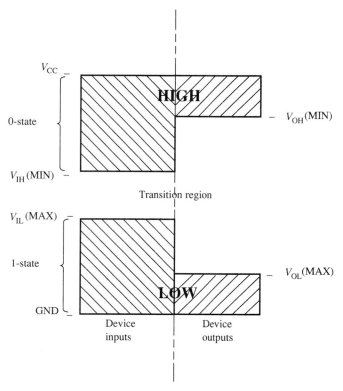

FIGURE 5-6
The negative logic convention relating an external 0-state to a high-level voltage range, and an external 1-state to a low-level voltage range.

logic convention to symbolically represent the family of devices. Today, most manufacturers of TTL and CMOS utilize either the positive logic convention or direct polarity indication to symbolically represent their devices. For this reason, we will concentrate our discussions primarily on the positive logic convention and direct polarity indication.

5-3-3 Direct Polarity Indication

Direct polarity indication provides an alternate method of relating high-level and low-level voltages on signal lines to the inputs and outputs of real devices. A diagram using direct polarity indication utilizes the presence or absence of the polarity symbol ($\triangleright$) to provide the relationship between the external logic levels of a signal and the internal logic states of 1 and 0 as shown in Fig. 5-7. The same relationship that exists for device inputs in Fig. 5-7 also applies to device outputs.

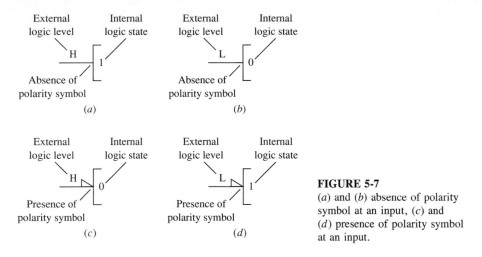

FIGURE 5-7
(*a*) and (*b*) absence of polarity symbol at an input, (*c*) and (*d*) presence of polarity symbol at an input.

Table 5-2 summarizes the relationship between the external voltage ranges H and L and the internal logic states for devices when the polarity symbol is absent and when the polarity symbol is present.

Direct polarity indication has not found wide acceptance in the academic environment, but in the last few years it has found wide acceptance in industry. This is primarily due to the availability of schematic capture programs designed to increase engineering productivity, and the wide variety of library parts utilizing direct polarity indication that are available for these programs.

> **Example 5-3.** Obtain the voltage truth table for the AND element shown in Fig. E5-3*a*.
>
> *Solution* Signal line *A* is an active low line, since it contains a polarity indicator, while signal lines *B* and *C* are active high lines. The operation of the AND element can be described by the following logic statement: if input *A* is low AND input *B* is high, then output *C* is high; else, output *C* is not high. Using this logic statement, the following voltage truth table can be written for the AND element.

TABLE 5-2
Direct polarity indication summary

External voltage range	Polarity symbol present	Internal logic state
H	No	1
L	No	0
H	Yes	0
L	Yes	1

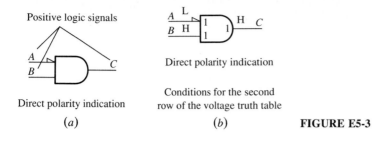

Positive logic signals

Direct polarity indication

(*a*)

Direct polarity indication

Conditions for the second
row of the voltage truth table

(*b*)

FIGURE E5-3

A	B	C
L	L	L
L	H	H
H	L	L
H	H	L

The conditions indicated in the second row of the voltage truth table are shown in Fig. E5-3*b*. Notice that the internal logic states for the specified external logic levels are also included for the conditions indicated in the second row.

The presence or absence of the traditionally-used negation symbol (○) provides the relationship between the external logic states of a signal and the internal logic states of 1 and 0. This is shown in Fig. 5-8 so that a comparison can be made with Fig. 5-7. It is important to understand the meaning of the polarity symbol and the negation symbol as illustrated in Figs. 5-7 and 5-8 respectively. When each polarity symbol in Fig. 5-7 is substituted by the negation symbol, the result is Fig. 5-8, using the positive logic convention. The top-down design process provides a tool for obtaining a diagram using either the positive logic

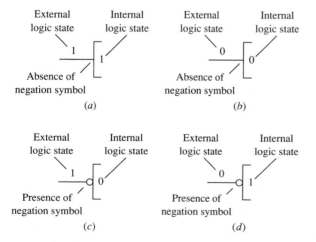

FIGURE 5-8
(*a*) and (*b*) absence of negation symbol at an input, (*c*) and (*d*) presence of negation symbol at an input.

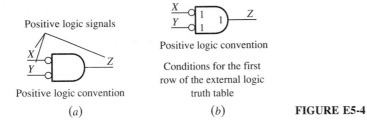

Positive logic signals

Positive logic convention

(a)

Positive logic convention

Conditions for the first
row of the external logic
truth table

(b)

FIGURE E5-4

convention or direct polarity indication. It is the designer's choice which diagram one prefers to use.

> **Example 5-4.** Obtain the voltage truth table for the AND element shown in Fig. E5-4*a*.
>
> **Solution** Negated input lines such as signal lines X and Y contain a negation indicator. Signal line Z is a nonnegated signal line. The operation of the AND element can be described as follows: If input X is 0 AND input Y is 0, then output Z is 1; else, output Z is not 1. Using this logic statement, the following external logic truth table can be written for the AND element and immediately converted to the voltage truth table shown to the right.

X	Y	Z	X	Y	Z
0	0	1	L	L	H
0	1	0	L	H	L
1	0	0	H	L	L
1	1	0	H	H	L

> The conditions indicated in the first row of the external logic truth table are shown in Fig. E5-4*b*. Notice that the internal logic states shown inside the symbol outline refer to the external logic states in the first row of the truth table. Keep in mind that the internal input logic states are negated, compared to the external input logic states, because of the presence of the negation symbol at each of the inputs.

5-4   USING DESIGN DOCUMENTATION TO OBTAIN FUNCTIONAL LOGIC DIAGRAMS

Step 3 in the top-down design process involves obtaining the design documentation for the circuits. Functional or basic logic diagrams are drawn using the design documentation. Functional logic diagrams show logic circuits in simple form. This form does not necessarily contain exact design information or represent a realizable physical form.

There are three different forms of logic diagrams. Logic diagrams can be prepared using the positive logic convention, the negative logic convention, or direct polarity indication. As we mentioned earlier, we will place our emphasis on diagrams using the positive logic convention and direct polarity indication.

The design documentation procedure provides an organized approach for obtaining logic circuit diagrams for Boolean functions using basic logic elements (gates). The designer can decide to obtain functional logic diagrams using either the positive logic convention or direct polarity indication. In the first step of the design documentation procedure (Step 3a) we obtain the logic layout part (LLP). The logic layout part is the same for either prepared form of diagram, as illustrated in the design documentation procedure shown in Fig. 5-9. In the second step of the design documentation procedure (Step 3b) we obtain the negation indication part (NIP), if a diagram is to be prepared using the positive logic convention, or the polarity indication part (PIP), if a diagram is to be prepared using direct polarity indication.

The logic layout part represents a Boolean function using only the available signals. This documentation part shows the AND and OR gates required for a circuit representing a function prior to adding negation symbols for a diagram using the positive logic convention, or polarity symbols to obtain a diagram using direct polarity indication. To write the Boolean function using only available signals simply means to write the function without using any overbars. The overbars are removed when writing the logic layout part of the design documentation for two reasons: (1) overbars are not used by our procedure when drawing the AND and OR gate structure required by the circuit, and (2) the available signal names

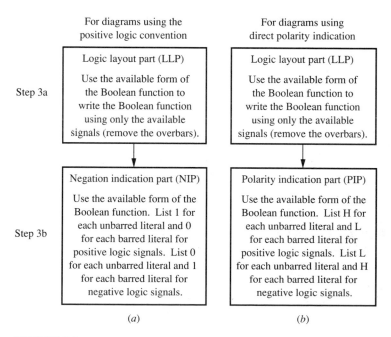

(a)   (b)

FIGURE 5-9
Design documentation procedure (*a*) for diagrams using the positive logic convention, (*b*) for diagrams using direct polarity indication.

used to label the inputs and outputs of the circuit are those signal names taken from the logic layout part of the design documentation.

The negation indication part of the design documentation provides the required negation symbols for a diagram using the positive logic convention, while the polarity indication part provides the required polarity symbols for a diagram using direct polarity indication. The need to fix signal mismatching problems also exists when the design documentation procedure is used. The correct convention of each signal is accounted for when the negation indication and polarity indication parts of the design documentation are obtained. If the logic convention of a signal is assumed to be positive logic and it is not, then an Inverter would have to be added along the signal line connecting the different logic conventions to complete the design.

> **Aside 5-1.** The division on a diagram between a negative logic section and a positive logic section is demonstrated in Fig. A5-1 for the logic signal $A[NL]$.
>
> Notice in the figure that a dotted line is used to mark the boundary between the negative logic and positive logic sections on the diagram. The signal line containing the negative logic signal A is changed to $\overline{A}$ just across the boundary on the positive logic side to account for the logic inversion caused by the different logic convention. On the output of the Inverter shown in Fig. A5-1 the signal is again shown as A, only now it is a positive logic signal. Using all positive logic signals, obtaining the circuit diagram, and then adding an Inverter to every signal line that contains a negative logic signal is one acceptable way to accomplish a design using negative logic signals.
>
> The design documentation procedure accomplishes the same goal by using the Boolean function and complementing each negative logic signal. If X is a negative logic signal and X occurs in a function, then 0 is substituted in the negation indication part rather than a 1, which would be required if the signal were a positive logic signal. If $\overline{X}$ occurs in the function, then 1 is substituted instead of 0 to account for the boundary condition change. Since all negative logic signals are complemented when following the procedure for the negation indication part and the polarity indication part of the design documentation, it is not necessary to add Inverters after the circuit is drawn.

Table 5-3 illustrates a function that contains complemented and uncomplemented available signals as well as positive and negative logic signals (as

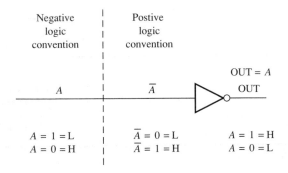

FIGURE A5-1

TABLE 5-3
Results of applying the design documentation procedure

Step 1:		$F = \overline{\sim B} \cdot \overline{\sim D} + \overline{A} \cdot \sim B \cdot \overline{C} + A \cdot \sim B \cdot \sim D$
Step 2:	Signal list:	$F, A, \sim B, C[\text{NL}], \sim D[\text{NL}]$
Step 3a:	Logic Layout Part:	$F = \sim B \cdot \sim D \quad + \quad A \cdot \sim B \cdot C \quad + \quad A \cdot \sim B \cdot \sim D$
Step 3b:	Negation Indication Part:	1 0 1 0 1 1 1 1 0
Step 3a:	Logic Layout Part:	$F = \sim B \cdot \sim D \quad + \quad A \cdot \sim B \cdot C \quad + \quad A \cdot \sim B \cdot \sim D$
Step 3b:	Polarity Indication Part:	H L H L H H H H L

observed from the signal list that accompanies the function). The design documentation procedure is simply an organized method for obtaining the layout steps for functional logic diagrams. If one masters the technique of obtaining the design documentation parts in Table 5-3, one can obtain the design documentation for any Boolean function or its complement in SOP form. Most of the time, designs will not require complemented available signals or negative logic signals, thus simplifying the design task.

5-4-1 Binary Adders

To illustrate the design documentation method for a useful circuit, consider the half adder truth table.

A0	B0	SUM0	C_OUT1
0	0	0	0
0	1	1	0
1	0	1	0
1	1	0	1

A half adder is the name given to a circuit that adds the two least significant bits at bit position 0 in an n-bit binary parallel adder. No carry in to bit 0 is required, but a carry out (C_OUT1) from bit 0 to the next significant bit, bit 1, is required. A block diagram for a 4-bit parallel adder is shown in Fig. 5-10a, and a block diagram for a half adder for bit 0 of this adder is shown in Fig. 5-10b. If the 4-bit parallel adder in Fig. 5-10a were designed without partitioning, it would require a truth table with 8 inputs and $2^8 = 256$ entries. An 8-bit parallel adder would require $2^{16} = 65,536$ entries, and a 16-bit parallel adder would require $2^{32} = 4,294,967,296$ entries. For a large number of inputs, it is easy to see that an iterative approach is not just desirable to keep the size of a problem manageable, but it can be mandatory to obtain a solution.

The Boolean functions for the half adder can be written in available form with their corresponding signal lists from the truth table.

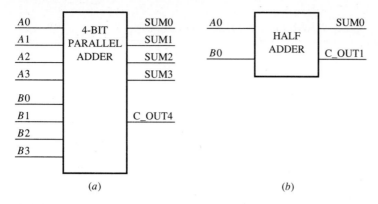

(a)   *(b)*

FIGURE 5-10
(*a*) Block diagram of a 4-bit parallel adder, (*b*) half adder.

SUM0 $= \overline{A0} \cdot B0 + A0 \cdot \overline{B0}$ C_OUT1 $= A0 \cdot B0$
Signal list: SUM0, A0, B0   Signal list: C_OUT1, A0, B0

The design documentation for diagrams using the positive logic convention and direct polarity indication is provided below.

Design documentation for diagrams using the positive logic convention

---

Logic layout part (LLP):	SUM0 $= A0 \cdot B0$	$+$	$A0 \cdot B0$		C_OUT1 $= A0 \cdot B0$		
Negation indication part (NIP):	1	0	1	1	0	1	1	1

Design documentation for diagrams using direct polarity indication

---

Logic layout part (LLP):	SUM0 $= A0 \cdot B0$	$+$	$A0 \cdot B0$		C_OUT1 $= A0 \cdot B0$		
Polarity indication part (PIP):	H	L H		H L	H	H H	

The design documentation represents the necessary information required to draw the functional logic diagrams for the Boolean functions of the half adder in a convenient form. When the functional logic diagrams are drawn for the negation indication parts, a 0 signifies the presence of a negation symbol (○), while a 1 signifies the absence of a negation symbol. For the polarity indication parts, an L signifies the presence of a polarity symbol (⊳), while an H signifies the absence of a polarity symbol.

In the construction of the functional logic diagrams for the half adder, the logic layout parts are used to draw the gates and interconnect them in the order of

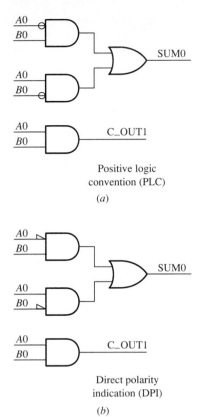

Positive logic
convention (PLC)

(*a*)

Direct polarity
indication (DPI)

(*b*)

FIGURE 5-11
Functional or basic logic diagrams, (*a*) using the positive logic
convention, (*b*) using direct polarity indication.

precedence of the logic operators. Negation symbols are added at the inputs and
outputs where 0s appear in the negation indication parts. The half adder circuit is
shown in Fig. 5-11*a* using the positive logic convention. Notice that the negation
symbols can be drawn as shown in the figure with the signal lines extending
through the negation symbols. While the design documentation is used to draw
the logic layout for the circuit, and the negation indication provides the location
of the negation symbols, the signal list is also used to indicate which signals
reside in the positive sections of the diagram, and which signals reside in the
negative logic section of the diagram. For the half adder shown in Fig. 5-11 all
the signals are positive logic signals, and so the whole diagram is labeled positive
logic convention (PLC).

The functional logic diagrams for the half adder are shown in Fig. 5-11*b*
using direct polarity indication. The gates are drawn first, and interconnected
using the logic layout parts in the order of precedence of the logic operators.
Next, polarity symbols are added at the inputs and outputs where Ls appear in
the polarity indication parts. When negative logic signals appear on a diagram
using polarity indication, they will be written as they appear in the signal list,
since a diagram using polarity indication does not have separate positive and
negative logic sections.

Example 5-5. In Fig. E5-5a, a 2-bit parallel binary adder block diagram is shown using a half adder and a full adder. The full adder has an additional carry-in input that is not required by the half adder; it can be used iteratively by simply adding additional stages.

1. Obtain the truth table for the full adder.
2. Obtain functional logic diagrams using the positive logic convention for the full adder using the top-down design process.

Solution The truth table for the full adder is generated as follows: Using each combination for the inputs, the sum and carry output of the full adder is exactly the

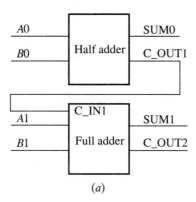

(a)

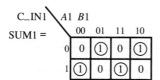

$$\text{SUM1} = m_1 + m_2 + m_4 + m_7$$
$$= \overline{\text{C\_IN1}} \cdot \overline{\text{A1}} \cdot B1 + \overline{\text{C\_IN1}} \cdot A1 \cdot \overline{B1} + \text{C\_IN1} \cdot \overline{\text{A1}} \cdot \overline{B1} + \text{C\_IN1} \cdot A1 \cdot B1$$

$$\text{C\_OUT2} = p_1 + p_2 + p_3$$
$$= A1 \cdot B1 + \text{C\_IN1} \cdot A1 + \text{C\_IN1} \cdot B1$$

(b)

FIGURE E5-5

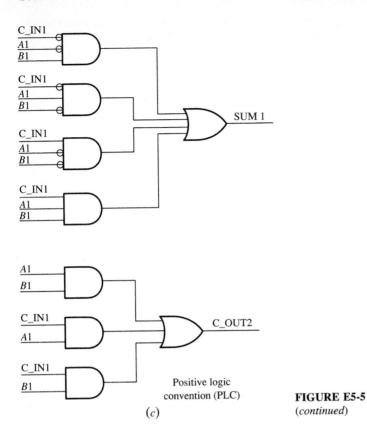

Positive logic
convention (PLC)

(*c*)

FIGURE E5-5
(*continued*)

same as the result of the binary addition operation on two 2-bit binary numbers $A1$ $A0$ and $B1$ $B0$ as indicated by the block diagram shown in Fig. E5-5a.

C_IN1	A1	B1	SUM1	C_OUT2
0	0	0	0	0
0	0	1	1	0
0	1	0	1	0
0	1	1	0	1
1	0	0	1	0
1	0	1	0	1
1	1	0	0	1
1	1	1	1	1

Minimum SOP expressions for SUM1 and C_OUT2 are obtained using the Karnaugh maps as shown in Fig. E5-5b. Step 1 and Step 2 of the top-down design process are shown below. Step 1 shows the Boolean functions in available form, while Step 2 shows the signal lists for the functions. Notice that the signals in both signal lists are represented as positive logic signals.

Step 1: $SUM1 = \overline{C\_IN1} \cdot \overline{A1} \cdot B1 + \overline{C\_IN1} \cdot A1 \cdot \overline{B1} + C\_IN1 \cdot \overline{A1} \cdot \overline{B1} + C\_IN1 \cdot A1 \cdot B1$
Step 2:   Signal list: SUM1, C_IN1, A1, B1

Step 1: $C\_OUT2 = A1 \cdot B1 + C\_IN1 \cdot A1 + C\_IN1 \cdot B1$
Step 2:   Signal list: C_OUT2, C_IN1, A1, B1

The design documentation, or Step 3, is provided below for diagrams using the positive logic convention.

Step 3a :   LLP: $SUM1 = C\_IN1 \cdot A1 \cdot B1 + C\_IN1 \cdot A1 \cdot B1 + C\_IN1 \cdot A1 \cdot B1 + C\_IN1 \cdot A1 \cdot B1$
Step 3b :   NIP: 1   0   0   1   0   1   0   1   0   0   1   1   1

Step 3a :   LLP: $C\_OUT2 = A1 \cdot B1 + C\_IN1 \cdot A1 + C\_IN1 \cdot B1$
Step 3b :   NIP: 1   1   1   1   1   1   1

The functional logic diagrams represent Step 4 in our design process, and these are drawn using the design documentation as shown in Fig. E5-5c. The diagrams use the positive logic convention, since the design documentation was written utilizing the negation indication part, and not the polarity indication part.

Functional logic diagrams represent the starting point for obtaining realizable logic diagrams — step 5 in the top-down design process. From functional logic diagrams, substitutions of symbols for real devices can easily be made in order to obtain diagrams with realizable or constructable circuits. The final design step, Step 6, involves obtaining detailed logic diagrams which (1) depict circuits as they are actually implemented; and (2) contain information for manufacturing or maintenance. Realizable logic diagrams and detailed logic diagrams will be presented in Chapter 6.

5-4-2 The Carry-Lookahead Concept

Circuits that are designed with modules and then cascaded have a ripple effect associated with them. One such circuit is a binary parallel adder using a half adder followed by full adders. The 5-bit parallel adder shown in Fig. 5-12, for

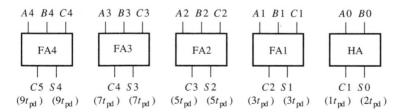

FIGURE 5-12
Five-bit parallel (ripple) adder with worst-case settling times shown in parentheses.

example, has nine gate delays (propagation delays). This is shown as $9t_{pd}$ in parentheses under the sum output, $S4$, of full adder 4 (FA4) in Fig. 5-12.

The abbreviation "t_{pd}" is used as a relative measure of the time it takes for a signal to propagate through a gate (average of t_{PLH} and t_{PHL} specified in the data book). The propagation delay time (low-to-high-level output or t_{PLH}) is the delay time between two specified references points (one on the input switching waveform and one on the output switching waveform) with the output changing from a low level to a high level. The propagation delay time (high-to-low-level output or t_{PHL}) is the delay time between two specified references points (one on the input switching waveform and one on the output switching waveform) with the output changing from a high level to a low level. This means that the final sum is not available or accurate until nine gate delays after the signals $A4$, $B4$, $A3$, $B3$, $A2$, $B2$, $A1$, $B1$, $A0$, and $B0$ are simultaneously made available at the inputs to the parallel adder. This represents a worst-case settling time of $9t_{pd}$ for the output of the parallel adder. Tracing the number of gates from the half adder through to each of the succeeding full adders results in the worst-case gate delays shown in parentheses under each of the output signal names. The circuits for the half adder and the full adder are shown in Figs. 5-11 and E5-5c respectively. Note that each full adder consists of two two-level circuits and therefore has two gate delays for each output. The half adder consists of one two-level circuit and one one-level circuit. The ripple effect is simply due to the accumulation of gate delays from the first carry-out of the half adder to the output $S4$ of FA4. In other words, the output of each full adder in the chain is dependent on the carry-outputs of the stages all the way back to the carry-output of the half adder.

By making the delay paths shorter from the output of the half adder to the carry-outputs of each full adder, a speed improvement can be achieved. The technique of using additional combinational logic to shorten the delay paths of the carry-outputs is referred to as "carry-lookahead". Ideally, if no more than two gate delays, $2t_{pd}$, are required to obtain the outputs for all carry-bits, then the worst-case settling time for the parallel adder would be only $4t_{pd}$, since the sum-bits will settle $2t_{pd}$ after the carry-bits settle.

Three things limit the usefulness of carry-lookahead when it is applied over a large number of stages. First, the carry-output of the half adder stage must be capable of driving the carry-input circuit of every succeeding stage. Second, each succeeding stage requires gates with an ever increasing number of inputs. Third, the gate count and hence the cost increases with each additional stage that is added. Because of these limiting factors, partial carry-lookahead is usually implemented over a limited number of stages. The mathematics for one popular implementation technique will now be presented for a 5-bit parallel (carry-lookahead) adder. The worst-case delay time for the 5-bit parallel (ripple) adder shown in Fig. 5-12 and a carry-lookahead version can then be compared.

To keep our presentation simple, we only need to modify the carry functions, since we can utilize the same sum functions that we previously developed for the half adder and the full adder. The sum function for stage i of the full adder is written as follows.

$$S_i = \overline{C_i} \cdot \overline{A_i} \cdot B_i \ + \ \overline{C_i} \cdot A_i \cdot \overline{B_i} \ + \ C_i \cdot \overline{A_i} \cdot \overline{B_i} \ + \ C_i \cdot A_i \cdot B_i$$

S_i is the sum bit for stage i and C_i is the carry-in bit for stage i. As soon as C_i is known for any stage, the gate delay to the output S_i is $2t_{pd}$ later for any stage, since the function S_i represents a two-level circuit.

The function for the carry-bit for the half adder is

$$C1 = A0 \cdot B0$$

$$= G0$$

$G0$ is called the carry-generate term for stage 0.

The carry function for stage i of the full adder is written as follows.

$$C_i + 1 = A_i \cdot B_i \ + \ C_i \cdot A_i \ + \ C_i \cdot B_i$$

From this function, the carry-bit for stage 1 can be written as

$$
\begin{aligned}
C2 &= A1 \cdot B1 \ + \ C1 \cdot A1 \ + \ C1 \cdot B1 \\
&= A1 \cdot B1 \ + \ C1 \cdot (A1 + B1) \\
&= G1 \ + \ C1 \cdot P1 \\
&= G1 \ + \ G0 \cdot P1
\end{aligned}
$$

$G1$ is the carry-generate term for stage 1. $P1$ is called the carry-propagate term for stage 1. Similarly, the carry-bits for stages 2, 3, and 4 can be written using simple two-input carry-generate terms, $G_i = A_i \cdot B_i$, and carry-propagate terms, $P_i = A_i + B_i$, as follows.

$$
\begin{aligned}
C3 &= A2 \cdot B2 \ + \ C2 \cdot A2 \ + \ C2 \cdot B2 \\
&= A2 \cdot B2 \ + \ C2 \cdot (A2 + B2) \\
&= G2 \ + \ C2 \cdot P2 \\
&= G2 \ + \ (G1 \ + \ G0 \cdot P1) \cdot P2 \\
&= G2 \ + \ G1 \cdot P2 \ + \ G0 \cdot P1 \cdot P2
\end{aligned}
$$

$$
\begin{aligned}
C4 &= A3 \cdot B3 \ + \ C3 \cdot A3 \ + \ C3 \cdot B3 \\
&= A3 \cdot B3 \ + \ C3 \cdot (A3 + B3) \\
&= G3 \ + \ C3 \cdot P3 \\
&= G3 \ + \ (G2 \ + \ G1 \cdot P2 \ + \ G0 \cdot P1 \cdot P2) \cdot P3 \\
&= G3 \ + \ G2 \cdot P3 \ + \ G1 \cdot P2 \cdot P3 \ + \ G0 \cdot P1 \cdot P2 \cdot P3
\end{aligned}
$$

$$
\begin{aligned}
C5 &= A4 \cdot B4 \ + \ C4 \cdot A4 \ + \ C4 \cdot B4 \\
&= A4 \cdot B4 \ + \ C4 \cdot (A4 + B4)
\end{aligned}
$$

$$= G4 + C4 \cdot P4$$

$$= G4 + (G3 + G2 \cdot P3 + G1 \cdot P2 \cdot P3 + G0 \cdot P1 \cdot P2 \cdot P3) \cdot P4$$

$$= G4 + G3 \cdot P4 + G2 \cdot P3 \cdot P4 + G1 \cdot P2 \cdot P3 \cdot P4$$

$$+ G0 \cdot P1 \cdot P2 \cdot P3 \cdot P4$$

In general the carry-lookahead term can be written for stage i in the following form.

$$C_i = G_{i-1} + G_{i-2} \cdot P_{-1} + G_{i-3} \cdot P_{i-2} \cdot P_{i-1} + \cdots$$
$$+ G_1 \cdot P_2 \cdot P_3 \cdot P_4 \cdot \cdots \cdot P_{i-1} + G_0 \cdot P_1 \cdot P_2 \cdot P_3 \cdot \cdots \cdot P_{i-1}$$

Each of the carry functions $C2$ through $C5$ has a worst-case settling time of $2t_{pd}$ from the time the carry-generate and carry-propagate terms are available. Each carry-generate term and carry-propagate term represents a one-level circuit with a worst-case delay of only $1t_{pd}$. A circuit diagram for a 5-bit parallel (carry-lookahead) adder is shown in Fig. 5-13. It can be noted that the carry-lookahead version of this 5-bit parallel adder has a settling time of $5t_{pd}$ compared to the settling time of $9t_{pd}$ for the 5-bit parallel (ripple) adder shown in Fig. 5-12. This speed improvement has not been made without a compromise, since the total gate count for the parallel (ripple) adder is only 40, while the total gate count for the parallel (carry-lookahead) adder is 46.

The 5-bit parallel adder shown in Fig. 5-13 can be redesigned to have a minimum worst-case settling time of $4t_{pd}$ by solving for carry functions that contain only A_i and B_i inputs. For example,

$$C3 = G2 + G1 \cdot P2 + G0 \cdot P1 \cdot P2$$

$$= A2 \cdot B2 + A1 \cdot B1 \cdot (A2 + B2) + A0 \cdot B0 \cdot (A1 + B1) \cdot (A2 + B2)$$

$$= A2 \cdot B2 + A1 \cdot B1 \cdot A2 + A1 \cdot B1 \cdot B2 + A0 \cdot B0 \cdot A1 \cdot A2$$

$$+ A0 \cdot B0 \cdot A1 \cdot B2 + A0 \cdot B0 \cdot B1 \cdot A2 + A0 \cdot B0 \cdot B1 \cdot B2$$

Since the last expression is in terms of the A_i and B_i inputs of the adder, the worst-case settling time for the $C3$ output is consequently only $2t_{pd}$. The sum output $S3$ that would use this output, therefore, would settle $2t_{pd}$ later for a total setting time of $4t_{pd}$ compared to $5t_{pd}$ using the first expression for $C3$. It is also quite obvious that the gate count, and, hence, the cost of implementing the last expression, is greater than the gate count for implementing the first expression, (the expression for $C3$ using carry-generate and carry-propagate terms).

Implementing lookahead circuitry for a larger number of stages provides realistic problems that are complicated by the three limiting factors discussed earlier. Lookahead circuitry is therefore usually implemented in groups of bits. The lookahead technique can then be applied again over the separate groups as they are cascaded.

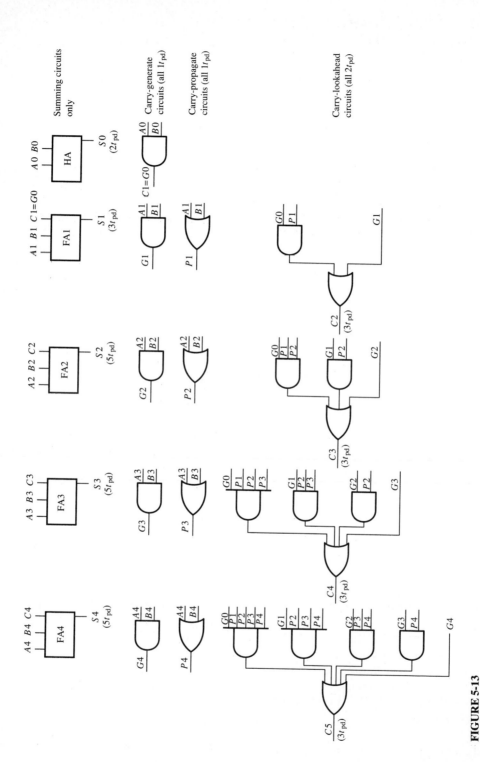

FIGURE 5-13

Five-bit parallel (carry-lookahead) adder with worst-case settling times shown in parentheses.

5-5   APPLYING THE TOP-DOWN DESIGN PROCESS TO OTHER REALISTIC PROBLEMS

In this section we present nine different practical design examples for gate level combinational design.

5-5-1   Binary Subtractors

Just like the half adder and the full adder, it is also desirable to specify and design a half subtractor and full subtractor so that larger subtractors can be built iteratively. If we designate $A0$ as a minuend bit, $B0$ as a subtrahend bit, DIFF0 the difference bit, and B_OUT1 as the borrow-out bit, then the following binary subtractions are possible.

A0	0	0	1	1
B0	0	1	0	1
DIFF0	0	1	1	0
B_OUT1	0	1	0	0

where the difference DIFF0 is the arithmetic operation $DIFF0 = A0 - B0$. A borrow bit B_OUT1 is an output that flags (signals) the next bit position, bit position 1, in a binary parallel subtractor that a borrow from position 1 is necessary. When a borrow is necessary, then 2 (binary 10) is added to the minuend bit, bit $A0$. The subtractor for bit position 1 would be a full subtractor with a borrow-input bit B_IN1 as illustrated in the 2-bit parallel binary subtractor block diagram shown in Fig. 5-14. An n-bit parallel subtractor has a half subtractor at the least significant bit position, and a full subtractor at all higher bit positions.

From the logic description of the half subtractor illustrated above, the truth table for the half subtractor can be written as follows.

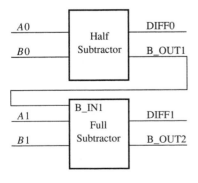

FIGURE 5-14

Block diagram of a 2-bit binary parallel subtractor using a half subtractor and a full subtractor.

A0	B0	DIFF0 (A0 − B0)	B_OUT1
0	0	0	0
0	1	1	1
1	0	1	0
1	1	0	0

Minimum SOP expressions for the outputs DIFF0 and B_OUT1 for the half subtractor and their signal lists for positive logic signals can be written as follows.

$\text{DIFF0} = \overline{A0} \cdot B0 + A0 \cdot \overline{B0}$ $\text{B\_OUT1} = \overline{A0} \cdot B0$
Signal list: DIFF0, A0, B0   Signal list: B_OUT1, A0, B0

From our previous discussions of the half adder, observe that the expression for the function SUM0 is the same as the expression for the function DIFF0. This results in the same functional logic diagram.

Using these Boolean functions the design documentation can be written as follows.

LLP:	$\text{DIFF0} = A0 \cdot B0$	$+$	$A0 \cdot B0$	$\text{B\_OUT1} = A0 \cdot B0$
NIP:	1 0	1	1 0	1 0 1

LLP:	$\text{DIFF0} = A0 \cdot B0$	$+$	$A0 \cdot B0$	$\text{B\_OUT1} = A0 \cdot B0$
PIP:	H L	H	H L	H L H

Using the design documentation for the half-subtractor the functional logic diagrams are drawn as shown in Fig. 5-15.

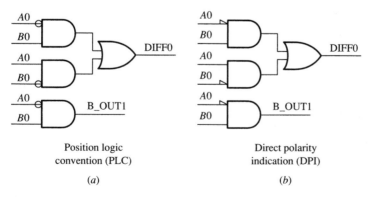

Position logic   Direct polarity
convention (PLC)   indication (DPI)

(a)   (b)

FIGURE 5-15
Functional logic diagrams for a half subtractor (a) using the positive logic convention and (b) using direct polarity indication.

Example 5-6. Obtain functional logic diagrams using direct polarity indication for a full subtractor. Use the top-down design process.

Solution The block diagram for the full-subtractor is shown in Fig. 5-14 and has three inputs B_IN1, A1, and B1. The output DIFF1 represents the arithmetic operation DIFF1 $= A1 - B1 - $ B_IN1. When a borrow out is necessary, (B_OUT2 $= 1$), then 2 (binary 10) is added to the minuend bit, bit A1.

Ċ   The truth table for the full subtractor can now be filled in as follows.

B_IN1	A1	B1	DIFF1	B_OUT2
0	0	0	0	0
0	0	1	1	1
0	1	0	1	0
0	1	1	0	0
1	0	0	1	1
1	0	1	0	1
1	1	0	0	0
1	1	1	1	1

Notice that the first four lines of the truth table are the same as the half subtractor. This suggests that a full subtractor circuit can be used as a half subtractor circuit if input B_IN1 is permanently tied to logic 0. Minimum SOP expressions for DIFF1 and B_OUT2 are obtained using the Karnaugh maps as shown in Fig. E5-6a.

Ċ   Minimum SOP expressions for the outputs DIFF1 and B_OUT2 for the full subtractor and their signal lists for positive logic signals can be written as follows.

$$\text{DIFF1} = m_1 + m_2 + m_4 + m_7$$
$$= \text{B\_IN1} \cdot \overline{A1} \cdot B1 + \overline{\text{B\_IN1}} \cdot A1 \cdot \overline{B1} + \text{B\_IN1} \cdot \overline{A1} \cdot \overline{B1} + \text{B\_IN1} \cdot A1 \cdot B1$$

$$\text{B\_OUT2} = p_1 + p_2 + p_3$$
$$= \overline{A1} \cdot B1 + \text{B\_IN1} \cdot \overline{A1} + \text{B\_IN1} \cdot B1$$

FIGURE E5-6a

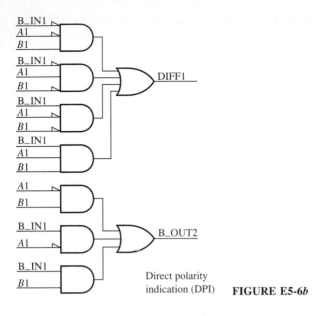

Direct polarity
indication (DPI)   **FIGURE E5-6*b***

DIFF1 $= \overline{\text{B\_IN1}} \cdot \overline{A1} \cdot B1 \ + \ \overline{\text{B\_IN1}} \cdot A1 \cdot \overline{B1} \ + \ \text{B\_IN1} \cdot \overline{A1} \cdot \overline{B1} \ + \ \text{B\_IN1} \cdot A1 \cdot B1$
Signal list: DIFF1, B_IN1, $A1$, $B1$

B_OUT2 $= \overline{A1} \cdot B1 + \text{B\_IN1} \cdot \overline{A1} + \text{B\_IN1} \cdot B1$
Signal list: B_OUT2, B_IN1, $A1$, $B1$

      The design documentation is provided below for diagrams using direct polarity
indication.

LLP:	DIFF1 $=$ B_IN1 $\cdot A1 \cdot B1$	$+$	B_IN1 $\cdot A1 \cdot B1$	$+$	B_IN1 $\cdot A1 \cdot B1$	$+$	B_IN1 $\cdot A1 \cdot B1$
PIP:	H L L H		L H L		H L L		H H H

LLP:	B_OUT2 $= A1 \cdot B1$	$+$	B_IN1 $\cdot A1$	$+$	B_IN1 $\cdot B1$
PIP:	H L H		H L		H H

      The functional logic diagrams for the full subtractor are shown in Fig. E5-6*b*
draw using direct polarity indication.

5-5-2   Binary Comparators

Using the iterative approach and the top-down design process, we will now show
the design of a comparator. A comparator is a combinational circuit that can be
used to compare the relative magnitudes of two binary numbers. First we will
consider a half comparator which can handle the two least-significant bits of the
binary numbers, then we will consider a full comparator that can be cascaded in
an iterative fashion. Figure 5-16 shows a block diagram of an *n*-bit comparator
using a half comparator and $n - 1$ cascadable full comparators.

The truth table below represents the logic description for the least significant bits of the *n*-bit comparator shown in Fig. 5-16.

$A0$	$B0$	OLES0 $A0 < B0$	OEQU0 $A0 = B0$	OGRE0 $A0 > B0$
0	0	0	1	0
0	1	1	0	0
1	0	0	0	1
1	1	0	1	0

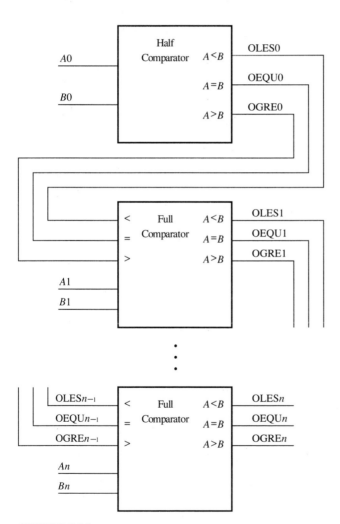

FIGURE 5-16
Block diagram of an *n*-bit comparator using a half comparator and $n - 1$ cascadable full comparators.

The signal names OLES0, OEQU0, and OGRE0 are short for output less for bit 0, output equal for bit 0, and output greater for bit 0 respectively. Minimum SOP expressions can be written directly from the truth table.

$OLES0 = \overline{A0} \cdot B0$	$OEQU0 = \overline{A0} \cdot \overline{B0} + A0 \cdot B0$	$OGRE0 = A0 \cdot \overline{B0}$
Signal list: OLES0, A0, B0	Signal list: OEQU0, A0, B0	Signal list: OGRE0, A0, B0

The design documentation for the half comparator can be written as follows.

LLP:	$OLES0 = A0 \cdot B0$	$OEQU0 = A0 \cdot B0$	+	$A0 \cdot B0$	$OGRE0 =$	$A0 \cdot B0$			
NIP:	1 0 1	1 0 0		1 1	1	1 0			
LLP:	$OLES0 = A0 \cdot B0$	$OEQU0 = A0 \cdot B0$	+	$A0 \cdot B0$	$OGRE0 =$	$A0 \cdot B0$			
PIP:	H L H	H L L		H H	H	H L			

The circuit diagrams for a half comparator in functional form are shown in Fig. 5-17. They are prepared for both the positive logic convention and direct polarity indication.

Example 5-7

1. Obtain the truth table for the full comparator illustrated in the block diagram shown in Fig. 5-16. Since each full comparator represents the same circuit, use the signal names for bit position 1 in the figure.
2. Using the top-down design process, obtain the functional logic diagrams for the full comparator using the positive logic convention.

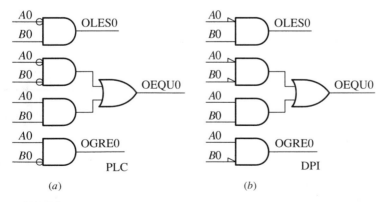

(a)   *(b)*

FIGURE 5-17
Functional logic diagrams for a half comparator *(a)* using the positive logic convention, *(b)* using direct polarity indication.

Solution The truth table description for the full comparator is shown below. The signal names ILES1, IEQU1, and IGRE1 are mnemonics for Input LESs for bit 1, Input EQUal for bit 1, and Input GREater for bit 1 respectively.

ILES1 $A0<B0$	IEQU1 $A0=B0$	IGRE1 $A0>B0$	$A1$	$B1$	OLES1 $A1\,A0<B1\,B0$	OEQU1 $A1\,A0=B1\,B0$	OGRE1 $A1\,A0>B1\,B0$
0	0	0	0	0	X	X	X
0	0	0	0	1	X	X	X
0	0	0	1	0	X	X	X
0	0	0	1	1	X	X	X
0	0	1	0	0	0	0	1
0	0	1	0	1	1	0	0
0	0	1	1	0	0	0	1
0	0	1	1	1	0	0	1
0	1	0	0	0	0	1	0
0	1	0	0	1	1	0	0
0	1	0	1	0	0	0	1
0	1	0	1	1	0	1	0
0	1	1	0	0	X	X	X
0	1	1	0	1	X	X	X
0	1	1	1	0	X	X	X
0	1	1	1	1	X	X	X
1	0	0	0	0	1	0	0
1	0	0	0	1	1	0	0
1	0	0	1	0	0	0	1
1	0	0	1	1	1	0	0
1	0	1	0	0	X	X	X
1	0	1	0	1	X	X	X
1	0	1	1	0	X	X	X
1	0	1	1	1	X	X	X
1	1	0	0	0	X	X	X
1	1	0	0	1	X	X	X
1	1	0	1	0	X	X	X
1	1	0	1	1	X	X	X
1	1	1	0	0	X	X	X
1	1	1	0	1	X	X	X
1	1	1	1	0	X	X	X
1	1	1	1	1	X	X	X

Although the table for the full comparator looks formidable, it is easy to fill in. Only 12 output entries must be determined. The remaining input entries result in don't care output conditions, since they represent input combinations that can not occur. For example, don't care outputs occur in the table for minterm 01100 which can not happen, since $A0 = B0$ cannot be true at the same time $A0 > B0$ is true. Each of the other don't care outputs represents a similar situation.

Minimum SOP expressions for outputs OLES1, OEQU1, and OGRE1 are obtained using the Karnaugh maps shown in Fig. E5-7a.

We show minimum SOP expressions for the outputs OLES1, OEQU1, and

OGRE1 for the full comparator below. The signal lists are written for positive logic signals.

OLES1 $= \overline{A1} \cdot B1 + ILES1 \cdot \overline{A1} + ILES1 \cdot B1$
Signal list: OLES1, ILES1, A1, B1

OEQU1 $= IEQU1 \cdot \overline{A1} \cdot \overline{B1} + IEQU1 \cdot A1 \cdot B1$
Signal list: OEQU1, IEQU1, A1, B1

OGRE1 $= IGRE1 \cdot \overline{A1} + IGRE1 \cdot A1 + A1 \cdot \overline{B1}$
Signal list: OGRE1, IGRE1, A1, B1

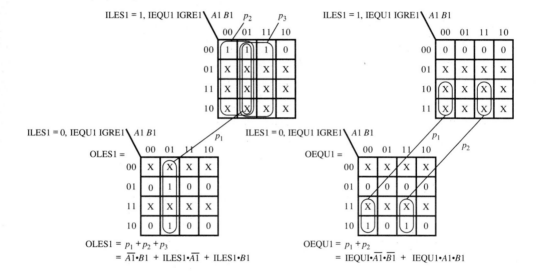

$$OLES1 = p_1 + p_2 + p_3$$
$$= \overline{A1} \cdot B1 + ILES1 \cdot \overline{A1} + ILES1 \cdot B1$$

$$OEQU1 = p_1 + p_2$$
$$= IEQU1 \cdot \overline{A1} \cdot \overline{B1} + IEQU1 \cdot A1 \cdot B1$$

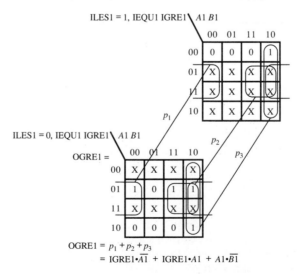

$$OGRE1 = p_1 + p_2 + p_3$$
$$= IGRE1 \cdot \overline{A1} + IGRE1 \cdot A1 + A1 \cdot \overline{B1}$$

FIGURE E5-7a

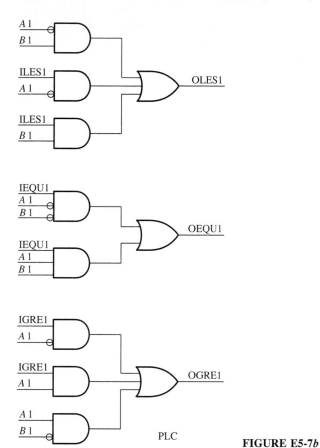

PLC

FIGURE E5-7b

The design documentation is provided below for diagrams using the positive logic convention.

LLP:	OLES1 $=$ $A1 \cdot B1$	$+$	ILES1 $\cdot A1$	$+$	ILES1 $\cdot B1$
NIP:	1 $\quad$ 0 $\quad$ 1		1 $\quad$ 0		1 $\quad$ 1

LLP:	OEQU1 $=$ IEQU1 $\cdot A1 \cdot B1$	$+$	IEQU1 $\cdot A1 \cdot B1$
NIP:	1 $\qquad$ 1 $\quad$ 0 $\quad$ 0		1 $\quad$ 1 $\quad$ 1

LLP:	OGRE1 $=$ IGRE1 $\cdot A1$	$+$	IGRE1 $\cdot A1$	$+$	$A1 \cdot B1$
NIP:	1 $\qquad$ 1 $\quad$ 0		1 $\quad$ 1		1 $\quad$ 0

The functional logic diagrams for the full comparator are shown in Fig. E5-7b. They are drawn using the positive logic convention.

5-5-3 Code Converters

The idea of converting from one binary code to a different binary code was presented in Chapter 2 and again in Chapter 3. The circuit that performs the conversion is called a code converter. The "from" code represents the inputs and the "to" code represents the outputs in a truth table representation. Karnaugh maps can then be used to obtain minimum SOP expressions, from which functional logic diagrams can be generated for the code converter.

> **Example 5-8.** In Example 3-7 in Chapter 3, the following Boolean functions were obtained for the outputs of a XS3 (excess 3) to BCD (binary coded decimal) code converter. $I3$, $I2$, $I1$, and $I0$ are the XS3 input bits from the most significant to the least significant bit, while $F3$, $F2$, $F1$, and $F0$ are the BCD output bits from the most significant to the least significant bit.
>
> $$F3 = I3 \cdot I2 + I3 \cdot I1 \cdot I0$$
>
> $$F2 = \overline{I2} \cdot \overline{I0} + \overline{I2} \cdot \overline{I1} + I2 \cdot I1 \cdot I0$$
>
> $$F1 = \overline{I1} \cdot I0 + I1 \cdot \overline{I0}$$
>
> $$F0 = \overline{I0}$$
>
> For the combined signal list
>
> $$\text{Signal list: } F3, F2, F1, F0, I3, I2, I1, I0$$
>
> obtain the design documentation and the functional logic diagrams for the XS3 to BCD code converter using direct polarity indication.
>
> ***Solution*** The design documentation for the XS3 to BCD code converter is written using the logic layout parts and the polarity indication parts as follows:

LLP:	$F3 = I3 \cdot I2$	+	$I3 \cdot I1 \cdot I0$				
PIP:	H	H H		H H H			
LLP:	$F2 = I2 \cdot I0$	+	$I2 \cdot I1$	+	$I2 \cdot I1 \cdot I0$		
PIP:	H	L L		L L		H H H	
LLP:	$F1 = I1 \cdot I0$	+	$I1 \cdot I0$				
PIP:	H	L H		H L			
LLP:	$F0 = I0$						
PIP:	H	L					

> The functional logic diagrams for the XS3 to BCD code converter using direct polarity indication are shown in Fig. E5-8.
>
> Since the polarity symbol must be drawn connecting a logic symbol when drawing the circuit for $F0$, a Buffer symbol is first drawn, then the polarity symbol is added in the usual fashion. If we were obtaining the diagram for $F0$ using the positive logic convention, the same procedure would be followed by drawing a Buffer symbol; however, the negation symbol would be added as indicated by the negation indication part.

5-5-4 Encoders

An Encoder is a combinational logic circuit that is designed to generate a binary output code for n different inputs. The number of bits required in the encoded

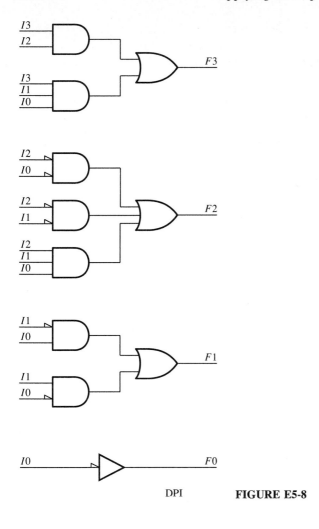

DPI   **FIGURE E5-8**

output m must satisfy the following relationship:

$$2^m \geq n$$

It is normally the case that only one of the n-inputs can be active at any time. In some cases, more than one input may be active; a special Encoder called a Priority Encoder must be designed to handle these cases. A Priority Encoder ensures that the output code responds to the highest priority input line if more than one input is active.

An Encoder for a telephone keypad similar to the one shown in Fig. 2-3 in Chapter 2 can be represented by the following truth table. Each switch on the keypad would have a separate input line to the Encoder that would represent a 0 when a key was pressed and a 1 when not pressed.

Keypad input switch	Binary coded output			
	W	X	Y	Z
KIS1	0	0	0	1
KIS2	0	0	1	0
KIS3	0	0	1	1
KIS4	0	1	0	0
KIS5	0	1	0	1
KIS6	0	1	1	0
KIS7	0	1	1	1
KIS8	1	0	0	0
KIS9	1	0	0	1
KIS0	1	0	1	0
KIS*	1	0	1	1
KIS#	1	1	0	0

The binary coded output functions for the Encoder are written below.

$$W = \overline{\text{KIS8}} + \overline{\text{KIS9}} + \overline{\text{KIS0}} + \overline{\text{KIS*}} + \overline{\text{KIS\#}}$$

$$X = \overline{\text{KIS4}} + \overline{\text{KIS5}} + \overline{\text{KIS6}} + \overline{\text{KIS7}} + \overline{\text{KIS\#}}$$

$$Y = \overline{\text{KIS2}} + \overline{\text{KIS3}} + \overline{\text{KIS6}} + \overline{\text{KIS7}} + \overline{\text{KIS0}} + \overline{\text{KIS*}}$$

$$Z = \overline{\text{KIS1}} + \overline{\text{KIS3}} + \overline{\text{KIS5}} + \overline{\text{KIS7}} + \overline{\text{KIS9}} + \overline{\text{KIS*}}$$

The combined signal list for the Encoder is: W, X, Y, Z, KIS1, KIS2, KIS3, KIS4, KIS5, KIS6, KIS7, KIS8, KIS9, KIS0, KIS*, KIS#.

The design documentation is written as follows.

LLP:	$W =$ KIS8 +	KIS9 +	KIS0 +	KIS* +	KIS#		
LIP:	1	0	0	0	0	0	
LLP:	$X =$ KIS4 +	KIS5 +	KIS6 +	KIS7 +	KIS#		
LIP:	1	0	0	0	0	0	
LLP:	$Y =$ KIS2 +	KIS3 +	KIS6 +	KIS7 +	KIS0 +	KIS*	
LIP:	1	0	0	0	0	0	
LLP:	$Z =$ KIS1 +	KIS3 +	KIS5 +	KIS7 +	KIS9 +	KIS*	
LIP:	1	0	0	0	0	0	

The functional logic diagrams using the positive logic convention for the Encoder of a telephone keypad are shown in Fig. 5-18. For a large number of keys the Encoder equations require many inputs, and the usual encoding technique is to interface the keyboard via a microcomputer as shown in Fig. 2-3.

Example 5-9. Design a Priority Encoder with four inputs such that when the highest priority input is active or true, the output is encoded as 001, when the second highest

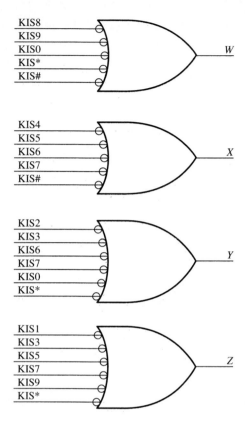

FIGURE 5-18
Functional logic diagrams for an Encoder of a telephone keypad.

priority input is active, the output is encoded as 010, when the third highest priority input is active, the output is encoded as 011, and finally when the fourth highest input is active, the output is encoded as 100. When none of the inputs are active, the output is encoded as 000. Obtain the design documentation for the design along with the functional logic diagrams using the positive logic convention.

Solution A truth table that describes the above design specification can be written as follows.

Minterms	$I3$	$I2$	$I1$	$I0$	$F2$	$F1$	$F0$
8,9,10,11,12,13,14,15	1	X	X	X	0	0	1
4,5,6,7	0	1	X	X	0	1	0
2,3	0	0	1	X	0	1	1
1	0	0	0	1	1	0	0
0	0	0	0	0	0	0	0

The representation 1XXX includes 11XX, 111X, and 1111; therefore, it has the highest priority and is encoded as 001. If inputs $I3$, $I2$, $I1$, and $I0$ were all active or true at the same time, the output code 001 would result. Similarly 01XX includes 011X, and 0111; therefore, it has the second highest priority, and is encoded as 010. If $I2$, $I1$, and $I0$ were all active at the same time, the output code 010 would result. $I1$ is at a higher priority than $I0$ since it includes 0011. Writing the truth table in this manner provides a convenient compact form. The minterm listing included with the truth table is not necessary, but is included as a reminder of the minterms represented by each input row.

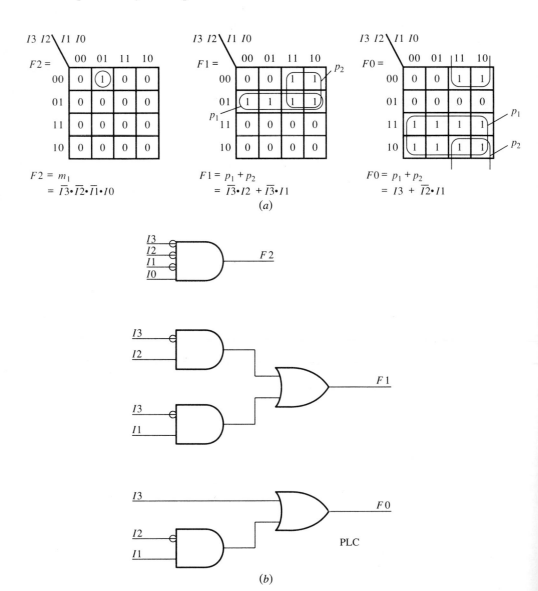

$$F2 = m_1$$
$$= \overline{I3} \cdot \overline{I2} \cdot \overline{I1} \cdot I0$$

$$F1 = p_1 + p_2$$
$$= \overline{I3} \cdot I2 + \overline{I3} \cdot I1$$
$$(a)$$

$$F0 = p_1 + p_2$$
$$= I3 + \overline{I2} \cdot I1$$

$$(b)$$

FIGURE E5-9

From the Karnaugh maps shown in Fig. E5-9*a* the following minimum SOP expressions are obtained:

$$F2 = \overline{I3} \cdot \overline{I2} \cdot \overline{I1} \cdot I0 \qquad F1 = \overline{I3} \cdot I2 \;+\; \overline{I3} \cdot I1 \qquad F0 = I3 \;+\; \overline{I2} \cdot I1$$

From these equations and the following combined signal list the following design documentation can be obtained.

Signal list: *F2, F1, F0, I3, I2, I1, I0.*

LLP:	$F2 = I3 \cdot I2 \cdot I1 \cdot I0$	$F1 = I3 \cdot I2 \;+\; I3 \cdot I1$	$F0 = I3 \;+\; I2 \cdot I1$
NIP:	1 0 0 0 1	1 0 1 0 1	1 1 0 1

The functional logic diagrams for the Priority Encoder are illustrated in Fig. E5-9*b*.

For an example of a different type of Encoder, refer to the shaft angle Encoder shown in Fig. 2-7 which encodes the shaft position in degrees into a binary code. The binary code most often used for this type of application is the "Gray code." The successive binary values contained in the Gray code only allow one bit to change between successive code values. For this reason the Gray code is a unit distance code.

5-5-5 Decoders

A circuit that converts a binary code applied to n-input lines to 2^n different output lines is called a n-to-2^n line Decoder or just a Decoder. A Decoder in effect translates each possible combination of values at its input to a distinct output line. A Decoder can be thought of as a minterm generator, since each input is decoded by using the functions for the minterms of the coded input variables.

A Decoder is often designed with an enable input that must be used to decode its input code. The enable input allows a Decoder to be cascadable. In addition, it also allows a Decoder to be used as a Demultiplexer. Demultiplexers will be discussed later. The truth table description for a 3-to-8 line decoder with an enable input is presented below.

EN	A	B	C	D0	D1	D2	D3	D4	D5	D6	D7
1	X	X	X	1	1	1	1	1	1	1	1
0	0	0	0	0	1	1	1	1	1	1	1
0	0	0	1	1	0	1	1	1	1	1	1
0	0	1	0	1	1	0	1	1	1	1	1
0	0	1	1	1	1	1	0	1	1	1	1
0	1	0	0	1	1	1	1	0	1	1	1
0	1	0	1	1	1	1	1	1	0	1	1
0	1	1	0	1	1	1	1	1	1	0	1
0	1	1	1	1	1	1	1	1	1	1	0

This Decoder decodes its input code when the enable input EN is 0 and does not decode its input code when the enable input is 1. When the enable input is 0 and a binary input code 010 is applied to $A \, B \, C$, then the output at $D2$ is 0 and all other outputs are 1. The other output lines operate in a similar manner as observed from the truth table.

Minimum SOP expressions for the 0s of the outputs can be written by inspection from the truth table.

$$\overline{D0} = \overline{EN} \cdot \overline{A} \cdot \overline{B} \cdot \overline{C}$$

$$\overline{D1} = \overline{EN} \cdot \overline{A} \cdot \overline{B} \cdot C$$

$$\overline{D2} = \overline{EN} \cdot \overline{A} \cdot B \cdot \overline{C}$$

$$\overline{D3} = \overline{EN} \cdot \overline{A} \cdot B \cdot C$$

$$\overline{D4} = \overline{EN} \cdot A \cdot \overline{B} \cdot \overline{C}$$

$$\overline{D5} = \overline{EN} \cdot A \cdot \overline{B} \cdot C$$

$$\overline{D6} = \overline{EN} \cdot A \cdot B \cdot \overline{C}$$

$$\overline{D7} = \overline{EN} \cdot A \cdot B \cdot C$$

Using all uncomplemented positive logic available signals, the design documentation for the 3-to-8 line Decoder is provided below for diagrams using the positive logic convention.

LLP:	$D0 = EN \cdot A \cdot B \cdot C$				
NIP:	0	0	0	0	0
LLP:	$D1 = EN \cdot A \cdot B \cdot C$				
NIP:	0	0	0	0	1
LLP:	$D2 = EN \cdot A \cdot B \cdot C$				
NIP:	0	0	0	1	0
LLP:	$D3 = EN \cdot A \cdot B \cdot C$				
NIP:	0	0	0	1	1
LLP:	$D4 = EN \cdot A \cdot B \cdot C$				
NIP:	0	0	1	0	0
LLP:	$D5 = EN \cdot A \cdot B \cdot C$				
NIP:	0	0	1	0	1
LLP:	$D6 = EN \cdot A \cdot B \cdot C$				
NIP:	0	0	1	1	0
LLP:	$D7 = EN \cdot A \cdot B \cdot C$				
NIP:	0	0	1	1	1

The functional logic diagrams for the 3-to-8 line Decoder are shown in Fig. 5-19. Like an ordinary Encoder, the design of a Decoder is relatively straightforward.

It is instructive to note that since the enable input EN must be 0 in order for the Decoder to decode its input code, the enable input line on each of the

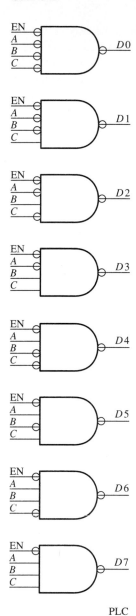

FIGURE 5-19
Functional logic diagrams for a 3-to-8 line Decoder with an enable
input.

functional logic diagrams contains a negation symbol. Designers often use the qualifying symbol indicated at the enable input as a memory aid of the logic state (or logic level if diagrams are prepared using direct polarity indication) necessary for decoder action to occur. In a similar manner, the qualifying symbol (a negation symbol is present in this case) on the output lines of each of the functional logic diagrams in Fig. 5-19 can be used as an indication of the logic state of an output line when the input code representing that line has been decoded.

Primarily due to the popularity of transistor-transistor logic NAND gates, which are used to implement functions with Decoders, each decoded output is a 0 logic state when its respective input code is decoded. Decoders are also generally designed to be enabled by a 0 logic state. Besides their usefulness in implementing Boolean functions, Decoders are especially useful in computer applications where they are used in designing memory elements to decode address locations.

5-5-6 Demultiplexers

A Demultiplexer (DX or DMUX) is a circuit that receives a signal on a single input line and directs that signal to one of 2^n possible output lines. Looking back at the truth table for the 3-to-8 line Decoder with an enable input EN, the enable input can be used as the single input line for a Demultiplexer. The input lines *A B C* contain a three-bit code used to select one of eight possible output lines. The input signal that is supplied to the enable input EN can thus be directed to a selected output. Since a Demultiplexer is simply a decoder with an enable input, the design of a Demultiplexer is the same as the design of a Decoder.

> **Example 5-10.** An application requiring a simple 1-to-4 line Demultiplexer is required. Obtain the design documentation for the Demultiplexer, and show functional logic diagrams for the device.
>
> **Solution** The block diagram for the device is shown in Fig. E5-10*a*, and its truth table is shown below.

EN	X	Y	D0	D1	D2	D3
0	X	X	0	0	0	0
1	0	0	1	0	0	0
1	0	1	0	1	0	0
1	1	0	0	0	1	0
1	1	1	0	0	0	1

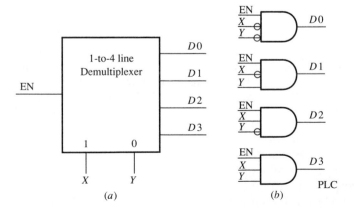

FIGURE E5-10

Notice that when the code 00 is present on the selection inputs X Y and $EN = 1$, the output line $D0 = EN = 1$, since a negation indicator is not shown on the enable input or on any of the output lines.

Minimum SOP expressions for the 1s of the outputs can be written by inspection from the truth table.

$$D0 = EN \cdot \bar{X} \cdot \bar{Y}$$

$$D1 = EN \cdot \bar{X} \cdot Y$$

$$D2 = EN \cdot X \cdot \bar{Y}$$

$$D3 = EN \cdot X \cdot Y$$

For available uncomplemented positive logic signals, the design documentation for the 1-to-4 line Demultiplexer using the positive logic convention is written below.

LLP:	$D0 = EN \cdot X \cdot Y$			
NIP:	1	1	0	0
LLP:	$D1 = EN \cdot X \cdot Y$			
NIP:	1	1	0	1
LLP:	$D2 = EN \cdot X \cdot Y$			
NIP:	1	1	1	0
LLP:	$D3 = EN \cdot X \cdot Y$			
NIP:	1	1	1	1

The functional logic diagrams for the 1-to-4 line Demultiplexer are shown in Fig. E5-10*b*.

Smaller Decoders and Demultiplexers can be connected together to obtain larger configurations. Fig. 5-20 illustrates in block diagram form how five 2-to-4 line Decoders, each with an enable input, can be connected to obtain a 4-to-16 line Decoder or a 1-to-16 line Demultiplexer. The disadvantage of this type of implementation is the longer propagation delay time required over an implementation designed as a single unit. The single-unit design would require gates with five inputs. A single-unit design has a propagation delay time of $1t_{pd}$, while the design shown in Fig. 5-20 has a propagation delay time of $2t_{pd}$. Observe when input enable signal ENABLE is low (the enable input enabled) and input signals $I3$ $I2 = $ L L, then output signal OD0 (output $D0$ of Decoder #1) is low, and this enables Decoder #2. If input signals $I1$ $I0 = $ H L, then output signal $O2$ (output $D2$ of Decoder #2) $= $ L, and all other Decoder outputs are high. When ENABLE $= $ H (the enable input disabled), output lines $O0$ through $O15$ are all high including all outputs of Decoder #1.

Example 5-11. Fig. E5-11*a* shows a block diagram representation for the 3-to-8 line Decoder (1-to-8 line Demultiplexer) shown in Fig. 5-19. Use two of the block diagram representations, and draw an implementation for a 4-to-16 line Decoder.

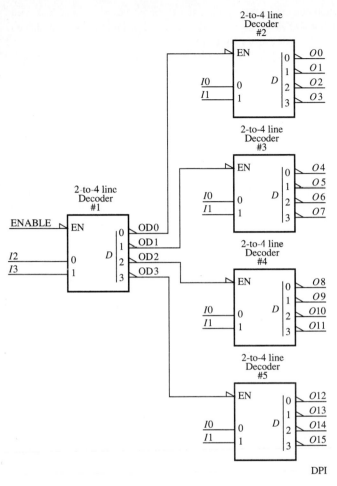

FIGURE 5-20
A 4-to-16 line Decoder or 1-to-16 line Demultiplexer configured from five 2-to-4 line Decoders, each with an enable input.

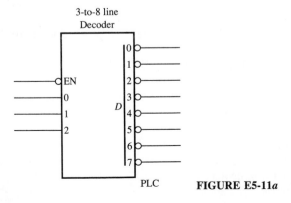

FIGURE E5-11a

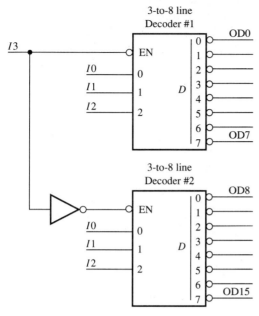

PLC   **FIGURE 5-11***b*

Solution No formal design procedure is used other than a knowledge of how the pieces work so a plausible solution can be tried. The design procedure most often involves breaking the problem down into several simpler problems that can be solved by the required devices. The implementation shown in Fig. E5-11*b* fits the requirements, since each decoder solves part of the problem. When Decoder #1 is enabled and Decoder #2 is disabled ($I3 = 0$), then $I2$ through $I0$ select output lines OD0 through OD7. When Decoder #2 is enabled and Decoder #1 is disabled ($I3 = 1$), then $I2$ through $I0$ select output lines OD8 through OD15. The resulting circuit is a 4-to-16 line decoder without an enable input.

5-5-7   Multiplexers (Data Selectors)

A Multiplexer (MUX) is a circuit that is used to direct one out of 2^n inputs to a single output. Another name for a multiplexer is a Data Selector, since n-input select lines are used to select one of the input signals and direct it to the output. A block diagram of a 2-to-1 line MUX with an enable input is shown in Fig. 5-21.

     The truth table for the 2-to-1 line MUX is shown below.

EN	DI0	DI1	SI0	F
1	X	X	X	0
0	0	X	0	0
0	1	X	0	1
0	X	0	1	0
0	X	1	1	1

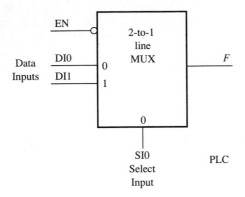

FIGURE 5-21

PLC   Block diagram of a 2-to-1 line Multiplexer (MUX) (also referred to as a 2-to-1 line Data Selector).

Since the enable input EN has a negation symbol at its input in the block diagram, 0 enables the device and 1 disables the device. Since the output line has no negation symbol, when the device is disabled, the output signal F is 0. When the device is enabled, the device output signal F follows the data input selected by the single select line. When the select input signal SI0 (select input 0) is 0, then the data input signal DI0 (data input 0) is selected and $F = $ DI0. When the select input signal SI0 is 1, then the data input signal DI1 (data input 1) is selected and $F = $ DI1. Now that the explanation of the functionality of a Multiplexer is understood, filling in the truth table is rather routine.

Example 5-12. Obtain the functional logic diagram for the 2-to-1 line Multiplexer represented by the truth table description shown above.

Solution Using the truth table for the 2-to-1 line Multiplexer, the Karnaugh map shown in Fig. E5-12a can be drawn.

From the Karnaugh map, the following minimum SOP expression is obtained.

$$F = \overline{EN} \cdot DI0 \cdot \overline{SI0} + \overline{EN} \cdot DI1 \cdot SI0$$

From this equation and the corresponding signal list the following design documentation can be obtained.

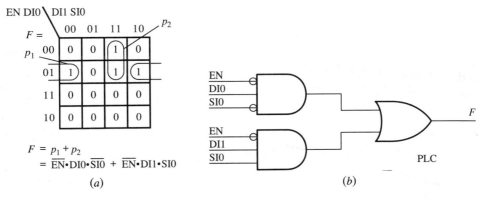

(a) (b)

FIGURE E5-12

Signal list: F, EN, DI0, DI1, SI0.

LLP:	$F =$	EN $\cdot$ DI0 $\cdot$ SI0		$+$	EN $\cdot$ DI1 $\cdot$ SI0			
NIP:		1	0	1	0	0	1	1

The functional logic diagram for the 2-to-1 line Multiplexer is in Fig. E5-12*b*.

The Boolean function for larger Multiplexers follows the same design process. However, since the number of inputs can become rather large, the Boolean function can be obtained by closely observing the 2-to-1 MUX and writing larger MUX configurations by inspection. The Boolean expression for a 4-to-1 MUX with an enable input can be written as follows:

$$F = \overline{EN} \cdot DI0 \cdot \overline{SI1} \cdot \overline{SI0} \;+\; \overline{EN} \cdot DI1 \cdot \overline{SI1} \cdot SI0 \;+\; \overline{EN} \cdot DI2 \cdot SI1 \cdot \overline{SI0}$$
$$+ \; \overline{EN} \cdot DI3 \cdot SI1 \cdot SI0$$

When EN $= 0$, the expression indicates that $F = $ DIn when SI1 SI0 $= n$ for $n = 0$ to 3 decimal. The design documentation for signal list: F, EN, DI0, DI1, DI2, DI3, SI1, SI0 is written as follows.

LLP:	$F =$	EN $\cdot$ DI0 $\cdot$ SI1 $\cdot$ SI0			$+$	EN $\cdot$ DI1 $\cdot$ SI1 $\cdot$ SI0			$+$	EN $\cdot$ DI2 $\cdot$ SI1 $\cdot$ SI0			$+$	EN $\cdot$ DI3 $\cdot$ SI1 $\cdot$ SI0		
NIP:		1	0	1	0	0	0	1	0	1	0	1	1	0	0	1	1	1

Figure 5-22*a* shows the functional logic diagram for a 4-to-1 MUX with an enable input, while Fig. 5-22*b* illustrates a block diagram for the circuit.

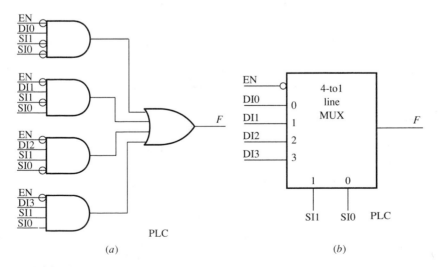

(*a*)   (*b*)

FIGURE 5-22
A 4-to-1 line Multiplexer (Data Selector) (*a*) functional logic diagram, (*b*) block diagram.

Example 5-13. Write a truth table description of the 4-to-1 line MUX shown in Fig. 5-22.

Solution The truth table description for the 4-to-1 line Multiplexer is shown below. Notice that the table contains seven inputs, and the complete table, if written without don't care inputs, would require 128 rows.

EN	DI0	DI1	DI2	DI3	SI1	SI0	F
1	X	X	X	X	X	X	0
0	0	X	X	X	0	0	0
0	1	X	X	X	0	0	1
0	X	0	X	X	0	1	0
0	X	1	X	X	0	1	1
0	X	X	0	X	1	0	0
0	X	X	1	X	1	0	1
0	X	X	X	0	1	1	0
0	X	X	X	1	1	1	1

From this truth table description one can easily see that the 4-to-1 line MUX functionally performs the same way as the 2-to-1 line MUX discussed earlier.

The Boolean function written for the 4-to-1 line Multiplexer can be checked by substituting the inputs into the Boolean expression on the right side of the function, one row at a time, and verifying that the output function F agrees with the output in the table. With practice, the Boolean function can be written directly from the truth table description.

Like Demultiplexers, smaller Multiplexers or Data Selectors can also be connected together to obtain larger configurations. The procedure is to use block diagrams for the available Multiplexers and determine how they can be connected to obtain a Multiplexer with a larger number of inputs. For example, using three 2-to-1 line Multiplexers, a 4-to-1 line Multiplexer can be constructed as shown in Fig. 5-23. This implementation has a propagation delay time of $4t_{pd}$ compared to the propagation delay time of $2t_{pd}$ for the single unit Multiplexer design shown in Fig. 5-22.

Example 5-14. Use four 8-to-1 line Multiplexers and one 4-to-1 line Multiplexer and show an implementation for a 32-to-1 line Multiplexer.

Solution The block diagram for the 32-to-1 line Multiplexer is shown in Fig. E5-14.

Observe in Fig. E5-14 that the 4-to-1 line Multiplexer selects each of the 8-to-1 line Multiplexers in turn by its select inputs. For select inputs SI4 SI3 = m = 0 0, MUX #1 is selected. MUX #2 is selected for select inputs SI4 SI3 = 0 1, and so on. When one of the 8-to-1 line MUXs is selected, then select inputs SI2 SI1 SI0 = n determine which data input line is selected for n = 0 0 0 through

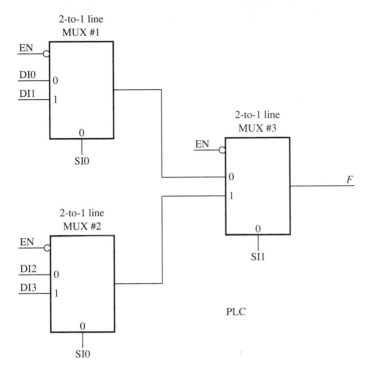

FIGURE 5-23
A 4-to-1 line Multiplexer constructed from three 2-to-1 line Multiplexers.

1 1 1 for each respective MUX. The overall data input selected is determined by concatenating the binary values for *m* and *n*.

5-5-8   Exclusive OR and Exclusive NOR

Exclusive ORs and Exclusive NORs are special two-input devices that are obtained from the definition of their respective truth table logic descriptions. The truth table description for these devices are listed as follows:

X	Y	F1		X	Y	F2
0	0	0		0	0	1
0	1	1		0	1	0
1	0	1		1	0	0
1	1	0		1	1	1

By definition, $F1$ is an Exclusive OR function, and its complement $F2$ is an Exclusive NOR function. The Karnaugh maps for $F1$ and $F2$ are shown in Fig. 5-24. Neither function can be reduced.

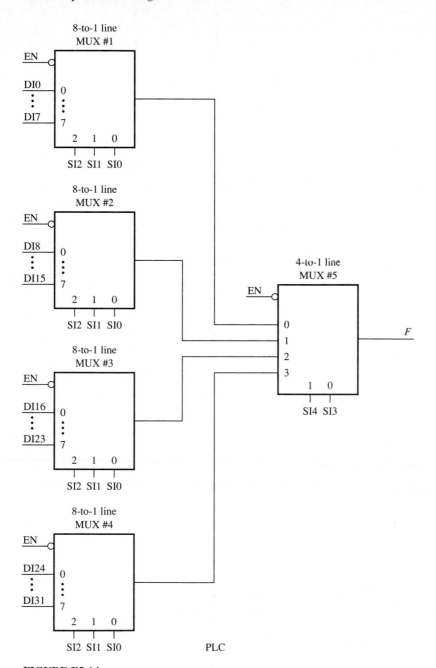

FIGURE E5-14

$F1 =$

XY			
00	01	11	10
0	①	0	①

$F1 = m_1 + m_2$
$F1 = \overline{X}{\cdot}Y + X{\cdot}\overline{Y}$

(a)

$F2 =$

XY			
00	01	11	10
①	0	①	0

$F2 = m_0 + m_3$
$F2 = \overline{X}{\cdot}\overline{Y} + X{\cdot}Y$

(b)

FIGURE 5-24
Karnaugh map for (a) the Exclusive OR Function, (b) the Exclusive NOR function.

The Boolean functions for the Exclusive OR and the Exclusive NOR can be written with the following signal lists.

$F1 = \overline{X}{\cdot}Y + X{\cdot}\overline{Y}$ $F2 = \overline{X}{\cdot}\overline{Y} + X{\cdot}Y$
Signal list: $F1, X, Y$ Signal list: $F2, X, Y$

The design documentation for each function can therefore be written as

LLP:	$F1 = X{\cdot}Y$	$+$	$X{\cdot}Y$	LLP:	$F2 = X{\cdot}Y$	$+$	$X{\cdot}Y$
NIP:	1 0		1 0	NIP:	1 0 0		1 1.

Using the design documentation, the functional logic diagram for each function is shown in Fig. 5-25a and b respectively.

The logic symbols shown in Fig. 5-26a and b are the symbols most often used to represent the Exclusive OR and Exclusive NOR functions. The logic operator $\oplus$, or the Exclusive OR operator, is used to represent an Exclusive OR of two Boolean signals (where $X \oplus Y = \overline{X}{\cdot}Y + X{\cdot}\overline{Y}$). The logic operator $\odot$, or the Exclusive NOR operator or Equivalence operator, is used to represent an Exclusive NOR of two Boolean signals (where $X \odot Y = \overline{X \oplus Y} = \overline{X}{\cdot}\overline{Y} + X{\cdot}Y$). The Exclusive NOR is also referred to as the Equivalence function, since its truth table outputs are 1 when its inputs are equal. On the other

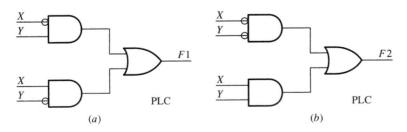

(a)   *(b)*

FIGURE 5-25
(a) Functional logic diagram for the Exclusive OR function, and (b) functional logic diagram for the Exclusive NOR function.

$$F1 = X \oplus Y$$
$$= \overline{X} \cdot Y + X \cdot \overline{Y}$$

$$F2 = \overline{X \oplus Y} = X \odot Y$$
$$= \overline{X} \cdot \overline{Y} + X \cdot Y$$

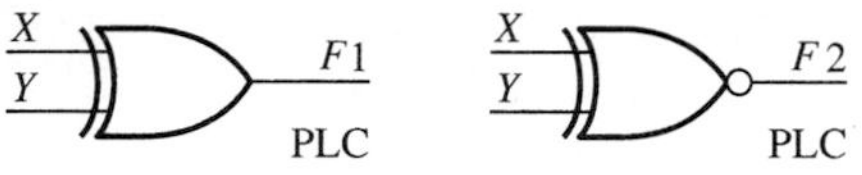

(a) PLC

(b) PLC

FIGURE 5-26
(a) Exclusive OR logic symbol,
(b) Exclusive NOR logic symbol

hand, the truth table outputs for the Exclusive OR function are 1 only when its inputs are mutually exclusive (when one of the inputs is 1 the other is 0).

Some designers like to hunt for patterns in design specifications represented either by Karnaugh maps or Boolean functions, and use the Exclusive OR and Exclusive NOR logic operators when those patterns are found. When logic diagrams are drawn for the Boolean functions, the symbols for the Exclusive OR and Exclusive NOR operators can then be substituted by the Exclusive OR and Exclusive NOR logic symbols illustrated in Fig. 5-26. For example, each of the following Boolean functions can be modified using either the Exclusive OR or the Exclusive NOR logic operators as illustrated below. Each of the functions has been presented earlier in the text.

$$SUM0 = \overline{A0} \cdot B0 + A0 \cdot \overline{B0} = A0 \oplus B0$$

$$SUM1 = \overline{C\_IN1} \cdot \overline{A1} \cdot B1 + \overline{C\_IN1} \cdot A1 \cdot \overline{B1} + C\_IN1 \cdot \overline{A1} \cdot \overline{B1}$$
$$+ C\_IN1 \cdot A1 \cdot B1$$
$$= \overline{C\_IN1} \cdot (A1 \oplus B1) + C\_IN1 \cdot (A1 \odot B1)$$

$$DIFF0 = \overline{A0} \cdot B0 + A0 \cdot \overline{B0} = A0 \oplus B0$$

$$DIFF1 = \overline{B\_IN1} \cdot \overline{A1} \cdot B1 + \overline{B\_IN1} \cdot A1 \cdot \overline{B1} + B\_IN1 \cdot \overline{A1} \cdot \overline{B1}$$
$$+ B\_IN1 \cdot A1 \cdot B1$$
$$= \overline{B\_IN1} \cdot (A1 \oplus B1) + B\_IN1 \cdot (A1 \odot B1)$$

$$OEQU0 = \overline{A0} \cdot \overline{B0} + A0 \cdot B0 = A0 \odot B0$$

$$OEQU1 = IEQU1 \cdot \overline{A1} \cdot \overline{B1} + IEQU1 \cdot A1 \cdot B1$$
$$= IEQU1 \cdot (A1 \odot B1)$$

$$F1 = \overline{I1} \cdot I0 + I1 \cdot \overline{I0} = I1 \oplus I0$$

When Exclusive OR and Exclusive NOR functions are used, the resulting Boolean functions contain more than two levels of logic, hence having more propagation delay than Boolean functions implemented in a SOP or POS form. In addition to adder and subtractor circuits, comparator circuits, code converter circuits, parity generator and parity checker circuits, error detection and error correction circuits can be implemented using Exclusive OR and Exclusive NOR devices. Some of these applications will be discussed further in Chapter 7.

Example 5-15

1. From the following truth table description, obtain the Boolean functions for a three-bit Binary-to-Gray Code code converter. Use the Exclusive OR and Exclusive NOR logic operators when appropriate. Obtain the logic diagrams for the Binary-to-Gray Code code converter.

Binary			Gray Code		
$B2$	$B1$	$B0$	$G2$	$G1$	$G0$
0	0	0	0	0	0
0	0	1	0	0	1
0	1	0	0	1	1
0	1	1	0	1	0
1	0	0	1	1	0
1	0	1	1	1	1
1	1	0	1	0	1
1	1	1	1	0	0

2. Using the same truth table description, obtain the Boolean functions for a three-bit Gray Code to Binary code converter. Use Exclusive OR and Exclusive NOR logic operators when appropriate. Obtain the logic diagrams for the Gray Code to Binary code converter.

Solution

1. The Karnaugh maps for the Gray Code outputs $G2$ through $G0$ are shown in Fig. E5-15a.

     Using the definition for the Exclusive OR function and all positive logic signals, the Boolean functions and their corresponding signal lists can be written as follows:

$G2 = B2$	$G1 = B2 \oplus B1$	$G0 = B1 \oplus B0$
Signal list: $G2, B2$	Signal list: $G1, B2, B1$	Signal list: $G0, B1, B0$

     The design documentation for the Binary-to-Gray Code code converter can be written as follows:

LLP:	$G2 = B2$	$G1 = B2 \oplus B1$	$G0 = B1 \oplus B0$
NIP:	1 1	1	1

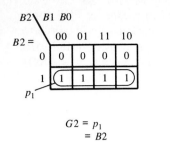

$$G2 = p_1$$
$$= B2$$

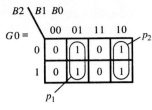

$$G1 = p_1 + p_2$$
$$= B2 \cdot \overline{B1} + \overline{B2} \cdot B1$$
$$= B2 \oplus B1$$

$$G0 = p_1 + p_2$$
$$= \overline{B1} \cdot B0 + B1 \cdot \overline{B0}$$
$$= B1 \oplus B0$$

(a)

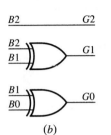

(b)

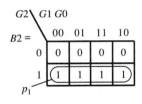

$$B2 = p_1$$
$$= G2$$

$$B1 = p_1 + p_2$$
$$= G2 \cdot \overline{G1} + \overline{G2} \cdot G1$$
$$= G2 \oplus G1$$
$$= B2 \oplus G1$$

$$B0 = m_1 + m_2 + m_4 + m_7$$
$$= \overline{G2} \cdot \overline{G1} \cdot G0 + \overline{G2} \cdot G1 \cdot \overline{G0}$$
$$\quad + G2 \cdot \overline{G1} \cdot \overline{G0} + G2 \cdot G1 \cdot G0$$
$$= \overline{G2} \cdot (G1 \oplus G0) + G2 \cdot (G1 \odot G0)$$
$$= \overline{G2} \cdot (G1 \oplus G0) + G2 \cdot \overline{(G1 \oplus G0)}$$
$$= G2 \oplus (G1 \oplus G0)$$
$$= G2 \oplus G1 \oplus G0$$
$$= B1 \oplus G0$$

(c)

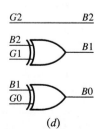

(d)

FIGURE E5-15

This is the only design documentation that is required, since the Exclusive OR functions are considered to be single logic elements in this case. The logic diagrams for the Binary-to-Gray Code code converter are shown in Fig. E5-15*b*.

2. If we consider *B*2 through *B*0 as outputs and *G*2 through *G*0 as inputs, we can fill in the Karnaugh Maps for a Gray Code to Binary code converter as shown in Fig. E5-15*c*.

The Boolean functions and their corresponding signal lists can be written as follows for the Gray Code to Binary code converter:

$B2 = G2$	$B1 = B2 \oplus G1$	$B0 = B1 \oplus G0$
Signal list: *B*2, *G*2	Signal list: *B*1, *B*2, *G*1	Signal list: *B*0, *B*1, *G*0

The design documentation for the Gray Code to Binary code converter can be written as follows:

LLP:	$B2 = G2$	$B1 = B2 \oplus G1$	$B0 = B1 \oplus G0$
NIP:	1 1	1	1

The logic diagrams for the Gray Code to Binary code converter are shown in Fig. E5-15*d*.

The Boolean functions for larger Binary-to-Gray Code code converters or Gray Code-to-Binary code converters can easily be obtained by observing the results for the three-bit converters obtained in Example 5-15.

The general Boolean functions can be written as follows for a Binary-to-Gray Code code converter.

$$G_n = B_n$$

$$G_{n-1} = B_n \oplus B_{n-1}$$

$$G_{n-2} = B_{n-1} \oplus B_{n-2}$$

$$\vdots$$

$$G_1 = B_2 \oplus B_1$$

$$G_0 = B_1 \oplus B_0$$

The general Boolean functions for a Gray Code to Binary code converter are listed as follows:

$$B_n = G_n$$

$$B_{n-1} = B_n \oplus G_{n-1}$$

$$B_{n-2} = B_{n-1} \oplus G_{n-2}$$

$$\vdots$$

$$B_1 = B_2 \oplus G_1$$

$$B_0 = B_1 \oplus G_0$$

Using these general functions, a converter of any size can be constructed.

5-5-9 Binary Multipliers

Consider a binary number represented by A where A is the multiplier and another binary number represented by B where B is the multiplicand. The arithmetic product can be obtained for two bits; $A = A1\ A0 = 11$ and $B = B1\ B0 = 11$ as follows. Note that this is the same procedure used in decimal multiplication.

```
  B    multiplicand      11   B1   B0   m bits
× A    multiplier       × 11  A1   A0   n bits
A × B  result            11
                         11
                       1001   R3   R2   R1   R0
```

A truth table for a Binary Multiplier using the multiplier bits $A1\ A0$ and the multiplicand bits $B1\ B0$ as inputs, and the four result bits $R3\ R2\ R1\ R0$ as outputs can be now be written and filled in. When an m-bit number is multiplied with an n-bit number, observe that the resulting product has $m + n$ bits.

A1	A0	B1	B0	R3	R2	R1	R0
0	0	0	0	0	0	0	0
0	0	0	1	0	0	0	0
0	0	1	0	0	0	0	0
0	0	1	1	0	0	0	0
0	1	0	0	0	0	0	0
0	1	0	1	0	0	0	1
0	1	1	0	0	0	1	0
0	1	1	1	0	0	1	1
1	0	0	0	0	0	0	0
1	0	0	1	0	0	1	0
1	0	1	0	0	1	0	0
1	0	1	1	0	1	1	0
1	1	0	0	0	0	0	0
1	1	0	1	0	0	1	1
1	1	1	0	0	1	1	0
1	1	1	1	1	0	0	1

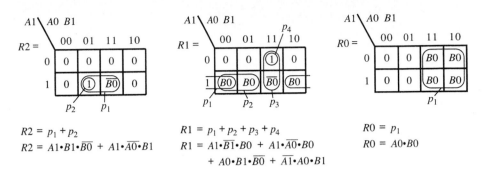

$R2 = p_1 + p_2$
$R2 = A1 \cdot B1 \cdot \overline{B0} + A1 \cdot \overline{A0} \cdot B1$

$R1 = p_1 + p_2 + p_3 + p_4$
$R1 = A1 \cdot \overline{B1} \cdot B0 + A1 \cdot \overline{A0} \cdot B0$
$\qquad + A0 \cdot B1 \cdot \overline{B0} + \overline{A1} \cdot A0 \cdot B1$

$R0 = p_1$
$R0 = A0 \cdot B0$

Signal list: $R2$, $A1$, $A0$, $B1$, $B0$

Signal list: $R1$, $A1$, $A0$, $B1$, $B0$

Signal list: $R0$, $A0$, $B0$

FIGURE 5-27
Karnaugh maps for a 2-bit by 2-bit Binary Multiplier.

Product bit $R3$ ($R3 = A1 \cdot A0 \cdot B1 \cdot B0$) is written from the truth table. Minimum SOP expressions for the remaining product bits $R2$, $R1$, and $R0$ are obtained using the Karnaugh maps shown in Fig. 5-27. The signal lists are written using all positive logic signals.

The design documentation for a 2-bit by 2-bit Binary Multiplier is written as follows:

LLP:	$R3 = A1 \cdot A0 \cdot B1 \cdot B0$
NIP:	1 1 1 1 1
LLP:	$R2 = A1 \cdot B1 \cdot B0 \quad + \quad A1 \cdot A0 \cdot B1$
NIP:	1 1 1 0 1 0 1
LLP:	$R1 = A1 \cdot B1 \cdot B0 \quad + \quad A1 \cdot A0 \cdot B0 \quad + \quad A0 \cdot B1 \cdot B0 \quad + \quad A1 \cdot A0 \cdot B1$
NIP:	1 1 0 1 1 0 1 1 1 0 0 1 1
LLP:	$R0 = A0 \cdot B0$
NIP:	1 1 1

The functional logic diagrams using the positive logic convention for a 2-bit by 2-bit Binary Multiplier are shown in Fig. 5-28.

Example 5-16. Design a Binary Multiplier using four multiplier bits and five multiplicand bits.

Solution Since the design would require $2^9 = 512$ truth table entries, the following approach is used.

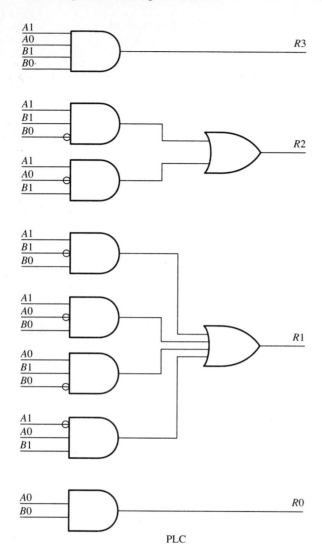

PLC

FIGURE 5-28
Functional logic diagrams for a 2-bit by 2-bit Binary Multiplier.

					B4	B3	B2	B1	B0	m multiplicand bits
					×	A3	A2	A1	A0	n multiplier bits
					P04	P03	P02	P01	P00	A0 partial products
			P14	P13	P12	P11	P10			A1 partial products
		C16	C15	C14	C13	C12	0			Adder No. 1 carry bits
		S16	S15	S14	S13	S12	S11	S10		Adder No. 1 sum bits
		P24	P23	P22	P21	P20				A2 partial products
	C27	C26	C25	C24	C23	0				Adder No. 2 carry bits
	S27	S26	S25	S24	S23	S22	S21	S20		Adder No. 2 sum bits
	P34	P33	P32	P31	P30					A3 partial products
C38	C37	C36	C35	C34	0					Adder No. 3 carry bits
S38	S37	S36	S35	S34	S33	S32	S31	S30		Adder No. 3 sum bits

In the multiplication carried out above, each P*XX* term represents a partial product ($P00 = A0 \times B0$). Recall from Chapter 1 that the AND operation of two binary bits is identical to the product of the binary bits, so $P00 = A0 \times B0 = A0 \wedge B0$, where $\wedge$ is used to represent the AND operation in expressions involving both arithmetic and logical operations. $P21 = A2 \wedge B1$, and $P34 = A3 \wedge B4$ are other examples of Boolean expressions of product terms. Notice that the partial products utilizing A0 are lined up with A0, while the partial products utilizing A1 are lined up with A1, and so forth. In other words, each row of partial products is shifted one bit position to the left, just like in decimal multiplication. After two rows of partial products are obtained, they are added column by column. When a column contains only one partial product, the partial product is simply recorded as the sum for that column. A carry bit is recorded in the next column when the previous column contains two or more partial products to be added. The first number associated with each sum or carry bit represents the adder number, while the second number represents the bit position, starting with bit 0. After the sum of two rows of partial products is obtained, the next row of partial products is obtained, and then added column by column to the previous sum. This process is continued until all the partial product rows are added, forming the final product.

Note that there are $m + n$ product bits in the final sum, the design requires $n - 1$ binary adders, and each binary adder contains m bits. The number of partial products and hence the number of two input AND gates required is $m \times n$. Figure E5-16 shows the design layout for the binary multiplier. The design contains 3 binary adders of five bits each and produces the required nine product bits in the final sum. To save space the 20 AND gates required to generate the partial product terms are not shown.

Larger Binary Multipliers can be designed using this strategy, but the propagation delays increase rapidly for a large number of bits, limiting usefulness in fast applications.

5-6 REVIEWING THE TOP-DOWN DESIGN PROCESS

In this section we present another realistic problem from the truth table description all the way through to the functional logic diagrams.

5-6-1 Binary to Seven-Segment Hexadecimal Character Generator

An excellent example demonstrating an occasion when a logic specification or description naturally results in a truth table tabulation is the binary to seven-segment hexadecimal character generator illustrated by the block diagram shown in Fig. 5-29. This type of character generator is often used in hand held calculators (such as the HP 16*C* and TI Programmer) and stand-alone microprocessor trainers. The available signals for the binary inputs are D (most significant bit, MSB), C, B, and A (least significant bit, LSB), and the available output signals are OA, OB, OC, OD, OE, OF, and OG as indicated on the block diagram.

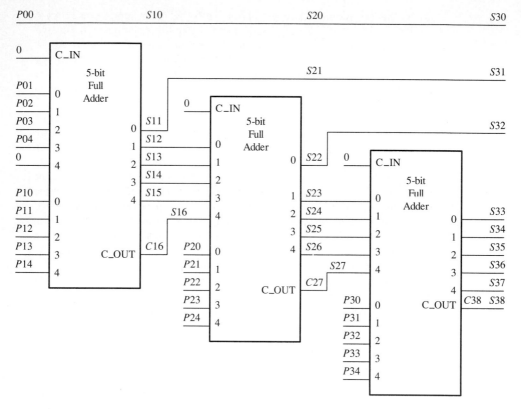

FIGURE E5-16

Each of the output signal lines of the character generator connect via a protective current-limiting resistor to the corresponding input signal lines of a seven-segment display device like the one shown in Fig. 5-30.

When an active or true signal (a logic 1) is present on an output line of the character generator, that signal line lights the respective segment. The sixteen input combinations ($2^4 = 16$), provided by the four input signals D, C, B, and A, re-

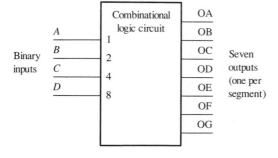

FIGURE 5-29

Block diagram of a binary to seven-segment hexadecimal character generator.

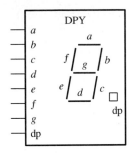

FIGURE 5-30
Block diagram of a seven-segment display device.

sult in sixteen different output combinations for the output signals OA, OB, OC, OD, OE, OF, and OG. These output signals drive the seven-segment display and produce the digits 0 through 9 and letters *A* through *F* shown in Fig. 5-31.

Example 5-17

1. Obtain the truth table for a character generator that will drive the seven-segment display device shown in Fig. 5-30.
2. Show Karnaugh maps for outputs OA through OG. Use the maps to obtain minimum SOP expressions for either the 1s or the 0s of the outputs.

Solution

1. The truth table for the character generator is shown in Fig. E5-17*a*. To represent the 16 hexadecimal characters on the output lines requires a minimum of four input bits or four input lines which contain the available signals *D, C, B,* and

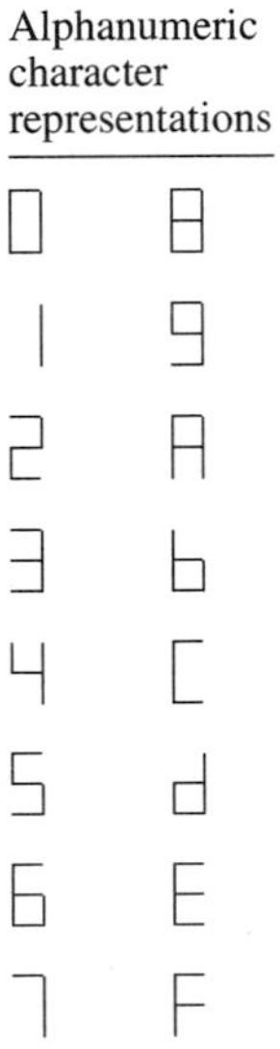

Alphanumeric
character
representations

FIGURE 5-31
List of desired alphanumeric character representations for the seven-segment display device.

A, respectively. For the binary input $D\ C\ B\ A\ =\ 0000$ every segment except segment g must be active to display the character 0; therefore, the binary output OA OB OC OD OE OF OG must be 1111110. For binary input 0001 only segment b and c must be active to display the character 1, so binary output OA OB OC OD OE OF OG must be 0110000. The truth table shown in Fig. E5-17a was generated using this line of thought for binary inputs 0000 through 1111.

2. The truth table for the character generator is used to plot the 3-variable maps shown in Fig. E5-17b. Reading the 3-variable maps with map-entered variables, the minimum SOP expressions can be obtained as illustrated in the figure. Larger 4-variable maps can also be used to obtain the minimum SOP expressions if desired.

It is instructive at this point to reflect on the strategy being used in our design process (first four steps illustrated again in Fig. 5-32). Note that we first used the design specifications (or descriptions) to obtain the Boolean functions (the top level of the design process, Step 1). As we move from the top level to a lower level, the design tasks usually become more detailed. This design method is similar to the strategy used in computer programming called top-down design. The method allows us to complete the design tasks of obtaining the Boolean functions in available forms before getting into further details. The details we are referring to involve obtaining the signal lists.

5-6-2   The Next Level Down in the Design Process, Step 2

Now that we have obtained the output functions for the character generator OA through OG in Example 5-17, the next step is to obtain their signal lists. Here is

Binary inputs				Seven outputs							Displayed characters
D	C	B	A	OA	OB	OC	OD	OE	OF	OG	
0	0	0	0	1	1	1	1	1	1	0	0
0	0	0	1	0	1	1	0	0	0	0	1
0	0	1	0	1	1	0	1	1	0	1	2
0	0	1	1	1	1	1	1	0	0	1	3
0	1	0	0	0	1	1	0	0	1	1	4
0	1	0	1	1	0	1	1	0	1	1	5
0	1	1	0	1	0	1	1	1	1	1	6
0	1	1	1	1	1	1	0	0	0	0	7
1	0	0	0	1	1	1	1	1	1	1	8
1	0	0	1	1	1	1	1	0	1	1	9
1	0	1	0	1	1	1	0	1	1	1	A
1	0	1	1	0	0	1	1	1	1	1	b
1	1	0	0	1	0	0	1	1	1	0	C
1	1	0	1	0	1	1	1	1	0	1	d
1	1	1	0	1	0	0	1	1	1	1	E
1	1	1	1	1	0	0	0	1	1	1	F

FIGURE E5-17a

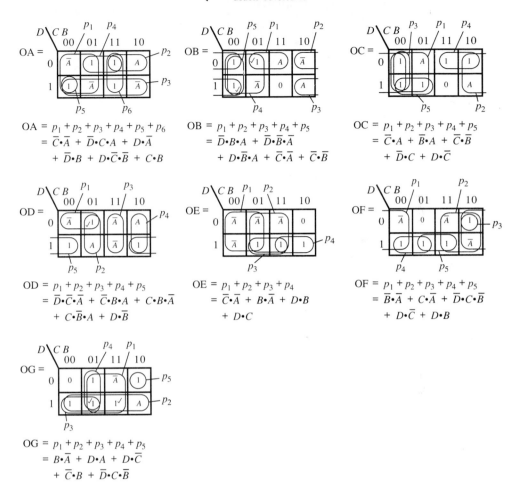

$\checkmark \equiv$ double-covered 1s

$OA = p_1 + p_2 + p_3 + p_4 + p_5 + p_6$
$= \overline{C} \cdot \overline{A} + \overline{D} \cdot C \cdot A + D \cdot \overline{A}$
$+ \overline{D} \cdot B + D \cdot \overline{C} \cdot \overline{B} + C \cdot B$

$OB = p_1 + p_2 + p_3 + p_4 + p_5$
$= \overline{D} \cdot B \cdot A + \overline{D} \cdot \overline{B} \cdot \overline{A}$
$+ D \cdot \overline{B} \cdot A + \overline{C} \cdot A + \overline{C} \cdot B$

$OC = p_1 + p_2 + p_3 + p_4 + p_5$
$= \overline{C} \cdot A + \overline{B} \cdot A + \overline{C} \cdot B$
$+ \overline{D} \cdot C + D \cdot \overline{C}$

$OD = p_1 + p_2 + p_3 + p_4 + p_5$
$= \overline{D} \cdot \overline{C} \cdot \overline{A} + \overline{C} \cdot B \cdot A + C \cdot B \cdot \overline{A}$
$+ C \cdot \overline{B} \cdot A + D \cdot \overline{B}$

$OE = p_1 + p_2 + p_3 + p_4$
$= \overline{C} \cdot A + B \cdot \overline{A} + D \cdot B$
$+ D \cdot C$

$OF = p_1 + p_2 + p_3 + p_4 + p_5$
$= \overline{B} \cdot \overline{A} + C \cdot \overline{A} + \overline{D} \cdot C \cdot \overline{B}$
$+ D \cdot \overline{C} + D \cdot B$

$OG = p_1 + p_2 + p_3 + p_4 + p_5$
$= B \cdot \overline{A} + D \cdot A + D \cdot \overline{C}$
$+ \overline{C} \cdot B + \overline{D} \cdot C \cdot \overline{B}$

FIGURES E5-17b

where we move to the next level down in the design process and get into more details. It is advantageous to choose the logic convention for each input (output) signal of a circuit to match the logic convention of its source (destination) so that signal mismatching problems do not occur. For this problem, it was verified that the source of the input signals D, C, B, and A originated from a positive logic circuit, so the input signals should be positive logic signals.

5-6-3 Common Cathode Type Seven-Segment Display

We also need to choose the logic convention for each output signal of our circuit to match the logic convention of its destination. We know that the output

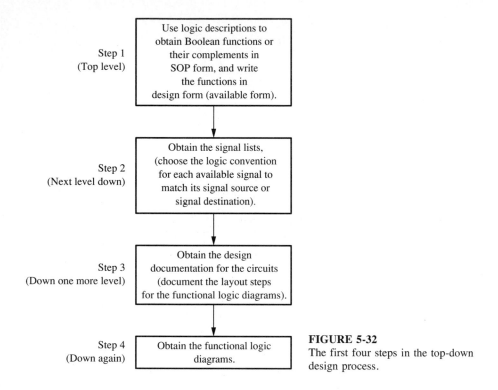

Step 1
(Top level) — Use logic descriptions to obtain Boolean functions or their complements in SOP form, and write the functions in design form (available form).

Step 2
(Next level down) — Obtain the signal lists, (choose the logic convention for each available signal to match its signal source or signal destination).

Step 3
(Down one more level) — Obtain the design documentation for the circuits (document the layout steps for the functional logic diagrams).

Step 4
(Down again) — Obtain the functional logic diagrams.

FIGURE 5-32
The first four steps in the top-down design process.

signals of our character generator connect to a seven-segment display. The type of seven-segment display we choose to use will dictate the declaration of the logic convention of each of the output signals OA through OG of our character generator. There are two types of seven-segment display devices. One type is a common cathode and the other type is a common anode. A common cathode type seven-segment display has all of its cathodes wired together as shown in the functional logic diagram in Fig. 5-33. It is usually necessary to add a resistor in series with each anode lead. The seven outputs of our generator, when connected to the seven series resistors and then to the display, require a high-level voltage to light each respective segment. If we choose to use a common cathode-type display, a high-level voltage corresponds to a 1 in the truth table, and we should declare each output signal as a positive logic signal.

5-6-4   Common Anode Type Seven-Segment Display

A common anode type of seven-segment display has all of its anodes wired together as shown in the functional logic diagram in Fig. 5-34. In this case it is usually necessary to add a resistor in series with each cathode lead. The seven outputs of our generator, when connected to the seven series resistors and then

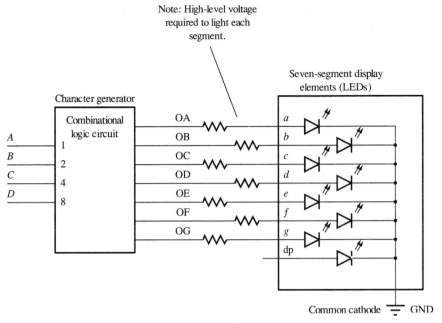

FIGURE 5-33
Connection for common cathode type seven-segment display.

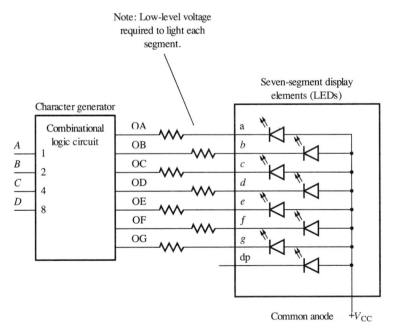

FIGURE 5-34
Connection for common anode type seven-segment display.

to the display, require a low-level voltage to light each respective segment. If we choose to use a common anode type display, a low-level voltage corresponds to a 1 in the truth table, and we should declare each output signal as a negative logic signal.

5-6-5 Moving Down One More Level to Step 3 in the Design Process

As illustrated in Fig. 5-32, after we obtain the required Boolean functions in available form and their signal lists, we are ready to move down one more level in the design process and obtain the design documentation for the circuits.

Example 5-18. Assume that the source for the input signals D, C, B, and A for the character generator originate from a positive logic circuit, and that the outputs are connected to a common anode type seven-segment display device like the one in Fig. 5-34. First, write the Boolean functions for the outputs OA through OG obtained in Example 5-17 for the character generator and their corresponding signal lists. Then write the design documentation for the circuits using the positive logic convention.

Solution For available signals D, C, B, and A declared as positive logic signals, and available signals OA through OG declared as negative logic signals, the Boolean functions for OA through OG and their combined signal list are written as follows:

$$OA = \overline{C} \cdot \overline{A} + \overline{D} \cdot C \cdot A + D \cdot \overline{A} + \overline{D} \cdot B + D \cdot \overline{C} \cdot \overline{B} + C \cdot B$$

$$OB = \overline{D} \cdot B \cdot A + \overline{D} \cdot \overline{B} \cdot \overline{A} + D \cdot \overline{B} \cdot A + \overline{C} \cdot \overline{A} + \overline{C} \cdot \overline{B}$$

$$OC = \overline{C} \cdot A + \overline{B} \cdot A + \overline{C} \cdot \overline{B} + \overline{D} \cdot C + D \cdot \overline{C}$$

$$OD = \overline{D} \cdot \overline{C} \cdot \overline{A} + \overline{C} \cdot B \cdot A + C \cdot B \cdot \overline{A} + C \cdot \overline{B} \cdot A + D \cdot \overline{B}$$

$$OE = \overline{C} \cdot \overline{A} + B \cdot \overline{A} + D \cdot B + D \cdot C$$

$$OF = \overline{B} \cdot \overline{A} + C \cdot \overline{A} + \overline{D} \cdot C \cdot \overline{B} + D \cdot \overline{C} + D \cdot B$$

$$OG = B \cdot \overline{A} + D \cdot A + D \cdot \overline{C} + \overline{C} \cdot B + \overline{D} \cdot C \cdot \overline{B}$$

Signal list: OA[NL], OB[NL], OC[NL], OD[NL], OE[NL], OF[NL], OG[NL], D, C, B, A

Using the information contained in the first two steps of the top-down design process, the design documentation for functions OA through OG can now be written for diagrams using the positive logic convention.

LLP:	OA = $C \cdot A$	+	$D \cdot C \cdot A$	+	$D \cdot A$	+	$D \cdot B$	+	$D \cdot C \cdot B$	+	$C \cdot B$
NIP:	0 0 0		0 1 1		1 0		0 1		1 0 0		1 1

LLP:	OB = $D \cdot B \cdot A$	+	$D \cdot B \cdot A$	+	$D \cdot B \cdot A$	+	$C \cdot A$	+	$C \cdot B$
NIP:	0 0 1 1		0 0 0		1 0 1		0 0		0 0

LLP:	$OC = C \cdot A$	+	$B \cdot A$	+	$C \cdot B$	+	$D \cdot C$	+	$D \cdot C$
NIP:	0 0 1		0 1		0 0		0 1		1 0

LLP:	$OD = D \cdot C \cdot A$	+	$C \cdot B \cdot A$	+	$C \cdot B \cdot A$	+	$C \cdot B \cdot A$	+	$D \cdot B$	
NIP:	0 0 0 0		0 1 1		1 1 0		1 0 1		1 0	

LLP:	$OE = C \cdot A$	+	$B \cdot A$	+	$D \cdot B$	+	$D \cdot C$	
NIP:	0 0 0		1 0		1 1		1 1	

LLP:	$OF = B \cdot A$	+	$C \cdot A$	+	$D \cdot C \cdot B$	+	$D \cdot C$	+	$D \cdot B$
NIP:	0 0 0		1 0		0 1 0		1 0		1 1

LLP:	$OG = B \cdot A$	+	$D \cdot A$	+	$D \cdot C$	+	$C \cdot B$	+	$D \cdot C \cdot B$
NIP:	0 1 0		1 1		1 0		0 1		0 1 0

5-6-6   Down Again, Step 4

Now that the design documentation has been obtained (see Example 5-18), the next step in the top-down design process is to use the documentation to obtain the functional logic diagrams. The functional logic diagrams for functions OA through OG are shown in Fig. 5-35a through g. The dotted line is used to delineate between the positive and negative logic regions on each diagram. The PLC region represents the positive logic convention region and the NLC region represents the negative logic convention region. Recall, a logic low level on the signal line just to the left of the dotted line (PLC region) represents a logic 0, and on the same signal line just to the right of the dotted line (NLC region), it represents a logic 1.

The signal list provides the following information: (1) a list of signals in design or available form in a Boolean function; and (2) the logic convention associated with each signal in a Boolean function.

The design documentation step assists the design process in the following ways: (1) it provides a method for handling both positive and negative logic signals that can occur in realistic designs; and (2) it provides a method for documenting the layout steps for functional logic diagrams.

Example 5-19. Write the design documentation for the following output signal OE of the character generator for a diagram using direct polarity.

$$OE = \overline{C} \cdot \overline{A} \; + \; B \cdot \overline{A} \; + \; D \cdot B \; + \; D \cdot C$$

Signal list: OE[NL], D, C, B, A

Obtain the functional logic diagram for output OE using direct polarity indication.

Solution The design documentation for output signal OE is written as follows.

LLP:	$OE = C \cdot A$	+	$B \cdot A$	+	$D \cdot B$	+	$D \cdot C$	
PIP:	L L L		H L		H H		H H	

The functional logic diagram for output OE using direct polarity indication is shown in Fig. E5-19. Notice that the negative logic signal OE is accompanied by its logic convention on the diagram when direct polarity indication is used.

To illustrate the design flexibility of the top-down design procedure, the following example provides both positive and negative logic signals in both complemented and uncomplemented form.

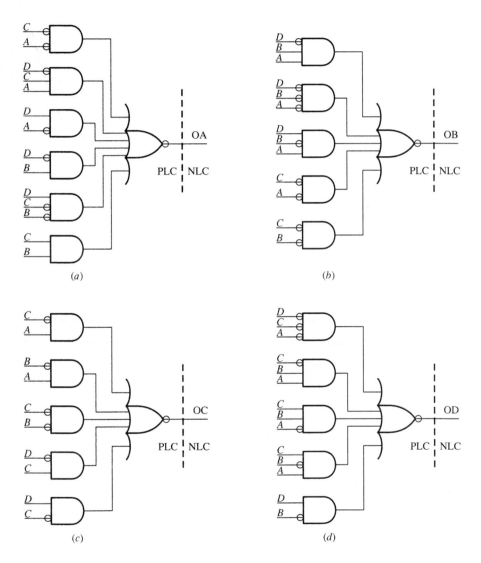

(a)   (b)

(c)   (d)

FIGURE 5-35
Functional logic diagrams for the Binary to Hexadecimal Character Generator (*a*) through (*d*).

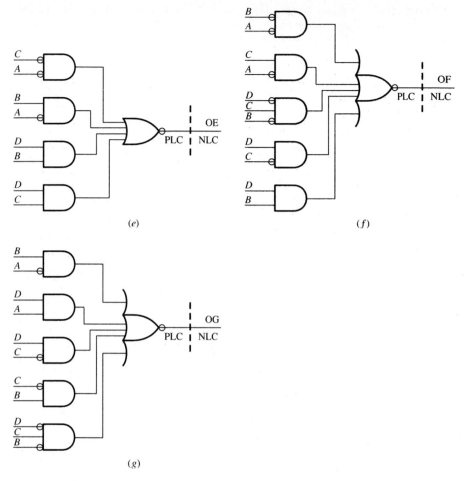

(e)

(f)

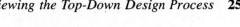

FIGURE 5-35 *(continued)*
Functional logic diagrams for the Binary to Hexadecimal Character Generator (e) through (g).

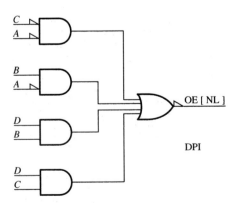

FIGURE E5-19

Example 5-20. Design a combinational logic circuit with the following truth table description. The signal list is: F, A, $\sim B$, $C[NL]$, $\sim D[NL]$. Notice that the available signals $\sim B$ and $\sim D$ are used in the truth table description.

A	~B	C	~D	F
0	0	0	0	1
0	0	0	1	0
0	0	1	0	1
0	0	1	1	0
0	1	0	0	1
0	1	0	1	1
0	1	1	0	0
0	1	1	1	0
1	0	0	0	1
1	0	0	1	0
1	0	1	0	1
1	0	1	1	0
1	1	0	0	0
1	1	0	1	1
1	1	1	0	0
1	1	1	1	1

Solution The Karnaugh map for the function F is illustrated in Fig. E5-20a. Notice that the complemented available signals $\sim B$ and $\sim D$ are used just like the uncomplemented available signals A and C along the perimeter of the map.

From the map, a minimum SOP expression for the 1s of the function and the corresponding signal list can be written in available form as:

$$F = \overline{\sim B} \cdot \overline{\sim D} \; + \; \overline{A} \cdot \sim B \cdot \overline{C} \; + \; A \cdot \sim B \cdot \sim D$$

Signal list: F, A, $\sim B$, $C[NL]$, $\sim D[NL]$

Notice that this is the Boolean function provided in Table 5-3. The design documentation is provided below for a diagram using the positive logic convention and also for a diagram using direct polarity indication.

LLP:	$F = \sim B \cdot \sim D$	+	$A \cdot \sim B \cdot C$	+	$A \cdot \sim B \cdot \sim D$			
NIP:	1 0 1		0 1 1		1 1 0			
LLP:	$F = \sim B \cdot \sim D$	+	$A \cdot \sim B \cdot C$	+	$A \cdot \sim B \cdot \sim D$			
PIP:	H L H		L H H		H H L			

Using the design documentation and signal list for function F, the diagram for the function F using the positive logic convention is shown in Fig. E5-20b. Notice that the available signals represented in the logic layout part are used to label the inputs and the output. The positive and negative logic signals are also clearly delineated via the dotted line.

The diagram for the function F using direct polarity indication is shown in Fig. E5-20c. In this case the available signals in the logic layout part are used to

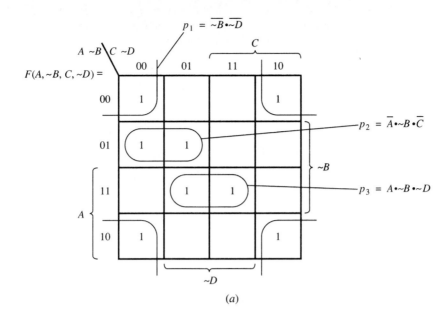

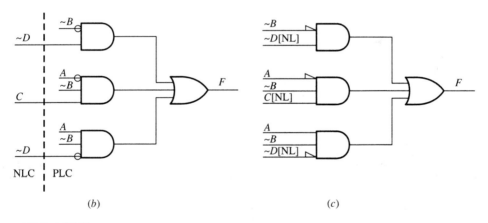

FIGURE E5-20

label the inputs and the output. The negative logic signals are then appended with the notation [NL].

An alternate and equally valid method would be to convert the negative logic signals C and $\sim D$ to the positive logic signals $\sim C$ and D in the truth table (see Fig. A5-1). This requires changing the C column to $\sim C$ by complementing every bit in the column, and changing the $\sim D$ column to D by complementing every bit in the column. Next, apply the top-down design process and obtain the gate level design using all positive logic signals. Draw the functional logic diagram. If preparing a diagram using the positive logic convention, extend the signal lines for

$\sim C$ and D into a negative logic region and label them C and $\sim D$. If preparing a diagram using direct polarity indication, relabel the signal lines $\sim C$ and D as $C[\text{NL}]$ or $C(\text{L})$ and $\sim D[\text{NL}]$ or $\sim D(\text{L})$, respectively.

REFERENCES

1. Peatman, J. B., *Digital Hardware Design,* McGraw-Hill, New York, 1980.
2. Clare, C. R., *Designing Logic Systems Using State Machines,* McGraw-Hill, New York, 1973.
3. Rhyne, Jr., V. T., *Fundamentals of Digital Systems Design,* Prentice-Hall, Englewood Cliffs, New Jersey, 1973.
4. Fletcher, W. I., *An Engineering Approach to Digital Design,* Prentice-Hall, Englewood Cliffs, New Jersey, 1980.
5. Dempsey, J. A., *Basic Digital Electronics with MSI Applications,* Addison-Wesley Pub., Reading, Mass., 1977.
6. Blakeslee, T. R., *Digital Design with Standard MSI and LSI,* John Wiley & Sons, New York, 1975.
7. Lee, S. C., *Digital Circuits and Logic Design,* Prentice-Hall, Englewood Cliffs, New Jersey, 1976.
8. IEC STANDARD Publication 617-12, *Graphic Symbols for Diagrams, Part 12: Binary logic elements,* International Electrotechnical Commission, Geneve, Swisse, 1983.
9. ANSI/IEEE Std 91-1984, *IEEE Standard Graphic Symbols for Logic Functions,* The Institute of Electrical and Electronic Engineers, Inc., New York, 1984.

PROBLEMS

Section 5-2   Gate-Level Combinational Logic Design

5-1. Write the following Boolean function in design form (available form) for the following available signals.

$$F = A \ + \ \overline{B} \cdot C$$

(a) $F, A, \sim B, C$
(b) $F, \sim A, \sim B, C$
(c) $F, A, \sim B, \sim C$
(d) $\sim F, \sim A, B, C$

5-2. List the available signals for the following Boolean function. Then write the normal form for the function.

$$F = \overline{\sim Y} \cdot Z \ + \ \overline{\sim X} \cdot \overline{\sim Y} \ + \ W \cdot \sim X \cdot \sim Y$$

5-3. Write the following Boolean function in available form for the following available signals.

$$\overline{F} = \overline{X} \cdot \overline{Y} \ + \ Y \cdot \overline{Z} \ + \ W \cdot \overline{Y} \cdot Z$$

(a) $F, \sim W, \sim X, Y, Z$
(b) $\sim F, W, X, \sim Y, Z$
(c) $F, W, \sim X, Y, Z$
(d) $\sim F, W, X, \sim Y, \sim Z$

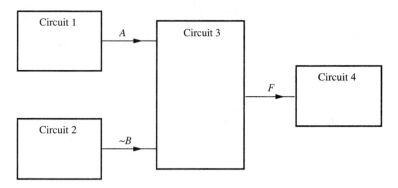

FIGURE P5-4

5-4. Apply the logic convention matching technique for the available signals shown in Fig. P5-4 to obtain the signal list for circuit 3 for each of the following logic conventions for circuits 1, 2, and 4.

	Circuit 1	Circuit 2	Circuit 4
(*a*)	PL	PL	PL
(*b*)	PL	PL	NL
(*c*)	NL	PL	PL
(*d*)	NL	NL	PL
(*e*)	PL	NL	NL

5-5. Apply the logic convention matching technique for the available signals shown in Fig. P5-5 to obtain the signal list for circuit 3 for each of the following logic conventions for circuits 1, 2, 4, and 5.

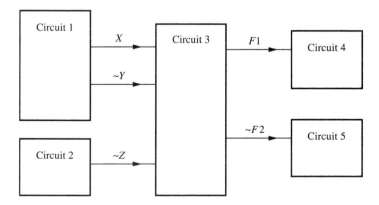

FIGURE P5-5

	Circuit 1	Circuit 2	Circuit 4	Circuit 5
(a)	NL	PL	PL	NL
(b)	PL	NL	NL	PL
(c)	PL	PL	PL	NL
(d)	NL	PL	PL	PL
(e)	NL	NL	NL	PL
(f)	PL	PL	NL	PL

Section 5-3 Logic Conventions and Polarity Indication

5-6. Given the following data, determine the high-level and low-level noise margins for a TTL 74ALS11. Repeat the calculations for a TTL 74AS11. Which device has the best noise margin?

	74ALS11 (TTL)	74AS11 (TTL)
$V_{IH}(MIN)$	2.0 Volts	2.0 Volts
$V_{IL}(MAX)$	0.8 Volts	0.8 Volts
$V_{OH}(MIN)$	2.5 Volts	2.5 Volts
$V_{OL}(MAX)$	0.4 Volts	0.5 Volts

5-7. Given the following data, determine the high-level and low-level noise margins for a CMOS 74ACT1011. Repeat the calculations for a CMOS 74AC1011. Which device has the best noise margin?

	74ACT1011 (CMOS)	74AC1011 (CMOS)
$V_{IH}(MIN)$	2.0 Volts	3.15 Volts
$V_{IL}(MAX)$	0.8 Volts	1.35 Volts
$V_{OH}(MIN)$	3.8 Volts	3.8 Volts
$V_{OL}(MAX)$	0.44 Volts	0.44 Volts

5-8. Obtain the voltage truth table for the logic elements shown in Fig. P5-8.

5-9. Obtain the voltage truth tables for U1, U2, and U3 in Fig. P5-9.

5-10. Obtain the external logic truth tables and the voltage truth tables for U1, U2, and U3 in Fig. P5-10.

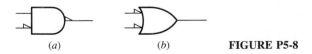

(a)   (b)   **FIGURE P5-8**

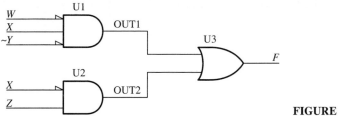

FIGURE P5-9

Section 5-4 Using Design Documentation to Obtain Functional Logic Diagrams

For the first four problems in this section, utilize the top-down design process to obtain the functional logic diagrams. Unless otherwise stated, also consider all available signals to be uncomplemented positive logic signals.

5-11. Design a circuit that provides at its output the 1's complement of each 3-bit binary number applied to its input.

5-12. Design a majority of 1s circuit such that the output F is 1 when a majority of the inputs $X\ Y\ Z$ are 1.

5-13. Design a minority of 1s circuit such that the output F is 1 when a minority of the inputs $A\ B\ C$ are 1.

5-14. Design a gate-level 2-bit full adder that adds $A1\ A0$, $B1\ B0$, and $C0$ to obtain $C2$ $S1\ S0$.

5-15. Design a 3-bit full adder using three 1-bit full adders.

5-16. Design a 4-bit full adder using four 1-bit full adders.

5-17. Design a 4-bit full adder using four 1-bit full adders and the carry-propagate technique discussed in Section 5-4.

Section 5-5 Applying the Top-Down Design Process to Other Realistic Problems

Unless otherwise stated, consider all available signals to be uncomplemented positive logic signals.

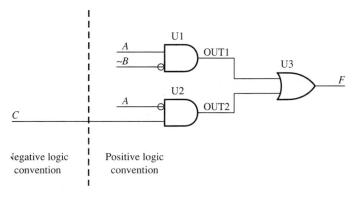

Negative logic convention ┊ Positive logic convention

FIGURE P5-10

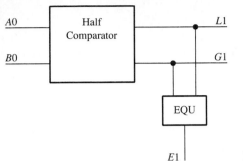

FIGURE P5-20

5-18. Design a 2-bit full subtractor using two 1-bit full subtractors.

5-19. Design a 3-bit full subtractor using borrow lookahead. Use the technique discussed in Subsection 5-4-2.

5-20. Design the half comparator represented by the block diagram shown in Fig. P5-20, where $L1$ represents $A0 < B0$, $G1$ represents $A0 > B0$, and $E1$ represents $A0 = B0$. Determine $E1$ from $L1 = G1 = 0$.

5-21. Design the full comparator represented by the block diagram shown in Fig. P5-21, where $L1$ represents $A0 < B0$, $G1$ represents $A0 > B0$, $L2$ represents $A1\ A0 < B1\ B0$, $G2$ represents $A1\ A0 > B1\ B0$, and $E2$ represents $A1\ A0 = B1\ B0$. Determine $E2$ from $L2 = G2 = 0$.

5-22. Design a gate-level 2-bit non-expandable magnitude comparator.

5-23. Design a BCD to XS3 code converter.

5-24. Design a 4-bit binary to BCD code converter.

5-25. Design a 3-to-8 line Decoder using two 2-to-4 line Decoders.

5-26. Design a Decimal to BCD encoder.

5-27. Design a 4-input priority encoder such that when input $I3$ is 1 it has priority over inputs $I2$, $I1$, and $I0$, and when $I2$ is 1 it has priority over inputs $I1$, and $I0$, etc. Provide an enable input for the encoder.

5-28. Show a design for a 3-bit Multiplexer using three 2-bit Multiplexers such that select lines $A\ B = 00$ select $I0$, $A\ B = 01$ select $I1$, and $A\ B = 1\ X$ (X represents a don't care input) select $I2$.

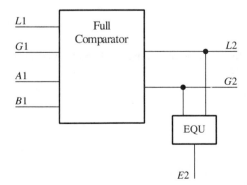

FIGURE P5-21

5-29. Using two 16-bit Multiplexers and one 2-bit Multiplexer, show a design for a 32-bit Multiplexer.

5-30. Design a circuit that converts each 3-bit binary number applied to its input to a binary number at its output, equivalent to the square of the corresponding input binary number.

5-31. Design a circuit that provides at its output the 2's complement of each 3-bit binary number applied to its input.

5-32. Repeat Prob. 5-31 using Inverters and a 3-bit Adder.

5-33. Design a circuit that converts each binary 4-bit number applied to its input lines to a BCD number at its output lines. To express a 4-bit input number in BCD, five output bits are required (the binary number 1100 converted to decimal is represented as 12 and when converted to BCD is represented as 1 0010).

5-34. Design a 2421 code to 2-out-of-5 code code converter.

5-35. Design a circuit with three inputs A B C that provides a 0 output for variable F when an even number of its inputs are 0.

5-36. Design a circuit with four inputs using Exclusive OR elements such that the output F is 1 when the binary number A B C D is odd, except when all of its inputs are 0.

5-37. Design a circuit to detect the equivalence of two 4-bit BCD numbers A and B by comparing the corresponding bits in each bit position and ANDing the results. Utilize the Equivalence operator to write the expressions, $A_i = B_i$, for each bit position.

5-38. Design a circuit to detect when a 4-bit binary number A is greater than another 4-bit binary number B by using the following algorithm. If the output $F(A > B) = 1$

then $A3 > B3$

   OR $(A3 = B3)$ AND $(A2 > B2)$

   OR $(A3 = B3)$ AND $(A2 = B2)$ AND $(A1 > B1)$

   OR $(A3 = B3)$ AND $(A2 = B2)$ AND $(A1 = B1)$ AND $(A0 > B0)$,

else the output F is not equal to 1. *Hint:* determine each expressions for $A_i > B_i$ and $A_i = B_i$ two bits at a time, using truth tables. Take advantage of the Equivalence operator in writing the expressions.

5-39. Design a circuit to detect when a 4-bit binary number A is less than another 4-bit binary number B by modifying the algorithm in Prob. 5-38 by replacing $>$ with $<$.

5-40. Design a circuit that provides the product of a 2-bit and a 3-bit number. The 2-bit number range is 0 through 2, while the 3-bit number range is from 0 through 4. All other output products are don't cares.

5-41. Design a circuit that provides the product of all 3-bit by 4-bit numbers. Use the technique illustrated in Example 5-16.

5-42. Design a circuit that will multiply two 2-bit numbers using four 2-input AND elements and two Half Adders.

5-43. Design a circuit that will divide two 2-bit numbers. Note that when n-bit numbers are divided, the quotient is $n/2$ bits and the remainder is n bits. Use one output quotient bit $Q0$ and two output remainder bits $R1$ and $R0$.

5-6   Reviewing the Top-Down Design Process

5-44. Design a BCD to seven segment code converter. Use the displayed characters 0 through 9 represented in Fig. E5-17a and let the displayed characters A through F represent don't care outputs. Assume that the converter is designed to drive a common cathode display.

5-45. Design a Binary to ASCII Hexadecimal character generator. Let the available inputs be A B C D and the available outputs be AK6 through AK0. The positive logic inputs range from 0000 to 1111 and the positive logic ASCII outputs range from 0 to F (see Table 2-1 in Chapter 2 for ASCII binary code values). From the language statement obtain

(a) the truth table

(b) the minimum SOP Boolean functions

(c) the design documentation

(d) the functional logic diagrams.

CHAPTER 6

OBTAINING REALIZABLE LOGIC DIAGRAMS USING SSI DEVICES

6-1 INTRODUCTION AND INSTRUCTIONAL GOALS

The design of gate-level logic circuits is usually carried out using functional logic diagrams as was demonstrated in Chapter 5. The next step in the top-down design process is to utilize off-the-shelf IC devices to obtain usable or realizable circuits. Our discussion begins with Section 6-2, Obtaining Realizable IC Circuits. This is a long section covering data book symbols and DeMorgan equivalent symbols for AND and OR elements. The concepts of equivalent signal lines and matching logic indicators are presented, providing a systematic method of converting functional logic diagrams to realizable logic diagrams. This involves using available small-scale integration (SSI) devices such as Buffers and Inverters in addition to AND, OR, NAND, and NOR gates. Equivalent signal lines and equivalent AND and OR element symbols for available devices are utilized to obtain realizable logic diagrams from functional logic diagrams in the same manner a system designer would carry out the task.

269

The discussion continues with Section 6-3, Implementing Circuits with NAND Gates and NOR Gates, where equivalent symbols and equivalent signal lines are used to realize logic diagrams with only NAND gates or NOR gates. Paralleled inputs and fixed-mode inputs provide a way to use larger gates (leftover or unused gates) in a useful manner in a circuit.

Section 6-4, Distributed Connections (Dot-AND and Dot-OR), and Section 6-5, Using AND-OR-Invert Elements, show how two special classes of logic circuits (open circuit outputs and AOI elements) can be used to obtain realizable diagrams. Section 6-6, Obtaining Boolean Functions from Diagrams Using the Positive Logic Convention, discusses how to analyze positive logic diagrams. Obtaining Boolean Functions from Diagrams Using Direct Polarity Indication, the topic of Section 6-7, presents signal names with level indication and also shows how to analyze diagrams using direct polarity indication. These techniques are useful for obtaining a Boolean function for an existing circuit or verifying that a designed circuit generates the required Boolean function.

This chapter should prepare you to

1. Obtain equivalent AND and OR element symbols for a given AND, OR, NAND, or NOR function.
2. Draw data book symbols and DeMorgan equivalent symbols for AND, OR, NAND, NOR, Buffer, and Inverter functions.
3. Obtain realizable logic diagrams using Boolean functions or functional logic diagrams.
4. Convert functional logic diagrams and realizable logic diagrams into equivalent realizable logic diagrams using different gate types.
5. Design realizable logic circuits using devices with open circuit outputs.
6. Design realizable logic circuits using AOI elements.
7. Obtain Boolean functions and their signal lists for functional logic diagrams using the positive logic convention.
8. Obtain Boolean functions and their signal lists for functional logic diagrams using direct polarity indication.

6-2   OBTAINING REALIZABLE IC CIRCUITS

In Chapter 5 the top-down design process was followed to obtain functional logic diagrams for Boolean functions. Functional logic diagrams are often used in the first phase of the design process. The actual devices used to realize a design do not have to be considered at this point in the design process.

The next step (step 5) in the design process is to choose actual logic elements that are realizable. All the logic elements used in functional logic diagrams can be realized by an IC design engineer at the transistor component level. Thus, at this level, switching transistors, such as those illustrated in Fig. 1-5 (Chapter 1), are used to design the logic elements. To a systems designer, however, realizable logic elements are only those elements that are available as off-the-shelf IC devices (devices currently available in IC manufacturer's data books that have been designed by an IC design engineer). In the following presentation we will restrict our discussions to the role of the systems designer and utilize only available devices in the realizable logic diagrams that are shown.

> **Definition 6-1.** For the purpose of our discussion, available devices are devices that are available as off-the-shelf IC devices.

Realizable logic diagrams may be obtained from functional logic diagrams by modifying the functional logic diagrams so that all the elements in the diagrams represent available devices. The AND and OR elements used in functional logic diagrams represent AND and OR operations specified by Boolean functions. The task is to choose appropriate gate devices to perform the operations represented by the AND and OR elements in our functional logic diagrams. First, we will consider available devices for AND elements.

6-2-1 2-Input AND Element Symbols

The matrix shown in Fig. 6-1 represents every possible combination for a 2-input AND element symbol. Normally the devices shown on the left side of the AND matrix are unavailable devices. The devices shown across the top of the AND matrix, however, are normally available devices. For easy reference, the names IC manufacturers provide for their available devices are specified directly under each of the available device symbols. Keep in mind that even though 2-input AND element symbols are shown in Fig. 6-1, IC manufacturers provide devices with additional inputs. Typically 2, 3, 4, and 8 inputs are common, but this also depends on the logic family being used, as well as the particular gate type. The AND element symbols used in the matrix in Fig. 6-1 for the AND and NAND gates are the symbols IC manufacturers use to represent these gates, that is, the data book symbols for these particular gates. The AND element symbols used in the matrix for the OR and NOR gates are often referred to as DeMorgan equivalent symbols for these respective gates. IC manufacturers show the OR and NOR gates in their data books using OR element symbols rather than DeMorgan equivalent AND element symbols.

6-2-2 Equivalent AND and OR Element Symbols

Before analyzing the realizable IC circuit forms in the body of the AND matrix in Fig. 6-1, we will first show how to obtain one of the AND element symbols and its equivalent OR element symbol for a simple logic expression. Each of the available AND element symbols, that is, the AND gate, OR gate, NAND gate, and NOR gate shown in Fig. 6-1, can also be represented as an equivalent OR element symbol. Consider the truth table shown in Fig. 6-2*a* for a 2-input

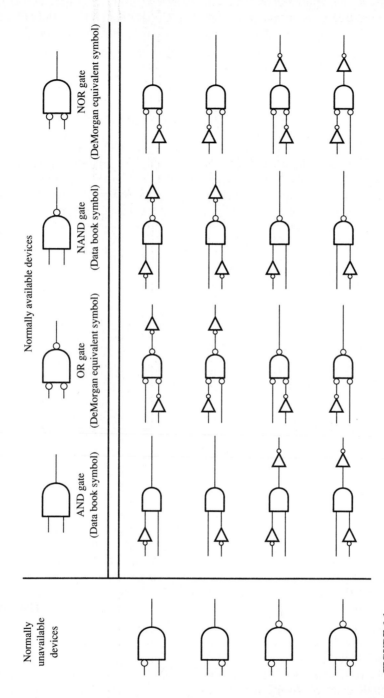

FIGURE 6-1
2-input AND element symbols.

A B	F
0 0	1
0 1	0
1 0	0
1 1	0

(a)

$F =$

A\B	0	1
0	1	0
1	0	0

(b)

FIGURE 6-2
(a) Truth table for 2-input NOR function, (b) Karnaugh map for 2-input NOR function.

NOR function (the term NOR is a contraction for NOT OR). The Karnaugh map description is shown in Fig. 6-2b.

From the truth table or the Karnaugh map, the Boolean expression for the 1s of the function F and its corresponding signal list is written as follows.

$$F = \overline{A} \cdot \overline{B}$$

Signal list: F, A, B

The design documentation for function F can be written as

LLP: $F = A \cdot B$
NIP: 1   0   0

Complementing both sides of function F and applying DeMorgan's theorem, or solving for the 0s of the function, results in the following complemented function.

$$\overline{F} = A + B$$

The design documentation for $\overline{F}$ can be written as follows.

LLP: $F = A + B$
NIP: 0   1   1

Using the design documentation for function F, the diagram for the positive logic 2-input NOR function can be drawn as shown in Fig. 6-3a. The AND element symbol shown in Fig. 6-3a is the AND element representation of the 2-input NOR function.

Using the design documentation for function $\overline{F}$, the diagram can also be drawn for the positive 2-input NOR function as shown in Fig. 6-3b. The latter

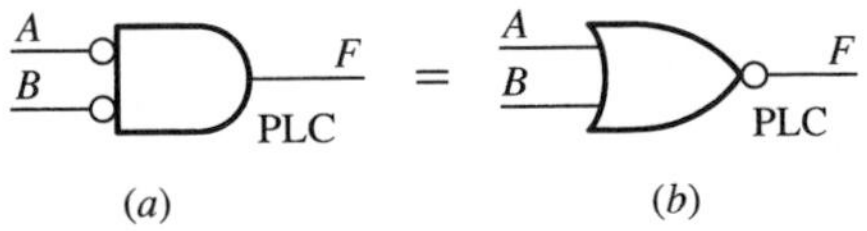

FIGURE 6-3
(a) AND element symbol for a positive logic 2-input NOR function, (b) OR element symbol for a positive logic 2-input NOR function.

symbol is the equivalent OR element symbol for the 2-input NOR function. The AND element symbol and the OR element symbol were obtained from the same logic description and, therefore, represent the same device. Notice that two different element symbols may be drawn to represent a positive logic 2-input NOR function.

In IC manufacturer's data books the device that generates the positive logic 2-input NOR function is represented symbolically using the OR element symbol shown in Fig. 6-3b, and is named a positive 2-input NOR gate. The fact that one of the element symbols in Fig. 6-3 has negation qualifying symbols (or polarity qualifying symbols if direct polarity indication is used) on all of its inputs, while the IC manufacturer's data book symbol does not, is the distinguishing feature between the two different symbols. Recognizing this fact, application software writers have begun to use the term DeMorgan equivalent symbols for symbols that contain negation or polarity qualifying symbols on all inputs. This applies to the symbols that represent the available logic AND, OR, NAND, NOR, Buffer, and Inverter functions. The AND element symbol shown in Fig. 6-3a is the DeMorgan equivalent symbol for the NOR function or NOR gate while the OR element symbol shown in Fig. 6-3b is the data book symbol for the NOR function or NOR gate.

> **Definition 6-2.** DeMorgan equivalent symbols for the available logic AND, OR, NAND, NOR, Buffer, and Inverter functions are symbols with negation qualifying symbols (or polarity qualifying symbols) on all their inputs.

Example 6-1. Obtain the AND element symbol and its equivalent OR element symbol for a 3-input AND function.

Solution The truth table for a 3-input AND function is shown in Fig. E6-1a, and its corresponding Karnaugh map is shown in Fig. E6-1b. The Boolean expression for the 1s of the function F and its signal list is written as follows.

$$F = A \cdot B \cdot C$$

Signal list: F, A, B, C

A	B	C	F
0	0	0	0
0	0	1	0
0	1	0	0
0	1	1	0
1	0	0	0
1	0	1	0
1	1	0	0
1	1	1	1

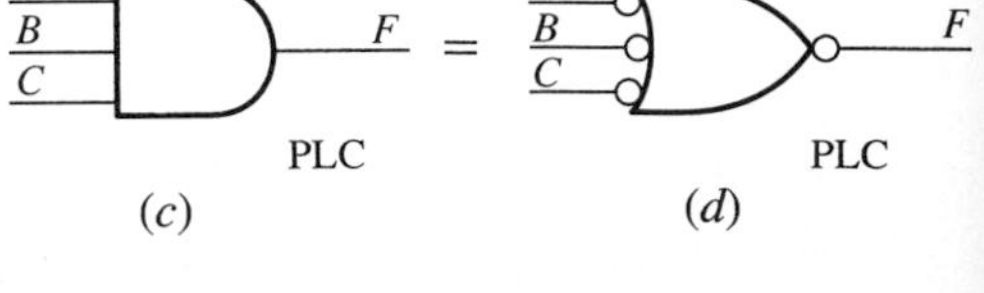

(a)

FIGURE E6-1

The Boolean expression for the 0s of the function F can be written as

$$\overline{F} = \overline{A} + \overline{B} + \overline{C}$$

Notice that the latter expression is also obtained by simply complementing both sides of the Boolean expression for the 1s of the function and applying DeMorgan's theorem.

The design documentation for the 1s of the function can be written as follows.

LLP: $F = A \cdot B \cdot C$
NIP: 1   1 1 1

From this design documentation the AND element symbol for a positive logic 3-input AND gate can be drawn as shown in Fig. E6-1c.

The design documentation for the 0s of the function can be written as follows.

LLP: $F = A + B + C$
NIP: 0   0   0   0

From this design documentation, we obtain the OR element symbol for the positive logic 3-input AND gate shown in Fig. E6-1d. The AND element symbol shown in Fig. E6-1c is the data book symbol, while the OR element symbol shown in Fig. E6-1d is the DeMorgan equivalent symbol for the 3-input AND function or AND gate.

Obtaining equivalent symbols for the 2-input NOR gate and the 3-input AND gate is a straightforward process as we have shown. This is also true for the NAND gate (the term NAND is a contraction for NOT AND) and the OR gate. Both are left as exercises for the student. It has become common practice by IC manufacturers to specify that the devices in their data books perform their specified functions in positive logic. This is one reason why diagrams are usually drawn using either the positive logic convention or direct polarity indication. Since the conversion between the positive logic convention and direct polarity indication is a simple substitution of the polarity symbol and the negation symbol, engineers can use either of the data book symbols (symbols drawn using the positive logic convention or direct polarity indication) rather routinely. In Fig. 6-4 we show a summary of available AND element symbols and their equivalent OR element symbols for positive logic AND, OR, NAND, and NOR gates for both distinctive-shape and rectangular-shape symbols.

6-2-3 Positive and Negative Logic Gate Types

The conversion from an AND gate using the positive logic convention to an OR gate using the negative logic convention is illustrated in Fig. 6-5. The result

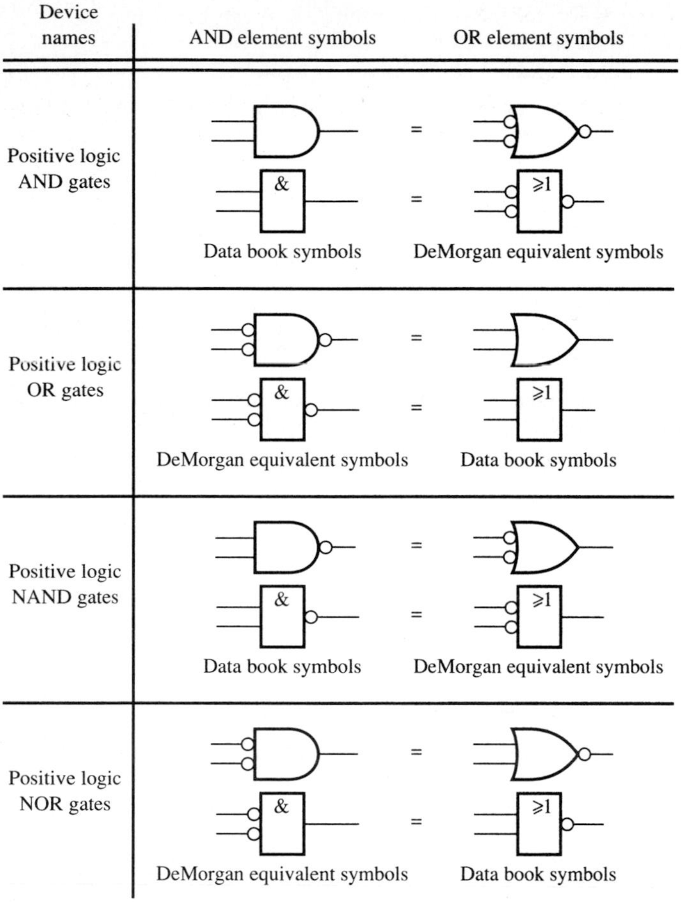

Device names	AND element symbols	OR element symbols
Positive logic AND gates		

FIGURE 6-4
Equivalent AND and OR element symbols for normally available devices.

$F = A \cdot B$
Signal list: F, A, B

$F = A \cdot B$
Signal list: $F[\text{NL}], A[\text{NL}], B[\text{NL}]$

LLP: $F = A \cdot B$
NIP: 1   1 1

LLP: $F = A \cdot B$
NIP: 0   0 0

Functional logic diagram

Functional logic diagram

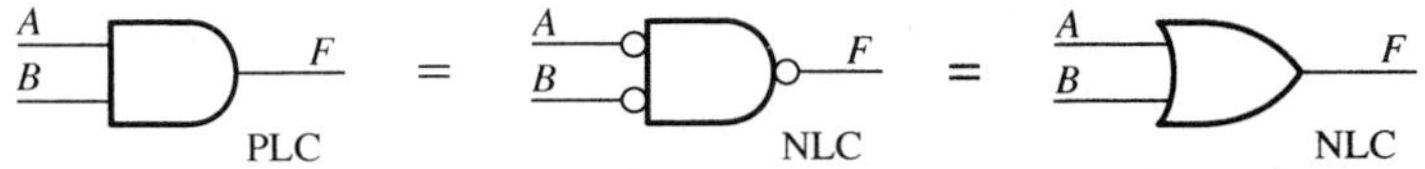

FIGURE 6-5
Converting from a positive logic AND gate to the equivalent negative logic gate (an OR gate).

of these conversions illustrates that a positive logic AND gate is the same device as a negative logic OR gate. The conversion of positive logic OR, NAND, and NOR gates to their equivalent negative logic gate types is left as an exercise for the student.

6-2-4 Using Equivalent AND and OR Element Symbols and Equivalent Signal Lines

One method of converting the symbols in a functional logic diagram to symbols that exist as off-the-shelf IC devices is by simply replacing each unavailable device symbol with the corresponding available device symbol and the necessary Inverters, as indicated in Fig. 6-1. This technique is valid, but a designer should not have to refer to Fig. 6-1 to obtain the replacement circuit. A logic designer needs to become intimately familiar with the available gates (AND, OR, NAND, and NOR) and their equivalent AND and OR element symbols shown in Fig. 6-4. Equipped with this information, it is quite routine to obtain equivalent representations for unavailable device symbols using equivalent AND and OR element symbols for available devices and equivalent signal lines. Equivalent input signal lines are shown in Figs. 6-6a and b, while equivalent output signal lines can be found in Figs. 6-6c and d.

The negation symbol may be drawn on the output side or the input side of an Inverter symbol as shown in Fig. 6-7. The Inverter symbol with the negation symbol on the output side is the data book symbol, while the Inverter symbol with the negation symbol on the input side is the DeMorgan equivalent symbol. For either Inverter symbol, the complementation function performed remains the same. Figure 6-7 shows a summary of equivalent Inverter and Buffer symbols for both distinctive-shape and rectangular-shape symbols. The data book symbol for a Buffer is drawn without negation symbols. The Buffer symbol with negation symbols (or wedges if direct polarity indication is being used) on both input and output sides (as shown in Fig. 6-7) is the DeMorgan equivalent symbol.

(a)   (b)

(c)   (d)

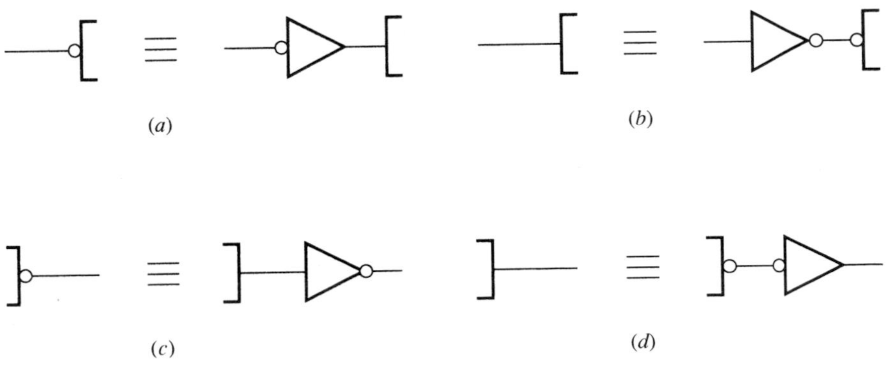

FIGURE 6-6
Equivalent signal lines (a), (b) for inputs, (c), (d) for outputs.

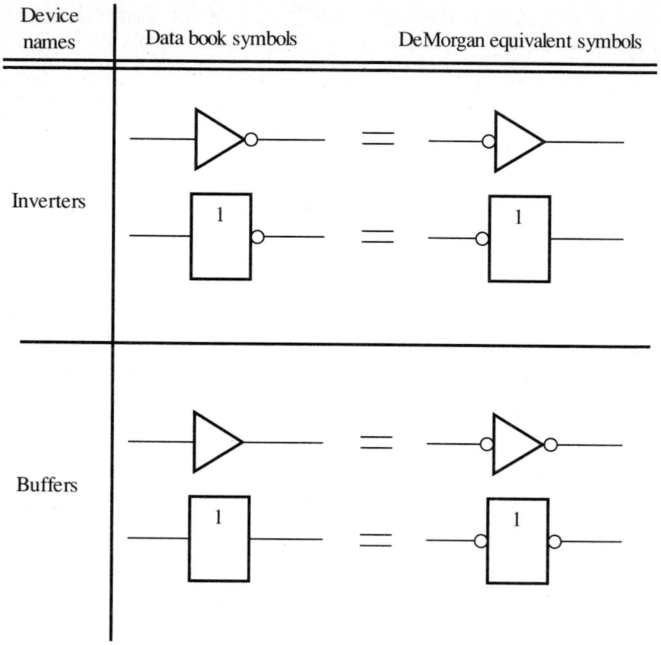

Device names	Data book symbols	DeMorgan equivalent symbols
Inverters		
Buffers		

FIGURE 6-7
Equivalent Inverter and Buffer symbols.

Keep in mind that it is not necessary, although it is good design practice, to match logic indicators on the signal lines between two logic elements. When possible, the logic indicators (the presence or absence of negation qualifying symbols if the positive logic convention is used, or the presence or absence of polarity qualifying symbols, if direct polarity indication is used) should match at both ends of each signal line connecting two logic elements as shown in Fig. 6-6. Notice that each of the signal lines for the replacement circuits in the body of the AND matrix in Fig. 6-1 is also drawn with matched logic indicators.

> **Definition 6-3.** The expression, matched logic indicators, refers to a signal line connecting two logic elements in which the logic indicator at the destination of the signal matches the logic indicator at the source of the signal.

A circle (negation symbol or a wedge, which is the polarity symbol, if using direct polarity indication) can be moved away from the input or output of an AND symbol as shown in Figs. 6-6a and c if we make the circle part of an Inverter symbol. If we add a circle at an input or output of an AND symbol as illustrated in Figs. 6-6b and d, we have created a signal mismatching problem, which we simply fix by adding an Inverter. Using equivalent signal lines and equivalent

AND and OR symbols for the normally available devices, as shown in Fig. 6-4, a designer can obtain a realizable IC logic circuit for any unavailable AND element device.

As an example, consider converting the functional logic diagram shown in Fig. 6-8 (represented by the following Boolean function and the corresponding signal list) into a realizable logic diagram.

$$\overline{F} = \overline{D} \cdot \overline{C} \cdot B \cdot \overline{A} \;+\; D \cdot C \cdot \overline{A} \;+\; D \cdot C \cdot B$$

Signal list: F[NL], D, C, B, A

Logic symbols 3 and 4 in the diagram in Fig. 6-8 appear to represent available devices; they need not be changed. Logic symbols 1 and 2 do not represent available devices. These symbols can be replaced with equivalent circuits representing available devices by using equivalent signal lines. To match the logic indicators (absence of circles) at the inputs to symbol 4 requires an AND element symbol for either an AND gate or a NOR gate for symbols 1 and 2, since the AND element symbols for these gates do not have circles on their outputs. If we were trying to choose symbols to reduce package count, we might replace symbol 2 with one like symbol 3, and, to do so, we would need to modify symbol 2 by moving the circle (negation symbol) at input A to make it part of an Inverter symbol as indicated by the equivalent signal lines shown in Fig. 6-6a.

If we were using a certain TTL (Transistor-Transistor Logic) technology, for instance STD TTL, LS, S, ALS, AS, or F (see Table 5-1), we might find 4-input AND gates available but 4-input NOR gates unavailable. If we choose to use an AND element symbol for a 4-input AND gate, we would need to modify symbol 1 to represent that AND element symbol. This requires moving each of the circles at inputs D, C, and A to make them a part of three separate Inverter symbols, that is, use the equivalent signal lines shown in Fig. 6-6a. OR gates with more than two inputs are not available for the TTL logic family, but 3-input NOR gates are. As you can tell from our discussion, when actual devices must be chosen, the decision to proceed one way or the other is based on a number of different criteria such as (1) actual available gates (since some gates may not be available for a particular technology), (2) number of desired packages (choosing similar gates allows fewer packages due to multiple gates in a package), and (3)

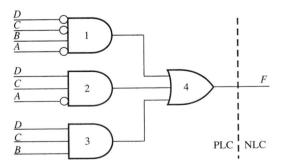

PLC | NLC

FIGURE 6-8
Functional logic diagram.

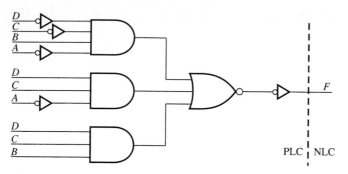

FIGURE 6-9
Realizable logic diagram for function *F*.

required speed (AND gates and OR gates tend to be slower than NAND gates and NOR gates, when fabricated in TTL).

In Fig. 6-9, we show a realizable logic diagram using Inverters, AND gates, and a NOR gate. The diagram is realizable because we can choose actual off-the-shelf devices from manufacturer's logic books for each of the symbols in the circuit.

After obtaining a realizable logic diagram, the last step (step 6) in the top-down design process illustrated in Fig. 5-1 is to obtain a detailed logic diagram like the one shown in Fig. 6-10 for function *F*. Designers generally use schematic-capture programs to obtain detailed logic diagrams, since these specialized computer programs provide an abundance of library parts to speed up the design process. OrCAD, Schema, and EE Designer (see the list of references at the end of this chapter) are examples of available schematic-capture programs that are designed to work on IBM PC XT/AT (or compatibles) or PS/2 type systems. (IBM PC XT/AT and IBM PS/2 are trademarks of International Business Machines Corporation.) After a diagram has been drawn on the computer monitor, it can be dumped to a printer or plotter to obtain a hard copy.

As can be observed from the detailed logic diagram in Fig. 6-10, only four Inverters are required. The dotted line from the output of U1C to U1D has been added to show that either one of these Inverters can be removed if we add the dotted signal line. Decisions like this often come up late in the design phase, when a designer realizes an Inverter is not needed and therefore can be used elsewhere. The detailed logic diagram shown in Fig. 6-10 shows (*a*) reference designators to identify each package, (*b*) each part value or type, and (*c*) pin numbers for input and output signals. The pin numbers for power and ground are shown in the accompanying power supply connection table. The unused spare devices in a package are usually shown in the diagram as indicated in Fig. 6-10. Information concerning available manufacturer's devices may be obtained from TTL and CMOS data books (see the list in the Reference section at the end of this chapter) or by accessing the library of available parts in a schematic-capture program. Detailed logic diagrams are necessary when circuits must be

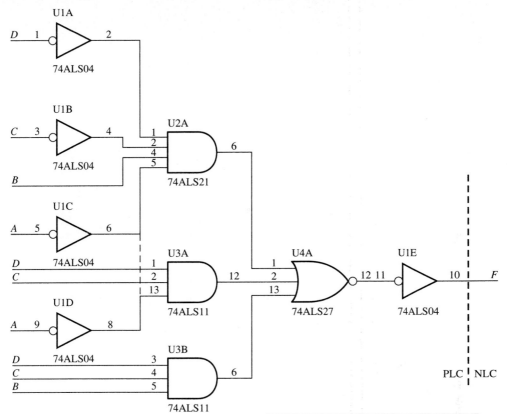

Spare devices

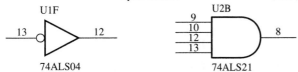

Reference designator	Part value	Pin numbers	
		+5 V	Ground
U1	74ALS04	14	7
U2	74ALS21	14	7
U3	74ALS11	14	7
U4	74ALS27	14	7

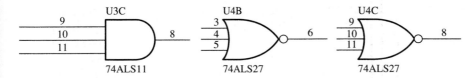

FIGURE 6-10
Detailed logic diagram for Function *F*.

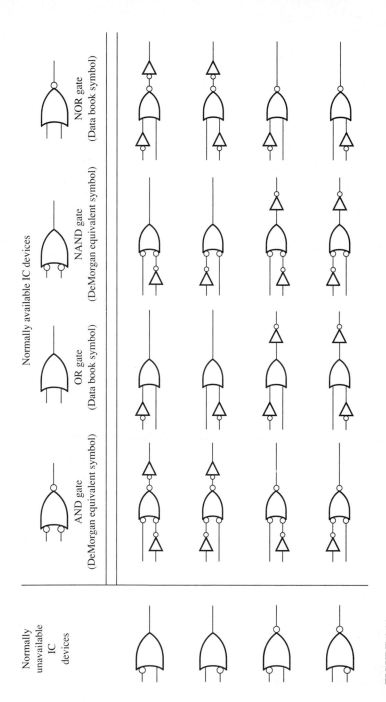

FIGURE 6-11

2-input OR element symbols.

constructed in the laboratory, and also when engineers must turn in their final documentation to Engineering Specs at the end of a project.

6-2-5   2-Input OR Element Symbols

Realizable IC circuit forms for 2-input OR element symbols are illustrated in Fig. 6-11. The OR matrix shown in Fig. 6-11 represents every possible combination for a 2-input OR element symbol. The devices shown on the left side of the OR matrix are normally unavailable devices; however, the devices shown across the top of the OR matrix are normally available off-the-shelf devices. For easy reference, the names IC manufacturers provide for their available devices are specified directly under each of the available device symbols. Even though only 2-input OR element symbols are shown in Fig. 6-11, keep in mind that IC manufacturers typically provide these devices with additional inputs (depending upon the logic family and gate types). The OR element symbols used in the matrix in Fig. 6-11 for the OR and NOR gates are the symbols IC manufacturers use to represent these gates, that is, the data book symbols for these particular gates. The OR element symbols used in the matrix for the AND and NAND gates are the DeMorgan equivalent symbols for these respective gates.

By knowing the available gates (AND, OR, NAND, and NOR) and their equivalent AND and OR element symbols, it is a simple process to obtain equivalent representations for unavailable device symbols using available device symbols and one or more Inverter symbols, as illustrated in the matrix in Fig. 6-11. Keep in mind that it is considered good design practice to match logic indicators at each end of every signal line connecting two logic elements.

A circle (negation symbol or a wedge which is the polarity symbol, if using direct polarity indication) can be moved away from the input or output of an OR symbol as shown in Figs. 6-6*a* and *c*, if we make the circle part of an Inverter symbol. If we add a circle at an input or output of an OR symbol as illustrated in Figs. 6-6*b* and *d*, we have created a signal mismatching problem which we simply fix by adding an Inverter. Using equivalent signal lines and equivalent AND and OR element symbols for the normally available devices shown in Fig. 6-4, a designer can obtain a realizable IC logic circuit for any unavailable OR element device.

Logic indicators on signal lines, connecting only two logic elements, can be matched or mismatched. The signal lines we are referring to are interior signal lines, since they are on the interior of the design. There are only two types of interior signal lines with matched logic indicators that can connect two logic elements together, and these are shown in Fig. 6-12. In Fig. 6-12*a*, the absence of the negation qualifying symbol at the source and the destination of

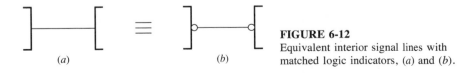

FIGURE 6-12
Equivalent interior signal lines with matched logic indicators, (*a*) and (*b*).

        (*a*)   (*b*)

FIGURE 6-13
Equivalent interior signal lines with
mismatched logic indicators, (*a*) and (*b*).

the signal provides a match, while in Fig. 6-12*b* the presence of the negation qualifying symbol at the source and destination of the signal also provides a match. Either one of these signal lines with matched indicators may be substituted for the other signal line.

There are also only two types of interior signal lines with mismatched logic indicators that can connect two logic elements. These are shown in Fig. 6-13. The short perpendicular line placed across the signal line, where mismatching occurs, is called the mismatched symbol. Either one of these signal lines with mismatched indicators may be substituted for the other signal line. Mismatched logic indicators are not necessary, but may aid in interpreting diagrams since logic indication mismatches are a common source of errors in logic design.

Equivalent interior signal lines are used in the following ways. Circles (wedges if working with direct polarity indication) may be added to both ends of a signal line that has no circles, or removed from both ends if they already exist. A circle (wedge) may be moved on a signal line from one end of the signal line to the other end. Making either one of these changes will not effect the functionality of a diagram. The ability to make these equivalent substitutions on interior signal lines provides the designer with additional flexibility in choosing available elements when obtaining realizable logic diagrams.

6-3   IMPLEMENTING CIRCUITS
WITH NAND GATES AND NOR GATES

Digital designers using TTL usually prefer to use NAND functions as opposed to using AND, OR, or NOR functions. This philosophy has good merit since TTL has a larger variety of NAND functions than AND, OR, and NOR functions. In this section we discuss how to implement circuits using both NAND and NOR functions.

6-3-1   Implementing Logic Circuits with NAND Gates

As we indicated earlier, NAND and NOR gates often have faster propagation delay times than AND and OR gates. In faster speed applications, when every nanosecond counts, faster gates must be used. The diagram for function F shown in Fig. 6-8 is shown again in Fig. 6-14. The following changes have been made to convert the functional logic diagram into a realizable logic diagram: The interior signal lines are drawn with circles at both ends, and the circles at the inputs have been moved away from the inputs and made a part of Inverter symbols.

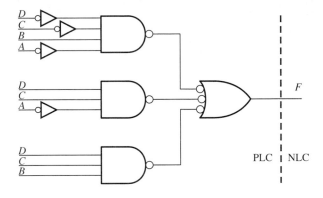

FIGURE 6-14

Realizable logic diagram implementation using NAND gates.

By making these simple equivalent signal line changes, and using the equivalent AND and OR element symbols for NAND gates, the functional logic diagram in Fig. 6-8 has been converted to a realizable logic diagram implementation using NAND gates.

The implementation using NAND gates shown in Fig. 6-14 is shown as a detailed logic diagram in Fig. 6-15. A function with n input signals will logically require at most only n Inverters on the input lines, since the outputs of those Inverters can be shared as needed by other inputs. In Fig. 6-15 only three Inverters are needed since the output of U1C may be shared, as shown in the figure. The power supply connection table and the spare gates are not included in Fig. 6-15 in order to keep the diagram simple.

6-3-2   Implementing Logic Circuits with NOR Gates

It is not necessary to obtain a Boolean function in product of sums form to obtain an implementation using NOR gates, as many digital design books suggest. The implementation using NOR gates shown in Fig. 6-16 is obtained by starting with the functional logic diagram in Fig. 6-8 and using equivalent signal lines and equivalent AND and OR element symbols for NOR gates. On each of the input signal lines that do not contain a circle we simply substitute the equivalent input signal line shown in Fig. 6-6b (an input signal line with an Inverter—Data book symbol—followed by a circle). On the output signal line we substitute the equivalent output signal line shown in Fig. 6-6d (an output signal line with a circle followed by an Inverter—DeMorgan equivalent symbol). The AND and OR element symbols that result from these simple signal line changes represent NOR gates. The circuit in Fig. 6-16, therefore, represents a realizable logic diagram implementation using NOR gates.

A corresponding detailed logic diagram is shown in Fig. 6-17. The 4-input NOR gate U2A used in Fig. 6-17 has a strobe input (pin 3) that we tied to +5 V so that the gate is always enabled, that is, so the gate always performs the function of a 4-input NOR gate. When the strobe input is tied to +5 V it is

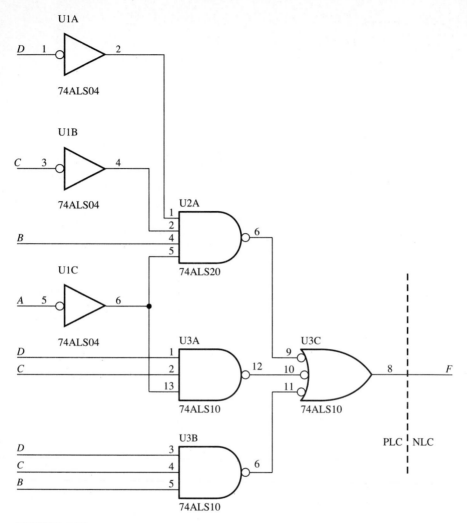

FIGURE 6-15
Detailed logic diagram implementation using NAND gates (ALS TTL devices).

being used as a fixed-mode input. Fixed-mode inputs will be covered in more detail a little later in this section.

Most circuits that utilize off-the-shelf TTL devices are implemented using either AND, NAND, or NOR gates since each of these types of gates is available with more than two inputs. As discussed earlier, OR gates are available with only two inputs.

Example 6-2

1. Obtain a realizable logic diagram implementation using NAND gates for the majority function represented by the following truth table.

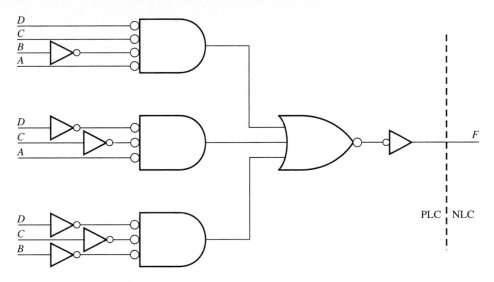

FIGURE 6-16
Realizable logic diagram implementation using NOR gates.

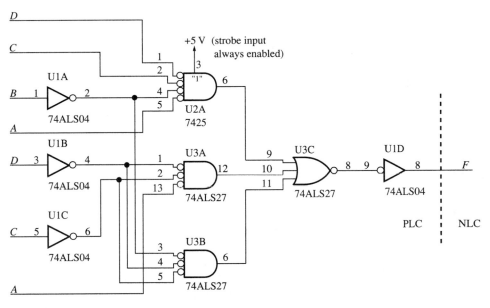

FIGURE 6-17
Detailed logic diagram implementation using NOR gates (ALS and STD TTL devices).

A	~B	C	F
0	0	0	0
0	0	1	0
0	1	0	0
0	1	1	1
1	0	0	0
1	0	1	1
1	1	0	1
1	1	1	1

2. Obtain a realizable logic diagram implementation using NOR gates for the majority function.

Solution

1. The Karnaugh map for the majority function is shown in Fig. E6-2a. From the Karnaugh map the following Boolean function, its signal list, and the corresponding design documentation can be written.

$$F = {\sim}B \cdot C \; + \; A \cdot C \; + \; A \cdot {\sim}B$$

$$\text{Signal list: } F, A, {\sim}B, C$$

LLP:	$F = {\sim}B \cdot C$		$+$	$A \cdot C$		$+$	$A \cdot {\sim}B$	
NIP:	1	1 1		1 1			1 1	

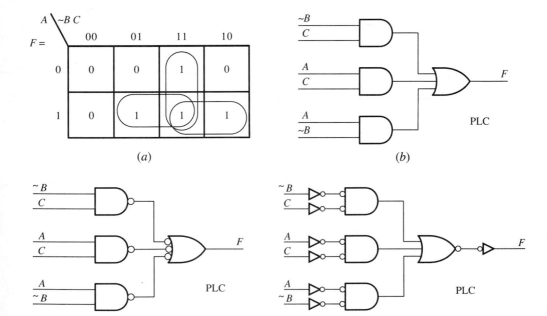

(a)   (b)

(c)   (d)

FIGURE E6-2

Using the design documentation, the functional logic diagram is drawn as shown in Fig. E6-2*b*. Using equivalent interior signal lines, an implementation using NAND gates for the majority function is shown in Fig. E6-2*c*.

2. Using equivalent input and output signal lines, an implementation using NOR gates for the majority function is shown in Fig. E6-2*d*.

When a diagram is converted from one form to another, one cannot always be assured that the gate types will exist with the proper number of inputs. For example, after checking in a data book we may find that the family of devices we prefer to use does not have 4-input NOR gates. At that point we back off and make another decision. Using equivalent signal lines for inputs *D*, *C*, and *A* of the 4-input AND element in Fig. 6-16 allows us to move each circle at these inputs and make them part of an Inverter. The signal line at input *B* in Fig. 6-16 is equivalent to a signal line without an Inverter. The equivalent diagram shown in Fig. 6-18 using a 4-input AND gate and three 3-input NOR gates illustrates the flexibility that a designer has when selecting available devices.

In the detailed logic diagram shown in Fig. 6-19, notice that there are four interior signal lines with mismatched logic indicators. A mismatched symbol has been placed across each of these lines as a reminder of each mismatched condition.

Using equivalent signal lines and equivalent AND and OR element symbols for the normally available devices provides us with a powerful tool when it comes time in the design process to choose the gate types to be utilized in the design. Once a functional logic diagram is obtained, we can substitute different gates without returning to the Boolean function to obtain an alternate form.

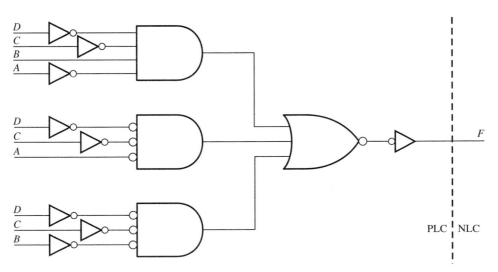

FIGURE 6-18
Realizable logic diagram implementation using NOR gates and an AND gate.

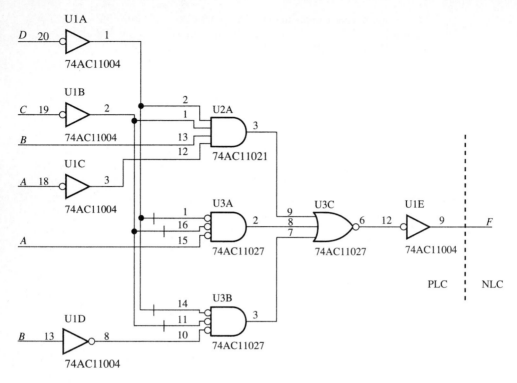

FIGURE 6-19
Detailed logic diagram implementation using NOR gates and an AND gate (advanced CMOS devices).

6-3-3 Using Gates with Paralleled and Fixed-Mode Inputs

If a gate with many inputs is available, it can be used with fewer inputs by paralleling one or more inputs as shown in Fig. 6-20. When a signal line has more than one input tied to it belonging to the same gate, then those inputs are paralleled inputs. The disadvantage of using paralleled input connections like the ones shown in Figs. 6-20*a* and *b* is twofold. First, the source of the input signal

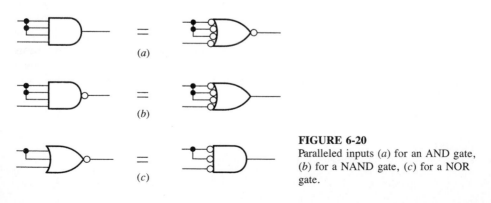

FIGURE 6-20
Paralleled inputs (*a*) for an AND gate, (*b*) for a NAND gate, (*c*) for a NOR gate.

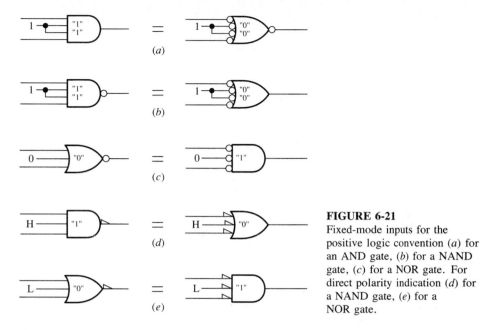

FIGURE 6-21
Fixed-mode inputs for the positive logic convention (*a*) for an AND gate, (*b*) for a NAND gate, (*c*) for a NOR gate. For direct polarity indication (*d*) for a NAND gate, (*e*) for a NOR gate.

for the paralleled inputs needs to provide more power than it would normally need to provide if it were connected to only one input. Finally, paralleled inputs can cause an increase in input capacitance in some logic families which results in an increase in noise coupling into the circuit.

The preferred method of using devices in a reduced input mode that does not require the source of the input signal to provide additional input power is shown in Fig. 6-21. The unused inputs are tied permanently to the logic level that will produce either (*a*) an internal logic 1-state for the AND element symbol of the device, or (*b*) an internal logic 0-state for the OR element symbol of the device. An internal logic 1-state for an AND element symbol and an internal logic 0-state for an OR element symbol allow the device to perform its represented function with the remaining signal inputs. The internal qualifying symbols "1" and "0" specify that the internal state always stands at its 1-state or 0-state respectively. Since the inputs with these qualifying symbols are fixed, they are called fixed-mode inputs. Knowing the internal fixed state and the external qualifying symbol, it is easy to determine the permanent assignment of each external logic state as shown in Figs. 6-21*a* through *c*. When fixed-mode inputs are used on diagrams using direct polarity indication, external logic levels are permanently assigned depending on the external qualifying symbol at the fixed-mode input as shown in Figs. 6-21*d* and *e*.

6-3-4   Universal Gates

Multiple inputs gates (both AND and OR gates) can also be reduced to a single signal input as shown in Fig. 6-22*a* using the preferred method of fixed-mode inputs. The resulting function is a Buffer. When multiple input NAND gates

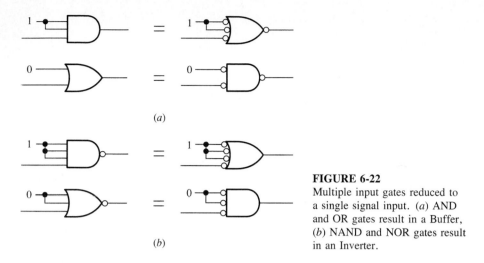

(a)

(b)

FIGURE 6-22
Multiple input gates reduced to
a single signal input. (*a*) AND
and OR gates result in a Buffer,
(*b*) NAND and NOR gates result
in an Inverter.

and multiple input NOR gates are reduced to a single signal input as shown in
Fig. 6-22*b*, we end up with the Inverter function. When unused NAND gates or
unused NOR gates are present in a circuit, these gates can be used productively
when the Inverter function is required. For this reason NAND gates and NOR
gates are often referred to as universal gates, since any switching circuit can be
implemented with only NAND gates or only NOR gates; that is, each type is
a functionally complete set. Refer to Comments on Functional Completeness in
subsection 1-8-4 for a more in-depth discussion of functionally complete sets.

6-3-5 Obtaining Larger Gate Functions

So far we have discussed how to utilize gates with more inputs than we need, but
what if we need a gate with more inputs than we either have access to or perhaps is
even available from IC manufacturers? In this case, a viable approach is to write
the Boolean function and partition the function according to the number of inputs
for the available devices. These partitioned functions can be used to construct the
larger gate function. The following 17-input NAND function is partitioned using
parenthesis so 8-input NAND gates can be use to build the larger function.

$$\overline{F} = (A \cdot B \cdot C \cdot D \cdot E \cdot G \cdot H \cdot I) \cdot (J \cdot K \cdot L \cdot M \cdot N \cdot O \cdot P \cdot Q) \cdot R$$

This 17-input NAND function is shown implemented in Fig. 6-23 using two
8-input NAND gates and three 3-input NOR gates. Two of the 3-input NOR gates
are used as Inverters by utilizing fixed-mode inputs. First, the functional logic
diagram for the 17-input NAND function is obtained; then equivalent signal lines
and equivalent AND and OR element symbols are used for the 8-input NAND
gates and 3-input NOR gates to obtain the diagram shown in Fig. 6-23.

Example 6-3. Many times certain devices may be available in a circuit as an unused
spare gate or Inverter. This occurs when multiple gates are available in a package

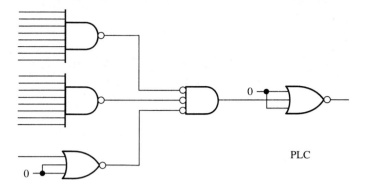

FIGURE 6-23
17-input NAND function.

and every gate is not currently used. Draw the circuits for the following situations. Assume that Inverters are available as needed.

(*a*) Draw a 3-input AND gate using a spare 3-input NAND gate.
(*b*) Draw a 3-input NAND gate using a spare 4-input AND gate.
(*c*) Draw a 2-input NOR gate using a spare 3-input NAND gate.
(*d*) Draw a 3-input OR gate using a spare 4-input NOR gate.

Solution

(*a*) First draw the 3-input AND gate (the function that is needed), and then use equivalent signal lines on the output to convert the AND gate to a circuit that uses the spare 3-input NAND gate as shown in Fig. E6-3*a*.

(*b*) The function that is needed is the 3-input NAND gate drawn in Fig. E6-3*b*. Using equivalent signal lines on the output and adding an input line, the NAND gate is converted to a circuit that uses the spare 4-input AND gate. The added input line is used as a fixed-mode input.

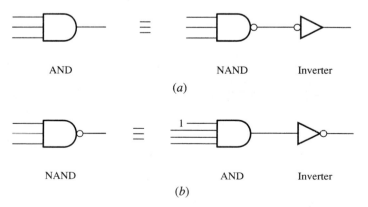

FIGURE E6-3

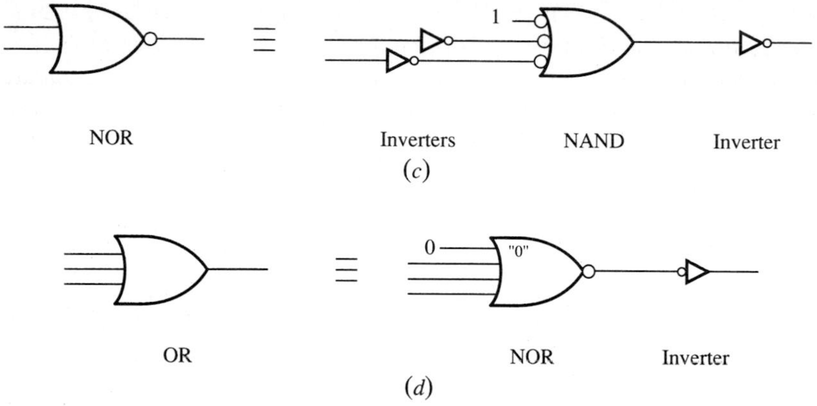

FIGURE E6-3

(c) First the 2-input NOR gate is drawn in data book form; then equivalent signal lines are drawn on both the inputs and the output to convert the NOR gate to a circuit that uses the spare 3-input NAND gate by adding an input as shown in Fig. E6-3c.

(d) Figure E6-3d shows an equivalent circuit that uses the spare 4-input NOR gate to obtain a 3-input OR gate.

In Figs. E6-3b, c, and d fixed-mode inputs are being used. Look closely at Fig. E6-3d. One of the NOR gate inputs has been tied to a logic 0 to allow the remaining three inputs to perform the NOR function. If the excess input had been tied to a logic 1, the output of the NOR gate would always be a logic 0, completely disabling the NOR gate. Care must be exercised when using fixed-mode inputs in order not to disable the function. Fixed-mode inputs must be connected to either a logic 0 or a logic 1 depending on the function being used. Floating lines generally pull to a logic 1 for TTL. If the spare 4-input NOR gate used in Fig. E6-3d was a TTL gate, an unconnected input would prevent the NOR gate from performing its function with the remaining inputs, and the output of the Inverter would be fixed at a logic 1. It helps to identify fixed-mode inputs on a schematic as we have illustrated in in Fig. E6-3d, but this is not always done.

6-4 DISTRIBUTED CONNECTIONS (DOT-AND AND DOT-OR)

Distributed connections are connections of certain logic elements used for the implementation of logic operations without the use of additional logic elements. Special logic circuits (L-type open-circuit outputs such as open-collector outputs) have the capacity to connect outputs with similar open-circuit outputs such that the distributed connection effectively represents an AND operation at the point of connection. Since the connection point is usually represented on a diagram as a dot, the AND operation can be appropriately referred to as a Dot-AND or Wired-AND. This type of circuitry is found in ICs using Transistor-Transistor Logic as shown in the simplified diagram in Fig. 6-24*a*. To review the basic switching operation performed by a bipolar transistor, refer to Fig. 1-5 in Chapter 1. The Dot-AND connection of two such ICs is shown in Fig. 6-24*b*. The rotated square with the line drawn below and touching its lower tip is the qualifying symbol used to denote an open-circuit output (L-type), where L-type means that the output is at a low level when the output transistor is turned on. Figure 6-24*b* illustrates that a pull-up resistor is required for the Dot-AND distributed connection to work properly. Usually the resistor is added externally, but it can also be supplied inside the IC. The rectangular-shape symbol for a Dot-AND distributed connection is shown in Fig. 6-24*c*.

Figure 6-25*a* illustrates an H-type open-circuit output that results in a Dot-OR or Wired-OR distributed connection. The circuit is the type found in Emitter-

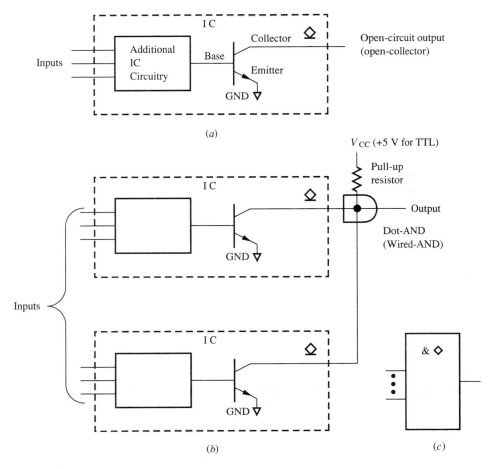

FIGURE 6-24
(*a*) Diagram for a device with a L-type open-circuit output, (*b*) a Dot-AND distributed connection formed by connecting two L-type open-circuit outputs to a pull-up resistor, (*c*) rectangular-shape symbol for a Dot-AND distributed connection.

Coupled Logic (ECL), one of the fastest bipolar transistor logic families available. In this case, a pull-down resistor is required for the Dot-OR distributed connection to work properly as shown in Fig. 6-25b. The output is at a high level when the output transistor is turned on. This is denoted by the qualifying symbol represented by a rotated square with a line drawn above and touching its upper tip, as illustrated in the diagram at the open-circuit output terminal of the IC. The

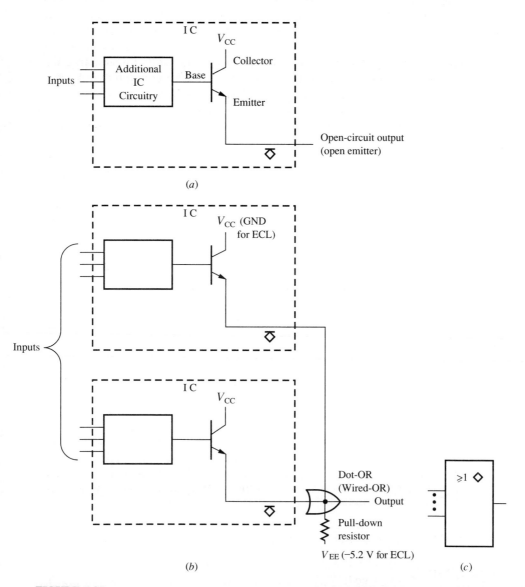

FIGURE 6-25
(a) Diagram for a device with a H-type open-circuit output, (b) a Dot-OR distributed connection formed by connecting two H-type open-circuit outputs to a pull-down resistor, (c) rectangular-shape symbol for a Dot-OR distributed connection.

rectangular-shape symbol for a Dot-OR distributed connection is shown in Fig. 6-25*c*.

Circuits using MOS transistors can also be designed with open-circuit outputs. Manufacturers of these ICs will generally use the qualifying symbols illustrated in Figs. 6-24 and 6-25 for an L-type or H-type output so that the user will understand which type of distributed connection (Dot-AND or Dot-OR) will be generated by using the IC. Manufacturers also provide in their data books the necessary information for determining the correct value for any required external resistors and the specified power supply voltage requirements for V_{CC} and V_{EE}.

The propagation delay times associated with ICs with open-circuit outputs (particularly open-collector outputs) tends to be much slower than the propagation delay times of ICs with normal active-device outputs (transistors used in place of pull-up or pull-down resistors). As a result, ICs with open-circuit outputs capable of being used to produce distributed connections are generally used in either slower applications or special applications that can produce a distributed AND function that might otherwise not be available when using ICs with normal active-device outputs. It should be noted that ICs with normal active-device outputs should never be connected together to perform distributed connections since these connections will surely damage, if not destroy, the ICs.

Open-circuit outputs can be used to perform Dot-AND and Dot-OR functions by using a distributed connection as we have discussed. However, special open-circuit outputs are also used in applications that require more current drive than a normal device output can supply. In these applications, the open-circuit outputs are usually called Drivers with open-collector outputs (nomenclature used by TTL manufacturers). These open-circuit output Drivers are only intended to be used with circuits requiring higher current applications, instead of in distributed-connection applications.

Example 6-4. For the functional logic diagram of the majority circuit shown in Fig. E6-2*b* in Example 6-2, show a realizable logic diagram using NAND gates with open-circuit outputs (TTL open-collector outputs).

Solution The procedure is to first obtain the functional logic diagram, as shown in Fig. E6-4*a*, and to then use equivalent signal lines and equivalent AND and OR element symbols to draw the circuit in a form that will accept ICs with open circuit outputs, as shown in Fig. E6-4*b*. The final realizable circuit is shown in Fig. E6-4*c* using ICs with open-collector outputs to perform the Dot-AND distributed connection. The Dot-AND distributed connection is represented by the OR form (DeMorgan equivalent symbol) shown in the figure.

6-5 USING AND-OR-INVERT ELEMENTS

After obtaining a functional logic diagram, a realizable logic diagram can sometimes be implemented using a circuit consisting of two or more AND elements feeding an OR element with a negated output. This type of circuit configuration is shown in Fig. 6-26, in two different configurations, and is called an AND-OR-

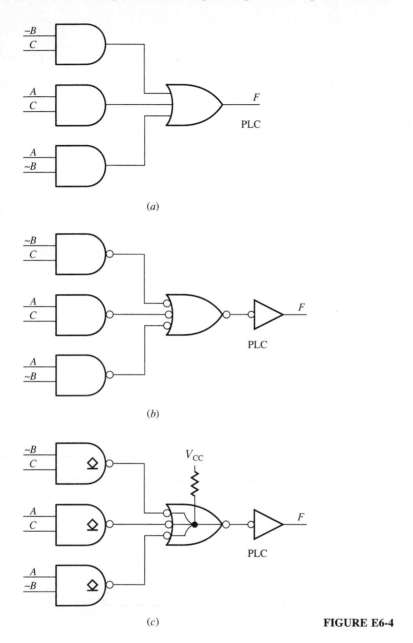

(a)

(b)

(c)

FIGURE E6-4

Invert (AOI) gate by IC manufacturers or an AND-OR-Invert element in IEEE Std 91. Figure 6-27a illustrates an AOI element using distinctive-shape symbols, while Fig. 6-27b shows the same configuration using a rectangular-shape symbol with qualifying symbols & and ≥1 to signify the respective logic functions that are performed by the individual rectangles making up the element.

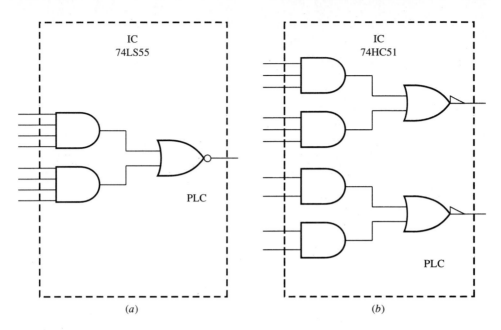

FIGURE 6-26
AND-OR-Invert elements (*a*) realizable logic diagram using the positive logic convention, (*b*) realizable logic diagram using direct polarity indication.

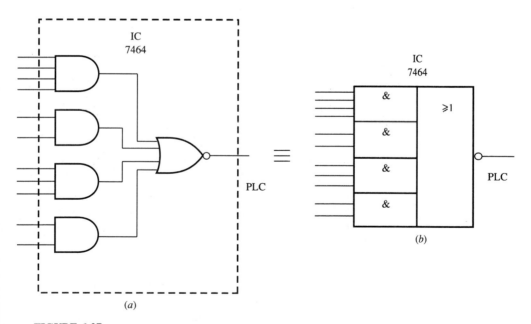

FIGURE 6-27
AND-OR-Invert element (*a*) drawn using distinctive-shape symbols, (*b*) drawn using a rectangular-shape symbol.

Several different configurations are available, as we have illustrated. Each differs by the number of AND elements and the number of inputs associated with each AND element. Some varieties of AND-OR-Invert elements also have expandable inputs to allow additional AND elements to be added. One advantage of using AOI elements to implement logic functions is package count reduction. Using AOI elements generally results in fewer packages than using separate AND gates, OR gates, and Inverters to implement the same circuitry.

The procedure for using AOI elements is to first obtain the functional logic diagram for the function being implemented and then to utilize manufacturer's data books to search for an AOI element that can be used to realize the circuit represented by the functional logic diagram. Since the gate circuitry of an AOI element cannot be changed, equivalent signal lines are changed on the functional logic diagram to organize the circuit so that an AOI element can be substituted in the circuit to implement the function, as illustrated in the following example.

Example 6-5. Design a realizable logic circuit diagram using an AOI element to implement the following Boolean function.

$$F(A, B, C, D) = \sum m(0, 5, 7, 11, 13, 15)$$

The signal list for the function is F, $\sim A$, B, $\sim C$, D. The function is mapped in Fig. E6-5a.

The available form of the function and its signal list is written as follows.

$$F = \sim A \cdot \overline{B} \cdot \sim C \cdot \overline{D} + B \cdot D + \overline{\sim A} \cdot \overline{\sim C} \cdot D$$

Signal list: F, $\sim A$, B, $\sim C$, D

The design documentation is written as

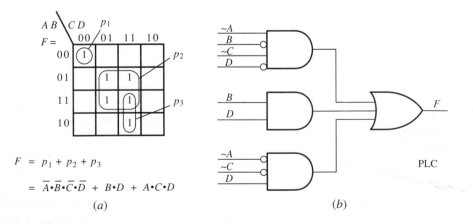

$$F = p_1 + p_2 + p_3$$

$$= \overline{A} \cdot \overline{B} \cdot \overline{C} \cdot \overline{D} + B \cdot D + A \cdot C \cdot D$$

(a)   (b)

FIGURE E6-5

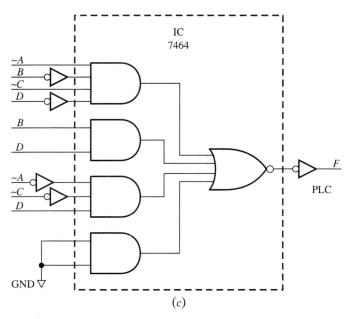

(c)

FIGURE E6-5

LLP:	$F = \sim A \cdot B \cdot \sim C \cdot D$					$+$	$B \cdot D$		$+$	$\sim A \cdot \sim C \cdot D$		
NIP:	1	1	0	1	0		1	1		0	0	1

The functional logic diagram for the function is shown in Fig. E6-5*b*.

The AOI element (IC 7464) shown in Fig. 6-27 has the required number of AND gates and AND-gate inputs for the circuit represented by the functional logic diagram. Equivalent signal lines are substituted, as shown in the Fig. E6-5*c* in the functional logic diagram. This allows the AOI element to be used to implement the function. The inputs to the unused 2-input AND gate of the AOI element are tied to logic 0 (ground) to disable the gate.

6-6 OBTAINING BOOLEAN FUNCTIONS FROM DIAGRAMS USING THE POSITIVE LOGIC CONVENTION

Suppose it is necessary to read someone else's logic diagrams in order to determine the Boolean functions represented by the diagrams. If a diagram is converted to a different form, are the two diagrams really equivalent? How does one check to see if a diagram obtained from the design documentation really represents the original Boolean function? In this section we will discuss how to obtain Boolean functions from diagrams using a single logic convention. The next section will discuss how to obtain Boolean functions from diagrams using direct polarity indication. The diagrams we present in this section contain available off-the-shelf devices, but the same technique is also valid for functional logic diagrams with unavailable devices.

6-6-1 Signal Flow through Circuits Using the Positive Logic Convention

Obtaining Boolean functions from diagrams using the positive logic convention is not difficult. The object is to start with the available signals at the inputs and proceed through the circuit, keeping track of each function being performed. The function each symbol performs is indicated by the outline of the symbol, if the diagram uses distinctive-shape symbols, or by the general qualifying symbols (refer to Figs. 6-4, and 6-7), if the diagram uses rectangular-shape symbols. When negation is performed, we should see either a negation-qualifying symbol or a partition between two different logic conventions. When a signal moves through a circuit in the direction of signal flow and encounters a negation symbol or a logic convention partition, the signal leaving the negation symbol or partition is negated with respect to the approaching signal, as illustrated by the examples shown in Fig. 6-28. We wrote the output signal for the Inverter symbol in Fig. 6-28*b* as $\sim\overline{A}$, that is, in available form, in order to preserve the identity of the available signal $\sim A$.

Note that the signal A on the negative logic section of the diagram in Fig. 6-28*d* is complemented when written on the positive logic section of the diagram. The reason may not be obvious until recalling that if the signal line labeled A goes low, for example, then that signal line must be a logic 1 on the negative

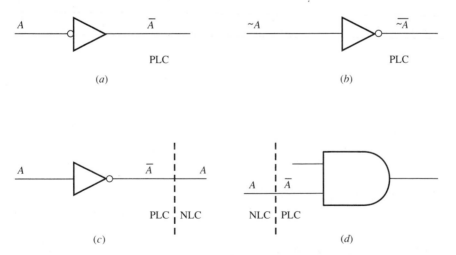

FIGURE 6-28
Examples of signals that experience logic negation, (*a*) through (*d*).

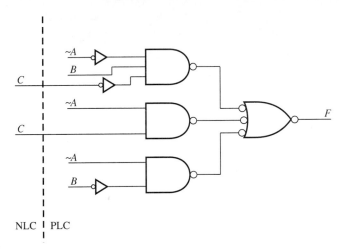

NLC ┊ PLC

FIGURE 6-29
Realizable logic diagram using the positive logic convention.

logic section of the diagram and a logic 0 on the positive logic section of the diagram. Since the signal line is labeled A on the negative logic section, it must of course be labeled $\overline{A}$ on the positive logic section to accommodate the logic convention change.

When several input signals are present on a diagram, each signal's path through the circuit must be considered in relation to every other signal and to the functions represented by the symbols in the circuit. Consider the diagram shown in Fig. 6-29.

The available output signal on the diagram in Fig. 6-29 is F. Following the signal flow for every input signal on the diagram shown in Fig. 6-29 from left to right, allowing each signal to be acted on by the functions represented by the symbols, and writing down the results as we go, we obtain the signals shown on the signal lines in Fig. 6-30. The Boolean expression for the output F of the circuit is written as follows.

$$\overline{\sim A \bullet B \bullet C \; + \; \sim A \bullet \overline{C} \; + \; \sim A \bullet \overline{B}}$$

This expression is now equated to the available output signal F as shown below.

$$F = \overline{\sim A \bullet B \bullet C \; + \; \sim A \bullet \overline{C} \; + \; \sim A \bullet \overline{B}}$$

Complementing both sides of the function results in the following sum of products expression for the complemented function, $\overline{F}$. The signal list is included to provide complete documentation.

$$\overline{F} = \overline{\sim A} \bullet B \bullet C \; + \; \sim A \bullet \overline{C} \; + \; \sim A \bullet \overline{B}$$

Signal list: $F, \sim A, B, C$[NL]

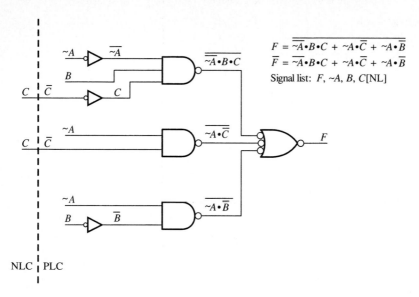

$$F = \overline{\sim A \cdot B \cdot C} + \sim A \cdot \overline{C} + \sim A \cdot \overline{B}$$

$$\overline{F} = \overline{\sim A \cdot B \cdot C} + \sim A \cdot \overline{C} + \sim A \cdot \overline{B}$$

Signal list: F, $\sim A$, B, C[NL]

FIGURE 6-30
Obtaining the Boolean function of a diagram that uses the positive logic convention.

After obtaining a Boolean function for a diagram, the intermediate signals that lead to the function should be removed from the diagram, since they have served their purpose.

Example 6-6. Obtain the Boolean function for the diagram shown in Fig. E6-6a.

Solution It is good design practice to match logic indicators on the signal lines between logic elements when drawing a logic diagram. The designer of the circuit shown in Fig. E6-6a obviously did not follow this recommendation. One way to approach this problem is to redraw the diagram and match the indicators. Another approach is to work directly from the diagram as presented. Following the latter approach, we show the signal flow in Fig. E6-6b. Across each signal line where a

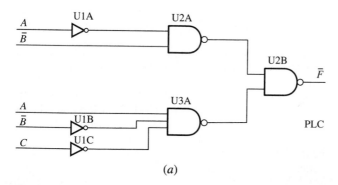

(a)

FIGURE E6-6

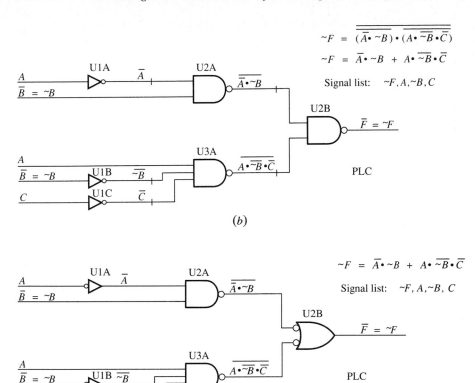

$$\sim F \ = \ \overline{(\overline{A} \bullet \sim B) \bullet (A \bullet \overline{\sim B} \bullet \overline{C})}$$

$$\sim F \ = \ \overline{A} \bullet \sim B \ + \ A \bullet \overline{\sim B} \bullet \overline{C}$$

Signal list: $\sim F, A, \sim B, C$

$$\overline{F} = \sim F$$

PLC

(b)

$$\sim F \ = \ \overline{A} \bullet \sim B \ + \ A \bullet \overline{\sim B} \bullet \overline{C}$$

Signal list: $\sim F, A, \sim B, C$

$$\overline{F} = \sim F$$

PLC

(c)

FIGURE E6-6   *(continued)*

logic indicator mismatch occurs, we place a mismatched symbol on the diagram in Fig. E6-6*b* to show that a mismatch exists. Since the designer used $\overline{B}$ as an input and $\overline{F}$ as an output, it should be obvious that $\overline{B}$ and $\overline{F}$ are both available signals which we prefer to represent as $\sim B$ and $\sim F$ respectively. By using $\sim B$ and $\sim F$, we can preserve the identity of these available signals when they must be complemented. After obtaining the expression for the output and equating it to the available output signal $\sim F$, DeMorgan's theorem is applied to the expression to obtain the Boolean function in sum of products form as shown in Fig. E6-6*b*. If we had complemented both sides of the expression and then applied DeMorgan's theorem to the expression on the right side of the function, we could have obtained the Boolean function in product of sums form.

    Figure E6-6*c* shows the same diagram drawn such that the logic indicators on the signal lines between the logic elements are all matched. This is accomplished using equivalent signal lines and equivalent AND and OR element symbols for the normally available devices. To provide matching indicators on the signal lines between the Inverters and the AND element symbols, the circles (negation symbols) on the outputs of the Inverters are moved to the inputs. The expression for the output is now obtained from the signal flow and equated to the available output signal $\sim F$. The output expression for the function $\sim F$ is shown in Fig. E6-6*c*.

6-7 OBTAINING BOOLEAN FUNCTIONS FROM DIAGRAMS USING DIRECT POLARITY INDICATION

Signal names entered on diagrams using direct polarity indication can be composed of two parts. The first part of such signal names consists of the usual abbreviation of a statement called the signal state, while the second part consists of an indication of which logic level corresponds to a signal state of 1. The syntax used for two part signal names is shown below.

Signal state	(Level indication)
A	(H)
$\overline{A}$	(L)
$\sim B$	(H)
$\overline{\sim B}$	(L)
C	(L)
$\overline{C}$	(H)
$\sim D$	(L)
$\overline{\sim D}$	(H)

The signal state and the level indication are then written as a single unit by putting the two parts together as A(H), $\overline{A}$(L), $\sim B$(H), $\overline{\sim B}$(L), C(L), $\overline{C}$(H), $\sim D$(L), and $\overline{\sim D}$(H) respectively.

6-7-1 Equivalent Signals

Notice that the signals are shown in sets of two. The signal state of the first signal name in each set is uncomplemented, while the signal state of the second signal name is complemented. Also notice that the level indication for each set differs by a level inversion. It can easily be shown that the signal names in each of the above sets are equivalent and may be represented by the following identities.

$$A(\text{H}) = \overline{A}(\text{L})$$

$$\sim B(\text{H}) = \overline{\sim B}(\text{L})$$

$$C(\text{L}) = \overline{C}(\text{H})$$

$$\sim D(\text{L}) = \overline{\sim D}(\text{H})$$

For the signal name A(H), A is the signal state and also the available signal name; however, for the signal name $\overline{A}$(L), $\overline{A}$ is the signal state, while the available signal name is A.

The signals $A(H)$ and $\bar{A}(L)$ are representations for the same signal. We can demonstrate this by writing the logic statement for each signal and then comparing the results as follows.

1. The signal $A(H)$ means: If the signal state $A = 1$, then the signal is high, else the signal is not high.
2. The signal $\bar{A}(L)$ means: If the signal state $\bar{A} = 1$ ($A = 0$), then the signal is low, else the signal is not low.

Organizing the logic statement for each signal into the following table format clearly shows that the two signals $A(H)$ and $\bar{A}(L)$ behave in the same manner and are therefore identical.

	Signal level high	Signal level low
$A(H)$	$A = 1$	$A = 0$
$\bar{A}(L)$	$\bar{A} = 0$ ($A = 1$)	$\bar{A} = 1$ ($A = 0$)

In a similar fashion we can also prove the other three identities listed above. These proofs are left as an exercise for the student.

The following statement applies to each of the four identities we have been discussing. "A signal name that can be derived by applying both logic negation [to the statement part] *and* level inversion [to the level indication part] to an existing signal name is equivalent to the existing signal name and therefore shall not be used to identify a different signal (see ANSI/IEEE Std 991)." As we will show, identities play a key role in writing Boolean functions for diagrams using direct polarity indication.

Obtaining Boolean functions from diagrams using direct polarity indication can be easier and lead to better understanding than obtaining Boolean functions from diagrams using a single logic convention. This is because a signal name using level indication may be written in more that one way on a particular signal line. It is easy to obtain Boolean functions for diagrams using direct polarity indication when the input signal names are chosen such that their level indications match the polarity indications at the inputs and outputs of the symbols in the circuits.

As you are now aware, the method we presented to obtain diagrams using direct polarity indication did not consider signal names with level indication. In Chapter 5 diagrams were drawn using direct polarity indication, and signal names were used without a level indication part as illustrated by the following available signal list.

Signal list: A, $\sim B$, $C[\text{NL}]$, $\sim D[\text{NL}]$

We wrote A to represent $A(H)$ and did not mention the level indication part of the signal name. We also wrote $\sim B$ to represent $\sim B(H)$. The signal name $C[\text{NL}]$

Available signals		Equivalent forms using level indication		
(a)	A	=	A (H)	=	$\overline{A}$ (L)
(b)	~B	=	~B (H)	=	$\overline{~B}$ (L)
(c)	C [NL]	=	C (L)	=	$\overline{C}$ (H)
(d)	~D [NL]	=	~D (L)	=	$\overline{~D}$ (H)

FIGURE 6-31
Available signals and their equivalent forms using level indication (*a*) through (*d*).

was written to indicate a negative logic signal, but it also represents C(L). The signal name ~D[NL], written to represent a negative logic signal, also represents ~D(L). Figure 6-31 shows each of the available signals for the previous signal list and its corresponding equivalent form using level indication.

When a signal name with level indication is used in a diagram using direct polarity indication, it is good practice to choose the level indication parts to agree with the polarity indication on the signal line where the signal name is placed. This may be accomplished by using one of the identities shown in Fig. 6-31.

Equivalent signals that are used on a diagram as an aid to determine the Boolean function of the diagram should be removed or crossed off after they have served their purpose. This is because it is not good practice to write a signal name on a signal line in multiple forms even though the forms are equivalent.

6-7-2   Signal Flow through Circuits Using Direct Polarity Indication

To obtain the Boolean function for a diagram using direct polarity indication, we start with the available signals at the inputs and proceed through the circuit, keeping track of each function being performed on the input signals. The function being performed by a symbol is indicated by the outline of the symbol, if the diagram uses distinctive-shape symbols, or by the general qualifying symbols (refer to Figs. 6-4 and 6-7), if the diagram uses rectangular-shape symbols. A symbol performs its function on the input signal names whose level indications are the same as the polarity indications at the inputs to the symbol. When a signal moves through a circuit in the direction of signal flow, an equivalent signal name may be chosen to allow the level indication of the signal to agree with the polarity indication at each input as illustrated by the examples shown in Fig. 6-32. When the level indication of a signal name agrees with the polarity indication at the input of an Inverter and the signal flows through an Inverter, the signal experiences a level inversion; that is, the statement part of the signal name remains the same, but its level indication part is inverted on the output of each Inverter as shown in Fig. 6-32.

Since the polarity indication at the input to the Inverter symbol is active low in Fig. 6-32*a*, the signal A is written in its equivalent form A(H). This is then converted to the equivalent form $\overline{A}$(L) so that the low level indication of the

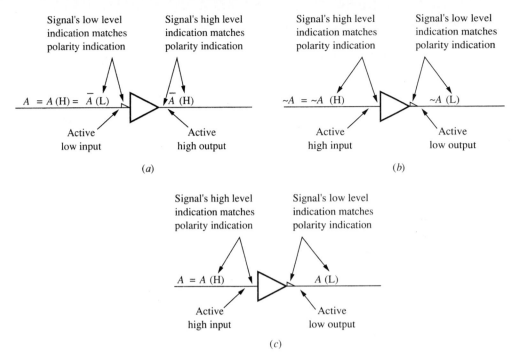

FIGURE 6-32
Examples of signals that experience level inversion, (*a*) through (*c*).

signal name matches the active low input of the Inverter symbol. Now the signal $\overline{A}$(L) may be thought of as moving through the Inverter symbol with the result that the signal $\overline{A}$ experiences only a level inversion. The level of the signal $\overline{A}$ on the output of the Inverter must exist at a high level to match the active high output of the Inverter symbol. Figure 6-32 shows each input signal rewritten in the form required by the polarity indication at the input of the symbol where it is used. With practice, it is not necessary to rewrite each input signal in its various equivalent forms, since this can be done by inspection.

While using the technique of matching the level indication of each signal with the polarity indication at each input and output, consideration must be given to the path of each input signal through the circuit, all the way to the output, in relation to every other signal and to the functions represented by the symbols in the circuit. Consider the realizable logic diagram using direct polarity indication shown in Fig. 6-33.

The available output signal on the diagram in Fig. 6-33 is $F = F$(H) $= \overline{F}$(L). Following the signal flow on the diagram for each input signal from left to right, allowing each signal to be acted on by the functions represented by the symbols, and writing down the results as we go, we obtain the signals shown on the signal lines in Fig. 6-34. In each case the level indication appended to a Boolean output expression is selected to agree with the logic indication at the

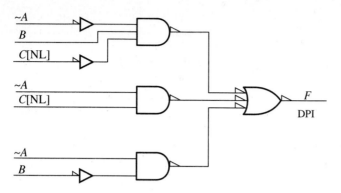

FIGURE 6-33
Realizable logic diagram using direct polarity indication.

output of the symbol. The Boolean expression $\overline{F}(L)$ for the output of the circuit is written as follows.

$$(\overline{\sim A} \cdot B \cdot C \; + \; \sim A \cdot \overline{C} \; + \; \sim A \cdot \overline{B})(L)$$

This expression is then equated to the available output signal $\overline{F}(L)$ to obtain the following Boolean function with level indication.

$$\overline{F}(L) = (\overline{\sim A} \cdot B \cdot C \; + \; \sim A \cdot \overline{C} \; + \; \sim A \cdot \overline{B})(L)$$

Rewriting both sides of the Boolean function using identities,

$$F(H) = (\overline{\sim A \cdot B \cdot C \; + \; \sim A \cdot \overline{C} \; + \; \sim A \cdot \overline{B}})(H)$$

so, $$F = (\overline{\sim A \cdot B \cdot C \; + \; \sim A \cdot \overline{C} \; + \; \sim A \cdot \overline{B}})$$

Complementing both sides of the function F results in the following sum of products expression for the complemented function $\overline{F}$. The signal list is also provided to assist in documenting the results.

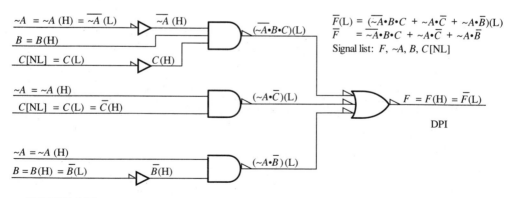

FIGURE 6-34
Obtaining the Boolean function of a diagram that uses direct polarity indication.

$$\overline{F} = \overline{\sim A} \cdot B \cdot C \;+\; \sim A \cdot \overline{C} \;+\; \sim A \cdot \overline{B}$$

Signal list: F, $\sim A$, B, C[NL]

Like any algebra, once the complete procedure is presented, a shortcut can usually be found. After forming the Boolean function using the output signal name with level indication and the Boolean expression with level indication, the level indication part can just be crossed out, that is, (L) in this case, on both sides of the equation. This is the procedure used in Fig. 6-34.

After obtaining the Boolean function for a diagram, the intermediate signals that lead to the function should be removed, having already served their purpose.

Example 6-7. Obtain the Boolean function for the diagram shown in Fig. E6-7*a*.

Solution It is good design practice to draw each logic element with the symbol that best depicts the logic function performed by the element in the system. This is achieved by choosing matching logic indicators on each signal line between the logic elements. The designer of the circuit in Fig. E6-7*a* unfortunately did not follow this recommendation. One way to approach the current problem is to redraw the diagram using equivalent symbols to effectively provide matching polarity indicators on each signal line as shown in Fig. E6-7*b*. Since all the signal lines in Fig. E6-7*b*

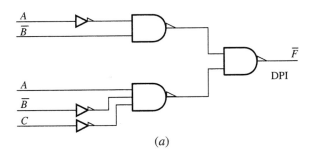

(*a*)

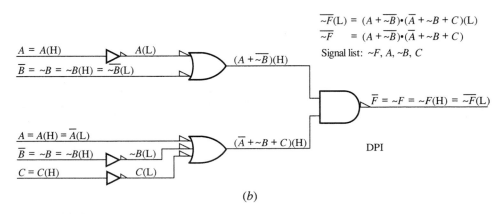

$$\overline{\sim F}(L) = (A + \overline{\sim B}) \cdot (\overline{A} + \sim B + C)(L)$$
$$\overline{\sim F} \;\; = (A + \overline{\sim B}) \cdot (\overline{A} + \sim B + C)$$

Signal list: $\sim F$, A, $\sim B$, C

(*b*)

FIGURE E6-7

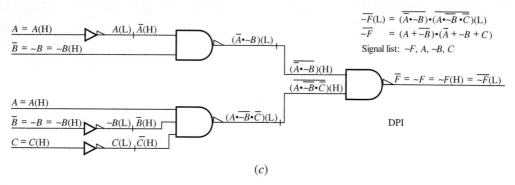

$$\sim\!\overline{F}(L) = \overline{(\overline{A}\bullet\sim\!B)}\bullet\overline{(A\bullet\sim\!B\bullet\overline{C})}(L)$$
$$\sim\!\overline{F} = (A + \sim\!\overline{B})\bullet(\overline{A} + \sim\!B + C)$$

Signal list: $\sim\!F$, A, $\sim\!B$, C

$\overline{F} = \sim\!F = \sim\!F(H) = \sim\!\overline{F}(L)$

DPI

(c)

FIGURE E6-7 (*continued*)

have matching logic indicators, the Boolean function is obtained as shown in Fig. E6-7*b* in a straightforward manner.

Another approach to the problem represented by the diagram in Fig. E6-7*a* is shown in Fig. E6-7*c*. Across each signal line where a logic indicator mismatch occurs, a mismatched symbol is used to show that a mismatch exists. The appropriate equivalent signal is written on each section of each mismatched signal line, as shown in Fig. E6-7*c* such that its level indication is consistent with the logic indicator at the output or the input of the logic element associated with the respective line. Notice that the problem of mismatched polarity indication is clearly undesirable. It is not a clean solution like the solution shown in Fig. E6-7*b* since both logic negation and level inversion must be used to obtain the Boolean function.

To help preserve the identity of the available signal $\sim\!B$ when the signal must be negated, observe that the available signal name $\sim\!B$ is used rather than $\overline{B}$.

As another example, let's find the Boolean function for the output for the diagram shown in Fig. 6-35 using signals with level indication. We can then convert the diagram into a diagram using the positive logic convention and find the Boolean function using signals without level indication. The resulting functions should be the same.

Referring to Fig. 6-35, it can be seen that the diagram has matching polarity indicators on all the signal lines between the logic symbols. This makes our job easier since the designer did a good job of organizing the drawing. Figure 6-36 shows the solution for the Boolean function for the diagram using direct polarity indication. Once the Boolean expression for the output is determined, it is equated to the output signal written in the form with the same level indication as the polarity indication at the output. The resulting Boolean function and its corresponding signal list is shown in Fig. 6-36. Note that an available input or output signal with a high level indication is a positive logic signal, while an available input or output signal with a low level indication is a negative logic signal.

The diagram using direct polarity indication shown in Fig. 6-35 is converted to a diagram using the positive logic convention as shown in Fig. 6-37. To ob-

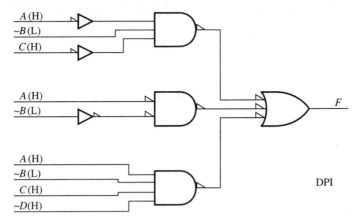

FIGURE 6-35
Realizable logic diagram using direct polarity indication.

tain the symbols for the diagram using a single logic convention requires substituting a circle (the negation symbol) for each wedge (the polarity symbol) in Fig. 6-35. Each signal name shown with level indication in Fig. 6-35 must be converted to its equivalent signal name without level indication. Listing the input and output signals without level indication on the diagram shown in Fig. 6-37 and recording the signal flow on each signal line throughout the circuit, we obtain the Boolean function and the signal list shown in Fig. 6-37. As you can observe from Figs. 6-36 and 6-37 the Boolean equations and their corresponding signal lists are the same as we would expect. This exercise shows that the two separate procedures presented for obtaining Boolean functions for diagrams using the positive logic convention and direct polarity indication are indeed consistent.

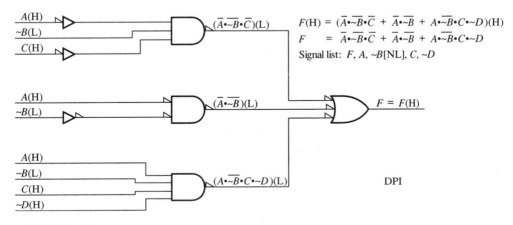

FIGURE 6-36
Obtaining the Boolean function of a diagram that uses direct polarity indication.

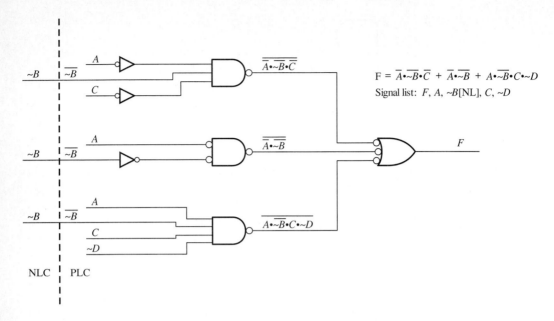

FIGURE 6-37
Obtaining the Boolean function of a diagram that uses direct polarity indication by first converting the diagram to a diagram that uses the positive logic convention.

REFERENCES

1. ANSI/IEEE Std 991-1986, *IEEE Standard for Logic Circuit Diagrams,* The Institute of Electrical and Electronic Engineers, New York, 1986.
2. ANSI/IEEE Std 91-1984, *IEEE Standard Graphic Symbols for Logic Functions,* The Institute of Electrical and Electronic Engineers, New York, 1984.
3. IEC STANDARD Publication 617-12, *Graphic Symbols for Diagrams, Part 12: Binary logic elements,* International Electrotechnical Commission, Geneva, 1983.
4. *The TTL Data Book Volume 1,* Texas Instruments, Dallas, Texas, 1984.
5. *Data Book, Low Power Schottky TTL ICs,* 3rd Ed, SGS-Semiconductor Corporation, Phoenix, AZ, June 1985.
6. *Bipolar Digital IC, ALSTTL,* Mitsubishi Electric Corporation, Marunouchi, Tokyo, 1985.
7. *Logic, TC74HC Series,* Toshiba America, Tustin, CA, March 1986.
8. *Logic Databook Volume II,* National Semiconductor Corporation, Santa Clara, CA, 1984.
9. *Advanced CMOS Logic Data Book,* Texas Instruments, Dallas, Texas, 1988.
10. *ALS/AS Logic Data Book,* Texas Instruments, Dallas, Texas, 1986.
11. *Fast Data Manual,* Signetics Corporation, Sunnyvale, CA, 1987.
12. *Schematic Design Tools, OrCAD/SDT III,* OrCAD Systems, Hillsboro, Oregon, 1987.
13. *Schema II User's Manual,* Omation, Richardson, Texas, 1987.
14. *EE Designer User's Guide,* Version 1.7, Visionics Corp., Sunnyvale, CA, 1987.
15. *EE Designer Reference Manual,* Visionics Corp., Sunnyvale, CA, 1987.

PROBLEMS

Section 6-2 Obtaining Realizable IC Circuits

6-1. Obtain the AND element symbol and the equivalent OR element symbol for the positive logic convention for a 2-input OR function, starting with the truth table for the function. Assume that the available input signals are A and B and the available output signal is F.

6-2. Obtain the AND element symbol and the equivalent OR element symbol using direct polarity indication for a 3-input NAND function, starting with the truth table for the function. Assume that the available input signals are X, Y, and Z and the available output signal is F.

6-3. Obtain the AND element symbol and the equivalent OR element symbol for the following 4-input NOR function.

$$F = \overline{A + B + C + D}; \text{ Signal list: } F, A, B, C, D$$

6-4. Obtain the data book symbol and the DeMorgan equivalent symbol for a 2-input AND function given the available input signals X and Y and the available output signal F.

6-5. Construct a table consisting of distinctive-shape data book symbols using direct polarity indication for 2-input AND, OR, NAND, and NOR gates.

6-6. Construct a table consisting of rectangular-shape data book symbols using the positive logic convention for 2-input AND, OR, NAND, and NOR gates.

6-7. Construct a table consisting of distinctive-shape DeMorgan equivalent symbols using the positive logic convention for 2-input AND, OR, NAND, and NOR gates.

6-8. Construct a table consisting of rectangular-shape DeMorgan equivalent symbols using direct polarity indication for 2-input AND, OR, NAND, and NOR gates.

6-9. Convert a positive logic OR gate to the equivalent negative logic gate type.

6-10. Convert a positive logic NAND gate to the equivalent negative logic gate type.

6-11. Convert a negative logic NOR gate to the equivalent positive logic gate type.

6-12. For each of the circuits shown in Fig. P6-12 draw an equivalent circuit such that the logic indicators on the signal lines between the logic elements are matched. Use equivalent Inverter symbols.

6-13. For each of the circuits shown in Fig. P6-13 draw an equivalent circuit such that the logic indicators on the signal lines between the logic elements are matched. Use equivalent Inverter symbols.

6-14. For the AND element symbol shown in Fig. P6-14 draw the circuits that result from the following substitutions, and name the normally available gate type for each circuit.

(*a*) Substitute an equivalent signal line for input 1.
(*b*) Substitute an equivalent signal line for input 2.
(*c*) Substitute equivalent signal lines for input 1 and output 3.
(*d*) Substitute equivalent signal lines for input 2 and output 3.

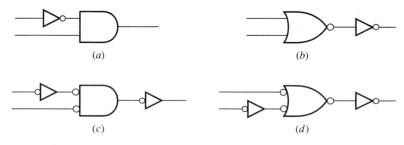

(*a*) (*b*)

(*c*) (*d*)

FIGURE P6-12

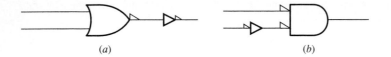

FIGURE P6-13

6-15. For the AND element symbol shown in Fig. P6-15 draw the circuits that result from the following substitutions, and name the normally available gate type for each circuit.
(*a*) Substitute an equivalent signal line for input 1.
(*b*) Substitute an equivalent signal line for input 2.
(*c*) Substitute equivalent signal lines for input 1 and output 3.
(*d*) Substitute equivalent signal lines for input 2 and output 3.

6-16. For the AND element symbol shown in Fig. P6-16 draw the circuits that result from the following substitutions, and name the normally available gate type for each circuit.
(*a*) Substitute an equivalent signal line for input 1.
(*b*) Substitute an equivalent signal line for input 2.
(*c*) Substitute equivalent signal lines for input 1 and output 3.
(*d*) Substitute equivalent signal lines for input 2 and output 3.

6-17. For the AND element symbol shown in Fig. P6-17 draw the circuits that result from the following substitutions, and name the normally available gate type for each circuit.
(*a*) Substitute an equivalent signal line for input 1.
(*b*) Substitute an equivalent signal line for i9put 2.
(*c*) Substitute equivalent signal lines for input 1 and output 3.
(*d*) Substitute equivalent signal lines for input 2 and output 3.

6-18. Use the AND element symbol shown in Fig. P6-14 and equivalent signal lines to draw an equivalent circuit using each of the following gates.
(*a*) an AND gate
(*b*) an OR gate
(*c*) a NAND gate
(*d*) a NOR gate

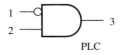

PLC

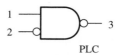

PLC

FIGURE P6-14   **FIGURE P6-15**

PLC

FIGURE P6-16

PLC

FIGURE P6-17

6-19. Use the AND element symbol shown in Fig. P6-15 and equivalent signal lines to draw an equivalent circuit using each of the following gates.
(*a*) an AND gate
(*b*) an OR gate
(*c*) a NAND gate
(*d*) a NOR gate

6-20. Use the AND element symbol shown in Fig. P6-16 and equivalent signal lines to draw an equivalent circuit using each of the following gates.
(*a*) an AND gate
(*b*) an OR gate
(*c*) a NAND gate
(*d*) a NOR gate

6-21. Use the AND element symbol shown in Fig. P6-17 and equivalent signal lines to draw an equivalent circuit using each of the following gates.
(*a*) an AND gate
(*b*) an OR gate
(*c*) a NAND gate
(*d*) a NOR gate

Section 6-3 Implementing Circuits with NAND Gates and NOR Gates

6-22. Draw each of the circuits shown in Fig. P6-22 as an implementation using NAND gates. Substitute only equivalent signal lines.

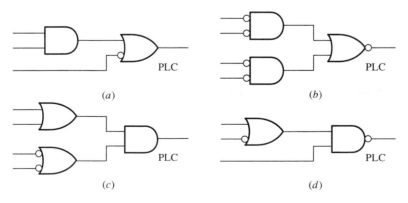

FIGURE P6-22

6-23. Draw each of the circuits shown in Fig. P6-22 as an implementation using NOR gates. Substitute only equivalent signal lines.

6-24. For each of the circuits shown in Fig. P6-22, obtain an equivalent circuit using direct polarity indication. Use only NAND gates and Inverters.

6-25. For each of the circuits shown in Fig. P6-22, obtain an equivalent circuit using direct polarity indication. Use only NOR gates and Inverters.

6-26. Draw each of the circuits shown in Fig. P6-22 as an implementation using both AND and OR gates. Substitute only equivalent signal lines.

6-27. Draw each of the circuits shown in Fig. P6-22 as an implementation using AND gates. Substitute only equivalent signal lines. Why is an implementation using AND gates generally not a good idea?

6-28. Draw each of the circuits shown in Fig. P6-22 as an implementation using OR gates. Substitute only equivalent signal lines. Why is an implementation using OR gates generally not a good idea?

6-29. For the function $F(X,Y,Z) = \sum m(1,2,4)$ with signal list: F,X,Y,Z solve for the SOP form for the 1s of the function. Obtain implementations using the positive logic convention for the following type of gates. Assume that Inverters are available.
   (*a*) NAND gates
   (*b*) NOR gates
   (*c*) AND and OR gates

6-30. For the function $F(X,Y,Z) = \sum m(1,2,4)$ with signal list: F,X,Y,Z, solve for the SOP form for the 0s of the function. Obtain implementations using the positive logic convention for the following type of gates. Assume that Inverters are available.
   (*a*) NAND gates
   (*b*) NOR gates
   (*c*) AND and OR gates

6-31. For the function $\sim F(\sim A,B,\sim C) = \sum m(0,1,2,3,5)$ with signal list: $\sim F,\sim A,B,\sim C$, solve for the SOP form for the 1s of the function. Obtain implementations using direct polarity indication for the following type of gates. Assume that Inverters are available.
   (*a*) NAND gates
   (*b*) NOR gates
   (*c*) AND and OR gates

6-32. For the function $\sim F(\sim A,B,\sim C) = \sum m(0,1,2,3,5)$ with signal list: $\sim F,\sim A,B,\sim C$, solve for the SOP form for the 0s of the function. Obtain implementations using direct polarity indication for the following type of gates. Assume that Inverters are available.
   (*a*) NAND gates
   (*b*) NOR gates
   (*c*) AND and OR gates

6-33. Using the positive logic convention, draw a 3-input AND gate as a 2-input AND gate, and a 3-input OR gate as a 2-input OR gate, using both AND and OR element symbols.
   (*a*) using paralleled inputs
   (*b*) using fixed-mode inputs

6-34. Using the positive logic convention, draw a 4-input NAND gate as a 3-input NAND gate, and a 4-input NOR gate as a 3-input NOR gate using both AND and OR element symbols.

(*a*) using paralleled inputs

(*b*) using fixed-mode inputs

6-35. Using direct polarity indication, draw a 3-input AND gate as a 2-input AND gate, and a 3-input OR gate as a 2-input OR gate using both AND and OR element symbols.

(*a*) using paralleled inputs

(*b*) using fixed-mode inputs

6-36. Using direct polarity indication, draw a 4-input NAND gate as a 3-input NAND gate, and a 4-input NOR gate as a 3-input NOR gate using both AND and OR element symbols.

(*a*) using paralleled inputs

(*b*) using fixed-mode inputs

Note: For Probs. 6-37 through 6-44 utilize the preferred method of using a device in a reduced input mode.

6-37. Show an implementation for a 7-input NOR gate using available 3-input OR gates. Inverters are also available.

6-38. Show an implementation for a 6-input NAND gate using available 4-input AND gates. Inverters are also available.

6-39. Show an implementation for a 6-input AND gate using available 3-input NAND gates. Inverters are also available.

6-40. Show an implementation for a 3-input OR gate using available 2-input NOR gates. Inverters are also available.

6-41. Show an implementation for a 4-input NOR gate using available 2-input NAND gates. Inverters are also available.

6-42. Show an implementation for a 5-input AND gate using available 2-input AND gates. Inverters are also available.

6-43. Show an implementation for a 2-input AND gate using each of the following spare gates. Assume Inverters are available.

(*a*) a 4-input NAND

(*b*) a 3-input NOR

(*c*) a 2-input OR

6-44. Show an implementation for a 3-input OR gate using each of the following spare gates. Assume Inverters are available.

(*a*) a 4-input AND

(*b*) a 3-input NOR

(*c*) a 4-input NAND

Section 6-4   Distributed Connections (Dot-AND and Dot-OR)

6-45. Design a circuit using a distributed connection that will implement the following function using TTL open collector NAND gates. Assume that normal active-device output Inverters are available.

$$F = A \cdot \overline{B} \ + \ \overline{A} \cdot B \ + \ \overline{C}$$

Signal list: F, A, B, C

6-46. Implement the following function using a distributed connection in the design. TTL open-collector AND gates are available in addition to normal active-device output Inverters. Use direct polarity indication in the diagram.

$$F = W \cdot X \cdot \overline{Y} \ + \ X \cdot \overline{\sim Z} \ + \ X \cdot \overline{Y} \ + \ W \cdot Y \cdot \overline{Z}$$

Signal list: $F, W, X, Y, \sim Z$

6-47. Design a circuit using a distributed connection that will implement the following function. Use the ECL open emitter device IC1 shown in Fig. P6-47 called a Quad 2-input NOR. Notice that one of the NOR symbols in the package has both complemented and uncomplemented outputs. IC2 also shown in Fig. P6-47 contains six Inverters that are functional when the negated single-input OR symbol is tied to a logic 1. IC2 is referred to by ECL manufacturers as a Hex Inverter with Enable.

$$F = A \cdot \overline{B} \ + \ \overline{A} \cdot B \ + \ \overline{C}$$

Signal list: F, A, B, C

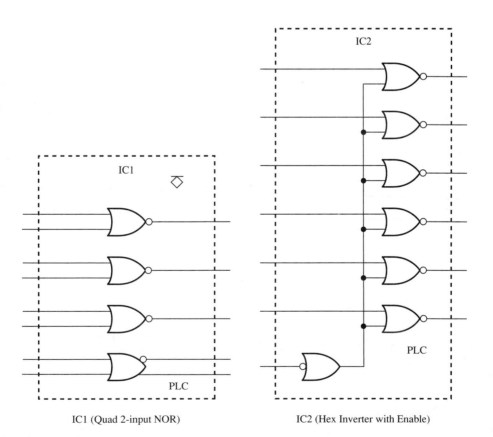

IC1 (Quad 2-input NOR)   IC2 (Hex Inverter with Enable)

FIGURE P6-47

Section 6-5 Using AND-OR-Invert Elements

6-48. Draw the rectangular-symbol for each of the AOI elements shown in Figs. P6-48a through c.

6-49. Design a circuit to implement the following truth table. Use one of the AOI elements in Fig. P6-48 and necessary Inverters to realize the design. Assume that all the signals are positive logic and uncomplemented.

A	B	C	F
0	0	0	1
0	0	1	0
0	1	0	1
0	1	1	0
1	0	0	0
1	0	1	1
1	1	0	1
1	1	1	0

6-50. Design a circuit to decode the all 0s and the all 1s patterns in a 4-bit binary number using one of the AOI elements shown in Fig. P6-48. Use Inverters as necessary. Draw the diagram using rectangular-shape symbols using direct polarity indication. Assume that all signals are uncomplemented positive logic signals.

6-51. Design a circuit to realize the following logic functions using the AOI element shown in Fig. P6-48c. Assume that all the signals are uncomplemented positive logic signals, and use Inverters as necessary.

$$\overline{F1} = \overline{X} \cdot \overline{Y} \cdot Z + X \cdot Y$$
$$\overline{F2} = \overline{Y} + X \cdot \overline{Z}$$

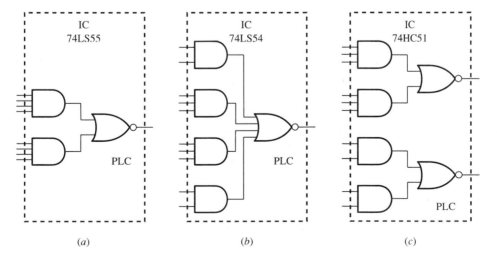

(a)   (b)   (c)

FIGURE P6-48

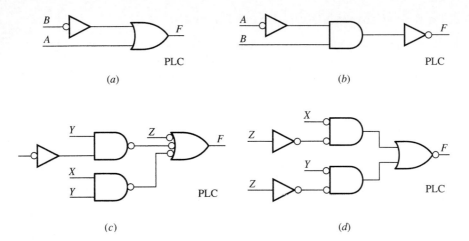

FIGURE P6-52

Section 6-6 Obtaining Boolean Functions from Diagrams Using the Positive Logic Convention

6-52. Obtain the Boolean function in SOP form and the signal list for each logic diagram shown in Fig. P6-52.

6-53. Obtain the Boolean function in SOP form and the signal list for each of the implementations shown in Fig. P6-53.

6-54. Obtain the Boolean function in SOP form and the signal list for each logic diagram shown in Fig. P6-54.

Section 6-7 Obtaining Boolean Functions from Diagrams Using Direct Polarity Indication

6-55. Write the logic statements for the signal names $\sim B$(H) and $\overline{\sim B}$(L). Make a table to show the logic value of the available signal for each signal name when the logic level of the signal is high, and when the logic level of the signal is low. What does the table indicate?

6-56. Write the logic statements for the signal names C(L) and $\overline{C}$(H). Make a table to show the logic value of the available signal for each signal name when the logic

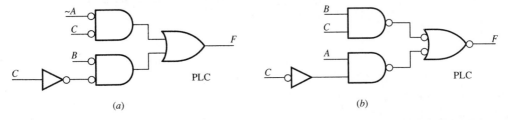

FIGURE P6-53

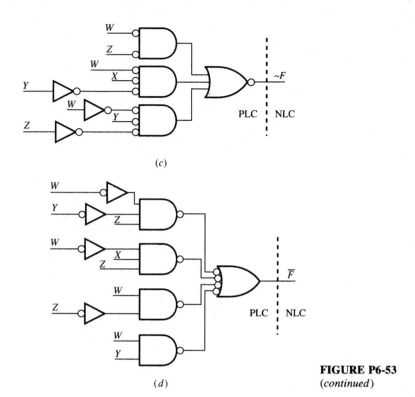

(c)

(d)

FIGURE P6-53
(*continued*)

level of the signal is high, and when the logic level of the signal is low. What does the table indicate?

6-57. Show that the signal names ~A(H) and ~A(L) are not identical.

6-58. Show that the signal name $\overline{\sim D}$(H) is equivalent to the signal name ~D(L) and therefore may not be used on the same diagram to identify a different signal.

6-59. Obtain the Boolean function in SOP form and the signal list for each of the implementations shown in Fig. P6-59.

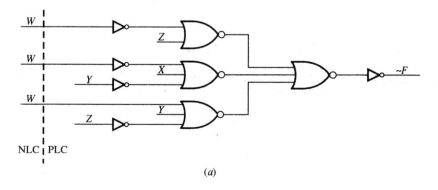

(a)

FIGURE P6-54

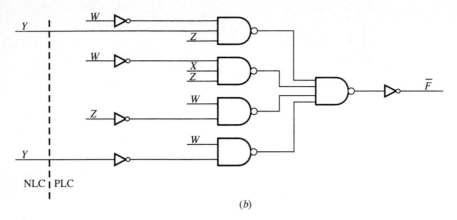

(b)

FIGURE P6-54
(*continued*)

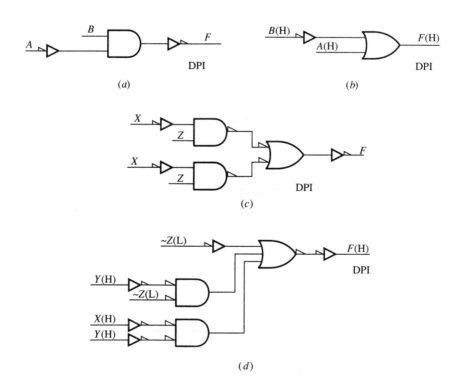

FIGURE P6-59

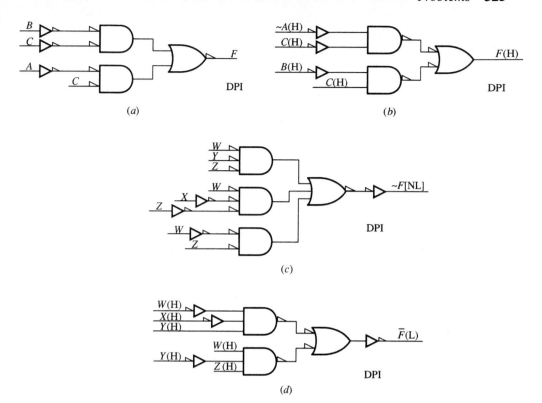

FIGURE P6-60

6-60. Obtain the Boolean function in SOP form and the signal list for each logic diagram shown in Fig. P6-60.

6-61. Obtain the Boolean function in SOP form and the signal list for each logic diagram shown in Fig. P6-61.

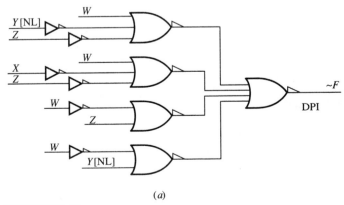

(a)

FIGURE P6-61

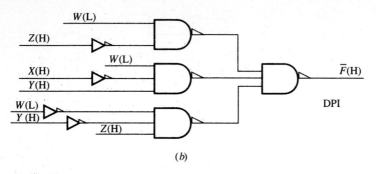

(*b*)

FIGURE P6-61
(*continued*)

IMPLEMENTING
LOGIC
FUNCTIONS
USING
MSI AND
PROGRAMMABLE
DEVICES

7-1 INTRODUCTION
AND INSTRUCTIONAL GOALS

This is the last chapter in Part II Combinational Logic Design. Until now we have only discussed designing circuits using small-scale integrated (SSI) devices. In this chapter we present medium-scale integrated (MSI) devices and programmable logic devices (PLDs) with up to a 800-gate equivalency. This does not imply that we can forget Boolean algebra and just pick a device that someone else has designed to save us time and effort. When function-specific MSI circuits are used as design blocks, it requires a different design approach. Some of the concepts we have presented will still be used, and newer design concepts will now be presented. These newer design concepts not only save us time and effort compared to designing strictly at the gate level, but they also broaden our appreciation and awareness of the many types of off-the-shelf MSI and programmable logic devices available to designers.

In Section 7-2, Implementing Logic Functions Using MSI Multiplexers, general purpose MSI Multiplexers are presented. The author prefers to use block diagrams that complement the suggested IEEE Multiplexer symbols. It is not unusual for a designer to formulate a design using simple block diagrams, and then use a schematic capture package with the recommended IEEE symbols to obtain the necessary detailed logic diagrams. This is what we do in most cases. In other cases, the recommended IEEE symbols are shown in order to emphasize important concepts. This way the exact details of the IEEE symbols are obtained after the design phase. The internal logic circuit of a Multiplexer was presented in Chapter 5. The theory associated with designing with off-the-shelf Multiplexers, Shannon's Expansion Theorem, is presented in this section.

A tabular method based on Shannon's Expansion Theorem is presented in Section 7-3, Designing with Multiplexers. Here a more efficient method is presented for Type 0, 1, 2, and 3 Multiplexer designs. Multiplexer trees are used when a function requires more inputs than are available with a single Multiplexer. The Multiplexer design approach must be modified when negative logic signals are present. Two different design approaches are presented to illustrate how to handle the problem of negative logic signals. The analysis of a Multiplexer design is also presented in this section. Since Multiplexers are used in bus-organized systems, Section 7-4, Additional Techniques for Designing with Multiplexers, provides a short introduction to this topic. The traditional use of Multiplexers and Demultiplexers is also presented in this section.

MSI Decoders are presented in Section 7-5, Implementing Logic Functions Using MSI Decoders. In this section a systems design divide-and-conquer approach is used to implement logic functions with Decoders. This technique utilizes the top down design approach for gate-level design presented in Chapter 5. The binary-to-hexadecimal character generator also presented in Chapter 5 is repeated using a Decoder design so that a comparison can be made. This section also shows how to analyze an existing Decoder design. Since Decoders are often used in bus-organized systems for address decoding, this topic is presented using a MSI Decoder design as well as a MSI comparator design.

In Section 7-6, Implementing Logic Functions Using Exclusive OR and Exclusive NOR Elements, we present the many different ways that special functions can be implemented using these MSI elements. Applications range from the simple controlled Buffer/Inverter application to the more complicated Parity Generator and Parity Checker applications using odd and even functions. Additional Boolean theorems are also presented in this section to provide a better understanding of Exclusive OR and Exclusive NOR functions.

Implementing Logic Functions Using Programmable Devices is the topic of Section 7-7. First an overview of the three basic programmable architectures: programmable read only memory (PROM), programmable array logic (PAL), and programmable logic array (PLA) devices, is presented. Each device is then discussed in depth. The binary-to-hexadecimal character generator is repeated using a PROM design to illustrate a single package implementation of a fairly complicated combinational design. The implementation uses an off-the-shelf PROM device. Next, we discuss perhaps the most used type of programmable device, the PAL. A design process is presented for PAL devices using first a simple PAL device with nonnegated outputs (active high outputs). Then the design is repeated using a commercially available device with negated outputs (active low outputs). The important task of choosing the correct forms for the equations for the different types of PAL outputs is stressed. Several different PAL devices are discussed, from simple combinational outputs to registered outputs including macrocell outputs with Multiplexers (to achieve versatility). Next, we discuss the architecture of PLA devices with programmable polarity fuses. PALs with polarity fuse capability are also discussed. The section ends with a discussion of multi-level PLD structures. These include Folded-NAND and Folded-NOR type devices. These PLDs can typically be used to replace up to three or four PAL or PLA conventional two-level PLDs.

Utilizing off-the-shelf PLDs to implement Boolean functions requires not only a computer system, but a software package and a programming unit. Section 7-8, Programming PALs Using PALASM, provides an introduction for first-time users of PALs. This section discusses the software package PALASM and provides several combinational logic examples. The design examples are completely worked out showing each step in the design process: the design file, the disassembled design file, the fuse map, the PAL circuit implementation (for the simple cases since this is really not necessary), and finally the JEDEC file. The author uses a programming unit to blow the fuses and test each of the PAL designs. These types of designs can be used as laboratory exercises if equipment is available.

The last section in this chapter, Section 7-9, Hazards in Combinational Logic Circuits, discusses the momentary error conditions called glitches that can occur in real combinational logic circuit implementations. Glitches are caused by static 1, static 0, dynamic 1 to 0, and dynamic 0 to 1 hazards as a result of delays in logic circuits. Examples of both function hazards and logic hazards are discussed. The last topic in this section contains a technique for designing logic hazard-free combinational circuits.

This chapter should prepare you to

1. Use Shannon's Expansion theorem to implement logic functions using Multiplexers.
2. Implement logic functions using a type 0, 1, 2, or 3 Multiplexer design.
3. Analyze existing Multiplexer designs, and obtain the Boolean functions being implemented.
4. Design Multiplexer trees to expand the capability of standard off-the-shelf Multiplexers.
5. Design with Multiplexers in a bus-organized system.
6. Implement logic functions using Decoders.
7. Design with Decoders and Identity Comparators in a bus-organized system.
8. Use Exclusive OR elements and Exclusive NOR elements to implement odd and even functions.
9. Design Parity Generator and Parity Checker circuits for error detection in communication links.

10. Design with different types of programmable logic devices (PLDs).
11. Design combinational logic circuits using programmable read only memory (PROM) devices.
12. Design combinational logic circuits using programmable array logic (PAL) devices.
13. Design combinational logic circuits using programmable logic array (PLA) devices.
14. Program PAL devices using PAL Assembly (PALASM).
15. Draw timing diagrams that illustrate the different types of hazards in combinational logic circuits.
16. Explain the difference between function hazards and logic hazards.
17. Design logic hazard-free combinational circuits.

7-2 IMPLEMENTING LOGIC FUNCTIONS USING MSI MULTIPLEXERS

Gate-level design for Multiplexers or Data Selectors was presented in Chapter 5, where it was also shown how larger Multiplexers can be built by leveraging existing multiplexer designs. In this section we will show how to implement Boolean functions using Multiplexers. Figure 7-1a shows a block diagram of an 8-to-1 line Multiplexer. Compare the block diagram with the IEEE (or IEC)

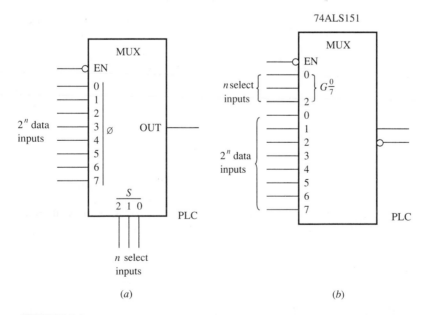

(a)   (b)

FIGURE 7-1
(a) Block diagram for an 8-to-1 line Multiplexer (MUX) (b) IEEE rectangular-shape symbol for an 8-to-1 line Multiplexer.

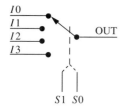

FIGURE 7-2
Switch equivalent for a 4-to-1 line Multiplexer (MUX).

rectangular-shape graphic symbol for the 8-to-1 line Multiplexer illustrated in Fig. 7-1b. Note: Read Section 4-2 (G (AND) Dependency) and Section 4-3 (Conventions for the Application of Dependency Notation in General) in *Overview of IEEE Standard 91-1984, Explanation of Logic Symbols* in Appendix A. These sections provide the information concerning the dependency notation shown in the IEEE graphic symbol for the Multiplexer.

Designers often use block diagrams to show the concepts of their logic, but then turn to schematic capture packages to draw realizable logic circuit diagrams, taking advantage of standard logic symbols in available data base libraries. The equation for the output of a Multiplexer with 2^n data inputs and n select inputs is

$$\text{OUT} = \sum_{i=0}^{2^n-1} Ii \cdot m_i \cdot \text{EN} \qquad (7\text{-}1)$$

where m_i is a minterm consisting of n select signals that are applied to the select inputs and Ii is a data input (used for applying 2^n data input signals) as illustrated in Fig. 7-1a. Equation 7-1 is provided with an enable input (which is sometimes referred to as the strobe input), EN. The enable input is used to enable the data and select inputs of the Multiplexer. Both the block diagram and the rectangular shape symbol shown in Fig. 7-1 have an enable or strobe input. If an enable input is not present for a device, then EN is not present in the equation of the device. Multiplexer Eq. 7-1, is written using the signal names labeled inside the block diagram of the device. For this reason the Multiplexer equation holds true whether or not a polarity symbol or a negation symbol is present at any of the inputs or output of the device.

Figure 7-2 illustrates the rotary switch equivalent of a 4-to-1 line Multiplexer without an enable input. The select inputs, $S1$ and $S0$, are decoded to determine the position of the movable contact of the rotary switch so the appropriate input Ii is steered to the output ($S1\ S0 = 00$ connects OUT to $I0$, $S1\ S0 = 01$ connects OUT to $I1$, $S1\ S0 = 10$ connects OUT to $I2$, and $S1\ S0 = 11$ connects OUT to $I3$).

7-2-1 Shannon's Expansion Theorem

Designing with Multiplexers revolves around applying a theorem called Shannon's Expansion Theorem. The theorem can be stated as follows.

T11a: $F(X1, \ldots, Xn) = F(0, X2, \ldots, Xn) \cdot \overline{X1} + F(1, X2, \ldots, Xn) \cdot X1$

T11b: $F(X1, \ldots, Xn) = [F(1, X2, \ldots, Xn) + \overline{X1}]$

$$\bullet [F(0, X2, \ldots, Xn) + X1]$$

The subfunction $F(0,X2, \ldots ,Xn)$ represents $F(X1, \ldots ,Xn)$ evaluated for $X1 = 0$. $F(1,X2, \ldots ,Xn)$ likewise represents $F(X1, \ldots ,Xn)$ evaluated for $X1 = 1$. The Expansion Theorem shows how any function can be expanded with respect to a particular variable. Theorem T11b is the dual of Theorem T11a. The subfunctions (the coefficients) of the literals $\overline{X1}$ and $X1$ contain $n - 1$ variables $(X2, \ldots ,Xn)$. Applying Shannon's Expansion Theorem again with respect to variable $X2$, we obtain

$$F(X1, \ldots, Xn) = F(0, 0, X3, \ldots, Xn) \bullet \overline{X1} \bullet \overline{X2} \ + \ F(0, 1, X3, \ldots, Xn)$$

$$\bullet \overline{X1} \bullet X2 \ + \ F(1, 0, X3, \ldots, Xn) \bullet X1 \bullet \overline{X2}$$

$$+ \ F(1, 1, X3, \ldots, Xn) \bullet X1 \bullet X2$$

The subfunctions of the product terms $\overline{X1} \bullet \overline{X2}$, $\overline{X1} \bullet X2$, $X1 \bullet \overline{X2}$, and $X1 \bullet X2$ now contain $n - 2$ variables $(X3, \ldots ,Xn)$. If the expansion process is continued with respect to all the variables ($X1$ through Xn), the resulting expression will be the canonical or standard sum of products expression for the function.

Shannon's Expansion Theorem, T11a, can be proven by perfect induction for all values of $X1$ by setting $X1 = 0$ to verify that the expression on the right is equivalent to the expression on the left. The process is continued setting $X1 = 1$ as shown below.

Setting $X1 = 0$ in T11a

$$F(0, \ldots, Xn) = F(0, X2, \ldots, Xn) \bullet \overline{0} \ + \ F(1, X2, \ldots, Xn) \bullet 0$$

$$= F(0, X2, \ldots, Xn) \bullet 1 \ + \ F(1, X2, \ldots, Xn) \bullet 0$$

$$= F(0, X2, \ldots, Xn)$$

Setting $X1 = 1$ in T11a

$$F(1, \ldots, Xn) = F(0, X2, \ldots, Xn) \bullet \overline{1} \ + \ F(1, X2, \ldots, Xn) \bullet 1$$

$$= F(0, X2, \ldots, Xn) \bullet 0 \ + \ F(1, X2, \ldots, Xn) \bullet 1$$

$$= F(1, X2, \ldots, Xn)$$

T11b, the dual of theorem T11a, is also true by the principle of duality.

Now observe that the expanded function for two variables $X1$ and $X2$ can be written in the same form as the Multiplexer Eq. 7-1.

$$F(X1, \ldots, Xn) = F(0, 0, X3, \ldots, Xn) \bullet \overline{X1} \bullet \overline{X2} \ + \ F(0, 1, X3, \ldots, Xn)$$

$$\bullet \overline{X1} \bullet X2 \ + \ F(1, 0, X3, \ldots, Xn) \bullet X1 \bullet \overline{X2}$$

$$+ \ F(1, 1, X3, \ldots, Xn) \bullet X1 \bullet X2$$

$$= I0 \bullet \overline{X1} \bullet \overline{X2} \ + \ I1 \bullet \overline{X1} \bullet X2 \ + \ I2 \bullet X1 \bullet \overline{X2} \ + \ I3 \bullet X1 \bullet X2$$

$$= I0 \bullet m_0 \ + \ I1 \bullet m_1 \ + \ I2 \bullet m_2 \ + \ I3 \bullet m_3$$

or the form

$$= \sum_{i=0}^{2^n-1} Ii \cdot m_i$$

where $m_i = m_i(X1, X2)$

The subfunctions $F(0,0,X3, \ldots ,Xn)$, $F(0,1,X3, \ldots ,Xn)$, $F(1,0,X3, \ldots ,$ $Xn)$, and $F(1,1,X3, \ldots Xn)$ provide the signals for inputs $I0$ through $I3$ of a 4 to 1 line Multiplexer. $X1$ and $X2$ provide the signals for the select inputs where $X1$ is the MSB and $X2$ is the LSB. This demonstrates that any Boolean function can be implemented with an appropriate size multiplexer provided that the subfunctions can be obtained.

The point of this development is mainly to show the theoretical basis for the procedures that will be presented for designing with Multiplexers. Actual design procedures are best carried out using a truth table or a Karnaugh map. Truth tables and Karnaugh maps usually provide a more efficient mechanism with which to evaluate the subfunctions of a specified Boolean function. Subfunction evaluation using truth tables and Karnaugh maps will be presented in the next section.

Example 7-1. Design a circuit using a Multiplexer to implement the following function by applying Shannon's Expansion Theorem T11a with respect to variables A and B.

$$F(A, B, C) = A + B \cdot \overline{C}$$

Signal list: F, A, B, C

Solution

$$F(A, B, C) = F(0,0,C) \cdot \overline{A} \cdot \overline{B} + F(0,1,C) \cdot \overline{A} \cdot B$$

$$= F(1,0,C) \cdot A \cdot \overline{B} + F(1,1,C) \cdot A \cdot B$$

for $F(A, B, C) = A + B \cdot \overline{C}$

$$F(0,0,C) = 0 + 0 \cdot \overline{C} = 0$$

$$F(0,1,C) = 0 + 1 \cdot \overline{C} = \overline{C}$$

$$F(1,0,C) = 1 + 0 \cdot \overline{C} = 1$$

$$F(1,1,C) = 1 + 1 \cdot \overline{C} = 1$$

A (MSB) and B (LSB) are the signals applied to the select inputs of a 4-to-1 line Multiplexer, 0 is applied to the $I0$ input line, $\overline{C}$ is applied to the $I1$ input line, 1 is applied to the $I2$ input line, and 1 is applied to the $I3$ input line as shown in Fig. E7-1. A logic 0 must be applied to the EN input line to enable the MUX, as shown in the figure.

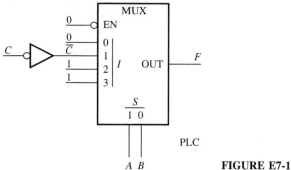

FIGURE E7-1

7-3   DESIGNING WITH MULTIPLEXERS

The circuit for a MSI Multiplexer cannot be changed; therefore, Karnaugh map minimization is not normally used to design with Multiplexers, except in a support capacity or to check to see if the function really requires a Multiplexer. When a Multiplexer is used to implement a logic function, that function does not need to be minimized in the normal manner; however, a minimized function consisting of only a single literal or a single product term would be more cost-effective using a gate-level design. Logic designers usually use truth tables to implement logic functions with Multiplexers. Truth tables can be used to obtain the subfunctions of Boolean functions in the same manner they were used to reduce the size of Karnaugh maps in Section 4-8 of Chapter 4. Of course, once the procedure is understood, the subfunctions of Boolean functions can also be obtained using a Karnaugh map description. To illustrate the process, the following function will be implemented using a Multiplexer beginning with a type 0 design through a type 3 design.

$$F(A, B, C, D) = \sum m(4, 5, 6, 7, 10, 14)$$

Signal list: *F, A, B, C, D*

The implementation for a type 0 Multiplexer design is obtained using either the minterm compact form of a Boolean function or its truth table representation. A type 0 MUX design required no signals in the truth table representing the function to be partitioned off. A type 1 MUX design requires one signal to be partitioned off, a type 2 MUX design requires two signals to be partitioned off, and so on. The independent signals represented in the function are applied to the select inputs. The characteristic numbers of the function that can be written by inspection from the minterm compact form of the function are applied to the data inputs. For the specified function listed above, the characteristic numbers are $f_0 = 0, f_1 = 0, f_2 = 0, f_3 = 0, f_4 = 1, f_5 = 1, f_6 = 1, f_7 = 1, f_8 = 0,$ $f_9 = 0, f_{10} = 1, f_{11} = 0, f_{12} = 0, f_{13} = 0, f_{14} = 1, f_{15} = 0$ (for a discussion of the characteristic numbers of a function refer to Section 1-8 in Chapter 1). The implementation for a type 0 Multiplexer design is shown in Fig. 7-3 using a 16-to-1 line Multiplexer.

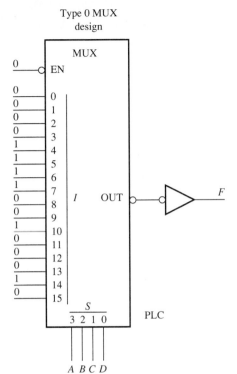

Type 0 MUX
design

FIGURE 7-3
Implementation for a type 0 Multiplexer
design for function $F(A, B, C, D) = \sum m(4, 5, 6, 7, 10, 14)$,
Signal list: F, A, B, C, D.

Notice in Fig. 7-3 that the logic state of each characteristic number is written on the corresponding input line such that the decimal subscript of the characteristic number matches the Multiplexer input with the same decimal value, e.g., for $f_5 = 1$, a 1 is written on input line I5 of the Multiplexer. Just to verify that we have not used a Multiplexer where a single line, a single Inverter, or single gate could be used, it is usually worthwhile to use a Karnaugh map to obtain a minimum expression for the function. For the function above, the minimum expression is $F(A, B, C, D) = \overline{A} \cdot B + A \cdot C \cdot \overline{D}$. The most significant input select line is 3, and the least significant input select line is 0 in the block diagram, or the rectangular-shape symbol for a 16-to-1 line Multiplexer. The most significant input signal in the function description or the truth table description must be applied to the most significant input select line of the Multiplexer, with the remaining least significant input signals arranged in descending order as shown in Fig. 7-3. An Inverter is included on the output of the Multiplexer since the function $F = $ OUT is required rather than $F = \overline{\text{OUT}}$.

7-3-1   Type 1 Multiplexer Design

Continuing with the same function, the truth table is now drawn and partitioned for a type 1 Multiplexer design as shown in Fig. 7-4a. The select inputs are

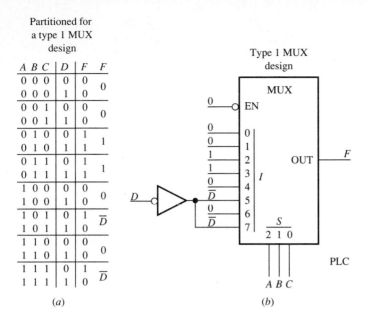

Partitioned for
a type 1 MUX
design

A B C	D	F	F
0 0 0	0	0	0
0 0 0	1	0	
0 0 1	0	0	0
0 0 1	1	0	
0 1 0	0	1	1
0 1 0	1	1	
0 1 1	0	1	1
0 1 1	1	1	
1 0 0	0	0	0
1 0 0	1	0	
1 0 1	0	1	$\overline{D}$
1 0 1	1	0	
1 1 0	0	0	0
1 1 0	1	0	
1 1 1	0	1	$\overline{D}$
1 1 1	1	0	

(a)

(b)

FIGURE 7-4
(a) Truth table partitioned for a type 1 MUX design, (b) implementation for a type 1 MUX design.

the independent signals A, B, and C, and the data inputs are the values (the subfunctions) represented in the reduced output column (the last column on the right) in the truth table. The reduced output column in Fig. 7-4a represents the output states as a function of the partitioned-off input signal D. The reduced outputs in the truth table consist of a set of subfunctions of the signal (or signals) that is partitioned off in the truth table. In this case, eight subfunctions are required where each subfunction is written as a function of D, that is, $F(D)$. The implementation for a type 1 Multiplexer design is shown in Fig. 7-4b using an 8-to-1 line Multiplexer. Notice that the most significant input signal in the truth table description is applied to the most significant input select line of the 8-to-1 line Multiplexer, and the remaining least significant input signals are arranged in descending order except input signal D (the least significant input signal), which was partitioned off. This procedure is generally considered more desirable than writing the minimum function and applying Shannon's Expansion Theorem T11a with respect to variables A, B, and C. This statement is especially true for complex functions, since the truth table approach requires the same effort whether the function is simple or complex.

7-3-2 Type 2 and Type 3 Multiplexer Designs

Figures 7-5a and b illustrate truth table partitioning for a type 2 and a type 3 Multiplexer design respectively. Partitioning off j of the n input signals in the

Partitioned for
a type 2 MUX
design

Partitioned for
a type 3 MUX
design

A B	C D	F	F
0 0	0 0	0	
0 0	0 1	0	0
0 0	1 0	0	
0 0	1 1	0	
0 1	0 0	1	
0 1	0 1	1	1
0 1	1 0	1	
0 1	1 1	1	
1 0	0 0	0	
1 0	0 1	0	$C \cdot \overline{D}$
1 0	1 0	1	
1 0	1 1	0	
1 1	0 0	0	
1 1	0 1	0	$C \cdot \overline{D}$
1 1	1 0	1	
1 1	1 1	0	

(a)

A	B C D	F	F
0	0 0 0	0	
0	0 0 1	0	
0	0 1 0	0	
0	0 1 1	0	B
0	1 0 0	1	
0	1 0 1	1	
0	1 1 0	1	
0	1 1 1	1	
1	0 0 0	0	
1	0 0 1	0	
1	0 1 0	1	
1	0 1 1	0	$C \cdot \overline{D}$
1	1 0 0	0	
1	1 0 1	0	
1	1 1 0	1	
1	1 1 1	0	

(b)

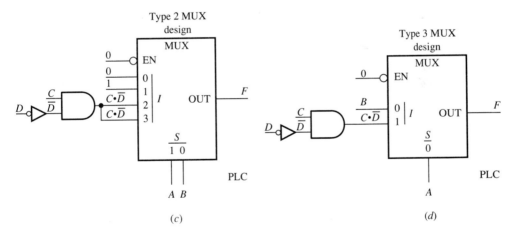

(c)

(d)

FIGURE 7-5
(a) and (b) Truth table partitioned for a type 2 and a type 3 MUX design, (c) and (d) implementations
for a type 2 and a type 3 MUX design.

truth table also requires partitioning off every k row in the table, where $k = 2^j$
for $j \geq 1$. The two input signals C and D are partitioned off for a type 2 MUX
design, while the three signals B, C, and D are partitioned off for a type 3 MUX
design. To make our job easier, the least significant inputs are partitioned off. The
type of MUX design simply agrees with the number of input signals partitioned
off in the truth table of the function. Figures 7-5c and d show the implementation
for a type 2 and type 3 MUX design. Karnaugh maps can be drawn to minimize

the subfunctions of the truth table if the minimum expressions cannot be obtained by inspection. The function represented by the bottom half of the truth table shown in Fig. 7-5*b* can be obtained from a three-variable Karnaugh using the input signals *B*, *C*, and *D*.

It should be obvious that it is possible for any function containing *n* input variables to be implemented with a Multiplexer design of type 0 (no input signals partitioned off) to type *n* − 1 (all input signals partitioned off except 1). Some functions require less external components for a particular type of MUX design, while other functions can require more external components for the same type. The procedure that is used to design with Multiplexers remains the same regardless of the complexity of the function being implemented. The criteria used in selecting onc design over another (a gate-level design compared to an MSI design in this case) usually hinges on either minimum space (less real estate on a printed circuit board) or lower expense. Designers most often choose a type 1 MUX design. Sometimes a design is chosen simply because it is easy to understand, as well as easy to implement. Multiplexer designs usually meet both of these criteria more so than gate-level designs (gate-level designs are also referred to as random logic, glue logic, or discrete logic designs).

In general, implementing a function with a larger Multiplexer requires fewer external components, but the Multiplexer itself can be a large device (typically a 16-to-1 line MUX is available in a 24-pin package, while an 8-to-1 line MUX, a dual 4-to-1 line MUX, and a quadruple 2-to-1 line MUX are each available in a smaller 16-pin package). Using a higher type MUX design results in a smaller MUX to implement a function, but a smaller MUX often requires more external discrete logic.

Example 7-2. Show a realizable implementation for the Boolean function plotted in the Karnaugh map shown in Fig. E7-2*a*:

1. using gate level logic,
2. using a type 1 MUX design, and
3. using a type 2 MUX design.

$A\,B \backslash C\,D$

$F =$	00	01	11	10
00	1	0	1	0
01	0	1	1	0
11	1	0	1	0
10	0	1	0	1

(*a*)

FIGURE E7-2

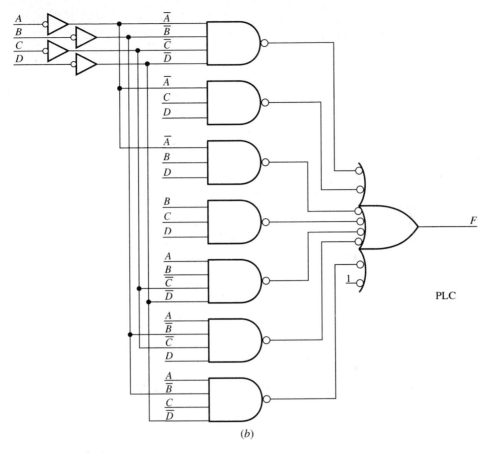

(b)

FIGURE E7-2 (*continued*)

Solution

1. The minimum SOP expression for the function is

$$F = \overline{A} \cdot \overline{B} \cdot \overline{C} \cdot \overline{D} \; + \; \overline{A} \cdot C \cdot D \; + \; \overline{A} \cdot B \cdot D \; + \; B \cdot C \cdot D$$

$$+ \; A \cdot B \cdot \overline{C} \cdot \overline{D} \; + \; A \cdot \overline{B} \cdot \overline{C} \cdot D \; + \; A \cdot \overline{B} \cdot C \cdot \overline{D}$$

For a signal list of F, A, B, C, D, Fig. E7-2b shows a realizable implementation for the function using NAND gates and Inverters. This implementation requires 5 IC packages. The circuit was drawn assuming that all the signals in the function were available signals. Equivalent signal lines were then substituted between each AND output and the corresponding OR input to convert each gate to a NAND gate implementation. All unavailable complemented inputs $\overline{A}$, $\overline{B}$, $\overline{C}$, and $\overline{D}$ were connected, and Inverters were then used to generate the unavailable complemented inputs via the available signals A, B, C, and D, respectively.

2. The signal D is partitioned off in the Karnaugh map shown in Fig. E7-2c. Each set of encircled output states is written as a function of the partitioned off input signal D, that is, $F(D)$. The results are entered in the reduced Karnaugh map

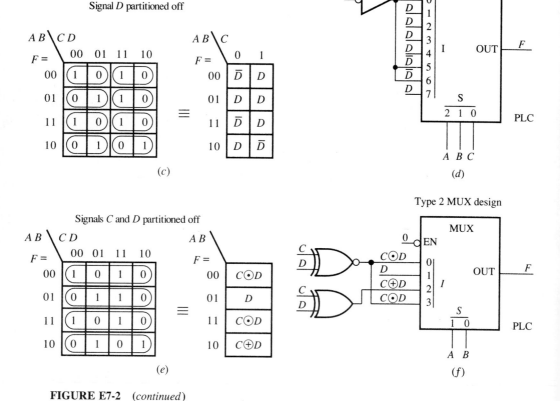

FIGURE E7-2   (*continued*)

shown in the figure (the output states are written as a function of the partitioned off signal D just like they would be in a truth table). The output states in the reduced map are used as the input signals to the 8-to-1 line Multiplexer in Fig. E7-2d. The nonpartitioned signals A, B, and C supply the select inputs where A is the most significant bit and C the least significant bit. The implementation requires only two packages.

3. In the Karnaugh map shown in Fig. E7-2e, input signals C and D are partitioned off as indicated by the four sets of encircled output states on the map. Each of these sets is written as a function of the partitioned-off inputs C and D, and the results are entered in the reduced map as shown in the figure. The output states in the reduced map are used as the input signals to the 4-to-1 line Multiplexer in Fig. E7-2f. The nonpartitioned signals A and B supply the select inputs where A is the most significant bit and B the least significant bit. The implementation shown requires three packages; however, since an Exclusive OR can be connected to function as an Inverter, the implementation can be realized using only two packages.

One can observe from Example 7-2 that the choice of variables as well as the number of variables used for the select inputs of a Multiplexer can affect the complexity and corresponding cost of the logic needed at the data inputs.

7-3-3   Analyzing a Multiplexer Design

Analyzing a Multiplexer design to obtain the Boolean function or truth table implemented by the Multiplexer is a simple process. Beginning with the Multiplexer equation, the signals driving the data inputs and the select inputs are substituted into the Multiplexer equation, and the equation is expanded to yield the Boolean function. The Boolean function in general will not be a minimum function. Expanding the function using Postulate P5a ($X + \overline{X} = 1$) allows one to obtain the minterm compact form. The process is illustrated below for the Multiplexer design shown in Fig. 7-6.

$$F(A, B, C) = \text{OUT} = \sum_{i=0}^{2^n - 1} Ii \cdot m_i \cdot \text{EN}$$

where $m_i = m_i(A, B)$, that is, $n = 2$

$$F(A, B, C) = \sum_{i=0}^{3} Ii \cdot m_i, \qquad \text{for EN} = 1$$

$$= 0 \cdot m_0 \;+\; \overline{C} \cdot m_1 \;+\; 1 \cdot m_2 + C \cdot m_3$$

$$= \overline{C} \cdot m_1 \;+\; 1 \cdot m_2 \;+\; C \cdot m_3$$

but $\qquad\qquad = \overline{C} \cdot \overline{A} \cdot B \;+\; 1 \cdot A \cdot \overline{B} \;+\; C \cdot A \cdot B \qquad A(\text{MSB}), B, C(\text{LSB})$

$$= \overline{A} \cdot B \cdot \overline{C} \;+\; A \cdot \overline{B} \cdot (C + \overline{C}) \;+\; A \cdot B \cdot C$$

$$= m_2 + m_5 + m_4 + m_7$$

$$= \sum m(2, 4, 5, 7)$$

where $\qquad\qquad m = m(A, B, C)$

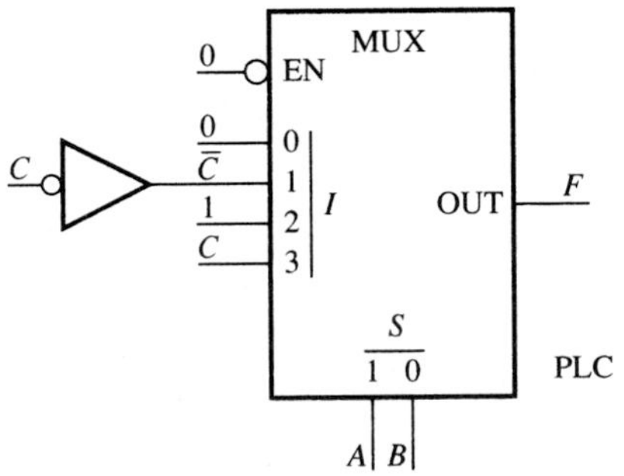

FIGURE 7-6
Multiplexer design.

7-4   ADDITIONAL TECHNIQUES FOR DESIGNING WITH MULTIPLEXERS

Boolean functions with a large number of inputs can be accommodated by constructing Multiplexer trees. Multiplexer trees (cascaded Multiplexers) can be used to implement functions that are not easily reduced or to implement functions with a larger than normal number of inputs (more inputs than normal off-the-shelf Multiplexers will accommodate). It is not necessary to write every binary entry for every signal in the truth table, since the reduced output column can be obtained by simply listing the binary values of the partitioned-off signal (or signals) and the output column so that these binary entries can be compared. The following example illustrates the procedure.

> **Example 7-3.** Show a Multiplexer tree implementation for the following six-variable Boolean function. All the signals are uncomplemented positive logic signals. Use a type 1 MUX design.
>
> $F(U, V, W, X, Y, Z)$
>
> $$= \sum m(0, 4, 9, 16, 20, 21, 24, 30, 37, 40, 42, 46, 49, 52, 56, 58, 62)$$

Solution A stripped-down version of the truth table is shown in Fig. E7-3a. Decimal values are provided for the minterms listed in the function so the output can be listed in the output column. The Z signal is listed in its normal alternating bit form for a type 1 MUX design; that is, Z is partitioned off in the table. Column Z and the output column provide all the information necessary to obtain each subfunction, $F(Z)$, required in the reduced output column.

Since there are 32 subfunctions in the reduced output column, a 32-to-1 line MUX tree is required for a type 1 MUX design. The four 8-to-1 line MUXs (U1 through U4) feeding one 4-to-1 line MUX (U5) as shown in Fig. E7-3b is one of

m	Z	F	F	m	Z	F	F	m	Z	F	F	m	Z	F	F
0	0	1	$\overline{Z}$	16	0	1	$\overline{Z}$	32	0		0	48	0		Z
1	1			17	1			33	1			49	1	1	
2	0		0	18	0		0	34	0		0	50	0		0
3	1			19	1			35	1			51	1		
4	0	1	$\overline{Z}$	20	0	1	1	36	0		Z	52	0	1	$\overline{Z}$
5	1			21	1	1		37	1	1		53	1		
6	0		0	22	0		0	38	0		0	54	0		0
7	1			23	1			39	1			55	1		
8	0		Z	24	0	1	$\overline{Z}$	40	0	1	$\overline{Z}$	56	0	1	$\overline{Z}$
9	1	1		25	1			41	1			57	1		
10	0		0	26	0		0	42	0	1	$\overline{Z}$	58	0	1	$\overline{Z}$
11	1			27	1			43	1			59	1		
12	0		0	28	0		0	44	0		0	60	0		0
13	1			29	1			45	1			61	1		
14	0		0	30	0	1	$\overline{Z}$	46	0	1	$\overline{Z}$	62	0	1	$\overline{Z}$
15	1			31	1			47	1			63	1		

FIGURE E7-3a

MUX tree using a type 1 MUX design

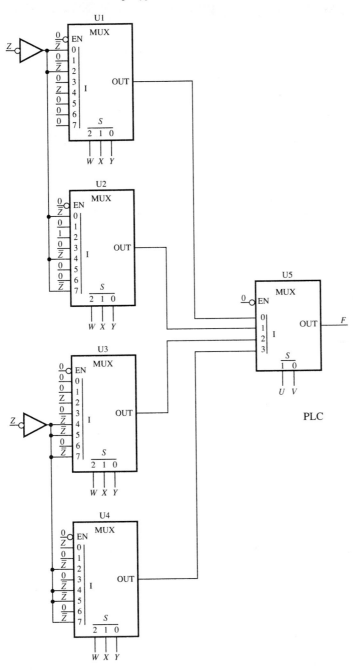

FIGURE E7-3b

several ways to connect existing off-the-shelf Multiplexers to obtain a 32-to-1 line MUX. Notice that the most significant select bit is select input 1 of U5, and the least significant select bit is select input 0 of U1 through U4. Signal U (the MSB in the truth table) is connected to the most significant select input, signal Y (the LSB in the truth table excluding the partitioned off signal) is connected to the least significant select bit, and the rest of the signals are connected to the select inputs in descending order of significance. Using the control bit ordering just described causes the I inputs to the 32-to-1 line MUX to be ordered from $I0$ through $I31$ beginning with the $I0$ input of U1 and ending with the $I7$ input of U4. This allows the subfunctions in the reduced output column to be listed in truth table order from the top input line of MUX (U1) to the bottom input line of MUX (U4) as illustrated in Fig. E7-3b.

To simplify loading calculations, manufacturers generally insure that their device will drive a specified number of other devices in the same technology. For TTL this refers to STD TTL, LS, S, ALS, AS, and F, while in CMOS this refers to AC, ACT, HC, and HCT (see Table 5-1). This number is called the fanout of the device, where fanout represents the maximum number of standard loads a device output can drive and still perform within specification. A standard load represents the current requirement of the input of another device in the same technology. If an Inverter in Fig. E7-3b had a fanout of 10 and each input that it is driving has one standard load, then two Inverters would be required as shown in the figure to reduce the loading requirement of a single Inverter to 10 or less. Standard TTL has a fanout of 10.

The ALS Inverter, the 74ALS04A, has the following published input and output current specifications.

$$I_{OH}(MAX) = -0.4 \text{ mA} \qquad I_{OL}(MAX) = 8 \text{ mA}$$

$$I_{IH}(MAX) = 20 \text{ } \mu A \qquad I_{IL}(MAX) = -0.1 \text{ mA}$$

The fanout is the smaller of the two ratios $I_{OH}(MAX)/I_{IH}(MAX) = 20$ and $I_{OL}(MAX)/I_{IL}(MAX) = 80$; therefore the fanout is 20. One Inverter would suffice to drive 20 standard loads in the circuit in Fig. E7-3b if the circuit were constructed using ALS technology devices.

Up to this point we have discussed implementing logic functions only with positive logic signals. If negative logic signals are present, one approach is to assume all positive logic signals and obtain the desired type MUX design. An Inverter is then added to every line that contains a negative logic signal as discussed in Aside 5-1 in Chapter 5. You may recall that Inverters are added to fix signal mismatching problems. This is the easiest method of handling negative logic signals. An alternate method is to convert each negative logic signal to an equivalent positive logic signal (complement the signal name and the values of the signal in the truth table). If a negative logic signal such as Y is changed to its positive logic equivalent $\sim Y$ in the truth table, then the signal mismatching problem has been fixed in the truth table, and an Inverter is no longer required when the design is completed.

Example 7-4. Obtain a type 1 MUX design for the following Boolean function.

$$F = \sum m(0, 3, 6) + d(5, 7)$$

Signal list: F, A, $B[NL]$, $C[NL]$

Solution The first method (method 1) we will use is to assume all positive logic signals and fix the signal mismatching problems by adding Inverters. The truth table for the function is shown in Fig. E7-4a. All the signals in the truth table are treated as positive logic signals and the table is partitioned for a type 1 MUX design. The don't cares in the truth table result in two separate choices for the subfunctions

A	B	C	F	F
0	0	0	1	$\overline{C}$
0	0	1	0	
0	1	0	0	C
0	1	1	1	
1	0	0	0	$C\bullet- = 0$ or C
1	0	1	X	
1	1	0	1	$\overline{C} + C\bullet- = 1$ or $\overline{C}$
1	1	1	X	

(X and – both
represent don't
cares)

(a)

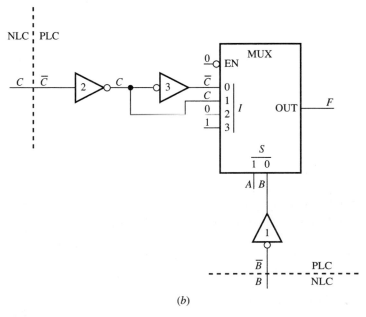

(b)

FIGURE E7-4

Original
Signal List:
$F, A, B[\text{NL}], C[\text{NL}]$

Modified
Signal list:
$F, A, \sim B, \sim C$

A	B	C	F
0	0	0	1
0	0	1	0
0	1	0	0
0	1	1	1
1	0	0	0
1	0	1	X
1	1	0	1
1	1	1	X

A	~B	~C	F	F
0	1	1	1	
0	1	0	0	~C
0	0	1	0	
0	0	0	1	$\overline{\sim C}$
1	1	1	0	
1	1	0	X	$\overline{\sim C} \cdot -\ = 0 \text{ or } \sim C$
1	0	1	1	
1	0	0	X	$\sim C + \overline{\sim C} \cdot -\ = 1 \text{ or } \sim C$

(c)

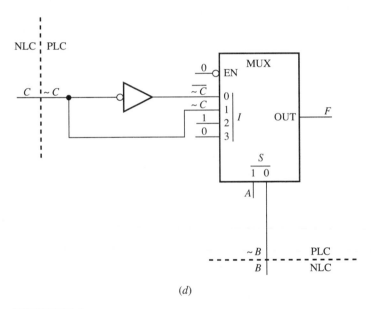

(d)

FIGURE E7-4

for minterms 2 and 3. The design is shown in Fig. E7-4*b* with Inverters (1 and 2) added to the negative logic signal lines *B* and *C*. Inverters 2 and 3 can obviously be replaced by a single line connecting the negative logic signal line *C* to the I0 input line of the MUX and a single Inverter connecting the negative logic signal line *C* to the *I*1 input line of the MUX. The resulting design thus requires only two Inverters.

The second method (method 2) is a little more clever since we will modify the truth table such that it actually represents all positive logic signals. The modified truth table is illustrated in Fig. E7-4*c*. First the truth table is generated for the function using both positive and negative logic signals as stated in the problem description. Next, a modified truth table is generated using only positive logic signals, thus fixing all signal mismatching problems. The implementation for the function is shown in Fig. E7-4*d*. For this type 1 MUX design only one Inverter is required. Since the minterms for the modified truth table have been altered, care must be exercised to insure that the required subfunctions connect to the correct I*i* MUX inputs as shown in the figure.

If one wishes to implement an equation, that is, a Boolean function, using a MUX implementation rather than random logic, there are three alternatives for obtaining the minterms of the Boolean function: (1) Expand the function into the canonical or standard form, then write the function in minterm compact form, (2) plot the Karnaugh map, or (3) fill in the output column of the truth table.

Example 7-5. Given the following Boolean function, obtain the minterms for the function: (1) by expanding the function and writing the minterm compact form, (2) by plotting the Karnaugh map, and (3) by filling in the output column of the truth table.

$$F = A \cdot \overline{B} + A \cdot \overline{C} + \overline{A} \cdot B \cdot C$$

Signal list: *F, A, B, C*

Implement a type 1 MUX design for the function using an IEEE rectangular-shape MUX symbol.

Solution

1. The function is expanded using the technique presented in Section 1-7 in Chapter 1 as follows using Postulate P5*a*, $X + \overline{X} = 1$ to supply the missing literals in each product term.

$$F = A \cdot \overline{B} + A \cdot \overline{C} + \overline{A} \cdot B \cdot C$$

$$= A \cdot \overline{B} \cdot (C + \overline{C}) + A \cdot (B + \overline{B}) \cdot \overline{C} + \overline{A} \cdot B \cdot C$$

$$= A \cdot \overline{B} \cdot C + A \cdot \overline{B} \cdot \overline{C} + A \cdot B \cdot \overline{C} + A \cdot \overline{B} \cdot \overline{C} + \overline{A} \cdot B \cdot C$$

$$= \Sigma \, m(5, 4, 6, 3)$$

or $$= \Sigma \, m(3, 4, 5, 6)$$

Shannon's Expansion Theorem T11*a* could also be used to expand the function to the canonical form, but this often requires more effort than using Postulate P5*a*.

2. The function when written in terms of its product terms can be expressed as

$$F = A \cdot \overline{B} + A \cdot \overline{C} + \overline{A} \cdot B \cdot C$$

or $$F = p_1 + p_2 + p_3$$

where $p_1 = A \cdot \overline{B}$, $p_2 = A \cdot \overline{C}$, and $p_3 = \overline{A} \cdot B \cdot C$.

Each product term can then be plotted on a three-variable Karnaugh as shown in Fig. E7-5*a*, thus identifying all the minterms for the function. This technique was presented in Example 3-3 in Chapter 3.

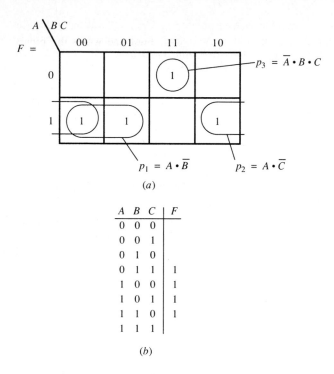

(a)

A	B	C	F
0	0	0	
0	0	1	
0	1	0	
0	1	1	1
1	0	0	1
1	0	1	1
1	1	0	1
1	1	1	

(b)

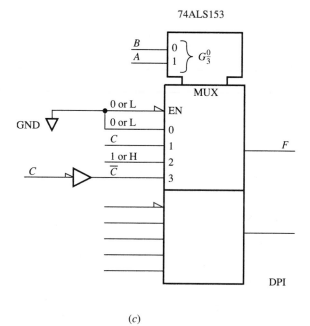

(c)

FIGURE E7-5

3. The technique of filling in the output column of the truth table from a Boolean function was suggested in Chapter 1, Section 1-4. First one writes the binary combinations for all the input variables in truth table format. Next, use each product term in the function to write the 1s of an uncomplemented function or the 0s of a complemented function in the output column. For example, product term $A \cdot \overline{B}$ causes a 1 in the output column for every combination of the input variables where $A = 1$ and $B = 0$. The other product terms are handled in a similar manner. After all the 1s are determined and entered in the output column as shown in Fig. E7-5*b*, 0s can then be entered in the designated vacant spots in the truth table.

A type 1 MUX design is shown in Fig. E7-5*c* using a rectangular-shape symbol for a 4-to-1 line MUX (a 74ALS153) in a diagram using direct polarity indication. The symbol for the 74ALS153 shows a common-control block with select lines that control two independent 4-to-1 line MUXs. Only the signal lines in the top MUX section need to be labeled, since the bottom section is identical to the top section. Note: Read Section 2.0 Symbol Composition in *Overview of IEEE Standard 91-1984, Explanation of Logic Symbols* in Appendix A for more information concerning the common-control block.

Notice that a second independent function could be implemented with the bottom half of the 74ALS153 device in Fig. E7-5*c*.

With the versatility and ease of design provided by Multiplexers or Data Selectors, they are considered one of the handiest design tools in a designer's bag of tricks. If a larger MUX is needed, then a Multiplexer tree can be constructed. Since Multiplexers are single output devices, this can be considered their major drawback. Implementing several functions requires several Multiplexers. Multiplexer designs, however, are usually easier to understand (compared to discrete logic circuits), and consequently easier to troubleshoot if a circuit board or system using a Multiplexer needs repair. In addition, Multiplexer designs often require fewer interconnections and take up less real estate; this is important from a reliability and cost perspective.

7-4-1   Other Uses for Multiplexers

The bus-organized system block diagram shown in Fig. 7-7*a* illustrates an application where Multiplexers are heavily used. A data bus consisting of 8 lines (a bus

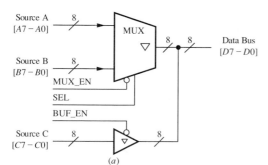

(*a*)

FIGURE 7-7

(*a*) Bus-organized system block diagram.

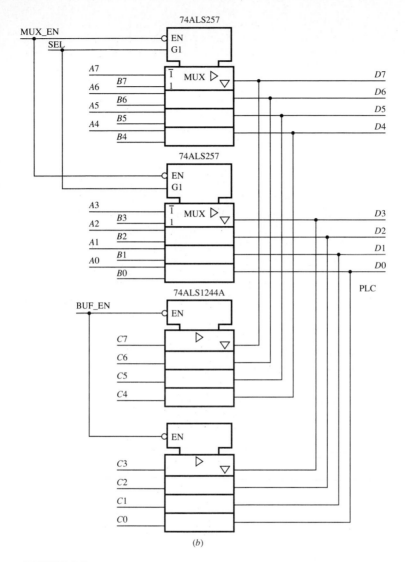

FIGURE 7-7
(b) Implementation for the system block diagram.

is a set of signal lines that provides a similar function) can accept 8 bits of data from three separate sources (either from source *A*, *B*, or *C*). The block diagram for the Multiplexer in the figure requires 2-to-1 line MUXs. The Multiplexer requires only one select input and has three-state outputs (denoted by the triangular symbol placed beside the output lines inside the symbol). The three-state outputs are controlled by a three-state enable input by signal MUX_EN. When the three-state enable input is enabled, MUX_EN = 0, the outputs perform normally; however,

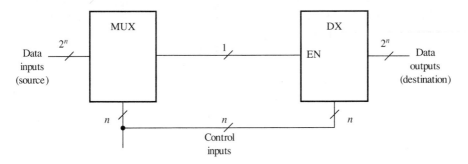

FIGURE 7-8
Block diagram for a communications link using a Multiplexer on the transmitting end (source) and a Demultiplexer on the receiving end (destination).

when disabled, the outputs take on a third high impedance state and thus have no logical significance. When the MUX outputs are in their high impedance state, $MUX\_EN = 1$, the three-state outputs of the Buffer can provide Source C data to the Data Bus by enabling the Buffer's three-state enable input using signal BUF_EN.

An implementation for the system block diagram in Fig. 7-7a is illustrated in Fig. 7-7b. The block labeled MUX is implemented using two packages containing quadruple 2-to-1 line MUXs with three-state outputs, while the Buffer block is implemented using a single package containing an octal buffer and line driver with three-state outputs.

One additional application where Multiplexers are frequently used is for data transmission or communication as illustrated in Fig. 7-8. The combination of a Multiplexer on the transmitting end of a communication link and a Demultiplexer on the receiving end of the link acts to reduce the number of signal lines required between the source and the destination. The length of the communication link is system dependent (it can be across a complicated IC, a PC board, or a communication network). The Multiplexer serves a traditional role in selecting a single signal from many inputs to be transmitted to the Demultiplexer. The Demultiplexer in turn uses the selected signal by means of its enable input and directs it to one of many outputs.

7-5   IMPLEMENTING LOGIC FUNCTIONS
USING MSI DECODERS

Since Decoders do not suffer from a lack of outputs like Multiplexers do, Decoder implementations are normally carried out for Boolean functions with multiple outputs. Function-specific MSI Decoders are readily available as off-the-shelf devices to perform such tasks as BCD-to-Decimal (4-to-10 line Decoder), XS3-to-Decimal (4-to-10 line Decoder), XS3 Gray-to-Decimal (4-to-10 line Decoder), and BCD-to-Seven Segment (4-to-7 line Decoder/Driver). It is a good idea to

check the data books for the logic family one intends to use to see if the MSI
Decoder needed already exists. This is a good rule to follow in general. There
are quite a few MSI and LSI devices available that need not be duplicated by
additional effort on the part of a system designer. Appendix B provides a list of
Digital Logic Devices from Texas Instruments. Browsing through a list of TTL
and CMOS devices such as this helps a designer recognize devices that may be
helpful in a design.

In this section we will discuss how to implement Boolean functions
using general purpose n-to-2^n line Decoders/Demultiplexers (DXs). The general form for a n-to-2^n line Decoder/Demultiplexer is illustrated by the 3-to-8
line Decoder/Demultiplexer shown in block diagram form in Fig. 7-9a. Recall
from subsection 5-5-6 in Chapter 5 that a Decoder with an enable input is also
a Demultiplexer. To construct Decoder/Demultiplexer trees, that is, cascaded
Decoders/Demultiplexers, requires an enable input, and so virtually all MSI
n-to-2^n line Decoders/Demultiplexers have one or more enable inputs. The rectangular-shape graphic symbol for a 3-to-8 line Decoder/Demultiplexer is shown
in Fig. 7-9b. This is the standard IEEE or IEC graphic symbol for a 74ALS138.
Notice that the 74ALS138 actually has three enable inputs (one nonnegated and
two negated) as illustrated by its rectangular-shape symbol. The enable signal
EN changes to the internal logic state of 1 only when all the signal lines that are
connected to the embedded AND gate provide an internal logic state of 1.

The equations for the outputs of a Decoder with n inputs and 2^n data outputs
is

$$Di = m_i \bullet EN, \qquad i = 0 \text{ to } 2^n - 1 \tag{7-2}$$

where m_i is a minterm consisting of n input signals that are applied to the data

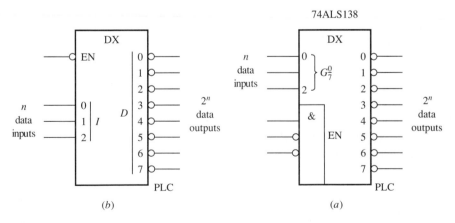

FIGURE 7-9
(a) Block diagram for a 3-to-8 line Decoder/Demultiplexer (DX or DMUX) (b) IEEE rectangular-shape
symbol for a 3-to-8 line Decoder/Demultiplexer.

inputs and EN is the enable (or strobe) input. Decoders with more than one enable input can be cascaded, or connected so that they have a common enable. Decoder equations expressed by Eq. 7-2 are written using the signal names labeled inside the block diagram of the device; therefore, the output equations hold true whether or not a polarity symbol or a negation symbol is present at any of the inputs or outputs of the device.

       With an external OR element connected to the appropriate outputs of a Decoder any Boolean function of n variables can be implemented. This is true since the 2^n outputs of a Decoder provide all the minterms of the n input signals applied to the data inputs of the Decoder. Standard off-the-shelf Decoders are available for $n = 2$, 3, and 4 data inputs. If k Boolean functions are to be implemented using a Decoder, k external OR elements can be required.

       Like Multiplexers, Decoders can easily be cascaded to obtain larger Decoders. If a larger Decoder is needed than is available, existing Decoders can be cascaded to form a Decoder tree (see Fig. 5-20 and Fig. E5-11b in Chapter 5 which show 4-to-16 line Decoder trees using smaller Decoders).

Example 7-6. Design a 4-bit binary-to-gray code code converter with the following truth table description (for more information about gray code refer to Section 2-8 in Chapter 2).

Binary number	Gray code
0000	0000
0001	0001
0010	0011
0011	0010
0100	0110
0101	0111
0110	0101
0111	0100
1000	1100
1001	1101
1010	1111
1011	1110
1100	1010
1101	1011
1110	1001
1111	1000

       Using binary signals $B3$ (MSB) $B2$ $B1$ $B0$ (LSB) and gray code signals $G3$ (MSB) $G2$ $G1$ $G0$ (LSB) we can write the Boolean functions for the outputs of the code converter in terms of the inputs as follows.

$$G3 = \sum m(8, 9, 10, 11, 12, 13, 14, 15) = B3$$

$$G2 = \sum m(4, 5, 6, 7, 8, 9, 10, 11)$$

$$G1 = \sum m(2, 3, 4, 5, 10, 11, 12, 13)$$

$$G0 = \sum m(1, 2, 5, 6, 9, 10, 13, 14)$$

where $m = m(B3, B2, B1, B0)$

Since there are four bits of input information required to determine the minterms, a 4-to-16 line Decoder is utilized as shown in Fig. E7-6. The design uses three external 8-input OR elements to implement the sum of products expressions for $G2$, $G1$, and $G0$. The OR element symbols shown in the figure are the DeMorgan equivalent symbols for NAND gates.

The negated outputs provided by n-to-2^n line Decoders allow us to use NAND gates with a large fanin (number of gate inputs). Recall that OR gates are only available with a fanin of 2. Using a Decoder for this problem only serves to illustrate the design approach. Since gray code is related to binary code by Exclusive OR operators, and vice versa, a simpler design can be obtained using MSI Exclusive OR devices as shown in Fig. 2-8 in Chapter 2. (Figure 2-8 illustrates a 3-bit binary-to-gray code code converter).

Notice in the design in Example 7-6 that there are the same number of 1s in each gray code function as there are 0s. When designing with Decoders, it is important to count the number of 1s and 0s in a function. If there are more 1s, the function is best implemented using the minterm compact form for the 1s of the

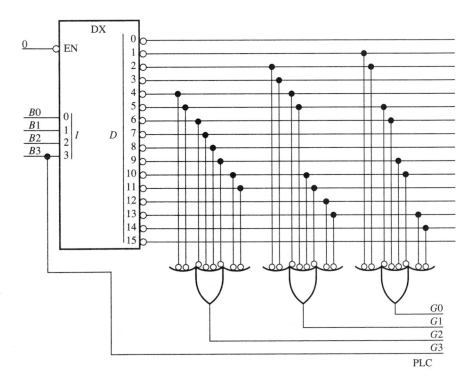

FIGURE E7-6

function; however, if there are more 0s, the function is best implemented using the minterm compact form for the 0s. A circuit implemented with fewer minterms requires gates with a smaller fanin and takes less wiring (or PC board traces). Package count can also be reduced in some cases by using fewer minterms to implement functions.

Consider the following function.

A	B	C	F
0	0	0	1
0	0	1	0
0	1	0	1
0	1	1	1
1	0	0	0
1	0	1	0
1	1	0	1
1	1	1	1

When designing with decoders, a divide-and-conquer approach can be used. Since a Decoder or Decoder tree will be utilized, a design can be represented in two parts as illustrated in Fig. 7-10. The figure suggests a hierarchal structure. The configuration inside the outer box, the system design, utilizes existing MSI devices. The configuration inside the inner box is the gate-level design. The system design involves choosing the Decoder and recognizing that an OR element is usually required to obtain a system output. The gate-level design uses the available output signals supplied by the decoder and the required system output to implement the necessary functions as discussed below.

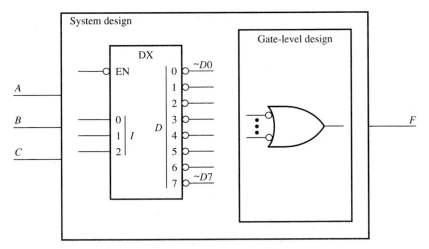

FIGURE 7-10
Designing with Decoders using the divide-and-conquer approach.

First write the Boolean expression using the 0s of the function, since there are only three 0s compared to five 1s.

$$\overline{F}(A, B, C) = \sum m(1, 4, 5)$$

$$\overline{F} = m_1 + m_4 + m_5$$

Applying Decoder Eq. 7-2:

$$Di = m_i, \qquad \text{for } i = 0 \text{ to } 7, \text{ and } EN = 1$$

so

$$\overline{F} = D1 + D4 + D5$$

Signal list: $F, \sim D1, \sim D4, \sim D5$
(signal list for the gate-level design)

Applying the top-down design process, but remembering that

$$\overline{F} = \overline{\sim D1} + \overline{\sim D4} + \overline{\sim D5}$$

(the available form for the gate-level design):

LLP:	F = ~D1 +	~D4 +	~D5
NIP:	0 0	0	0

Using the design documentation, the logic diagram is drawn as shown in Fig. 7-11.

Counting the 1s and 0s and implementing the Boolean expression, that results in the smallest number of minterms, is important when implementing functions using Decoders. The design steps used in the above discussion also serve to illustrate that the top-down design process is always a viable tool for the gate-level design part of a problem.

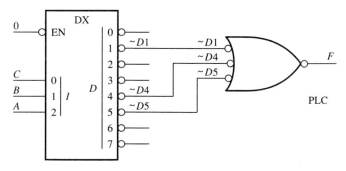

FIGURE 7-11
Implementation for the 0s of the function using a Decoder design.

7-5-1 Analyzing a Decoder Design

Analyzing a Decoder design to obtain the Boolean functions that are being implemented is not difficult. The hardest part is usually determining the order of the Decoder data input signals and the corresponding order of the Decoder data output signals. If a diagram uses direct polarity indication, the output functions are written with level indication by analyzing the circuit. The level indication is then removed to obtain the final functions.

> **Example 7-7.** Obtain the minterm compact forms for the outputs $F1$ and $F2$ in Fig. E7-7.
>
> **Solution** Notice that the circuit uses 74ALS139 dual 2-to-4 line Decoders/Demultiplexers in a useful arrangement to perform the function of a single 3-to-8 line Decoder. The Decoder symbol used in the figure is an alternate IEEE symbol. The inputs are ordered X(MSB) Y Z(LSB). The top Decoder is enabled when X is low, and disabled when X is high. The bottom Decoder is disabled when X is low, and enabled when X is high. When the top Decoder is enabled, the data outputs are $D0$ through $D3$ in the order shown on the symbol. When the Bottom Decoder is enabled, the data outputs are $D4$ through $D7$ in order from top to bottom. The overall 3-to-8 line Decoder has the equation $Di = m_i$, $i = 0$ to $2^n - 1$, since there is no enable input. Each Decoder data output connected to an OR element symbol represents a minterm in the output expression that corresponds to the Decoder data output subscript. When a data output Di inside the Decoder is 1, then the output

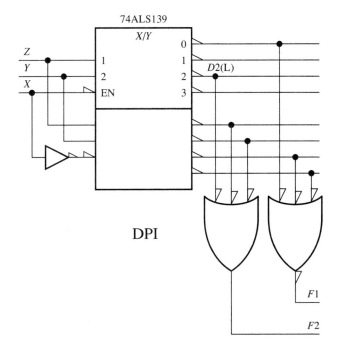

FIGURE E7-7

signal just to the right of the polarity indicator is Di(L) as shown in Fig. E7-7 for the data output $D2$. Using the same technique for each of the other outputs, one can write the output Boolean functions as shown below (see Section 6-7 in Chapter 6 for a discussion of level indication, that is, (H) and (L)).

$$F2(H) = (D2 + D4 + D5)(H)$$

$$F2 = D2 + D4 + D5$$

$$F2(X, Y, Z) = \sum m(2, 4, 5)$$

and
$$F1 = F1(H) = \overline{F1}(L)$$

$$\overline{F1}(L) = (D0 + D6 + D7)(L)$$

$$\overline{F1} = D0 + D6 + D7$$

$$\overline{F1}(X, Y, Z) = \sum m(0, 6, 7)$$

Remember the Binary-to-Hexadecimal Character Generator design in Chapter 5? Realizing the functional logic diagrams in Fig. 5-35 using NAND gates and Inverters requires a total of 19 IC packages. To illustrate the value of the Decoder design approach, Example 7-8 provides a Decoder solution for the same problem.

Example 7-8. Use a Decoder to design a binary-to-hexadecimal character generator as represented by the truth table in Fig. E7-8a. The outputs of the character generator

Binary inputs				Character generator outputs							Hexadecimal character displayed
D	C	B	A	OA	OB	OC	OD	OE	OF	OG	
0	0	0	0	1	1	1	1	1	1	0	0
0	0	0	1	0	1	1	0	0	0	0	1
0	0	1	0	1	1	0	1	1	0	1	2
0	0	1	1	1	1	1	1	0	0	1	3
0	1	0	0	0	1	1	0	0	1	1	4
0	1	0	1	1	0	1	1	0	1	1	5
0	1	1	0	1	0	1	1	1	1	1	6
0	1	1	1	1	1	1	0	0	0	0	7
1	0	0	0	1	1	1	1	1	1	1	8
1	0	0	1	1	1	1	1	0	1	1	9
1	0	1	0	1	1	1	0	1	1	1	A
1	0	1	1	0	0	1	1	1	1	1	b
1	1	0	0	1	0	0	1	1	1	0	C
1	1	0	1	0	1	1	1	1	0	1	d
1	1	1	0	1	0	0	1	1	1	1	E
1	1	1	1	1	0	0	0	1	1	1	F

(a)

FIGURE E7-8

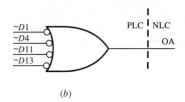

(b)

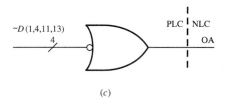

(c)

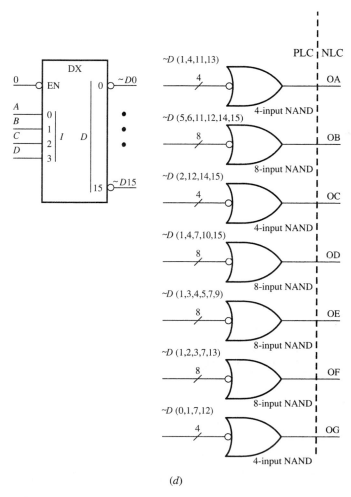

(d)

FIGURE E7-8

are to be connected via current limiting resistors to a common anode seven-segment display. Assume that the inputs D through A are positive logic signals.

Solution The problem can be solved by the divide-and-conquer approach illustrated in Fig. 7-10. Since each of the generator outputs contain fewer 0s than 1s, the following output functions are written using the 0s of the functions. The Decoder Eq. 7-2 is then applied, assuming that the Decoder is enabled (EN = 1).

$$\overline{OA}(D,C,B,A) = \Sigma\ m(1,4,11,13) = D1 + D4 + D11 + D13$$

$$\overline{OB}(D,C,B,A) = \Sigma\ m(5,6,11,12,14,15) = D5 + D6 + D11 + D12$$
$$+ D14 + D15$$

$$\overline{OC}(D,C,B,A) = \Sigma\ m(2,12,14,15) = D2 + D12 + D14 + D15$$

$$\overline{OD}(D,C,B,A) = \Sigma\ m(1,4,7,10,15) = D1 + D4 + D7 + D10 + D15$$

$$\overline{OE}(D,C,B,A) = \Sigma\ m(1,3,4,5,7,9) = D1 + D3 + D4 + D5 + D7 + D9$$

$$\overline{OF}(D,C,B,A) = \Sigma\ m(1,2,3,7,13) = D1 + D2 + D3 + D7 + D13$$

$$\overline{OG}(D,C,B,A) = \Sigma\ m(0,1,7,12) = D0 + D1 + D7 + D12$$

The combined signal list is OA[NL] through OG[NL], $\sim$D0 through $\sim$D15. Notice that the system outputs OA[NL] through OG[NL] are all negative logic signals (this is required for driving a common anode-type seven-segment display), and that the outputs of the Decoder are all negated output signals. Since all the Boolean functions have the same form, we only need to solve for one output to see the pattern for all the other system outputs.

Writing the system output for $\overline{OA}$ in available form we obtain:

$$\overline{OA} = \overline{\sim D1} + \overline{\sim D4} + \overline{\sim D11} + \overline{\sim D13}$$

so,

LLP:	$\overline{OA}$ = $\sim D1$ +	$\sim D4$ +	$\sim D11$ +	$\sim D13$
NIP:	1 0	0	0	0

Notice that 1 is written for $\overline{OA}$ in the negation indication part of the design documentation, since output OA is a negative logic signal. It is also valid to assume that output OA is a positive logic signal, obtain the circuit, and then add an external Inverter to signal line OA to fix the signal mismatching problem. Using the design documentation, the logic diagram is drawn for output OA as shown in Fig. E7-8b.

To simplify our diagrams, the output circuit for OA is shown in an abbreviated form in Fig. E7-8c.

The Decoder design for the Binary-to-Hexadecimal Character Generator is shown in Fig. E7-8d. Each of the outputs has the same form as the OA output. This implementation, however, requires only 7 IC packages and results in a savings of 12 IC packages over the gate-level design. The Decoder design is more desirable because it requires less real estate on a PC board. It is interesting to note that a type 1 Multiplexer design of the Binary-to-Hexadecimal Character Generator would require one Multiplexer for each output function. Two additional packages of Inverters (method 1) or only one package of Inverters (method 2) are required to complete the Multiplexer design. See Example 7-4 for an explanation of the two methods for handling negative logic signals. The methods are general and apply to all positive logic devices as repeated below for Decoders.

Since a positive logic Decoder (or any positive logic device) requires positive logic signals, negative logic signals can be handled in the following ways: (method 1) assume that the negative logic signals are positive logic signals, obtain the circuit, and add Inverters to the negative logic signal lines, or (method 2) modify the truth table using only positive logic signals, and obtain the circuit (do not add Inverters, since they are not needed). Method 2 can require fewer Inverters and is therefore preferred in a tight situation (when PC board real estate is critical).

7-5-2   Other Uses for Decoders

When a Decoder enable input is used for the data input then the function being performed by the Decoder is the Demultiplexer function. This application was illustrated in Fig. 7-8 where a Multiplexer is used to select one signal among many inputs, and a Demultiplexer is used to connect the selected signal by means of its enable input to one of many outputs. This application is often used in the telephone industry to connect the telephone line of a caller at one location to the line at the intended destination.

System designers can use Decoders to decode the binary address issued to a specific input or output device (I/O device) or memory device (read only memory—ROM, random access memory—RAM, or a register—a single group of storage locations, or bits) in a bus-organized system. A memory or store is simply an array of registers that are grouped together. The address bus contains the information necessary to locate all memory and I/O devices prior to sending or receiving data in a bus-organized digital system. Address-decoding devices perform the function of selecting a specific memory or I/O device in the system based on the binary number for the device appearing on the address bus. The actual data is transferred over bidirectional data lines on the data bus. A simple application of an address-decoding device connected to 5 bits of the address bus for decoding one specific memory or I/O device (device $A15$(MSB) $A14$ $A13$ $A12$ $A11 = 10100$) is illustrated by the block diagram in Fig. 7-12.

Typically, the most significant address bits are used in the location process (decoding process), and the least significant bits are reserved for communication with the selected device (either memory or I/O) after contact has been established. If all the allotted address-locator bits are used to address a particular device, the decoding scheme is referred to as absolute addressing (or fully decoded); otherwise, partial addressing is being used. Since partial addressing uses fewer address bits, less decoding is required. Figure 7-13 shows one possible circuit implementation for the system block diagram illustrated in Fig. 7-12. In this application

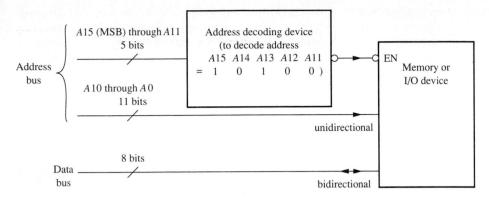

FIGURE 7-12
Address decoding device for a bus-organized system.

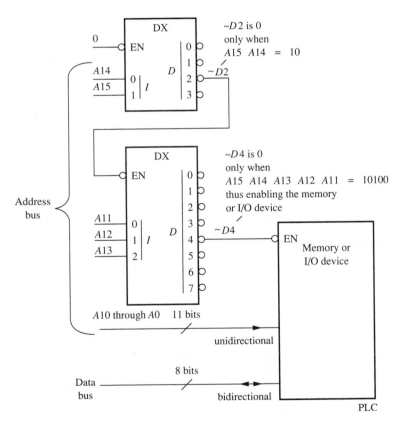

FIGURE 7-13
Implementation using Decoders for the address-decoding device shown in Figure 7-12.

the address-decoding device is referred to as simply an address decoder. It should be noted that IC chip designers also use address-decoders to decode extensively every individual register or storage location on a memory chip.

7-5-3 Using a Comparator for Address Decoding

The address-decoder shown in Fig. 7-13 uses absolute addressing; however, the memory or I/O device is hardwired (permanently connected) to a particular decoder output. An alternate and somewhat more useful decoding scheme that provides the opportunity to change the address of the memory or I/O device without rewiring the circuit is shown in Fig. 7-14. In this case an MSI 8-bit Identity Comparator is used with one set of inputs connected to a switch pack, and the other set connected to the appropriate address bits. When the single-pole single-throw switches in the switch pack are set to the device-select number ($A15$(MSB) $A14$ $A13$ $A12$ $A11$ = 10100) as shown in the figure, the output of the comparator is a logic 0.

The symbol for the 74ALS520, 8-bit Identity Comparator, shown in Fig. 7-14 is the recommended IEEE symbol. The 74ALS520 features 20 kΩ internal

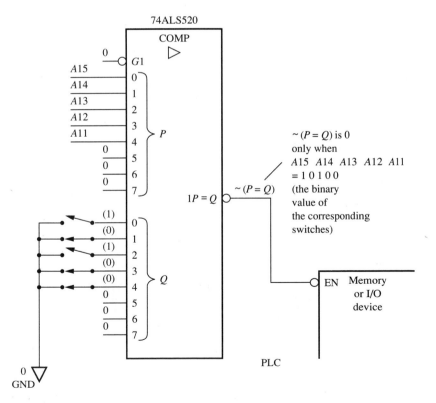

FIGURE 7-14
Implementation using an Identity Comparator for the decoding device shown in Figure 7-12.

pull-up resistors on the Q inputs for switch applications such as we are using. Therefore, when one of the single-pole single-throw switches is in the open position, the corresponding Q-input line of the Comparator is at a logic 1. The state of each switch is shown in parentheses in the figure. By simply changing the switch positions, the circuit can access the memory or I/O device at a different address. This allows different memory and I/O devices to use the same address-decoding scheme by simply setting the switches to a different device-select number without rewiring the circuit.

7-6 IMPLEMENTING LOGIC FUNCTIONS USING EXCLUSIVE OR AND EXCLUSIVE NOR ELEMENTS

In this section we present applications utilizing Exclusive OR and Exclusive NOR elements.

7-6-1 2's Complement Adder/Subtractor

Exclusive OR and Exclusive NOR elements or gates are packaged MSI devices that find a number of uses in implementing special functions. For example, Exclusive OR gates can be used as shown in Fig. 7-15 with a four-bit full adder to complement the B input when the subtraction operation is required. When the SUB input is 1, the B input bits are complemented (1's complement) by the Exclusive OR gates. Adding 1 to the least significant bit of the MSI 4-bit full adder, while supplying the 1's complement of the B input, results in $A - B$,

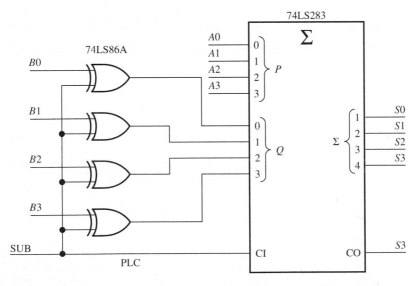

FIGURE 7-15
2's complement Adder/Subtractor using a 4-bit Adder with Exclusive OR gates.

or subtraction by addition of the 2's complement of the subtrahend (see Section 2-6 in Chapter 2). When the SUB input is 0, 0 is added to the least significant bit and the Exclusive OR gates act as noninverting buffers. The operation performed by the circuit is $A + B$. The 74LS283 has full internal carry lookahead across all four bits, thus providing each carry term in 10 ns (for the carry lookahead concept see subsection 5-4-2 in Chapter 5). The symbol used for the 74LS283 is the suggested IEEE symbol.

7-6-2 Odd and Even Functions

Consider the truth table shown in Fig. 7-16a. This table represents all possible binary numbers for five input variables and a unique function that identifies the odd number of 1s (1,3, or 5) in each binary number. The minterms for the binary numbers are also listed to aid in plotting the numbers in the five-variable map

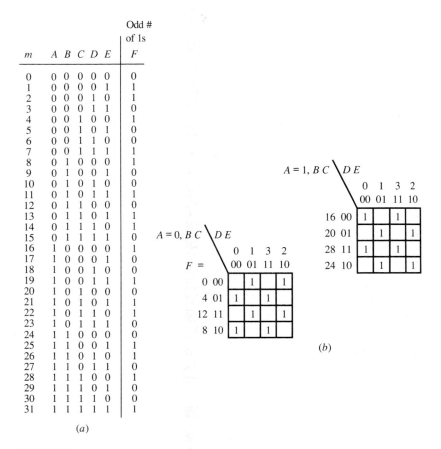

m	A	B	C	D	E	F
0	0	0	0	0	0	0
1	0	0	0	0	1	1
2	0	0	0	1	0	1
3	0	0	0	1	1	0
4	0	0	1	0	0	1
5	0	0	1	0	1	0
6	0	0	1	1	0	0
7	0	0	1	1	1	1
8	0	1	0	0	0	1
9	0	1	0	0	1	0
10	0	1	0	1	0	0
11	0	1	0	1	1	1
12	0	1	1	0	0	0
13	0	1	1	0	1	1
14	0	1	1	1	0	1
15	0	1	1	1	1	0
16	1	0	0	0	0	1
17	1	0	0	0	1	0
18	1	0	0	1	0	0
19	1	0	0	1	1	1
20	1	0	1	0	0	0
21	1	0	1	0	1	1
22	1	0	1	1	0	1
23	1	0	1	1	1	0
24	1	1	0	0	0	0
25	1	1	0	0	1	1
26	1	1	0	1	0	1
27	1	1	0	1	1	0
28	1	1	1	0	0	1
29	1	1	1	0	1	0
30	1	1	1	1	0	0
31	1	1	1	1	1	1

(Odd # of 1s heading above column F)

(a)

FIGURE 7-16
(a) Truth table for a function that identifies an odd number of 1s in a binary number, (b) five-variable map with the odd number of 1s identifier function plotted.

shown in Fig. 7-16b. To assist in plotting the minterm numbers in the map, decimal numbers equivalent to the binary entries along the tops and sides of the maps are added. Using the decimal numbers, the minterm locations (cells) where the 1s of the function are placed in the map are easily identified. A 1 (or 0) is placed in the intersecting cell in the map where a decimal row number and a decimal column number adds up to the specified minterm value.

Observing the map of the odd number of 1s identifier function (an odd function), one sees that a checkerboard pattern exists, suggesting that the function cannot be reduced in the normal manner. The checkerboard pattern illustrates that half of the minterms in a five-variable function have an odd number of 1s. For an odd function of n variables there are $2^n/2$ binary numbers with an odd number of 1s. We will now demonstrate how to obtain the expression for the odd function in the five-variable map in Fig. 7-16 using the map reduction technique presented in Section 4-8 in Chapter 4. The definitions of the Exclusive OR function $F1(X,Y) = \overline{X} \cdot Y + X \cdot \overline{Y}$ and the Exclusive NOR (also referred to as the Equivalence, or Coincidence) function $F2(X,Y) = \overline{X} \cdot \overline{Y} + X \cdot Y$ are listed below in terms of the Exclusive OR operator $\oplus$ and the Exclusive NOR operator $\odot$. Since we will use the above operator definitions, including their commutative and associative properties, these are also listed.

Definitions:

$$X \oplus Y = \overline{X} \cdot Y + X \cdot \overline{Y}$$

$$X \odot Y = \overline{X} \cdot \overline{Y} + X \cdot Y$$

Properties:

P1: $\quad X \oplus Y = \overline{X \odot Y}$ $\qquad\qquad\qquad\qquad$ complement

P2: $\quad X \odot Y = \overline{X \oplus Y}$

P3: $\quad X \oplus Y = Y \oplus X$ $\qquad\qquad\qquad\qquad\quad$ commutative

P4: $\quad X \odot Y = Y \odot X$

P5: $\quad (X \oplus Y) \oplus Z = X \oplus (Y \oplus Z) = X \oplus Y \oplus Z$ $\qquad$ associative

P6: $\quad (X \odot Y) \odot Z = X \odot (Y \odot Z) = X \odot Y \odot Z$

P7: $\quad (X \oplus Y) \odot Z = X \oplus (Y \odot Z) = X \oplus Y \odot Z$

Properties P1 through P7 can be easily verified by perfect induction.

Continuing in Fig. 7-17, the map reduction technique is used to systematically reduce the five-variable map to a one-variable map.

In Fig. 7-17a, two variables D and E are entered into the map as map-entered variables and removed as map variables (the variables along the perimeter of the map) as shown in the figure. After mapping D and E into the map, this leaves variables A, B, and C as map variables. The same procedure is performed

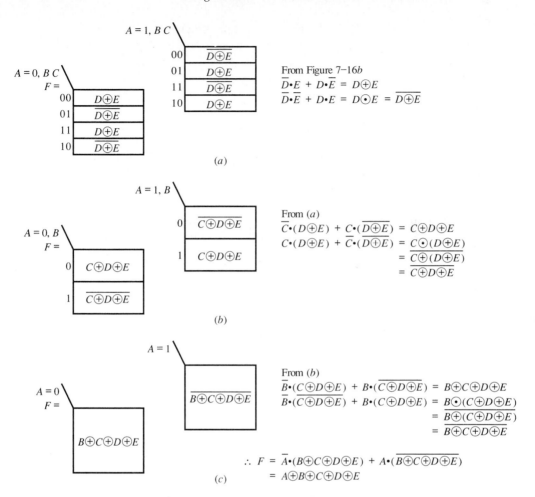

FIGURE 7-17
Solving the odd number of 1s identifier function using the map reduction technique, (*a*) through (*c*).

again in Fig. 7-17*b* with variable *C* using the definitions for the Exclusive OR and Equivalence operations. In Fig. 7-17*c* variable *B* is entered into the map leaving only variable *A* as a map variable. The five-variable odd function can now be written in terms of a set of Exclusive OR operations on the input variables as

$$F = A \oplus B \oplus C \oplus D \oplus E$$

The general form for an odd function (odd number of 1s identifier function) for *n* variables (or bits) can be written as follows:

$$FO = X1 \oplus X2 \oplus \cdots \oplus Xn, \quad \text{where } n \geq 2 \tag{7-3}$$

The general form for an even function (even number of 1s identifier function) for *n* variables is just the complement of Eq. 7-3.

$$FE = \overline{X1 \oplus X2 \oplus \cdots \oplus Xn}, \quad \text{where } n \geq 2 \tag{7-4}$$

Odd and even functions can be implemented utilizing the complement, commutative, and associative properties of the Exclusive OR/Exclusive NOR operators. Some additional properties for the Exclusive OR and Exclusive NOR operators that will allow odd and even functions to be implemented in a variety of different ways (using combinations of Exclusive OR gates, Exclusive NOR gates, and Inverters) are presented below.

Consider an odd number of variables, such as three, in an Exclusive OR relationship. By using the associative property and grouping expressions two at a time, we can write

$$A \oplus B \oplus C = (A \oplus B) \oplus C$$
$$= \overline{(A \oplus B) \odot C}$$
$$= \overline{A} \oplus (B \odot C)$$
$$= A \odot (B \odot C)$$

so
$$= A \odot B \odot C$$

For an odd number of variables, the following property holds true.

P8: $X1 \oplus X2 \oplus \cdots \oplus Xn = X1 \odot X2 \odot \cdots \odot Xn$ for n odd

Consider an even number of variables, such as four, in an Exclusive OR relationship.

$$A \oplus B \oplus C \oplus D = (A \oplus B) \oplus (C \oplus D)$$
$$= \overline{(A \oplus B) \odot (C \oplus D)}$$
$$= \overline{(A \oplus B \odot C) \oplus D}$$
$$= (A \oplus B \odot C) \odot D$$
$$= \overline{A \oplus (B \odot C \odot D)}$$
$$= \overline{A \odot (B \odot C \odot D)}$$

so
$$= \overline{A \odot B \odot C \odot D}$$

For an even number of variables, the following property holds true.

P9: $X1 \oplus X2 \oplus \cdots \oplus Xn = \overline{X1 \odot X2 \odot \cdots \odot Xn}$ for n even

Other useful properties using Exclusive OR and Exclusive NOR operators are obtained as follows.

$$\overline{A} \oplus B = (\overline{A}) \oplus B = \overline{(\overline{A})} \cdot B + (\overline{A}) \cdot \overline{B}$$
$$= A \cdot B + \overline{A} \cdot \overline{B}$$
$$= A \odot B$$

$$\overline{A} \odot B = (\overline{A}) \odot B = \overline{(\overline{A})} \cdot \overline{B} \ + \ (\overline{A}) \cdot B$$

$$= A \cdot \overline{B} \ + \ \overline{A} \cdot B$$

$$= A \oplus B$$

$$\overline{A} \oplus \overline{B} = (\overline{A}) \oplus (\overline{B}) = \overline{(\overline{A})} \cdot (\overline{B}) \ + \ (\overline{A}) \cdot \overline{(\overline{B})}$$

$$= A \cdot \overline{B} \ + \ \overline{A} \cdot B$$

$$= A \oplus B$$

$$\overline{A} \odot \overline{B} = (\overline{A}) \odot (\overline{B}) = \overline{(\overline{A})} \cdot \overline{(\overline{B})} \ + \ (\overline{A}) \cdot (\overline{B})$$

$$= A \cdot B \ + \ \overline{A} \cdot \overline{B}$$

$$= A \odot B$$

From this exercise we can list the following properties.

P10: $\overline{A} \oplus B = A \oplus \overline{B} = A \odot B$

P11: $\overline{A} \odot B = A \odot \overline{B} = A \oplus B$

P12: $\overline{A} \oplus \overline{B} = A \oplus B$

P13: $\overline{A} \odot \overline{B} = A \odot B$

The Exclusive OR and Exclusive NOR Properties P1 through P13 provide us with additional ways to design and analyze logic diagrams utilizing Exclusive OR and Exclusive NOR gates.

One of many possible circuits for a five-variable odd function is shown in Fig. 7-18a using Exclusive OR gates. Figure 7-18b shows an equivalent circuit implementation using Exclusive NOR gates. The circuits in Fig. 7-18c and d are also equally valid circuits for a five-variable odd function. The circuit in Fig. 7-18a was drawn from the five-variable odd function represented by Eq. 7-3. Each of the other circuits in Fig. 7-18 was obtained using the following three-step procedure: (1) write the five-variable odd function using Exclusive OR operators, (2) rewrite the five-variable odd function using one or more of the Exclusive OR or Exclusive NOR operator properties P1 through P13, and (3) draw the new circuit.

Example 7-9. Analyze the circuit implementation in Fig. E7-9 to determine the functions at outputs 1 through 5. Use the Exclusive OR and Exclusive NOR operator properties to convert each output expression to an odd or even function (identified by the forms of Eqs. 7-3 and 7-4). Specify whether each function is an odd or even function.

Solution

$\text{OUT1} = A \odot B = \overline{A \oplus B}$ an even function

$\text{OUT2} = D \odot E \odot F = \overline{D \oplus E} \odot F = D \oplus E \odot F$

$\quad\quad = \overline{D \oplus E \oplus F}$ an even function

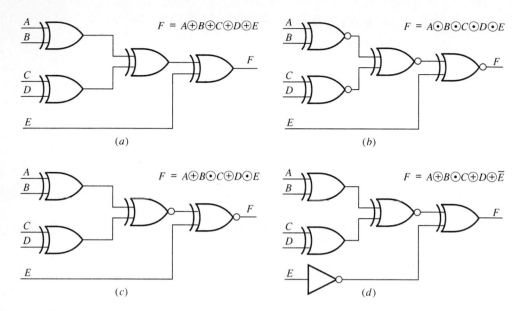

$$F = A \oplus B \oplus C \oplus D \oplus E$$

(a)

$$F = A \odot B \odot C \odot D \odot E$$

(b)

$$F = A \oplus B \odot C \oplus D \odot E$$

(c)

$$F = A \oplus B \odot C \oplus D \oplus \overline{E}$$

(d)

FIGURE 7-18
Different circuit implementations for a five-variable odd function.

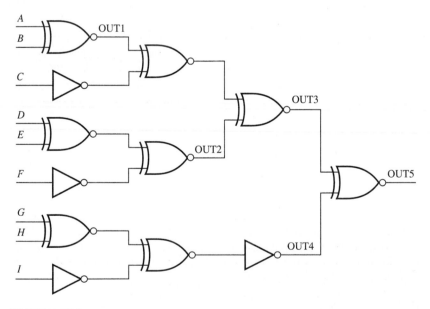

FIGURE E7-9

$$\text{OUT3} = \text{OUT1} \odot \overline{C} \odot \text{OUT2} = \overline{A \oplus B} \odot \overline{C} \odot \overline{D \oplus E \oplus F}$$

$$= \overline{A \oplus B \oplus C} \odot \overline{D \oplus E \oplus F}$$

$$= A \oplus B \oplus C \odot D \oplus E \oplus F$$

$$= \overline{A \oplus B \oplus C \oplus D \oplus E \oplus F} \qquad \text{an even function}$$

$$\overline{\text{OUT4}} = G \odot H \odot \overline{I} = \overline{G \oplus H} \odot \overline{I} = \overline{G \oplus H \oplus I}$$

$$\text{OUT4} = G \oplus H \oplus I \qquad \text{an odd function}$$

$$\text{OUT5} = \text{OUT3} \odot \text{OUT4}$$

$$= \overline{A \oplus B \oplus C \oplus D \oplus E \oplus F} \odot \overline{G \oplus H \oplus I}$$

$$= A \oplus B \oplus C \oplus D \oplus E \oplus F \oplus G \oplus H \oplus I \qquad \text{an odd function}$$

7-6-3 Designing Parity Generator and Parity Checker Circuits

Consider the application illustrated in Fig. 7-19. The figure shows a block diagram for a scheme to detect and report single bit errors that can occur in a communication link between a source and a destination. Such errors, although not frequent, are due to random noise changing one of the transmission bits from a 1 to a 0 or from a 0 to a 1. In the illustration, a circuit called a Parity Generator is used to interrogate the information bits and provide an output parity bit that is included in the transmission bits sent to the destination. A circuit called a Parity Checker is used to interrogate the transmission bit pattern received at the destination. Odd and even functions are used to implement the parity generation and parity checking circuits required in Fig. 7-19 as discussed below.

The parity bit (output PG of the Parity Generator in Fig. 7-19) is used to insure that either an odd number of 1s or an even number of 1s make up the transmission bit pattern. To transmit a bit pattern that uses even parity (even number of 1s), the Parity Generator must provide an output of 1 when the information bit pattern contains an odd number of 1s, so the number of 1s in the transmission

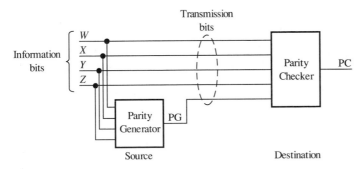

FIGURE 7-19
Single bit error detection scheme.

bit pattern is even. In other words, a Parity Generator used for transmitting even parity must be implemented using an odd function. If a logic 1 is used to signal an error at the destination for an even parity system, the Parity Checker must be implemented using an odd function.

To transmit a bit pattern that uses odd parity (odd number of 1s), the Parity Generator must provide an output of 1 when the information bit pattern contains an even number of 1s, so the number of 1s in the transmission bit pattern is odd. In this case, a Parity Generator used for transmitting odd parity must be implemented using an even function. If a logic 1 is used to signal an error at the destination for an odd parity system, the Parity Checker must be implemented using an even function.

> **Example 7-10.** Design a Parity Generator and a Parity Checker circuit to send four information bits to a destination using the single bit error detection scheme illustrated in Fig. 7-19 for an odd parity system (odd number of 1s make up the transmission bit pattern).
>
> **Solution** The Parity Generator at the source must generate a 1 when an even number of 1s is present in the information bit pattern, and a 0 when an odd number is present. This requires a four-variable even function to implement the Parity Generator. Using the even function (Eq. 7-4) for output PG for four variables results in
>
> $$PG = \overline{W \oplus X \oplus Y \oplus Z} \ = W \oplus X \odot Y \oplus Z$$
>
> If a 1 is used to signal an error at the destination, a five-variable even function must be used to implement the Parity Checker to check for odd parity. If an odd number of bits are present, everything is fine and the Parity Checker output PC is 0; however, if an even number of bits is present, an error is assumed to have occurred during transmission and PC is 1. Using the even function (Eq. 7-4) for output PC for five variables results in
>
> $$PC = \overline{W \oplus X \oplus Y \oplus Z \oplus (PG)} \ = W \oplus X \oplus Y \oplus Z \odot (PG)$$
>
> The Parity Generator and Parity Checker circuits necessary for transmitting and detecting five bits with odd parity are shown in Fig. E7-10.

Parity generation and parity checking are so common that special off-the-shelf MSI 9-bit odd/even Parity Generator/Checker devices are available to assist

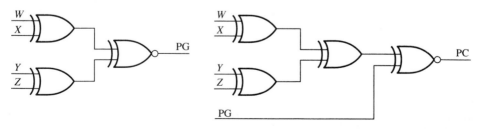

FIGURE E7-10

the designer in performing this task (see Appendix B (Digital Logic Devices) for available families). With the background we have provided in this section, it should be relatively easy to understand and utilize these special purpose devices.

The Hamming code is one of the most useful codes for detecting and correcting data transmission errors of one or more bits. This scheme involves using multiple parity bits to locate a bit position in error so that it can be corrected. The parity generation circuits and parity checking circuits for the Hamming code are also implemented using odd and even functions. Exclusive OR and Exclusive NOR gates are naturally used in the implementation of these circuits. We will not pursue the study of these specialized circuits in this text (see Hill and Peterson or Johnson and Karim in the list of references).

7-7   IMPLEMENTING LOGIC FUNCTIONS USING PROGRAMMABLE DEVICES

In this section we discuss the three basic architectures for programmable logic devices (PLDs). These are the programmable read only memory (PROM), the programmable array logic (PAL—a registered trademark of Monolithic Memories, Inc.), and the programmable logic array (PLA). Each of these configurations consist of AND elements followed by OR elements, so that the sum of products forms of Boolean functions can be implemented directly. A programmable device is a semicustom device since a user can tailor it to his or her particular need.

Figure 7-20 shows the generalized programmable logic device architectures for PROMs, PALs, and PLAs. The basic differences between these three types of PLDs are whether the AND and OR arrays are fixed or programmable. Other types of programmable architectures are constantly being introduced by manufacturers; however, these newer architectures are typically enhanced variations of these three basic types. The last topic we will present in this section will be multi-level PLDs which provide higher density and more functionality with a slightly different architecture.

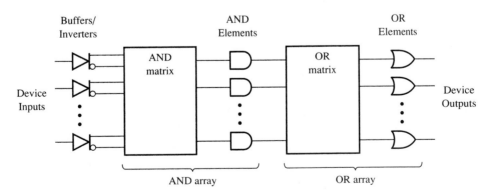

FIGURE 7-20
Generalized programmable logic device (PLD) architecture (for combinational PLDs).

Notice that each device type contains a set of Buffers/Inverters to provide both uncomplemented (or true) and complemented signals to the AND array. The AND and OR elements are shown with only one input for ease of drawing (these elements often contain as many as 32 inputs or more in some devices). Table 7-1 is a list of each combinational PLD and the type of AND and OR array it contains.

From a designer's point of view, the PROM with the fixed AND array and programmable OR array is the least versatile programmable device. The PAL with the programmable AND array and fixed OR array is a moderately versatile programmable device. The PLA with both programmable AND array and OR array is the most versatile programmable device, but is also typically the most expensive, a little harder to program, and a little slower.

In this presentation we will only discuss fuse programmable devices (bipolar technology). Fuse programmable devices have fuses that are blown during the programming process rendering the device one-time programmable. Once a fuse is blown, it is permanent. Rather than using fuses in the AND and OR matrices, some manufacturers offer ultraviolet (UV) erasable CMOS devices, while other manufacturers offer floating-gate electrically erasable CMOS devices. These devices simply mimic fuses. Electrical charge can be stored by a CMOS device (representing an intact or unblown fuse) or removed or erased (representing a blown fuse), either by light or electrically by the tunneling effect. Since fuses are conceptually easier to understand, they will be used throughout our discussions. PLDs utilizing fuses arrive from the manufacturer with all fuses intact, and selected fuses are blown as required during the programming process. PLDs utilizing UV or electrically-erasable CMOS devices arrive from the manufacturer erased, and charge is added to selected devices during the programming process. The UV and the electrically-erasable devices can be effectively used during the development phase of a project. During this phase mistakes are generally made and can be easily corrected by simply erasing the device, correcting the mistake, and reprogramming the device. A fused PLD must often be discarded when a programming mistake is make, unless an unblown section can be used to reprogram the device. Sometimes an application requires a device that can be erased while installed in the circuit. Electrically-erasable devices are the choice in this case.

TABLE 7-1
Combinational PLDs

Device	AND array	OR array
PROM	fixed	programmable
PAL	programmable	fixed
PLA	programmable	programmable

After a PLD is programmed and tested, it may be desirable to mask program the device. A mask is a data pattern that is created and used in the manufacturing process by the manufacturer. After the specification for a device is finalized by the user, a master device or computer data file is created and sent to the manufacturer. The manufacturer creates a mask to perform the same function as the fuses. Mask devices are generally less expensive than programmable devices when large quantities are needed. Once the mask(s) are created and the devices are manufactured they are no longer programmable. Errors detected at this stage of the process are expensive to fix. Fused, UV erasable, and electrically erasable devices are often called field-programmable devices since they can be programmed by designers outside the respective company or in the field. Field-programmable devices are the norm in a company that designs high tech digital equipment. The following three items are required to program PLDs: (1) a software package written for the desired devices, (2) a device programmer, and (3) a computer that will interface with the device programmer and will run the software package.

One might asked why use PLDs since SSI and MSI devices are readily available. It is desirable to use programmable devices for the following reasons:

1. to decrease PC board real estate (less PC board area means less cost) by reducing the package count (compared to only SSI perhaps 15 to 20 packages can be replaced by 1 package, compared to SSI/MSI perhaps 4 to 8 packages can be replaced by 1 package)
2. to shorten design time (compared to application specific—or custom—ICs (ASICs), PLD implementations can be achieved faster)
3. to allow design changes (reprogramming PLDs is less time consuming than redesigning a complete PC board using random logic or MSI logic devices)
4. to improve reliability (fewer packages means fewer interconnections and thus greater reliability)

Figure 7-21 is a summary of the circuit symbology adopted by several programmable device manufacturers for PLDs. The symbology is provided to help designers understand the PLD circuit configurations provided in data books. The intersection of two lines in the fuse matrix covered with an X as illustrated in Fig. 7-21a is a programmable connection (see the simplified intact fuse representation). The intersection of two lines without an X illustrated in Fig. 7-21b is an open connection or no connection (after programming). An intersection with a dot illustrates a permanent connection as shown in Fig. 7-21c. The AND and OR elements in Figs. 7-21d and e are shown as single-input devices when in fact they have multiple inputs. The symbology in Fig. 7-21f with the X inside the AND element illustrates that all input term fuses are intact (connected) and the resulting product term is 0 (in other words this product term can be ignored in the sum of products for the output). The symbology in Fig. 7-21g illustrates that all the input term fuses are blown (no connection) and results in a product term of 1. The OR element that this output is connected to is also a 1,

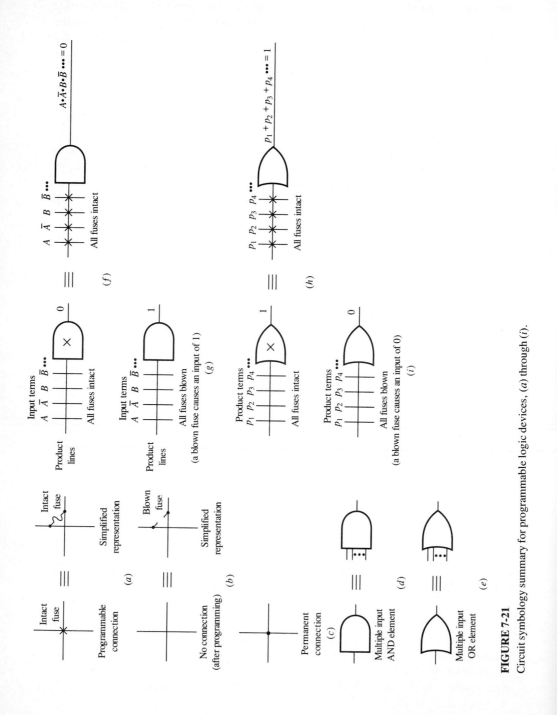

FIGURE 7-21

Circuit symbology summary for programmable logic devices, (*a*) through (*i*).

and is useless for generating an equation. The symbology in Fig. 7-21*h* with the X inside the OR element illustrates that all the product term fuses are intact and results in a sum term of 1. All open connections to an OR element, that is, all product term fuses blown, as shown in Fig. 7-21*i* result in a sum term of 0 at the output of the OR element.

7-7-1   Programmable Read Only Memory (PROM)

The diode matrix (consisting of sets of metallized rows and columns, connected at their crosspoints by diodes and fuses) was introduced in the early 1960s. This was the first programmable integrated circuit logic device. Integrated circuit designers later added an input Decoder and output Buffers, creating the first programmable read only memory (PROM). PROMs are typically used for code converters, character generators, data storage tables, and program stores. Notice that in each of these applications, tabular data rather than Boolean equations is generally used.

    The circuit diagram for a simple 3-input, 4-output PROM is shown in Fig. 7-22. A PROM this size is too small to be commercially available, but we can use it to illustrated the circuitry of a common off-the-shelf PROM. Notice that the AND array is fixed, and the OR array is programmable—the identifying feature of a PROM.

    Notice that the Buffers/Inverters and the fixed AND array for the PROM in Fig. 7-22 represent the circuit for a Decoder (see the Decoder circuit in Fig. 5-19 in Chapter 5). For n inputs this decoder provides all 2^n minterms at the outputs of the AND elements. Since these outputs are supplied to the OR elements via the OR array, PROMs can be used for implementing Boolean equations by simply generating and utilizing the truth tables that represent the equations. The number of programmable fuses in the OR fuse matrix is 2^n (one output for each minterm decoded) times m (the number of outputs) or $2^n \times m$. In many applications where PROMs are used, the inputs are address bus bits and the outputs are data bus bits in a bus-organized structure.

    The following example illustrates the single package solution of the binary to hexadecimal character generator first presented in Section 5-6 of Chapter 5, and, again, in Section 7-5 of this chapter. If you will recall, the solution took 19 IC packages using random logic and 7 IC packages using a Decoder design. The PROM solution is provided below in Example 7-11 and requires only 1 IC package or at most 2 IC packages, if additional current is needed to drive the LED display.

    Example 7-11. Show a design using a PROM to implement the binary to hexadec-imal character generator illustrated by the truth table repeated in Fig. E7-11*a*. The outputs of the character generator are to be connected via current limiting resistors to a common anode seven-segment display. Assume that the inputs *D* through *A* are positive logic signals.

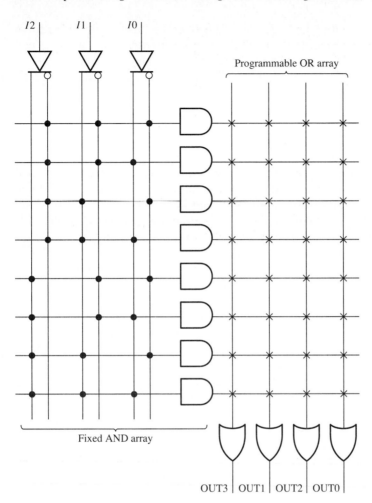

FIGURE 7-22
Simple 3-input, 4-output (8 × 4) programmable read only memory (PROM).

Solution Since the character generator must drive a common anode seven-segment display, the output signals OA through OG are negative logic signals. To fix the signal mismatching problem, we can assume that these negative logic signals are positive logic signals, obtain the PROM design, and add an Inverter at each output. This would require two packages for the implementation; one package for the PROM, plus one package of Inverters. If we complement the outputs in the truth table shown in Fig. E7-11*b* (recall this changes the available negative logic signals to positive logic signals), and use the complemented outputs, then only one package is necessary for the implementation.

When the PROM is programmed for the positive logic output signals ~OA through ~OG in the truth table, the fuses are blown where 0s appear, and left intact where 1s appear. The PROM circuit representation in Fig. E7-11*c* shows where the

the fuse-map information required to program the PROM. When the fuse-map is generated in this form it is called manual fuse-map entry.

Since this problem is very simple, a standalone PROM programming unit (such as a Data I/O 29A with a UniPack2 socket module) can easily be used to manually enter the data 03, 9F, 25, 0D, 99, 49, 41, 1F, 01, 09, 11, C1, 63, 85, 61, and 71 (Hex) for the 16 addresses 00 through 0F (Hex) and instruct the programming unit to blow the fuses. A diagram for the character generator using a 32 × 8 PROM is shown in Fig. E7-11*d*. This is one of the smallest off-the-shelf fuse programmable PROMs available, and is sufficiently large (a larger PROM could also be used) to handle our truth table. The symbol shown in the figure is one of the recommended IEEE graphic symbols for a PROM. Notice that the 27S19 has three-state outputs controlled by the enable input EN.

The 27S19 bipolar PROM will supply a maximum low-level output current of 16 mA ($I_{OH}(MAX) = 16$ mA). If this is not adequate for the seven-segment display, an additional package of octal Buffers can be used as illustrated in Fig. E7-11*e*. The 74ALS244A can supply a maximum low-level output current of 48 mA.

7-7-2   Programmable Array Logic (PAL)

The programmable [AND] array logic device or PAL device was developed and patented by John Birkner at Monolithic Memories, Inc. in 1976. PAL and PLA devices are generally used to implement functions in equation form as opposed to tabular form. Designers primarily choose programmable array logic (PAL) or programmable logic array (PLA) devices to reduce the real estate on a PC board.

Binary inputs				Character generator outputs							Hexadecimal character displayed
D	*C*	*B*	*A*	OA	OB	OC	OD	OE	OF	OG	
0	0	0	0	1	1	1	1	1	1	0	0
0	0	0	1	0	1	1	0	0	0	0	1
0	0	1	0	1	1	0	1	1	0	1	2
0	0	1	1	1	1	1	1	0	0	1	3
0	1	0	0	0	1	1	0	0	1	1	4
0	1	0	1	1	0	1	1	0	1	1	5
0	1	1	0	1	0	1	1	1	1	1	6
0	1	1	1	1	1	1	0	0	0	0	7
1	0	0	0	1	1	1	1	1	1	1	8
1	0	0	1	1	1	1	1	0	1	1	9
1	0	1	0	1	1	1	0	1	1	1	A
1	0	1	1	0	0	1	1	1	1	1	b
1	1	0	0	1	0	0	1	1	1	0	C
1	1	0	1	0	1	1	1	1	0	1	d
1	1	1	0	1	0	0	1	1	1	1	E
1	1	1	1	1	0	0	0	1	1	1	F

(a)

FIGURE E7-11

Binary inputs				Character generator outputs							Hexadecimal character displayed	Complemented outputs						
D	C	B	A	OA	OB	OC	OD	OE	OF	OG		~OA	~OB	~OC	~OD	~OE	~OF	~OG
0	0	0	0	1	1	1	1	1	1	0	0	0	0	0	0	0	0	1
0	0	0	1	0	1	1	0	0	0	0	1	1	0	0	1	1	1	1
0	0	1	0	1	1	0	1	1	0	1	2	0	0	1	0	0	1	0
0	0	1	1	1	1	1	1	0	0	1	3	0	0	0	0	1	1	0
0	1	0	0	0	1	1	0	0	1	1	4	1	0	0	1	1	0	0
0	1	0	1	1	0	1	1	0	1	1	5	0	1	0	0	1	0	0
0	1	1	0	1	0	1	1	1	1	1	6	0	1	0	0	0	0	0
0	1	1	1	1	1	1	0	0	0	0	7	0	0	0	1	1	1	1
1	0	0	0	1	1	1	1	1	1	1	8	0	0	0	0	0	0	0
1	0	0	1	1	1	1	1	0	1	1	9	0	0	0	0	1	0	0
1	0	1	0	1	1	1	0	1	1	1	A	0	0	0	1	0	0	0
1	0	1	1	0	0	1	1	1	1	1	b	1	1	0	0	0	0	0
1	1	0	0	1	0	0	1	1	1	0	C	0	1	1	0	0	0	1
1	1	0	1	0	1	1	1	1	0	1	d	1	0	0	0	0	1	0
1	1	1	0	1	0	0	1	1	1	1	E	0	1	1	0	0	0	0
1	1	1	1	1	0	0	0	1	1	1	F	0	1	1	1	0	0	0

(b)

FIGURE E7-11

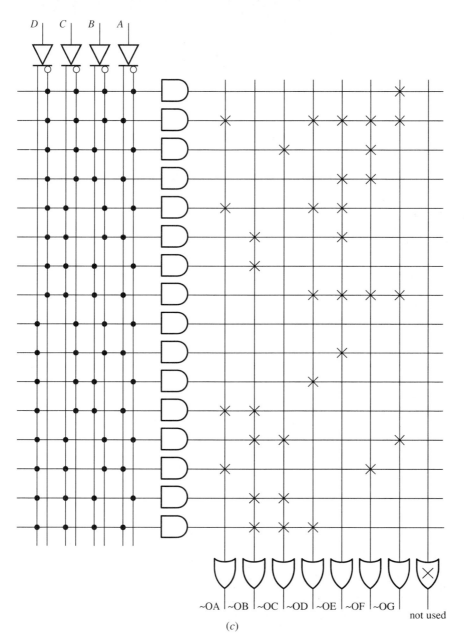

(c)

FIGURE E7-11

Consider the case where a PROM is used with more inputs than are required; there can be a lot of unused logic circuitry in the device. Unused logic circuitry can represent more PC board area, and more power to operate. For instance, in Example 7-11 the character generator implementation uses only 50 percent of the capacity of the 32 × 8 PROM. Now consider the case where a PROM designer

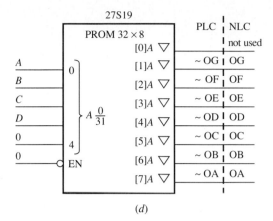

(d)

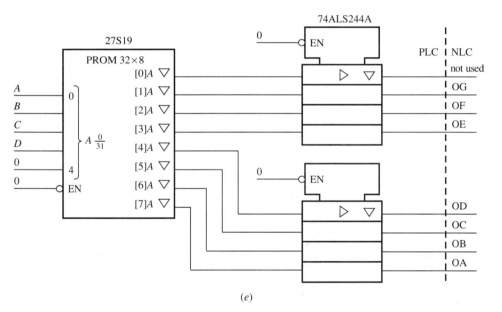

(e)

FIGURE E7-11

adds an input to the design, each additional input doubles the number of AND elements in the fixed AND array and doubles the number of fuses required in the programmable OR matrix. As a result, very large PROM circuits are required to implement functions with many inputs. This can require PROMs that are either not available or that are overly expensive compared to other programmable alternatives such as the PAL and the PLA.

Because the PAL and the PLA both have a programmable AND array, additional inputs are easily added by the PAL or PLA designer without doubling their sizes. PALs and PLAs are provided with a large number of inputs (up to 22 inputs are common for PALs and somewhat fewer inputs for PLAs). This allows Boolean equations with just a few inputs or a very large number of inputs to be implemented using either PALs or PLAs. A simple 3-input, 4-output PAL is illustrated in Fig. 7-23. As a comparison, a PROM provides all the minterms for a limited number of input signals; however, a PAL or a PLA provides product terms (programmable by the user) for a few to a large number of input signals.

Notice that the PAL in Fig. 7-23 has a programmable AND array and a fixed OR array. The number of AND element inputs for our simple PAL is $2n$ where n is the number of external signal (or AND array) inputs. (Note: For simple PALs the maximum number of external signal inputs and AND array inputs is the same, but this is not so for the more complex PALs with inaccessible internal feedback AND array inputs.) Each external signal input is supplied to every AND element in both true and complemented form. The number of product terms available for each OR element output is fixed during the manufacturing process and cannot be changed by the user. This holds down the PAL cost compared to the PLA. Each of the outputs in Fig. 7-23 is limited to only two product terms. Different part types have different OR gate configurations (number of product terms available), and the task of choosing a particular PAL for an implementation rests with the designer. Also, different part types are available with either active low (negated) or active high

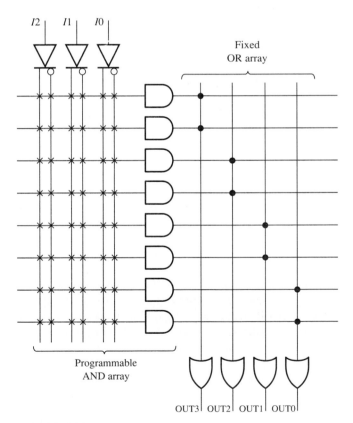

FIGURE 7-23
Simple 3-input, 4-output programmable array logic (PAL).

TABLE 7-2
PAL nomenclature

PAL *ii t oo* (or PALiitoo)

ii represents the maximum number of AND array inputs

t represents the type of outputs

combinational:
  H is active high
  L is active low
  P is programmable polarity
  C is complementary

registered:
  R is registered
  RP is registered with programmable polarity
  V is versatile, that is, programmable output macrocells

oo represents the maximum number of dedicated
(combinational, or registered) or programmed
(combinational and registered) outputs

(nonnegated) outputs. PAL devices have a special numbering system that assists in identifying a particular device. This special nomenclature is illustrated in Table 7-2 for the more common device types.

As an example, the PAL16L8 has a maximum number of 16 external signal (or AND array) inputs, active low (negated) outputs, and a maximum number of 8 dedicated combinational outputs. The logic diagram for the PAL16L8 is shown in Fig. 7-24. We see from the figure that this device can implement 8 Boolean equations with up to 10 inputs, and each equation can contain up to 7 product terms. Up to 6 of the outputs can be programmed as inputs by disabling three state buffers. Programmable input/output or I/O pins gives a PAL much more flexibility in implementing logic functions than a PROM. Because of the programmable I/O pins, the 16L8 PAL can be used to implement two logic functions with a maximum of 16 inputs, and each output can contain a maximum of 7 product terms. For combinational logic implementations of Boolean equations, the 16L8 is quite versatile since it can be used to implement a range of different sizes of equations.

Notice that there are no Xs marking the programmable locations in the AND matrix in Fig. 7-24. Manufacturers generally publish their PLD logic diagrams without the Xs in the fuse matrix so that a user can insert an X in each location where an intact fuse is desired. For small devices this allows PLD logic diagrams to be useful when generating specific functions by manual entry. For larger devices the absence of the Xs keeps the logic diagram from becoming cluttered.

Example 7-12

1. Show a simple PAL design for the four functions mapped in Fig. E7-12*a*. Use the simple 3-input, 4-output PAL in Fig. 7-23 if the equations will fit. If the equations

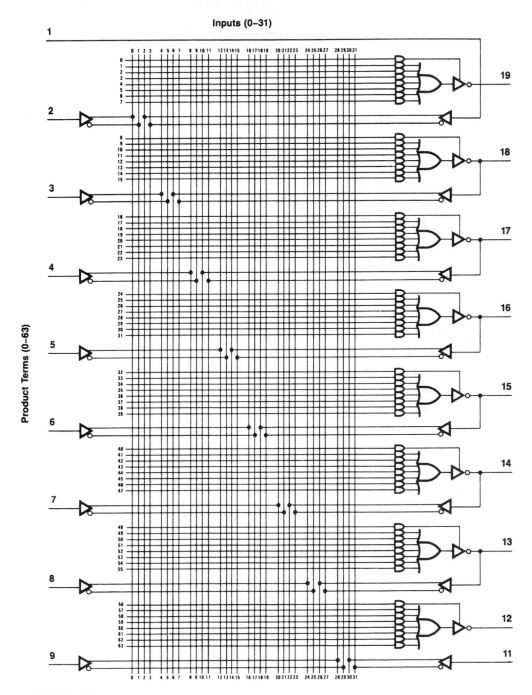

FIGURE 7-24
Logic Diagram PAL16L8. (Courtesy National Semiconductor)

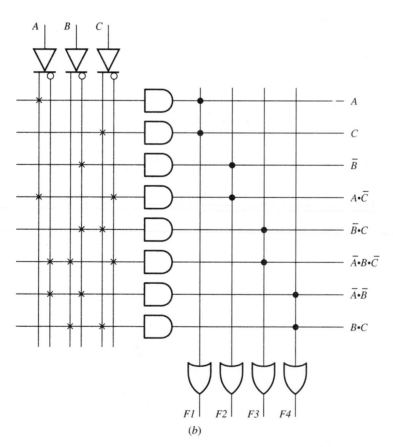

$F1 =$

A\B C	00	01	11	10
0	0	1	1	0
1	1	1	1	1

$F2 =$

A\B C	00	01	11	10
0	1	1	0	0
1	1	1	0	1

$F3 =$

A\B C	00	01	11	10
0	0	1	0	1
1	0	1	0	0

$F4 =$

A\B C	00	01	11	10
0	1	1	1	0
1	0	0	1	0

(a)

(b)

FIGURE E7-12

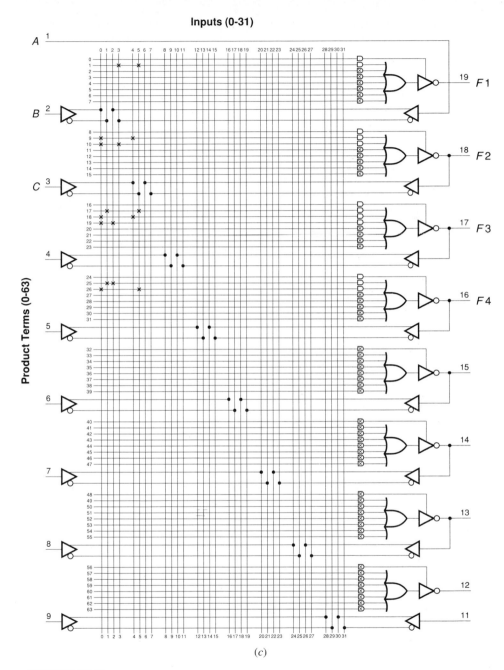

FIGURE E7-12
Logic Diagram PAL16L8. (Logic diagram adapted courtesy of National Semiconductor)

will not fit, draw a PAL large enough. Generate the fuse-map information for the equations manually. Assume that all the signals are positive logic signals, and that each signal is available in its true form (uncomplemented form).

2. Use a PAL16L8 to implement the four functions mapped in Fig. E7-12a. Generate the fuse-map information for the equations manually.

Solution

1. Utilizing the AND/OR circuit architecture of a PAL requires the designer to write the Boolean equations in SOP form. If the PAL has nonnegated or active high outputs, the Boolean equations must be written for the 1s of the functions. If the PAL has negated or active low outputs, the Boolean equations must be written for the 0s of the functions.

    Note that the PAL in Fig. 7-23 has nonnegated outputs. This tells us that the Boolean equations must be written in SOP form using the 1s of the functions to conform to the PAL architecture. These equations are obtained from the Karnaugh maps in Fig. E7-12a and are listed below.

$$F1 = A + C$$
$$F2 = \overline{B} + A \cdot \overline{C}$$
$$F3 = \overline{B} \cdot C + \overline{A} \cdot B \cdot \overline{C}$$
$$F4 = \overline{A} \cdot \overline{B} + B \cdot C$$

The PAL in Fig. 7-23 has 3 inputs, 4 outputs, and each output can contain up to 2 product terms. Since each of the SOP expressions has only 3 inputs and no more than 2 product terms, the equations will fit in this PAL. The PAL implementation for the equations is shown in Fig. 7-12b. By listing the product terms on the output lines of the AND elements, as shown in the figure, the fuse map or fuse pattern for the AND matrix can easily be generated manually. Simply provide an intact fuse for each literal in each product term.

2. Since the PAL16L8 has up to 16 AND array inputs (or external inputs since this is a simple PAL), up to 8 outputs, and up to 7 product terms in each output, this is more than sufficient to satisfy the functions in Fig. E7-12a. The active low or negated outputs of the 16L8, however, require SOP expressions for the 0s of functions to conform to the PAL AOI (AND-OR-Invert) architecture. The SOP expressions for the 0s of the functions are obtained by solving the Karnaugh maps in Fig. E7-12a.

$$\overline{F1} = \overline{A} \cdot \overline{C}$$
$$\overline{F2} = B \cdot C + \overline{A} \cdot B$$
$$\overline{F3} = \overline{B} \cdot \overline{C} + B \cdot C + A \cdot B$$
$$\overline{F4} = A \cdot \overline{B} + B \cdot \overline{C}$$

These or equivalent SOP expressions can also be obtained by first solving for the 1s of the functions, complementing both sides of the functions, and finally applying DeMorgan's theorem to the functions. This approach normally requires minimizing the resulting expressions algebraically, usually an unappealing task.

A 16L8 implementation for the functions in Fig. E7-12*a* is shown in Fig. E7-12*c*. Notice that we have placed Xs in the PAL16L8 logic diagram as required to satisfy the equations. Manual fuse-map entry is rather easy to do for small functions, but becomes quite tedious for large functions. Prior to the introduction of software programs to assist the designer with this task, fuse-map information was manually coded on a form, and then entered into a programming unit to blow the fuses.

Because a PAL or PLA device has a programmable AND matrix, functions that depend on one set of inputs (like the functions in Example 7-12) as well as functions that do not depend on the same inputs can be easily implemented in the same PAL or PLA. In other words, the PAL in Fig. E7-12*c* could be used to implement additional Boolean functions with a different set of signal inputs since there are still a maximum of 10 inputs and 4 outputs that are unused. Notice that it is possible to implement functions with different sets of signal inputs in a PROM, but this is not as easy to accomplish manually due to the fixed AND matrix. PALs and PLAs are often chosen over PROMs because functions can be easily implemented in equation form.

As an example of a registered output (recall that flip-flops or single bit registers will be discussed in Chapter 8), the logic diagram for a PAL16R4 is shown in Fig. 7-25. Notice that PAL16R4 has a maximum number of 16 AND array inputs (this is a more complex PAL and so it only has 12 external signal inputs since four of the AND array inputs do not connect to external pins), and a maximum number of four dedicated registered outputs (the rectangles that contain the letter *D* are single bit registers). Notice that the device also has four dedicated combinational logic outputs.

An example of a versatile PAL, the PAL22V10, is shown in Fig. 7-26. This PAL has a maximum number of 22 AND array inputs, and a maximum number of 10 independently programmable output macrocells (individually programmable by the user as combinational or registered for maximum flexibility). Each macrocell contains a flip-flop, a 4-to-1 line Multiplexer, and a 2-to-1 line Multiplexer. Programming the select inputs of the 4-to-1 line MUX provide the following four types of outputs: (1) a registered active low output (for MUX select signals 0 0), (2) a registered active high output (for MUX select signals 0 1), (3) a combinational active low output (for MUX select signals 1 0), or (4) a combinational active high output (for MUX select signals 1 1).

7-7-3   Programmable Logic Array (PLA)

The first [Field] Programmable Logic Array (PLA) was introduced in 1975. A simple 3-input, 4-output PLA with a programmable OR array is shown in Fig. 7-27. The number of product terms available for each output of the PLA is programmable by the user. Product terms can, therefore, be shared by PLA outputs, but product terms cannot be shared by PAL outputs due to their fixed OR array. Because a PLA has two programmable arrays compared to only one

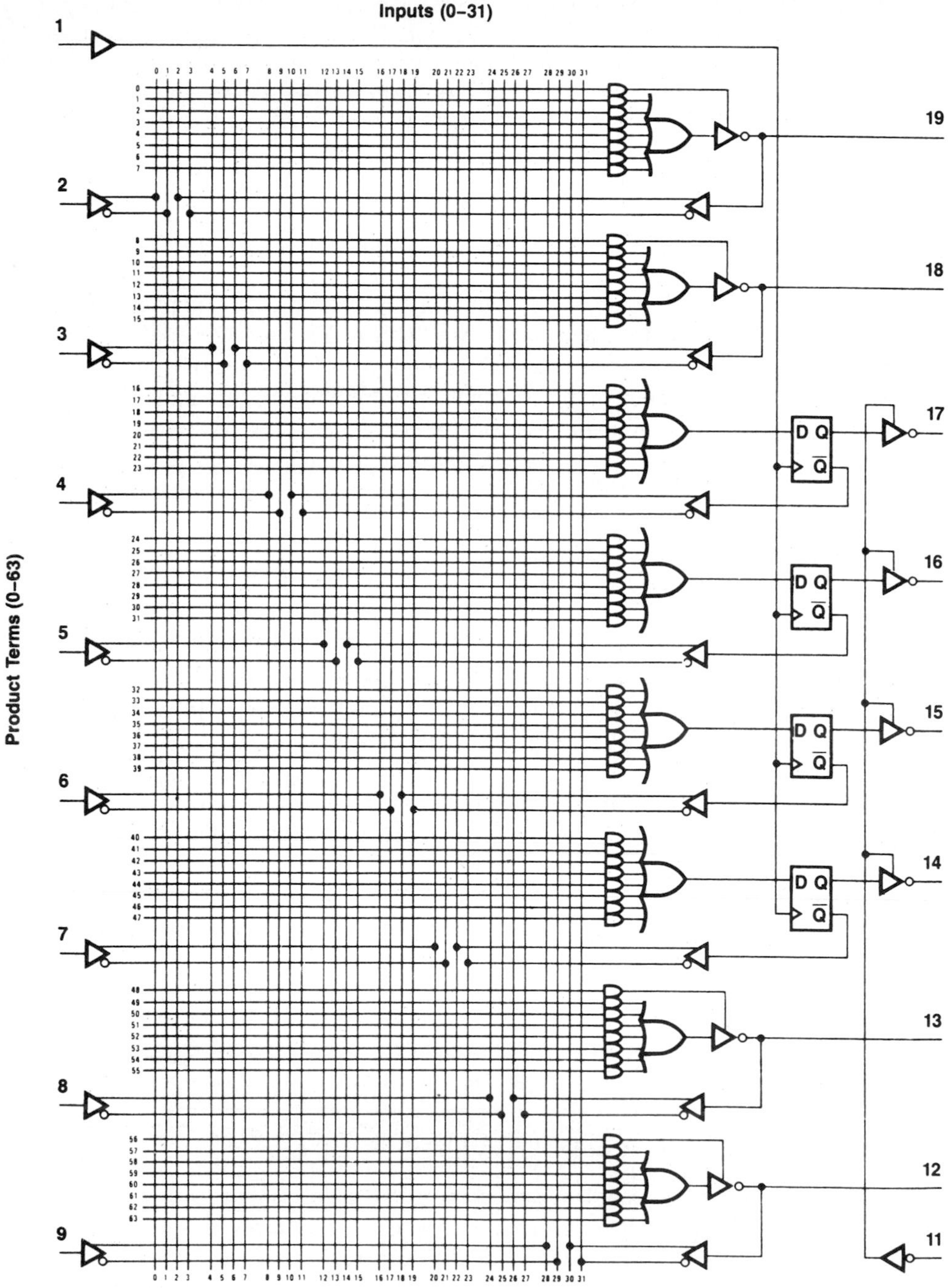

FIGURE 7-25
Logic Diagram PAL16R4. (Courtesy National Semiconductor)

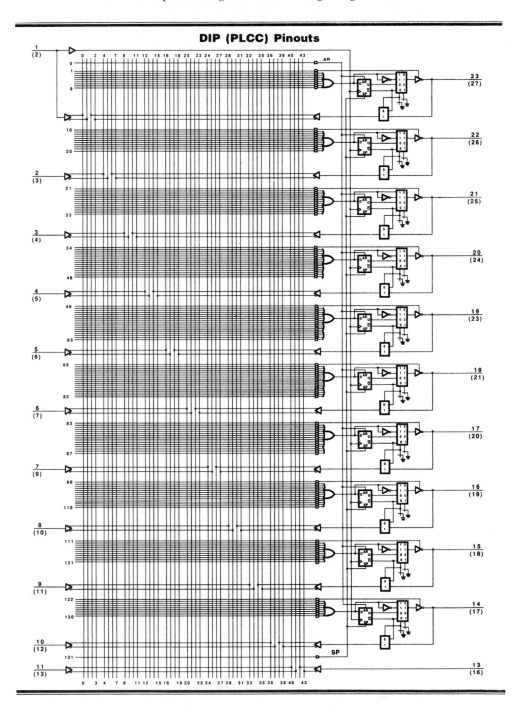

FIGURE 7-26
Logic Diagram PALC22V10. (Copyright ©Advanced Micro Devices, Inc., 1988. Reprinted with permission of copyright owner. All rights reserved.)

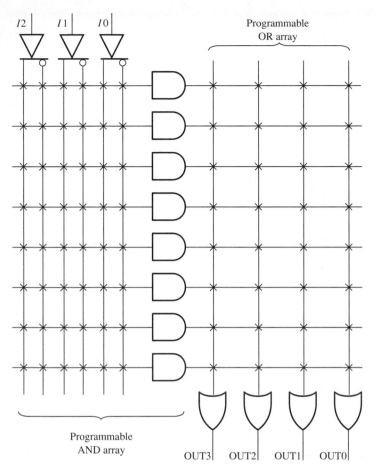

FIGURE 7-27
Simple 3-input, 4-output programmable logic array (PLA).

programmable array for both the PROM and the PAL, PLAs are typically a little slower (t_{pd} is longer) than PROMs and PALs.

To choose a particular PLA, it is necessary to use manufacturers' data books and look up the information concerning a particular type. This is because no industry standard PLA nomenclature exists. For example, the programmable logic diagram shown in Fig. 7-28 is the logic diagram for a Signetics PLS153 (field programmable logic array).

Typically the numbering system used for PLAs does not give any hint of their architecture size (signal inputs, product terms, or signal outputs) or their output types (combinational or registered). PLAs shown in manufacturers' data books utilize the notation illustrated in Table 7-3 to indicate their architectural size. A separate column in the data book indicates their output types.

Field-Programmable Logic Array (18 × 42 × 10)   PLS153

FPLA LOGIC DIAGRAM

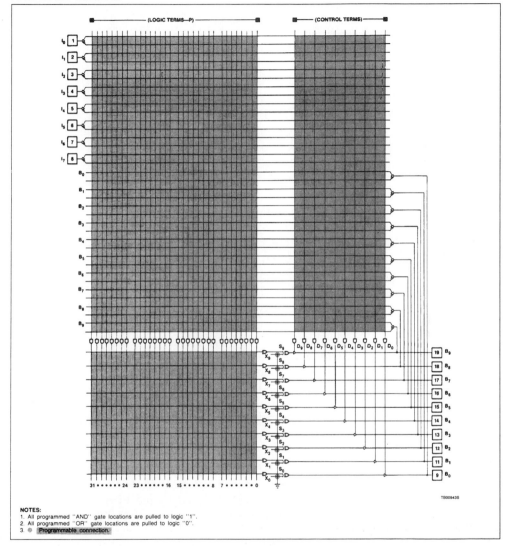

NOTES:
1. All programmed "AND" gate locations are pulled to logic "1".
2. All programmed "OR" gate locations are pulled to logic "0".
3. ● Programmable connection.

TB00943S

FIGURE 7-28
Logic Diagram PLS153. (Courtesy Signetics Company, a division of North American Philips Corporation)

TABLE 7-3
PLA architectural size

$i \times p \times o$

i represents the maximum number of signal inputs

p represents the maximum number of product terms (some of these product terms can be three state output control terms)

o represents the maximum number of signal outputs

Notice in Fig. 7-28 that the PLS153 has a programmable AND array and a programmable OR array. With both arrays programmable, this type of device can easily share product terms in the OR array, as we indicated earlier. The architectural size of this particular PLA is $18 \times 42 \times 10$ and its outputs are combinational. Its complexity allows it to compete with low to medium complexity PAL devices such as the 16L8 and the 16P8.

PLAs are often provided with programmable polarity fuses so that the output polarity of the PLA can be programmed by the user. This is achieved for the PLS153 by including an Exclusive OR element in the signal path on the output side of the PLA circuit as illustrated in Fig. 7-29. Notice that one input to the Exclusive OR element contains a programmable input (polarity fuse). A blown polarity fuse causes the Y input to the Exclusive OR element to be a 1. Since $X \oplus Y = X \oplus 1 = \overline{X}$, a blown polarity fuse causes the Exclusive OR element to act like an Inverter. When the polarity fuse is left intact, $(Y = 0)$, $X \oplus Y = X \oplus 0 = X$, the Exclusive OR element acts like a Buffer. Some PAL devices also have programmable output polarities. For example, the 16P8 has Exclusive OR elements with programmable polarity fuses, while the 18P8 has Exclusive NOR elements with programmable polarity fuses. The 22V10 also has polarity fuse capability. This is realized by programming the fuses in the macrocells. When designing with PLDs with programmable polarity capability, a designer can obtain SOP equations using either the 1s or 0s of the functions. Polarity fuses can therefore be used to simplify the designers task by conforming the PLD architecture to fit the equations. This is achieved by changing the architecture of the outputs with the programmable polarity fuses.

The two programmable array structures provided by the PLA architecture allows designers to implement relatively complex functions with PLA devices.

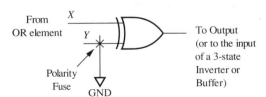

From OR element X Y Polarity Fuse   GND   To Output (or to the input of a 3-state Inverter or Buffer)

FIGURE 7-29
Exclusive OR element with a programmable polarity fuse.

For example, two commercially successful VLSI circuits, the Intel 8086 and the Motorola 68000 microprocessors, use PLAs as building blocks in their designs.

7-7-4   Multi-Level PLDs

A slightly different approach to the two programmable array structures of the traditional PLA are the Folded-NAND and Folded-NOR single programmable array structures. Each NAND of a Folded-NAND structure or NOR of a Folded-NOR structure can generate a one-level function. A single level of logic can therefore be implemented with these devices in a more efficient manner than with the two-level logic structures of PROMs, PALs, or PLAs. Multi-level logic can be generated by simply cascading multiple terms without any signal lines leaving the chip. Using a single programmable array, n levels of logic can be generated with n passes through the array. Both of these logic structures, therefore, can implement Boolean equations in SOP or POS form with two passes through the array, or with additional passes higher complexity multi-level logic functions can be implemented within the device. One Folded-NAND or Folded-NOR type device typically has the capability to replace the logic circuitry of 3 or 4 two-level PAL or PLA devices.

The logic circuit for a XL78C800 Folded-NOR or multi-level PLD is shown in Fig. 7-30. This particular device is only available in Electrically Erasable (E^2) CMOS technology. The XL78C800 PLD can contain from 1 to 42 levels of logic and is pin compatible with 24 pin PAL devices (PALs are typically available in 20 and 24 pin packages). The architectural size of the XL78C800 is 20 $\times$ 42 $\times$ 10. The device has a gate equivalent logic complexity of 600 to 800 gates.

Notice from the XL78C800 PLD logic diagram that the device also contains macrocells on 10 of the outputs. Each macrocell contains 3 programmable Multiplexers, 3 polarity control elements (represented by diamonds in the logic diagram), and a single-bit storage register that can be programmed for either D, T, or J-K operation (we will discuss each of these in the next chapter). Each macrocell can be programmed independently in 64 ways, thus providing a very versatile device. In addition the XL78C800 contains latches (latches will be presented in Chapter 8) on 8 of the inputs. These latches can temporally store input data or act transparently (the output of the latch simply follows the input—hence the latch is transparent). The XL78C800 like the PAL22V10 can be used to implement functions requiring both combinational and registered outputs.

Typically multi-level PLDs have more functionally than traditional PALs and PLAs. Because of their Folded-NAND or Folded-NOR architectures they can be used to implement functions that require one or more PALs or PLAs, thus requiring less real estate in the process. The quest for more and more circuitry in a smaller space is the primary reason for using such high density programmable devices in designs.

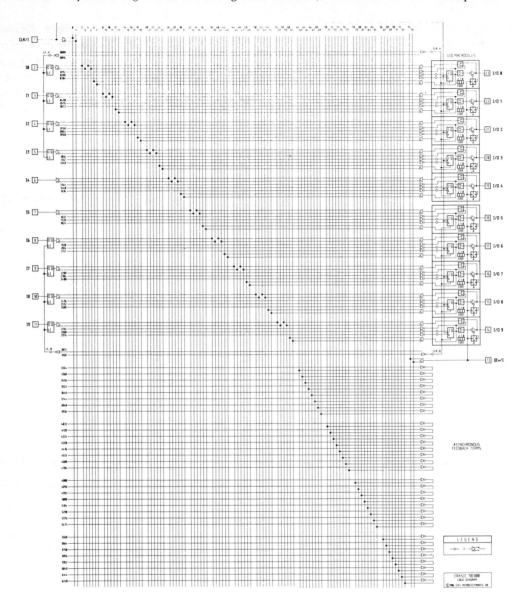

FIGURE 7-30
XL78C800 Logic Diagram. (The XL78C800 has been reproduced with the express authorization of EXEL Microelectronics, Inc.)

7-8 PROGRAMMING PALS USING PALASM

To design a combinational logic circuit using either a PAL or a PLA usually involves the following steps: (1) obtain a set of SOP equations (the designer can minimize the equations or this task can be done by a software package), (2) choose the device that the equations will fit into (PALs require more of a choice because of the fixed OR array—some software programs will assist with this task by providing choices), (3) use an ASCII editor (in nondocument mode) to enter the equations in a design file or design specification file in the form required by the software package, (4) run the software package and debug the design file, and (5) use the Joint Electronic Devices Engineering Council (JEDEC) standard fuse-map file created by the software package (from the design file) to download the fuse-map file in a programming unit to blow the fuses for the device specified in the design file. Boolean equations are the primary entry method for software programs; however, some software programs will also allow other entry methods such as truth tables, state machines, schematics, and waveforms.

There are a number of different software programs on the market that will run on IBM PC XT/AT or IBM PS/2 type systems. PAL Assembly (PALASM), Programmable Logic Analysis by National (PLAN)—very similar to PALASM, Automated Map And Zap Equations (AMAZE), Advanced Boolean Expression Language (ABEL), Universal Compiler for Programmable Logic (CUPL), and PLDesigner are perhaps the most visible programs available. IBM PC XT/AT and IBM PS/2 are trademarks of International Business Machines Corporation. PALASM is a trademark of Monolithic Memories, Inc. (A Wholly Owned Subsidiary of Advanced Micro Devices). PLAN is a trademark of National Semiconductor Corporation. AMAZE is a trademark of Signetics. ABEL is a trademark of Data I/O Corporation. CUPL is a trademark of Logical Devices Inc. PLDesigner is a trademark of Minc Incorporated. Each of these software programs assist the designer by eliminating manual fuse-map entry. Using one of these programs, a designer can enter the Boolean equations in a design file in the syntax required by the particular package. The designer then has the option of using the Boolean equations as specified in the design file or allowing the equations to be minimized by the software package (usually by an algorithm related to the Quine-McCluskey method, as presented in Chapter 4). The software package can then be used to generate the fuse-map information in JEDEC format (the JEDEC file) to be downloaded to a programmer unit. In the case of PALASM this means the design file is assembled, but in the case of ABEL, CUPL, and PLDesigner the design file is compiled. In addition to these normal features, each of these programs also has special features that can help the designer when working with PLDs. The software package PLDesigner will be discussed in Chapter 9 when it is utilized to obtain designs for synchronous sequential logic circuits.

7-8-1 Design Examples

The following examples will illustrate the procedure for obtaining the JEDEC fuse-map file for PALs using the software package PALASM. In each of the following examples we used PALASM V2.23. PALASM Software User Documentation is contained in the *PAL Device Data Book*. The software package and documentation for PALASM is available from a local sales office or by contacting Advanced Micro Device's main office (see reference 19 at the end of this chapter). The examples provided below will help to show the relative ease of using programmable devices to implement Boolean equations. After the JEDEC file is obtained, the file is downloaded into a programmer unit (such as a Data I/O 29A with a LogicPak socket module and the necessary adapters) so that the specified fuses can be blown for the device.

Example 7-13. Show the process of obtaining a PAL implementation for the function mapped in Fig. E7-13*a* and the corresponding signal list using the software package PALASM. Show the contents of the design file using a 10H8 (see Appendix C). Show the contents of the disassembled design file, the fuse map, the circuit diagram for the fuse map that is generated, and the fuse-map information in JEDEC format.

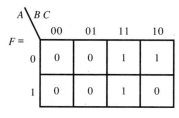

Signal list: *F, A, B, C*

(*a*)

```
;Design file (Input file in PALASM terminology)

TITLE                  Single function
PATTERN                singlef.pds(file name)
REVISION               A
AUTHOR                 R. S. Sandige
COMPANY                The Tech Team
DATE                   1/16/90

CHIP                   singlef               PAL10H8

;PINS   1    2    3    4    5    6    7    8    9    10
        A    B    C    NC   NC   NC   NC   NC   NC   GND


;PINS   11   12   13   14   15   16   17   18   19   20
        NC   NC   NC   NC   NC   NC   NC   NC   F    VCC


EQUATIONS

;Equation for the 1s of the function.

F  =  /A*B + B*C
```
                              (*b*)

FIGURE E7-13

```
TITLE                   Single function
PATTERN                 singlef.pd (file name)
REVISION                A
AUTHOR                  R. S. Sandige
COMPANY                 The Tech Team
DATE                    1/16/90
```

CHIP   SINGLEF   PAL10H8

```
   A   B   C    NC     NC     NC     NC     NC     NC     GND    NC    NC
      NC     NC     NC     NC     NC     NC   F     VCC
```

EQUATIONS
     F = B * C + /A * B

SIMULATION

(c)

```
                        PAL10H8
                        SINGLEF

                              11  1111 1111 2222 2222 2233
                     0123 4567 8901 2345 6789 0123 4567 8901

                  0  X--- X-00 --00 --00 --00 --00 --00 ----
                  1  X--X --00 --00 --00 --00 --00 --00 ----
```

TOTAL FUSES BLOWN: 36

(d)

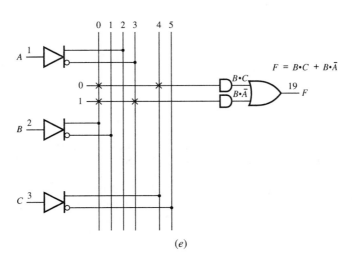

$F = B \cdot C + B \cdot \bar{A}$

(e)

FIGURE E7-13

```
PAL10H8
SINGLEF*
QP20*
QF320*
G0*F0*
L0000 01110111111111111111*
L0020 01101111111111111111*
C045A*
1521
```

(f)

FIGURE E7-13

Solution The 10H8 has a maximum of 10 signal inputs on pins 1 through 9 plus 11. There are a maximum of 8 active high outputs (nonnegated outputs) on pins 12 through 19. Pin 10 is GND and pin 20 is V_{CC}. Each output of the 10H8 has only 2 product terms (this is a rather small PAL). First we must generate the design file. The contents of the design file for the PAL10H8 is shown in Fig. E7-13*b*. Notice that lines that start with a semicolon are comment lines. The labels (such as TITLE, PATTERN, REVISION, AUTHOR, COMPANY, and DATE) for each of the other lines are specified, and the designer just uses them as a template and fills in the right side of the template to help keep records of the design, if desired. The chip name, PAL type, pin signal names, and the equation are the actual design specifications for the problem. Notice that the forward slash "/" is used for the complement operator, and the asterisk "*" is used for the AND operator. The equal sign and the plus sign represent the equal operator and the OR operator as they normally do.

　　After generating the design file, the designer usually executes AutoRun. This performs the Syntax check, Expand, Minimize, and Assemble (and Simulate if simulation data is provided—we will not discuss simulation). After executing AutoRun the designer usually elects to obtain a disassembled design file (execute Disassemble TRE in PALASM terminology). The disassembled design file (View Data—Disassembled JEDEC in PALASM terminology) contains the actual equation(s) that will be implemented in the fuse map. Figs. E7-13*c*, and *d* show the printouts of the disassembled design file, and a partial listing of the fuse map (only the portion where some of the fuses are blown for the function—in all other locations not shown, the fuses are left intact). A partial circuit diagram for the 10H8 with the intact fuses manually entered using the fuse map is shown in Fig. E7-13*e*. Notice that the Expand and Minimize routines utilize the equation for the 1s of the function since the PAL architecture has active high outputs. The fuse map consists of horizontally numbered lines (product lines) and vertically numbered lines (input lines). Using a partial logic diagram or the manufacturer's logic diagram of the 10H8, the intact fuses are simply drawn as shown in Fig. E7-13*e* with the same pattern as represented in the fuse map. The equation for the output *F* is the same equation as specified in the disassembled design file. A simulation file can be used to check the design, but this procedure will not be presented.

　　The partial circuit diagram illustrates that the fuse map provides the necessary information to blow the correct fuses and leave the correct fuses intact to implement the desired equation. The fuse-map information in JEDEC format (View Data—JEDEC Fuse Data in PALASM terminology) is shown in Fig. E7-13*f*. A file containing the fuse-map information in this standard form can be used with a programming unit to blow the fuses. By comparing the fuse map information in JEDEC format with the fuse map in Fig. E7-13d, one can see that an X (intact fuse) is represented as a 0 in JEDEC format and a – (blown fuse) is represented as a 1 in JEDEC format. Each letter O in the fuse map in Fig. E7-13d represents no programmable connection; that is, a fuse does not exist for the 10H8 in these locations.

Example 7-14. Repeat Example 7-13 for a 16L8 PAL. Show the contents of the design file using a 16L8. Show the contents of the disassembled design file, the fuse map, the circuit diagram for the fuse map that is generated, and the fuse-map information in JEDEC format.

Solution By executing the Expand and Minimize routines in PALASM, the designer can write the equations for either the 1s or the 0s of the functions. PALASM automatically selects the correct forms of the functions to fit the architecture of the PAL specified in the design file. The 10H8 and the 16L8 PALs are both 20-pin PALs with the same normal input and output pins. The 16L8 has a significantly larger circuit and is therefore capable of implementing much larger designs than our simple function that contain only two product terms. The 16L8 has active low outputs, while the 10H8 has active high outputs. The point is simply to show the design process using two different types of outputs, and any PAL that the equation will fit in will serve our purpose. By replacing "PAL10H8" with "PAL16L8" in the design file for the 10H8 in Fig. E7-13b, we obtain the design file for the PAL16L8 as shown in Fig. E7-14a.

Even though the equation is correct in the design file for the 16L8, the equation does not fit the 16L8 architecture because of the active low outputs. The equation must be obtained for the 0s of the function to fit the architecture. The equation that PALASM uses to generate the fuse map is shown in the disassembled design file in Fig. E7-14b. Notice that PALASM has obtained the correct form of the equation (the equation for the 0s of the function) to generate the fuse map. If we had bypassed the Expand and Minimize routines, we would be required to enter the correct form of the equation in the design file. Otherwise, PALASM would generate the following error message: "The polarity of the output on the left side of the equation should be opposite to the polarity in the pin list."

Figures E7-14c and d show a partial listing of the fuse map (only the portion where some of the fuses are blown for the function—in all other locations the fuses are left intact), and a partial circuit diagram for the 16L8 with the intact fuses manually entered using the fuse map. The equation for the output F is the same equation as specified in the disassembled design file.

The fuse-map information in JEDEC format is shown in Fig. E7-14e.

Example 7-15. Show a design using a PAL to implement the binary to hexadecimal character generator illustrated by the truth table in Fig. E7-11b. The outputs of the character generator are to be connected via current-limiting resistors to a common anode seven-segment display. Assume that the inputs D through A are positive logic signals. Use PALASM to show the stages of development provided by this software package.

```
;Design file

TITLE                   Single function
PATTERN                 singlef.pds(file name)
REVISION                A
AUTHOR                  R. S. Sandige
COMPANY                 The Tech Team
DATE                    1/16/90

CHIP                    singlef                    PAL16L8

;PINS   1    2    3    4    5    6    7    8    9    10
        A    B    C    NC   NC   NC   NC   NC   NC   GND


;PINS   11   12   13   14   15   16   17   18   19   20
        NC   NC   NC   NC   NC   NC   NC   NC   F    VCC


EQUATIONS

;Equation for the 1s of the function.

F = /A*B + B*C
```

(a)

```
TITLE                   Single function
PATTERN                 singlef.pds(file name)
REVISION                A
AUTHOR                  R. S. Sandige
COMPANY                 The Tech Team
DATE                    1/16/90

CHIP   SINGLEF                PAL16L8

 A   B   C    NC     NC     NC     NC     NC     NC     GND     NC     NC
     NC      NC     NC     NC     NC     NC  F     VCC

EQUATIONS
    /F = A * /C + /B

SIMULATION
```

(b)

FIGURE E7-14

Solution Our strategy will be to let the PALASM software package minimize the equations. Since the Expand and Minimize routines will select the correct form of each equation, we can simply write the canonical or standard SOP form for either the 1s or the 0s of each function. Since the functions ~OA through ~OG (or /OA through /OG in PALASM terminology) all have less 1s than 0s, we have written

PAL16L8
SINGLEF

```
                        11  1111  1111  2222  2222  2233
            0123  4567  8901  2345  6789  0123  4567  8901

   0  ----  ----  ----  ----  ----  ----  ----  ----
   1  --X-  -X--  ----  ----  ----  ----  ----  ----
   2  -X--  ----  ----  ----  ----  ----  ----  ----
```

TOTAL FUSES BLOWN: 93

(c)

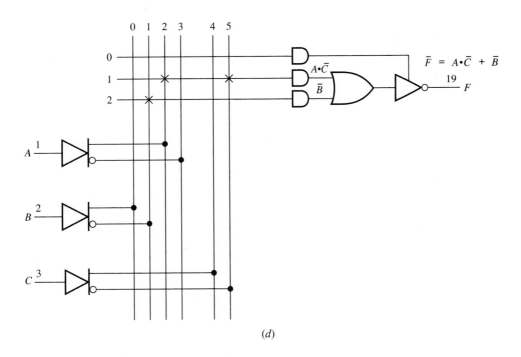

(d)

PAL16L8
SINGLEF*
QP20*
QF2048*
G0*F0*
L0000 11111111111111111111111111111111*
L0032 11011011111111111111111111111111*
L0064 10111111111111111111111111111111*
C0BCE*
21B7

(e)

FIGURE E7-14

the equations for the 1s for the functions in the design file shown in Fig. E7-15a. The PAL chosen for the implementation is the 16L8.

After executing AutoRun, disassemble design file is executed. The disassembled design file contains the actual equations that PALASM uses to construct the fuse map for the implementation. The contents of the disassembled design file is shown in Fig. E7-15b. Since the 16L8 PAL has active low outputs, the equations for the functions /OA through /OG must be solved in SOP form for the 0s of the functions. Notice that the Expand and Minimize routines were executed by AutoRun, and they correctly solved for the 0s of the functions as indicated in Fig. E7-15b in the equation list.

The fuse map for the design is shown in Fig. E7-15c. Notice that all the outputs except one were used. None of the fuses were blown for the unused output (the output connected to pin 19). If desired, this output could be used to implement a function with a different set of inputs.

The fuse-map information in JEDEC format is shown in Fig. E7-15d. The interested student might want to verify the equations by entering the intact fuses on a blank 16L8 logic diagram like the one shown in Fig. 7-24. This exercise is really not necessary, but it does provide a better understanding of the logic diagram, and the fuse map (either fuse-map format can be used).

The final proof of the operation of the programmed 16L8 device is to verify that the design actually works in the circuit application it was intended for. To verify our design after we programmed the fuses using a programming unit, we used the PLD Workshop Board designed by Data I/O Corporation (see the reference section at the end of this chapter for more information). This board allows a user to monitor and control the digital signals supplied to the pins of two 20-pin TEXTOOL sockets. Two different 20-pin PALs can be placed in the sockets and the functions implemented by the PALs can be verified.

Example 7-16. The following truth table represents the design specification for an 8-to-3 line Priority Encoder. The design specifies all input and output signals to be uncomplemented positive logic signals. Obtain a 16P8 PAL design for this Priority Encoder design specification. Use PALASM to show the stages of development provided by this software package. A MSI octal Priority Encoder, 74LS348, exists as indicated in Appendix B (Digital Logic Devices), but upon further investigation we find that its logic is complemented, compared to what we need for our design. We need nonnegated inputs and outputs rather than negated inputs and outputs. To use the 74LS348 in our design would require three packages (all input and output signals would need to be complemented using Inverters), and our goal is to use only one package for the design.

I7	I6	I5	I4	I3	I2	I1	I0	F2	F1	F0
1	X	X	X	X	X	X	X	1	1	1
0	1	X	X	X	X	X	X	1	1	0
0	0	1	X	X	X	X	X	1	0	1
0	0	0	1	X	X	X	X	1	0	0
0	0	0	0	1	X	X	X	0	1	1
0	0	0	0	0	1	X	X	0	1	0
0	0	0	0	0	0	1	X	0	0	1
0	0	0	0	0	0	0	1	0	0	0

;Design file

```
TITLE        Binary to Hex Char Gen
PATTERN      BHCG.pds(file name)
REVISION     A
AUTHOR       R. S. Sandige
COMPANY      The Tech Team
DATE         1/18/90

CHIP         BHCG              PAL16L8

;PINS   1    2    3    4    5    6    7    8    9    10
        NC   NC   NC   NC   NC   D    C    B    A    GND

;PINS   11   12   13   14   15   16   17   18   19   20
        NC   /OA  /OB  /OC  /OD  /OE  /OF  /OG  NC   VCC

EQUATIONS

;Equations are written in canonical or standard form
;for the 1s of the functions.

/OA = /D*/C*/B*A + /D*C*/B*/A + D*/C*/B*A + D*C*/B*A
/OB = /D*C*/B*A + /D*C*B*/A + D*/C*B*A + D*C*/B*/A + D*C*B*/A + D*C*B*A
/OC = /D*/C*B*/A + D*/C*B*/A + D*C*B*/A + D*C*B*A
/OD = /D*/C*/B*A + /D*C*/B*/A + /D*C*B*/A + D*/C*B*/A + D*C*B*A
/OE = /D*/C*/B*A + /D*/C*B*A + /D*C*/B*A + /D*C*B*A + /D*C*B*A + D*/C*/B*A
/OF = /D*/C*/B*A + /D*/C*B*/A + /D*/C*B*A + /D*C*/B*A + D*C*/B*/A + D*C*/B*A + D*/C*/B*A
/OG = /D*/C*/B*/A + /D*/C*/B*A + /D*/C*B*A + /D*C*B*A + D*C*/B*/A + D*C*/B*A
```

(a)

FIGURE E7-15

```
TITLE          Binary to Hex Char Gen
PATTERN        BHCG.pds(file name)
REVISION       A
AUTHOR         R. S. Sandige
COMPANY        The Tech Team
DATE           1/18/90


CHIP   BHCG                        PAL16L8

        NC      NC      NC      NC      NC   D   C   B   A      GND       NC    /OA
       /OB    /OC    /OD    /OE    /OF  /OG     NC     VCC

EQUATIONS
    OA  =  D   *   /A   +   B   *   C   +   /D   *   C   *   A   +   D
            *   /B   *   /C   +   /C   *   /A   +   /D   *   B
    OB  =  /C   *   /A   +   D   *   /B   *   A   +   /D   *   /B   *
           /A   +   /D   *   B   *   A   +   /D   *   /C
    OC  =  D   *   /C   +   A   *   /B   +   /D   *   C   +   /D   *
           A   +   /D   *   /B
    OD  =  C   *   A   *   /B   +   D   *   /C   *   A   +   /D   *   /
           C   *   /A   +   C   *   /A   *   B   +   D   *   /B   +   /
           D   *   /C   *   B
    OE  =  D   *   C   +   /A   *   /C   +   B   *   D   +   B   *   /A
    OF  =  D   *   /C   +   /D   *   C   *   /B   +   C   *   /A   +
           /B   *   /A   +   D   *   B
    OG  =  /D   *   C   *   /A   +   /C   *   B   +   D   *   /C   +
           D   *   A   +   /D   *   C   *   /B   +   D   *   B

SIMULATION
```

(b)

FIGURE E7-15 *(continued)*

Solution The 16P8 PAL we use for this design is shown in Fig. E7-16a. Notice that this PAL uses Exclusive OR elements (with polarity fuses), followed by Inverters allowing each PAL output to be either active high (polarity fuse blown) or active low (polarity fuse intact).

The Xs in the truth table represent don't cares. The Boolean functions for this design are written directly from the truth table specification and are shown in the Design file in Fig. E7-16b. Solving the equations by Karnaugh maps would require a four-variable map, a six-variable map, and a seven-variable map. Notice that a designer does not have to write the equations in either standard SOP or minimum SOP form. The minimization task can be provided by the PALASM software package for any SOP form the equations are written in.

The minimized equations are shown in the disassembled design file in Fig. E7-16c. Notice that the equation for $F0$ is solved for the 1s of the function, while the equations for $F1$ and $F2$ are solved for the 0s of the functions. This is possible since the polarity fuses allow the PAL16P8 outputs to be programmed as either active high or active low.

A partial fuse map for the implementation is shown in Fig. E7-16d. The entries not shown represent intact fuses. The fuse map includes the conditions of

```
PAL16L8
BHCG

                      11 1111 1111 2222 2222 2233
      0123 4567 8901 2345 6789 0123 4567 8901

 0    XXXX XXXX XXXX XXXX XXXX XXXX XXXX XXXX      32 ---- ---- ---- ---- ---- ---- ---- ----
 1    XXXX XXXX XXXX XXXX XXXX XXXX XXXX XXXX      33 ---- ---- ---- ---- ---- X--- -X-- X---
 2    XXXX XXXX XXXX XXXX XXXX XXXX XXXX XXXX      34 ---- ---- ---- ---- X--- -X-- ---- X---
 3    XXXX XXXX XXXX XXXX XXXX XXXX XXXX XXXX      35 ---- ---- ---- ---- -X-- -X-- ---- -X--
 4    XXXX XXXX XXXX XXXX XXXX XXXX XXXX XXXX      36 ---- ---- ---- ---- ---- X--- X--- -X--
 5    XXXX XXXX XXXX XXXX XXXX XXXX XXXX XXXX      37 ---- ---- ---- ---- X--- ---- -X-- ----
 6    XXXX XXXX XXXX XXXX XXXX XXXX XXXX XXXX      38 ---- ---- ---- ---- -X-- -X-- X--- ----
 7    XXXX XXXX XXXX XXXX XXXX XXXX XXXX XXXX      39 XXXX XXXX XXXX XXXX XXXX XXXX XXXX XXXX

 8    ---- ---- ---- ---- ---- ---- ---- ----      40 ---- ---- ---- ---- ---- ---- ---- ----
 9    ---- ---- ---- ---- -X-- X--- ---- -X--      41 ---- ---- ---- ---- X--- -X-- ---- ----
10    ---- ---- ---- ---- ---- -X-- X--- ----      42 ---- ---- ---- ---- ---- ---- -X-- X---
11    ---- ---- ---- ---- X--- -X-- ---- ----      43 ---- ---- ---- ---- -X-- X--- ---- ----
12    ---- ---- ---- ---- X--- ---- ---- X---      44 ---- ---- ---- ---- -X-- ---- ---- X---
13    ---- ---- ---- ---- -X-- X--- -X-- ----      45 ---- ---- ---- ---- -X-- ---- -X-- ----
14    ---- ---- ---- ---- X--- ---- X--- ----      46 XXXX XXXX XXXX XXXX XXXX XXXX XXXX XXXX
15    XXXX XXXX XXXX XXXX XXXX XXXX XXXX XXXX      47 XXXX XXXX XXXX XXXX XXXX XXXX XXXX XXXX

16    ---- ---- ---- ---- ---- ---- ---- ----      48 ---- ---- ---- ---- ---- ---- ---- ----
17    ---- ---- ---- ---- X--- -X-- ---- ----      49 ---- ---- ---- ---- ---- -X-- ---- -X--
18    ---- ---- ---- ---- -X-- X--- -X-- ----      50 ---- ---- ---- ---- X--- ---- -X-- X---
19    ---- ---- ---- ---- ---- X--- ---- -X--      51 ---- ---- ---- ---- -X-- ---- -X-- -X--
20    ---- ---- ---- ---- ---- ---- -X-- -X--      52 ---- ---- ---- ---- -X-- ---- X--- X---
21    ---- ---- ---- ---- X--- ---- X--- ----      53 ---- ---- ---- ---- -X-- -X-- ---- ----
22    XXXX XXXX XXXX XXXX XXXX XXXX XXXX XXXX      54 XXXX XXXX XXXX XXXX XXXX XXXX XXXX XXXX
23    XXXX XXXX XXXX XXXX XXXX XXXX XXXX XXXX      55 XXXX XXXX XXXX XXXX XXXX XXXX XXXX XXXX

24    ---- ---- ---- ---- ---- ---- ---- ----      56 ---- ---- ---- ---- ---- ---- ---- ----
25    ---- ---- ---- ---- X--- X--- ---- ----      57 ---- ---- ---- ---- X--- ---- ---- -X--
26    ---- ---- ---- ---- ---- -X-- ---- -X--      58 ---- ---- ---- ---- ---- X--- X--- ----
27    ---- ---- ---- ---- X--- ---- X--- ----      59 ---- ---- ---- ---- -X-- X--- ---- X---
28    ---- ---- ---- ---- ---- ---- X--- -X--      60 ---- ---- ---- ---- X--- -X-- -X-- ----
29    XXXX XXXX XXXX XXXX XXXX XXXX XXXX XXXX      61 ---- ---- ---- ---- ---- -X-- ---- -X--
30    XXXX XXXX XXXX XXXX XXXX XXXX XXXX XXXX      62 ---- ---- ---- ---- -X-- ---- X--- ----
31    XXXX XXXX XXXX XXXX XXXX XXXX XXXX XXXX      63 XXXX XXXX XXXX XXXX XXXX XXXX XXXX XXXX

TOTAL FUSES BLOWN:    1321
```

(*c*)

FIGURE E7-15 (*continued*)

```
PAL16L8
BHCG*
QP20*
QF2048*
G0*F0*
L0256  111111111111111111111111111111111*
L0288  111111111111111011011111111011*
L0320  111111111111111111101101111111*
L0352  111111111111110111101111111111*
L0384  111111111111110111111111110111*
L0416  111111111111111011011110111111*
L0448  111111111111110111111101111111*
L0512  111111111111111111111111111111*
L0544  111111111111110111101111111111*
L0576  111111111111111101101110111111*
L0608  111111111111111110111111111011*
L0640  111111111111111111111110111011*
L0672  111111111111110111111101111111*
L0768  111111111111111111111111111111*
L0800  111111111111110111011111111111*
L0832  111111111111111111101111111011*
L0864  111111111111110111101111011111*
L0896  111111111111111111111101111011*
L1024  111111111111111111111111111111*
L1056  111111111111111111011110110111*
L1088  111111111111110111101111110111*
L1120  111111111111111011011111111011*
L1152  111111111111111110111011110111*
L1184  111111111111111011111111110111*
L1216  111111111111110111011011011111*
L1280  111111111111111111111111111111*
L1312  111111111111111011110111111111*
L1344  111111111111111111111111011011*
L1376  111111111111111011011111111111*
L1408  111111111111111011111111110111*
L1440  111111111111111011111110111111*
L1536  111111111111111111111111111111*
L1568  111111111111111111011111111011*
L1600  111111111111111011111111011011*
L1632  111111111111111011111110111011*
L1664  111111111111111011111101110111*
L1696  111111111111111011011111111111*
L1792  111111111111111111111111111111*
L1824  111111111111110111111111111011*
L1856  111111111111111111011101111111*
L1888  111111111111110110111110111*
L1920  111111111111110111101110111111*
L1952  111111111111111111011111111011*
L1984  111111111111110111111101111111*
CAB0B*
5856
```

                                (d)

FIGURE E7-15   (*continued*)

Logic Diagram   16P8A

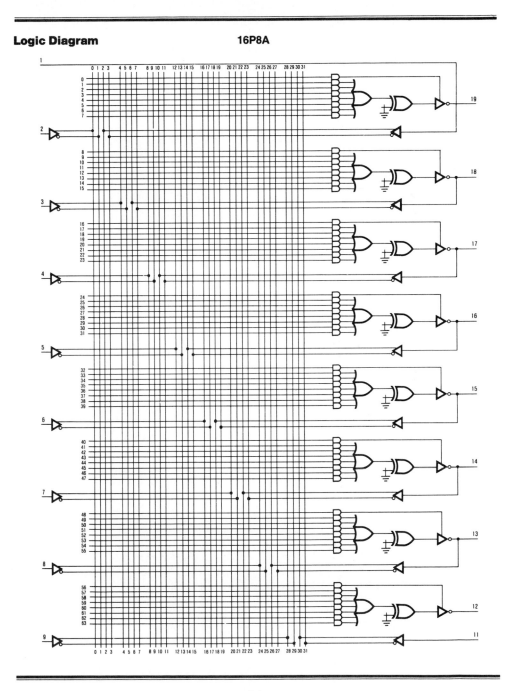

(*a*)

FIGURE E7-16

```
;Design file

TITLE         Priority Encoder
PATTERN       pencode.pds(file name)
REVISION      A
AUTHOR        R. S. Sandige
COMPANY       The Tech Team
DATE          1/24/90

CHIP          pencode        PAL16P8

;PINS  1    2    3    4    5    6    7    8    9    10
       I7   I6   I5   I4   I3   I2   I1   I0   NC   GND

;PINS  11   12   13   14   15   16   17   18   19   20
       NC   NC   NC   NC   NC   NC   F0   F1   F2   VCC

EQUATIONS

;Equations for the 1s of the functions.

F2 = I7 + /I7*I6 + /I7*/I6*I5 + /I7*/I6*/I5*I4
F1 = I7 + /I7*I6 + /I7*/I6*/I5*/I4*I3 + /I7*/I6*/I5*I4*/I3*I2
F0 = I7 + /I7*/I6*I5 + /I7*/I6*/I5*/I4*I3 + /I7*/I6*/I5*I4*/I3*/I2*I1
```

(b)

FIGURE E7-16 (continued)

```
TITLE          Priority Encoder
PATTERN        pencode.pds(file name)
REVISION       A
AUTHOR         R. S. Sandige
COMPANY        The Tech Team
DATE           1/24/90

CHIP   PENCODE                    PAL16P8

    I7   I6   I5   I4   I3   I2   I1   I0   NC    GND    NC    NC
         NC   NC   NC   NC   F0   F1   F2   VCC

EQUATIONS

    F0  =  /I6  *   I5  +  /I6  *  /I4  *   I3  +  /I6  *  /
           I4  *  /I2  *   I1  +   I7
   /F1  =  /I7  *  /I6  *   I5  +  /I7  *  /I6  *  /I3
               *  /I2  +  /I7  *  /I6  *   I4
   /F2  =  /I7  *  /I6  *  /I5  *  /I4

SIMULATION
```
                              (c)

FIGURE E7-16 *(continued)*

the polarity fuses and confirms the selected forms of the equations that were implemented. Notice that the user does not need to consider the polarity of any of the outputs since PALASM automatically take care of this task.

   The contents of the JEDEC file for our Priority Encoder design is shown in
Fig. E7-16*e*. Downloading the JEDEC file into a programmer unit allows us to
blow the fuses and obtain the PAL implementation for our design.

7-9 HAZARDS IN COMBINATIONAL LOGIC CIRCUITS

A hazard is a momentary error condition (a spurious signal or glitch) that occurs
in the output signal of a combinational logic circuit. Hazards normally occur in
combinational logic circuit implementations of Boolean functions; however, these
momentary error conditions do not prevent these circuits from performing their
specified functions. Since Boolean functions are implemented with real circuits
that have delays, propagation delays are responsible for the momentary error
conditions which are called hazards. If all the elements including the interconnect
wiring used in combinational logic circuits had 0 delay time, hazards would not
exist. The output signals of combinational logic circuits are normally not available
to use until all the outputs have settled (reached their final steady state values).
A combinational hazardous output signal can cause a problem if it is used to
drive a particular type of sequential logic circuit input, as illustrated in Fig. 7-31
(sequential logic circuits will be discussed beginning in Chapter 8).

PAL16P8
PENCODE

```
           11   1111 1111 2222 2222 2233
0123 4567  8901 2345 6789 0123 4567 8901

 0 ---- ----  ---- ---- ---- ---- ---- ----
 1 -X-X -X--  -X-- ---- ---- ---- ---- ----
 2 XXXX XXXX  XXXX XXXX XXXX XXXX XXXX XXXX
 3 XXXX XXXX  XXXX XXXX XXXX XXXX XXXX XXXX
 4 XXXX XXXX  XXXX XXXX XXXX XXXX XXXX XXXX
 5 XXXX XXXX  XXXX XXXX XXXX XXXX XXXX XXXX
 6 XXXX XXXX  XXXX XXXX XXXX XXXX XXXX XXXX
 7 XXXX XXXX  XXXX XXXX XXXX XXXX XXXX XXXX

 8 ---- ----  ---- ---- ---- ---- ---- ----
 9 -X-X X---  ---- ---- ---- ---- ---- ----
10 -X-X ----  ---- -X-- ---- ---- ---- ----
11 -X-X X---  ---X ---- ---- ---- ---- ----
12 XXXX XXXX  XXXX XXXX XXXX XXXX XXXX XXXX
13 XXXX XXXX  XXXX XXXX XXXX XXXX XXXX XXXX
14 XXXX XXXX  XXXX XXXX XXXX XXXX XXXX XXXX
15 XXXX XXXX  XXXX XXXX XXXX XXXX XXXX XXXX

16 ---- ----  ---- ---- ---- ---- ---- ----
17 -X-- X---  ---- ---- ---- ---- ---- ----
18 -X-- --X-  ---- ---- ---- ---- ---- ----
19 -X-- --X-  ---- -X-- -X-- ---- ---- ----
20 --X- ----  ---- ---- ---X ---- ---- ----
21 XXXX XXXX  XXXX XXXX XXXX XXXX XXXX XXXX
22 XXXX XXXX  XXXX XXXX XXXX XXXX XXXX XXXX
23 XXXX XXXX  XXXX XXXX XXXX XXXX XXXX XXXX

       OUTPUT PINS:   11111111
                      23456789
     POLARITY FUSE:   XXXXX-XX
TOTAL FUSES BLOWN:       329
```

(d)

PAL16P8
PENCODE*
QP20*
QF2056*
G0*F0*

```
L0000 1111111111111111111111111111111*
L0032 1010101011011111111111111111111*
L0256 1111111111111111111111111111111*
L0288 1010010100111111111111111111111*
L0320 1010101011011101011111111111111*
L0352 1010101111111111111111111111111*
L0512 1111111111111111111111111111111*
L0544 1011011111111111111111111111111*
L0576 1011111110110111111111111111111*
L0608 1011111110110111111111111111111*
L0640 1101111111011111111111010111111*
L2048 00100000*
C2B1B*
614F
```

(e)

FIGURE E7-16 *(continued)*

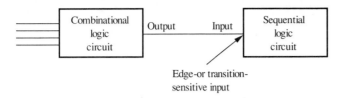

FIGURE 7-31
Hazard detection circuit.

Notice that the combinational circuit output signal is driving a sequential logic circuit input that is edge- or transition-sensitive. Edge- or transition-sensitive inputs are inputs that are sensitive to a 1 to 0 transition or a 0 to 1 transition. If a hazard is present in the combinational output signal used to drive such an input, the spurious signal or glitch present in the output signal will trigger or actuate the edge-sensitive input. Hazards are classified as either static hazards or dynamic hazards as illustrated by the timing or waveform diagrams in Fig. 7-32.

A static 1 hazard occurs as shown in Fig. 7-32*a* when the output signal condition should remain at 1 (but momentarily goes to 0) when one or more of the combinational logic circuit input signals change value. A static 0 hazard occurs (see Fig. 7-32*b*) when the output signal condition should remain at 0 (but momentarily goes to 1), when one or more of the combinational logic circuit input signals change value. When an output signal of a combinational logic circuit changes from 1 to 0 (0 to 1) due to a change in the input signal values, and one or more glitches occur as shown in Fig. 7-32*c* and *d*, a dynamic hazard is present.

7-9-1 Function Hazards

Hazards are classified as either function or logic. A hazard that can be caused in the output signal of a function, when more than one input signal is changed due to the way the function is defined, is called a function hazard. A function hazard can be spotted in the Karnaugh map of a function when more than one input signal is changed and the function change cannot be covered by the same cube. In a combinational logic circuit the designer has no control over function

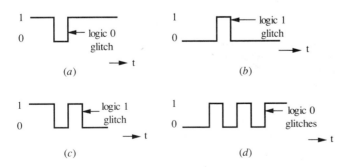

FIGURE 7-32
(*a*) Static 1 hazard, (*b*) Static 0 hazard, (*c*) Dynamic 1-to-0 hazard, and (*d*) Dynamic 0-to-1 hazard.

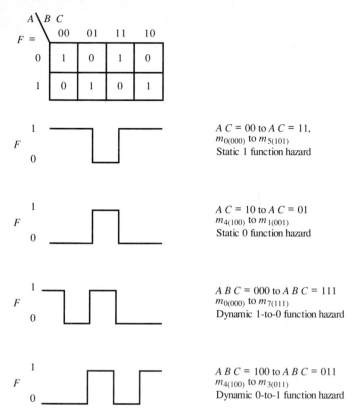

FIGURE 7-33
Function hazards—hazards that result from two or more input signals changing values at the same time.

hazards. Figure 7-33 illustrates the occurrence of static and dynamic function hazards.

7-9-2 Logic Hazards

A hazard that results in the output signal of a function when only one input signal changes due to delays in the particular circuit used to implement the function is called a logic hazard. Logic hazards can be eliminated by adding extra product terms in the Boolean equations implemented by the circuit (this requires adding more logic element in the implementation). To prevent logic hazards, a designer must recognize that a logic hazard can occur and add the necessary circuitry (additional logic) to prevent the hazard. Figure 7-34a illustrates a design specification with a single logic hazard.

The single logic hazard represented in Fig. 7-34a exists for any realization of the minimized logic function. Figure 7-34b shows a functional logic diagram and a plausible explanation for a static 1 logic hazard to be present. Figure 7-34c shows how the logic hazard is eliminated, and Fig. 7-34d shows the functional

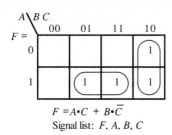

$$F = A\cdot C + B\cdot\overline{C}$$

Signal list: F, A, B, C

(a)

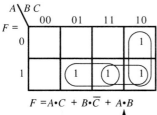

Consider the negated input to have additional
delay, when C changes from 1 to 0, Y changes to
0 before Z changes to 1, causing F to momentarily
change to 0.

$C = 1$ to $C = 0$

$m_{7(111)}$ to $m_{6(110)}$

Static 1 logic hazard

(b)

$$F = A\cdot C + B\cdot\overline{C} + A\cdot B$$

Since A and B do not change, only C changes, product
term $A\cdot B$ keeps F at 1 when C changes therefore
eliminating the static 1 logic hazard.

(c)

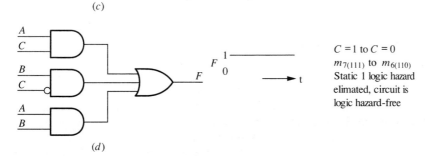

$C = 1$ to $C = 0$

$m_{7(111)}$ to $m_{6(110)}$

Static 1 logic hazard
elimated, circuit is
logic hazard-free

(d)

FIGURE 7-34
(a) Function containing a logic hazard, (b) circuit implementation showing a static 1 logic hazard (c)
adding a product term to eliminate the static 1 logic hazard, and (d) a logic hazard-free circuit.

logic diagram for a circuit that eliminates the logic hazard and thus results in a logic hazard-free circuit.

7-9-3 Designing Logic Hazard-Free Combinational Circuits

To eliminate logic hazards and obtain a logic hazard-free circuit use the following two-step procedure: (1) obtain a minimum covering of the 1s (0s) of the function, and (2) add a product term to cover each occurrence of adjacent 1s (0s) that are not already contained in the same p-subcube (r-subcube) in the minimum covering of the function. Following this procedure insures that the output signal for any realizable circuit implementation of the function will not momentarily have its value changed due to a change in value of any single input signal. This technique eliminates all static and dynamic logic hazards for the function being implemented (see McCluskey in the references at the end of this chapter). The best method to eliminate both function hazards and logic hazards is to simply wait until the hazards settle (die out). After the signals become stable, the signals may be used. This concept is the basis of synchronous sequential design that is introduced in the next chapter, Chapter 8.

Example 7-17. Design a logic hazard-free circuit to implement the following function. Use a 16P8 PAL for the implementation.

$$F(A,B,C,D) = \sum m(0,2,4,5,6,7,8,10,11,15)$$

Signal list: F, A, B, C, D

Solution First we map the function as shown in Fig. E7-17a to obtain a minimum covering for the 1s of the function.

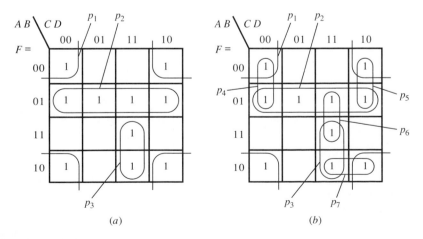

(a)   (b)

FIGURE E7-17

Logic Diagram   **16P8A**

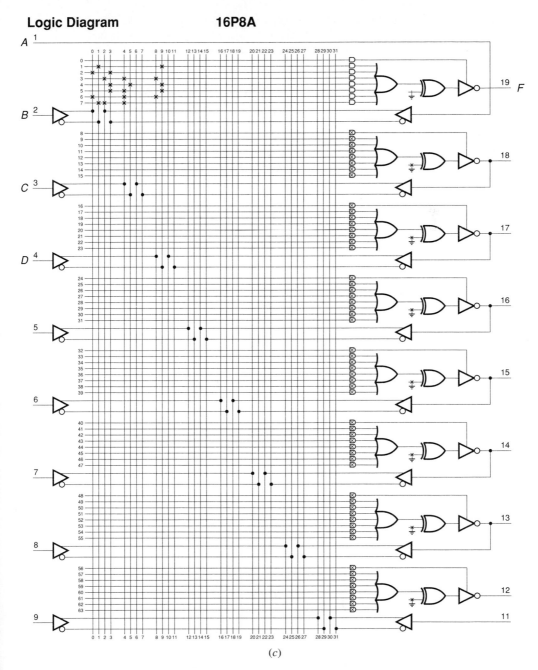

(c)

FIGURE E7-17 (*continued*)

Next we observe from the map that the adjacent 1s at minterm locations (1,4), (2,6), (7,15), and (10,11) are not covered in the same p-subcube. Adding product terms to cover these adjacent 1s results in a logic hazard-free function. The covering of the 1s of the function for a logic hazard-free design is shown in Fig. E7-17*b*.

For a minimum covering of the function only product terms p_1, p_2, and p_3 are required; however, we must add product terms p_4, p_5, p_6, and p_7 to obtain a logic hazard-free design as represented by the following function.

$$F = p_1 + p_2 + p_3 + p_4 + p_5 + p_6 + p_7$$

$$F = \overline{B}\cdot\overline{D} + \overline{A}\cdot B + A\cdot C\cdot D + \overline{A}\cdot\overline{C}\cdot\overline{D} + \overline{A}\cdot C\cdot\overline{D}$$

$$+ B\cdot C\cdot D + A\cdot\overline{B}\cdot C$$

An implementation for the logic hazard-free function is shown in Fig. E7-17*c*. If a minimum equation was already entered in the PAL, the additional product terms used to provide a logic hazard-free design can simply be added to the design, resulting in a relatively easy design change (reprogram the PAL). If a minimum equation was implemented using random logic, the additional product terms would require a complete redesign.

Notice that the 16P8 PAL has just enough product terms (7) to allow the logic hazard-free function to fit in the PAL. If more product terms were required, one could use a 16C1 which has a maximum of 16 product terms (see Appendix C).

REFERENCES

1. ANSI/IEEE Std 991-1986, *IEEE Standard for Logic Circuit Diagrams,* The Institute of Electrical and Electronic Engineers, New York, 1986.
2. ANSI/IEEE Std 91-1984, *IEEE Standard Graphic Symbols for Logic Functions,* The Institute of Electrical and Electronic Engineers, New York, 1984.
3. IEC STANDARD Publication 617-12, *Graphic Symbols for Diagrams, Part 12: Binary logic elements,* International Electrotechnical Commission, Geneve, Swisse, 1983.
4. Kampel, I., *A Practical Introduction to the New Logic Symbols,* 2d ed., Butterworths, London, 1986.
5. Mann, F. A., *Using Functional Logic Symbols and ANSI/IEEE Std 91-1984,* Texas Instruments, Dallas, Texas, 1987.
6. Hill, J. H., and Peterson, G. R., *Introduction to Switching Theory & Logic Design,* 3d ed., John Wiley & Sons, New York, 1971.
7. Johnson, E. L., and Karim, M. A., *Digital Design, A Pragmatic Approach,* Prindle, Weber & Schmidt, Boston, Mass., 1987.
8. McCluskey, E. J., *Logic Design Principles,* Prentice-Hall, Englewood Cliffs, New Jersey, 1986.
9. Edwards, H. E., *The Principles of Switching Circuits,* The M.I.T. Press., Cambridge, Mass., 1973.
10. Roth, Jr., C. H., *Fundamentals of Logic Design,* 3d ed., West Publishing, St. Paul, Minnesota, 1985.
11. Fletcher, W. I., *An Engineering Approach to Digital Design,* Prentice-Hall, Englewood Cliffs, New Jersey, 1980.
12. Blakeslee, T. R., *Digital Design with Standard MSI and LSI,* John Wiley & Sons, New York, 1975.
13. Rhyne, Jr., V. T., *Fundamentals of Digital Systems Design,* Prentice-Hall, Englewood Cliffs, New Jersey, 1973.
14. *Semiconductor Products, Circuit Design Tools and Support, Master Selection Guide,* Texas Instruments, Dallas, Texas, 1988.

15. *Programmable Logic Data Book,* Texas Instruments, Dallas, Texas, 1988.
16. *Programmable Logic Design Guide,* Rev. 1, National Semiconductor Corporation, Santa Clara, CA, 1986.
17. *Bipolar/MOS Memories Data Book,* Advanced Micro Devices, Sunnyvale, CA, 1986.
18. *PAL Device Handbook,* Advanced Micro Devices, Sunnyvale, CA, 1988.
19. *PAL Device Data Book,* Advanced Micro Devices (main office: 1-408-732-2400), Sunnyvale, CA, 1988.
20. *PLD Workshop Board,* Rev. A, Data I/O Corp., Redmond, WA, 1985.
21. Brayton, R. K., Hachtel, G. D., McMullen, C. T., and Sangiovanni-Vincentelli, A. L., *Logic Minimization Algorithms for VLSI Synthesis,* Kluwer Academic, Boston, Mass., 1984.
22. *CMOS Data Book,* Cypress Semiconductor Corp., San Jose, CA, 1988.
23. *Programmable Logic Data Manual,* Signetics Corp. (a subsidiary of U.S. Philips Corp.), Sunnyvale, CA, 1987.
24. *E² Data Book,* Exel Microelectronics, San Jose, CA, 1988.
25. *AMAZE PC/MS DOS User's Manual,* Release 1.65, Signetics Corp. (a subsidiary of U.S. Philips Corp.), Sunnyvale, CA, 1988.
26. *ABEL MS-DOS Guide,* ABEL 3.0, Data I/O Corp., Redmond, WA, 1988.
27. *CUPL User's Guide,* Logical Devices, Fort Lauderdale, FL, 1989.
28. *My First PAL Design,* Logical Devices, Fort Lauderdale, FL, 1989.
29. *PLDesigner User's Manual, Programmable Logic Device Design Program,* Version 1.2, Minc Incorporated, Colorado Springs, Colorado, 1988.

PROBLEMS

Section 7-2   Implementing Logic Functions
Using MSI Multiplexers

7-1. Design a circuit using a 2-to-1 line Multiplexer to implement each of the following functions. The signal list for each function is F, A, B. Use gates and Inverters if necessary. Apply Shannon's Expansion Theorem T11a.

(*a*) $F(A,B) = A \cdot B$ with respect to variable B
(*b*) $F(A,B) = A + B$ with respect to variable A
(*c*) $F(A,B) = A + B$ with respect to variable B
(*d*) $F(A,B) = A \oplus B$ with respect to variable A

7-2. Design a circuit using a 4-to-1 line Multiplexer to implement each of the following functions. The signal list for each function is F, A, B. Use gates and Inverters if necessary. Apply Shannon's Expansion Theorem T11a.

(*a*) $F(A,B) = A \cdot B$ with respect to variables A and B
(*b*) $F(A,B) = A + B$ with respect to variables A and B

7-3. Use Shannon's Expansion Theorem T11a to design a 2-to-1 line Multiplexer circuit to implement each of the following functions. The signal list for each function is F, X, Y, Z. Use gates and Inverters if necessary. In each case obtain the expansion with respect to variable X.

(*a*) $F(X,Y,Z) = X + \bar{Y} \cdot Z$
(*b*) $F(X,Y,Z) = X \cdot Y + \bar{X} \cdot Z$
(*c*) $F(X,Y,Z) = \sum m(0,2,4,5)$
(*d*) $F(X,Y,Z) = X + \bar{Y} \cdot Z$

7-4. Repeat Prob. 7-3 using a 4-to-1 line MUX by obtaining the expansion with respect to variables X and Y in each case.

7-5. Design a circuit to implement the following function using an 8-to-1 line Multiplexer. Apply Shannon's Expansion Theorem. Use gates and Inverters if necessary.

$$F(A,B,C,D) = \sum m(0,2,8,9,10)$$

Signal list: F, A, B, C, D

7-6. Repeat Prob. 7-5 using a 4-to-1 line Multiplexer.

Section 7-3   Designing with Multiplexers

7-7. Design a type 0 Multiplexer circuit for the following function.

$$F(A,B,C) = \sum m(2,5,7)$$

Signal list: F, A, B, C

7-8. Repeat Prob. 7-7 using a type 1 Multiplexer circuit implementation.
7-9. Repeat Prob. 7-7 using a type 2 Multiplexer circuit implementation.
7-10. Which one of the following implementations requires fewer symbols to draw the function.

$$F(A,B,C,D) = \sum m(0,2,8,9,10)$$

Signal list: F, A, B, C, D

(a) a gate-level design
(b) a type 3 MUX design
7-11. Repeat Prob. 7-10 for
(a) a gate-level design
(b) a type 2 MUX design
7-12. Repeat Prob. 7-10 for
(a) a gate-level design
(b) a type 1 MUX design
7-13. Obtain a type 1 MUX design for each of the following functions.
(a) $F1(A,B,C) = \sum m(1,3,4,5,6)$, Signal list: $F1$, A, B, C
(b) $F2(X,Y,Z) = \sum m(0,2,4,5,7)$, Signal list: $F2$, X, Y, Z
(c) $F3(A,B,C,D) = \sum m(2,6,7,8,9,12,13,14,15)$,
    Signal list: $F3$, A, B, C, D
7-14. Repeat Prob. 7-13 for a type 0 MUX design.
7-15. Obtain a type 1 MUX design for each of the Karnaugh maps shown in Fig. P7-15.
7-16. Obtain a type 2 MUX design for each of the Karnaugh maps shown in Fig. P7-15.
7-17. Obtain the Boolean function in minterm compact form for each of the circuits shown in Fig. P7-17.
7-18. Analyze the Multiplexer circuits shown in Fig. P7-18 and obtain their output functions in minterm compact form.

Section 7-4   Additional Techniques
for Designing with Multiplexers

7-19. Design a circuit to implement the following function using a type 1 MUX design.

$$F(A,B,C,D,E) = \sum m(0,5,7,11,15,16,18,25,29)$$

Signal list: F, A, B, C, D, E

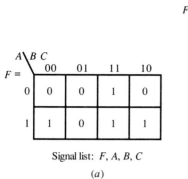

A B \ C D
$F =$

	00	01	11	10
00	0	0	1	0
01	1	0	0	1
11	1	0	0	1
10	0	1	0	0

A \ B C
$F =$

	00	01	11	10
0	0	0	1	0
1	1	0	1	1

Signal list: *F, A, B, C*

(a)

Signal list: *F, A, B, C, D*

(b)

FIGURE E7-15

     (a) use a 16-to-1 line Multiplexer
     (b) use two 8-to-1 line MUXs and one 2-to-1 line MUX to construct a 16-to-1 line
     MUX tree

7-20. Use four 2-to-1 line MUXs and one 4-to-1 line MUX to design a type 1 MUX
design for the following function.

$$F(W, X, Y, Z) = \sum m(5, 7, 13, 14, 15)$$

Signal list: *F, W, X, Y, Z*

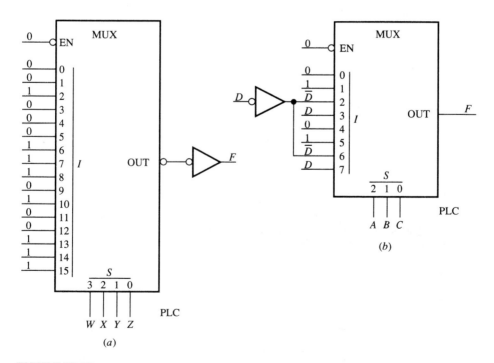

FIGURE P7-17

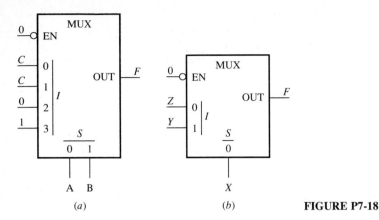

FIGURE P7-18

7-21. Implement each of the following functions using a type 1 MUX design.

$$F1(X,Y,Z) = \sum m(1,3,4,5,7)$$

$$F2(X,Y,Z) = \sum m(1,2,3,4,6)$$

Signal list: $F1$, $F2$, X[NL], Y, Z

(a) use method 1 (assume all positive logic signals and fix signal mismatching problems by adding Inverters)

(b) use method 2 (modify the truth table so that it represents all positive logic signals)

7-22. Obtain the minterms for the following minority function.

$$F(A,B,C) = \overline{A} \cdot \overline{B} \; + \; \overline{B} \cdot \overline{C} \; + \; \overline{A} \cdot \overline{C}$$

(a) by expanding the function and writing the minterm compact form

(b) by plotting the Karnaugh map

(c) by filling in the output column of the truth table

7-23. Design a circuit to implement each of the following functions using IEEE rectangular-shape symbols. Prepare each diagram using direct polarity indication (consult data books if necessary).

(a) $F(A,B,C) = B \cdot C \; + \; A \cdot C \; + \; A \cdot B$, Signal list: F, A, B, C; use a type 1 MUX design

(b) $F(A,B) = A \odot B$, Signal list: F, A, B[NL]; use a type 1 MUX design

(c) $F(W,X,Y,Z) = W \cdot \overline{X} \cdot Y \; + \; Z$, Signal list: F, W, X, Y, Z; use a type 1 MUX design

Section 7-5   Implementing Logic Functions Using MSI Decoders

7-24. Use a Decoder or Decoder tree and an external gate to implement each of the following functions.

(a) $F1(A,B,C) = \sum m(2,5,7)$, Signal list: $F1$, A, B, C

(b) $F2(A,B,C,D) = \sum m(3,6,9,12,14,15)$, Signal list: $F2$, A, B, C, D

(c) $F3(W,X,Y,Z) = \Sigma\, m(0,5,8,9,10,11)$, Signal list: $F3$, W, X, Y, Z

(d) $F4(V,W,X,Y,Z) = \Sigma\, m(1,5,19,23,31)$, Signal list: $F4$, V, W, X, Y, Z

7-25. Implement the following set of functions using a Decoder. Utilize OR elements with the smallest possible number of inputs (fanin); that is, count 1s and 0s and use the smallest number. Use the divide-and-conquer approach.

$$F1(A,B,C) = \Sigma\, m(0,1,2)$$

$$F2(A,B,C) = \Sigma\, m(1,2,3,4,5,7)$$

$$F3(A,B,C) = \Sigma\, m(2,3,5,7)$$

Signal list: $F1$, $F2$, $F3$, A, B, C

7-26. Use a Decoder to design the XS3-to-BCD code converter represented by the following truth table. Use external elements with the smallest possible fanin; that is, count the 1s and 0s. The signal list for the code converter is $F3$, $F2$, $F1$, $F0$, $I3$, $I2$, $I1$, $I0$.

XS3				BCD			
$I3$	$I2$	$I1$	$I0$	$F3$	$F2$	$F1$	$F0$
0	0	1	1	0	0	0	0
0	1	0	0	0	0	0	1
0	1	0	1	0	0	1	0
0	1	1	0	0	0	1	1
0	1	1	1	0	1	0	0
1	0	0	0	0	1	0	1
1	0	0	1	0	1	1	0
1	0	1	0	0	1	1	1
1	0	1	1	1	0	0	0
1	1	0	0	1	0	0	1

7-27. Design a Decoder circuit to perform the function of the BCD-to-XS3 code converter represented by the truth table below. Use external elements with the smallest possible fanin. The signal list for the code converter is $F3$, $F2$, $F1$, $F0$, $I3$, $I2$, $I1$, $I0$.

BCD				XS3			
$I3$	$I2$	$I1$	$I0$	$F3$	$F2$	$F1$	$F0$
0	0	0	0	0	0	1	1
0	0	0	1	0	1	0	0
0	0	1	0	0	1	0	1
0	0	1	1	0	1	1	0
0	1	0	0	0	1	1	1
0	1	0	1	1	0	0	0
0	1	1	0	1	0	0	1
0	1	1	1	1	0	1	0
1	0	0	0	1	0	1	1
1	0	0	1	1	1	0	0

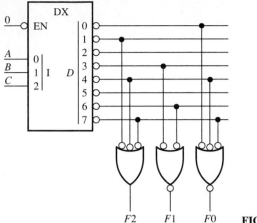

FIGURE P7-28

7-28. Obtain the Boolean functions for the Decoder Design in Fig. P7-28 in minterm compact form.

7-29. Design a Decoder circuit to implement the following functions. Use gates with the smallest possible fanin.

$$F1(X, Y, Z) = X \cdot \bar{Y} + X \cdot Z$$

$$F2(X, Y, Z) = Y \cdot Z + X \cdot \bar{Y} \cdot Z$$

Signal list: $F1[NL]$, $F2$, X, $Y[NL]$, Z

(a) use method 1 (assume all positive logic signals and fix signal mismatching problems by adding Inverters)

(b) use method 2 (modify the truth table so that it represents all positive logic signals)

7-30. Implement a Decoder design for the 2421 to BCD code converter represented by the truth table shown below. The code converter signal list is $F3[NL]$, $F2[NL]$, $F1[NL]$, $F0[NL]$, $I3$, $I2$, $I1$, $I0$. Use the design method that generally provides the simplest circuit.

2421				BCD			
$I3$	$I2$	$I1$	$I0$	$F3$	$F2$	$F1$	$F0$
0	0	0	0	0	0	0	0
0	0	0	1	0	0	0	1
0	0	1	0	0	0	1	0
0	0	1	1	0	0	1	1
0	1	0	0	0	1	0	0
1	0	1	1	0	1	0	1
1	1	0	0	0	1	1	0
1	1	0	1	0	1	1	1
1	1	1	0	1	0	0	0
1	1	1	1	1	0	0	1

7-31. Design an address decoding circuit to decode each of the following addresses, first using a decoder(s) and finally using gate-level logic. Why is the Decoder approach usually preferred?
(*a*) *A*15 *A*14 *A*13 = 101
(*b*) *A*15 *A*14 *A*13 *A*12 *A*11 = 11011
(*c*) *A*15 *A*14 *A*13 *A*12 *A*11 *A*10 = 001001

Section 7-6   Implementing Logic Functions Using Exclusive OR and Exclusive NOR Elements

7-32. Plot the odd function (odd number of 1s identifier function)

$$FO(A, B, C, D) = \sum m(1, 2, 4, 7, 8, 11, 13, 14)$$

using a Karnaugh map to obtain the function in terms of Exclusive OR operations on the input variables.

7-33. Plot the even function (even number of 1s identifier function)

$$FE(X, Y, Z) = \sum m(0, 3, 5, 6)$$

using a Karnaugh map, and obtain the function in terms of Exclusive OR operations on the input variables.

7-34. Obtain three different, but equivalent, implementations for each of the following functions using Exclusive OR and Exclusive NOR gates and Inverters.
(*a*) $F1 = X \oplus Y \oplus Z$, Signal list: *F*1, *X*, *Y*, *Z*
(*b*) $F2 = W \oplus X \oplus Y \oplus Z$, Signal list: *F*2, *W*, *X*, *Y*, *Z*
(*c*) $F3 = \overline{A \oplus B \oplus C \oplus D}$, Signal list: *F*3, *A*, *B*, *C*, *D*

7-35. For each circuit shown in Fig. P7-35, determine if the output function is odd or even by writing the output function in the form of Eq. 7-3 or Eq. 7-4.

7-36. A circuit will be constructed to send three information bits from a source to a destination using a single-bit error detection scheme (a single parity bit). Design a Parity Generator for the source and a Parity Checker for the destination to signal or flag an error (let the output be 1 when an error occurs) for
(*a*) an even parity system (even transmission bit pattern)
(*b*) an odd parity system (odd transmission bit pattern)

7-37. Design a Parity Generator/Parity Checker circuit such that the same circuit can perform the task of either parity generation or parity checking. A system will utilize your design to transmit seven information bits plus a parity bit. Design your circuit so that it can also be used in either an odd or even parity system.

Section 7-7   Implementing Logic Functions Using Programmable Devices

7-38. Draw a circuit for a simple PROM with four inputs and four outputs similar to the simple 3-input, 4-output PROM shown in Fig. 7-22. How many fuses are required for the simple 3-input, 4-output PROM? How many fuses are required for the 4-input, 4-output PROM?

7-39. Show an implementation for a XS3-to-BCD code converter (see Prob. 7-26 for the truth table and the signal list). Use a simple 4-input, 8-output PROM with the

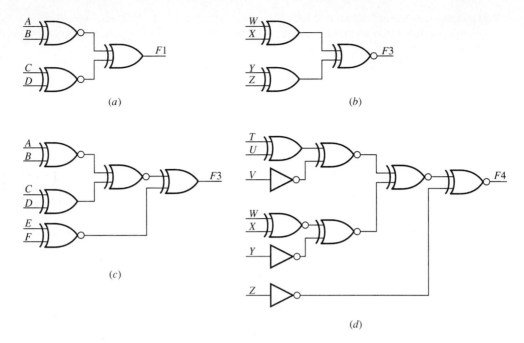

FIGURE P7-35

inputs and outputs labeled like the logic diagram illustrated in Fig. P7-39. The Xs that indicate programmable connections in the logic diagram are not shown so that they can be manually added in the OR matrix where they are needed. List each hexadecimal address and the corresponding data required to program the OR matrix using a standalone programming unit. Show a rectangular diagram or IEEE graphic symbol representation for a 32 × 8 PROM (27S19) for the code converter implementation.

7-40. Repeat Prob. 7-39 for a BCD-to-XS3 code converter (see Prob. 7-27 for the truth table and the signal list).

7-41. Show a PROM implementation for the 2421 to BCD code converter in Prob. 7-30 using the input and output signals labeled like the logic diagram illustrated in Fig. P7-39. Pay attention to the signal list: $F3[NL]$, $F2[NL]$, $F1[NL]$, $F0[NL]$, $I3$, $I2$, $I1$, $I0$. Use the design method for handling the negative logic signals that generally provides the simplest circuit. The Xs that indicate programmable connections in the logic diagram in Fig. P7-39 are not shown so that they can be manually added in the OR matrix where they are needed. List each hexadecimal address and the corresponding data required to program the OR matrix using a standalone programming unit. Show a rectangular diagram or IEEE graphic symbol representation for a 32 × 8 PROM (27S19) for the code converter implementation.

7-42. Show a design for the following Boolean functions using a simple 4-input, 4-output PAL like the one illustrated in Fig. P7-42. Generate the fuse map information for the equations manually. If the PAL in Fig. P7-42 were commercially available, write the PAL nomenclature that could be used to describe it.

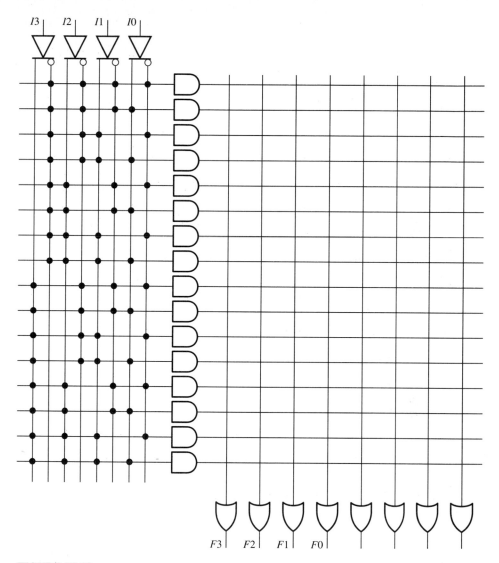

FIGURE P7-39

$$F1(A, B, C, D) = \sum m(6, 7, 9, 11, 12, 13)$$

$$F2(A, B, C, D) = \sum m(0, 2, 3, 4, 5, 10, 11, 13, 15)$$

$$F3(A, B, C, D) = \sum m(2, 3, 6, 7, 10, 11, 14, 15)$$

Signal list: $F1$, $F1$, $F3$, A, B, C, D

7-43. Repeat Prob. 7-42 using the simple 4-input, 4-output PAL illustrated in Fig. P7-43.

7-44. Implement the following Boolean equations; that is, manually generate the fuse-map information using a simple 8-input, 4-output PAL like the one illustrated in

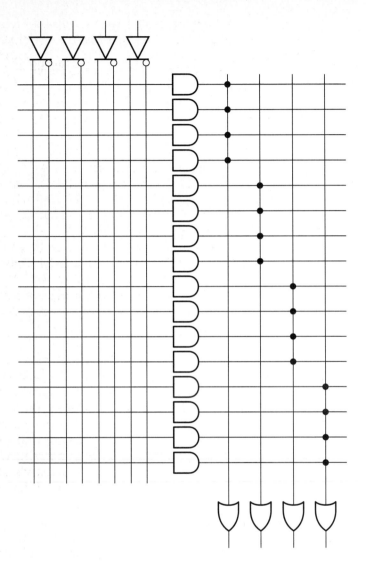

FIGURE P7-42

Fig. P7-44. If the PAL in Fig. P7-44 were available as an off-the-shelf device, write the PAL nomenclature for the device.

$$\overline{F1}(X, Y, Z) = X \ + \ \overline{Y} \cdot Z$$

$$F2(W, X, Y, Z) = W \cdot \overline{X} \cdot Y \ + \ Z$$

$$F3(A, B, C) = A \cdot C \ + \ B \cdot C \ + \ A \cdot B$$

Signal list: $F1$, $F2$, $F3$, W, X, Y, Z, A, B, C

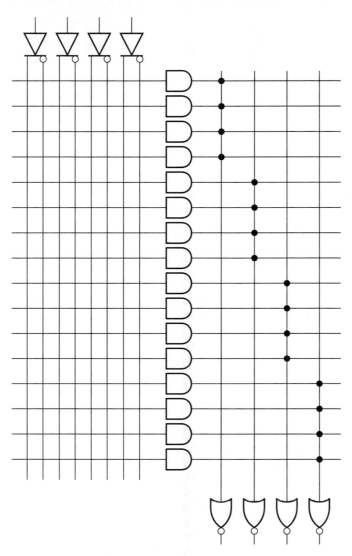

FIGURE P7-43

7-45. Design a circuit to implement a BCD-to-2421 code converter. The signal list is $F3$, $F2$, $F1$, $F0$, $I3$, $I2$, $I1$, $I0$. Use a simple PAL like the one illustrated in Fig. P7-43. Manually generate the fuse map for the PAL.

7-46. Implement the Boolean equations in Prob. 7-42 using the following off-the-shelf fuse programmable PALs. Make a copy of the necessary PAL logic diagrams; use manual fuse-map entry. (Logic diagrams for commercially available PALs are presented in Appendix C).

    (*a*) 10H8 if the equations will fit, otherwise use a 12H6

    (*b*) 10L8 if the equations will fit, otherwise use a 14L4.

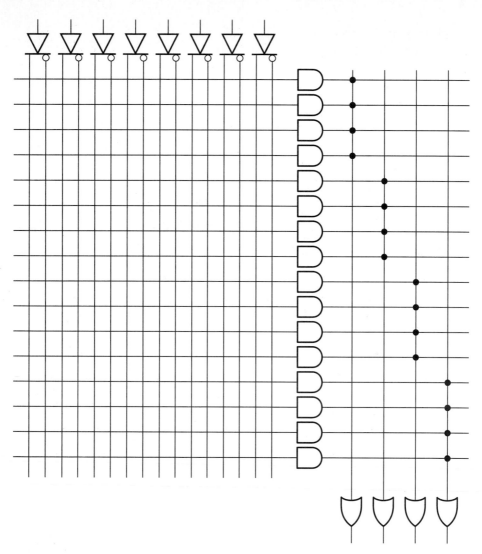

FIGURE P7-44

7-47. Show a PAL14H4 implementation for the 2421-to-BCD code converter in Prob. 7-30. The code converter signal list is $F3[NL]$, $F2[NL]$, $F1[NL]$, $F0[NL]$, $I3$, $I2$, $I1$, $I0$. Use method 2 for handling negative logic signals (modify the truth table using only positive logic signals). Make a copy of the 14H4 PAL logic diagram; use manual fuse-map entry. (Logic diagrams for commercially available PALs are presented in Appendix C).

7-48. Show an implementation for the following equations using the simple PLA shown in Fig. P7-48. Notice that the PLA circuit has polarity fuses that are programmable. Program each polarity fuse to fit the SOP form of the output function with the

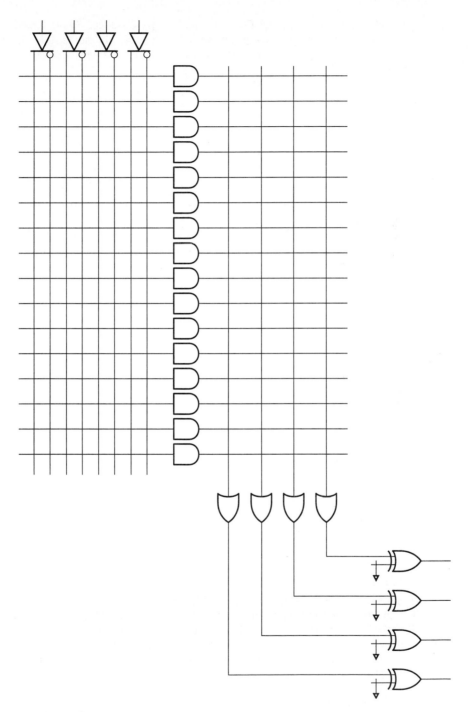

FIGURE P7-48

fewest product terms. Use manual fuse-map entry taking advantage of any product term sharing allowed by the PLA programmable OR array. If the PLA in Fig. P7-48 were available as an off-the-shelf device, write the PLA architectural size for the device.

$$F1(A,B,C) = \sum m(2,5,6)$$

$$F2(A,B,C,D) = \sum m(1,4,5,6,7,8,9,12,13,14,15)$$

$$F3(A,B,C,D) = \sum m(0,1,2,3,4,6,8,9,10,11,12,14) + d(13,15)$$

$$F4(A,B,C,D) = \sum m(0,2,4,5,12,13)$$

Signal list: $F1$, $F2$, $F3$, A, B, C, D

7-49. Implement a BCD-to-84-2-1 code converter (see Table 2-2, Common Binary Codes in Chapter 2) using the simple PAL shown in Fig. P7-49. The signal list is $F3$, $F2$, $F1$, $F0$, $I3$, $I2$, $I1$, $I0$. Use manual fuse-map entry, and program the polarity fuses so that all the output functions are implemented with the fewest product terms. If the PAL in Fig. P7-49 were available as an off-the-shelf device, write the PAL nomenclature that could be used for the device.

Section 7-8   Programming PALs Using PALASM

Work Examples 7-13 through 7-16 in Subsection 7-8-1 using the PALASM software program. For additional practice examples, use PALASM to implement any of the combinational logic functions presented in the text throughout Parts I and II.

Section 7-9   Hazards in Combinational Logic Circuits

7-50. State a procedure that can be used to eliminate all static and dynamic logic hazards in a circuit.

7-51. Eliminate all logic hazards that can result if the following functions are implemented. List a minimum covering and a logic hazard-free covering first for the 1s and then for the 0s for each of the following functions.
(a) $F(A,B,C) = \sum m(1,3,4,5)$, Signal list: F, A, B, C
(b) $F(X,Y,Z) = \sum m(1,2,3,6)$, Signal list: F, X, Y, Z
(c) $F(X,Y,Z) = \sum m(0,2,3,4,6)$, Signal list: F, X, Y, Z

7-52. Determine the number of logic hazards the circuit shown in Fig. P7-52 could contain. What product terms are necessary to eliminate these logic hazards? Write a logic hazard-free function for the circuit.

7-53. Will a circuit implementing each of the following functions for a minimum covering for the 1s contain logic hazards, if so, how many? What product terms must be added to each function to remove the logic hazards. Repeat the problem for a minimum covering of the 0s.
(a) $F(A,B,C,D) = \sum m(0,1,2,5,6,7,15)$, Signal list: F, A, B, C, D
(b) $F(W,X,Y,Z) = \sum m(0,1,5,7,8,9,14,15)$, Signal list: F, W, X, Y, Z
(c) $F(A,B,C,D) = \sum m(2,3,8,9,10,11)$, Signal list: F, A, B, C, D

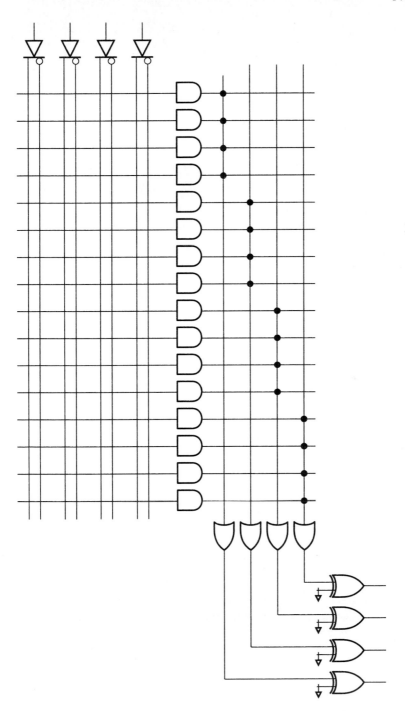

FIGURE P7-49

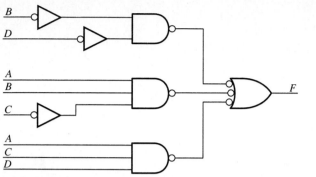

FIGURE P7-52

PART
III

SEQUENTIAL LOGIC DESIGN

SEQUENTIAL LOGIC CIRCUITS AND BISTABLE MEMORY DEVICES

8-1 INTRODUCTION AND INSTRUCTIONAL GOALS

This is the first chapter in Part III, Sequential Logic Design. In this chapter we present the memory side of logic analysis. Bistable memory devices (often called flip-flops) are introduced along with various tools and techniques that are used to analyze and design synchronous and asynchronous sequential logic circuits. As in Chapter 7, we will present many new concepts. Bistable devices including latches, master-slave flip-flops, and edge-triggered flip-flops are discussed and analyzed in detail, thus providing an in-depth presentation of their operating characteristics.

In Section 8-2, Combinational Logic Circuits versus Sequential Logic Circuits, the classification of logic circuits is presented. In this section we discuss block diagram models representing combinational logic circuits using worst-case delays, synchronous sequential logic circuits using edge-triggered D-type memory devices, and asynchronous sequential logic circuits using propagation delays. External input signal restrictions are emphasized for synchronous and asynchronous circuits to illustrate the difference between these two classes of sequential circuits.

Latches, the simplest form of bistable memory device or flip-flop, are presented in Section 8-3, The Basic Bistable Memory Devices. Circuit models for latches are analyzed using delay models. In this section, characteristic or present state/next state tables, characteristic equations, state diagrams, ASM charts, transition maps, flow maps, and timing diagrams are presented for the purpose of analyzing the behavior of latch circuits. A flow table, for example, is used to demonstrate the critical race condition of an asynchronous cross-coupled *S-R* NOR gated latch circuit, while a timing diagram is used to illustrate the result of debouncing a switch with a latch circuit.

The topic of Section 8-4 is Additional Bistable Memory Devices. Three major groups of memory devices are presented in this section. These include latches, master-slave (pulse-triggered) flip-flops with or without data lockout, and edge-triggered flip-flops. For each group, *S-R*, *D*, *J-K* and *T* circuit types are presented, and the behavior of each circuit is investigated. Recommended IEEE logic symbols are presented for each type of latch and flip-flop. Timing diagrams are used to show setup and hold time requirements for the different groups of bistable devices. Delay models, flow maps, and state diagrams are used as tools to analyze the circuit behavior of edge-triggered flip-flops.

We present shortened characteristic tables for analyzing bistable devices in a circuit and excitation tables for designing with bistable devices in Section 8-5, Reduced Characteristic and Excitation Tables for Bistable Devices.

The last section in this chapter, Section 8-6, Metastability and Synchronization Using Bistables, discusses the metastable state and how it can occur as a result of violating the data setup or hold time specification of a bistable. A basic synchronizer circuit, which is often used to synchronize asynchronous external input signals with the system clock signal, is also presented.

This chapter should prepare you to

1. Make a sketch showing the classification of logic circuits.
2. Draw block diagrams for synchronous and asynchronous logic circuits, labeling the blocks and all the signals.
3. Draw timing diagrams showing the input signal restrictions for synchronous or clock mode, fundamental mode, and pulse mode sequential logic circuits.
4. Obtain the characteristic equations of latch circuits using delay models.
5. Obtain the PS/NS table, state diagram, ASM chart, Karnaugh map, transition map, flow map, and timing diagram for basic latch circuits using their characteristic equations.
6. Explain the operation of basic latch circuits using one or more of the tools listed in number 5.
7. Make a sketch showing the three major groups of bistable devices.
8. Obtain the characteristic equations of gated latch circuits using delay models.

9. Obtain the PS/NS table, state diagram, ASM chart, Karnaugh map, transition map, flow map, and timing diagram for gated latch circuits using their characteristic equations.

10. Draw timing diagrams that show the data setup and hold time requirements for gated transparent latch circuits, master-slave flip-flops with and without data lockout, and edge-triggered flip-flops.

11. Draw the recommended IEEE logic symbols for each of the three groups of bistable devices for types *S-R*, *D*, *J-K*, and *T*.

12. Obtain the characteristic equations of edge-triggered flip-flops using delay models and show their behavior using one or more of the tools listed in number 5.

13. Draw circuits to show how negative edge-triggered *J-K* and *D*-type flip-flops can be used as positive edge-triggered flip-flops and vice versa.

14. Draw circuits to show how edge-triggered *J-K* flip-flops can be used as *S-R* and *T* flip-flops.

15. List the reduced characteristic tables for bistable types *S-R*, *D*, *J-K*, and *T* and explain their operation.

16. List the excitation tables for the bistable types *S-R*, *D*, *J-K*, and *T*.

17. Explain what a metastable state is and how it can occur.

18. Draw a basic synchronizer circuit and explain its operation by means of a timing diagram.

8-2   COMBINATIONAL LOGIC CIRCUITS VERSUS SEQUENTIAL LOGIC CIRCUITS

All logic circuits can be classified as either combinational or sequential as illustrated in Fig. 8-1. A circuit is classified as combinational if it has the property that its outputs are determined totally by its external inputs. A circuit is classified as sequential if it has the property that its outputs are determined not only by its external inputs but also by the past history of the circuit. The class of circuits called combinational (or combinatorial) is shown on the left side of Fig. 8-1, and the class of circuits called sequential is shown on the right side of Fig. 8-1. In this and the following chapters, we will present sequential logic circuit analysis and design. Our primary emphasis in this chapter will be placed on analysis, and our primary emphasis in Chapters 9 and 10 will be placed on design.

8-2-1   Model for Combinational Logic Circuits

Figure 8-2 illustrates a block diagram model for combinational logic circuits. The implementation of any combinational logic function results in a circuit whose output signals $f1, f2, \ldots, fm$ are totally determined by the external input signals to the circuit $I1, I2, \ldots, In$. At any instant of time, the output signals $F1, F2, \ldots,$ Fm of the combinational logic circuit are available as functions of the external

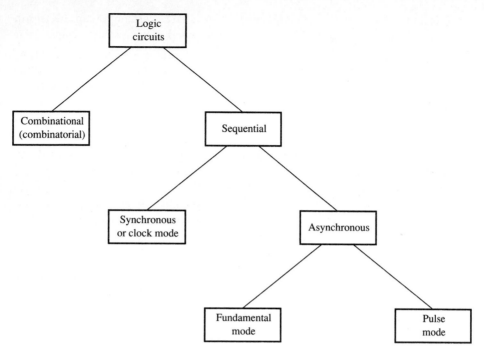

FIGURE 8-1
The classification of logic circuits.

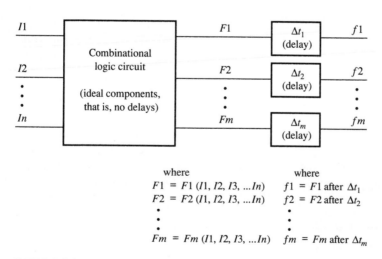

FIGURE 8-2
Block diagram model for combinational logic circuits.

input signals $F1(I1,I2, \ldots, In)$, $F2(I1,I2, \ldots, In)$, $\ldots$, $Fm(I1,I2, \ldots, In)$, assuming no delays (propagation delays).

After an appropriate period of time, represented by each lumped delay time Δt_1, Δt_2, $\ldots$, Δt_m, the output signals $f1 = F1$, $f2 = F2$, $\ldots$, $fm = Fm$ are available. Lower case letters are used for the output signals $f1$ through fm to emphasize their dynamic nature; that is, the values of these signals can change during time interval Δt. For a given set of input signal values, the output signals $f1, f2, \ldots, fm$ settle or become stable at the steady state values of $F1, F2, \ldots, Fm$ after the specified delay times Δt_1, Δt_2, $\ldots$, Δt_m. The combinational logic circuit is then stable.

The box in the model labeled combinational logic circuit in Fig. 8-2 is assumed to contain ideal components, that is, components with no propagation delays. The accumulative propagation delays or lumped delays of the circuit components are represented by the delays Δt_1, Δt_2, $\ldots$, Δt_m, lumped at the outputs represented by the separate smaller boxes in the model. These are worst-case delays through the longest delay path from the inputs to each output. Each time the input signals change and then settle or become stable by reaching a steady state value, the output signals become stable a short time later (depending on the lumped propagation delay times associated with each output). The model in Fig. 8-2 is important when one is required to determine settling time for output signals in a combinational logic circuit. For an application of this model, refer to Fig. 5-12 for the settling time for a five-bit parallel ripple adder, and Fig. 5-13 for the settling time for a five-bit parallel carry lookahead adder. In each of these figures, the worst-case settling times of the circuit output signals are shown in parentheses. For each adder circuit, the sum is not valid until all the output signals have settled or become stable. These examples illustrate the validity of the block diagram model for combinational logic circuits shown in Fig. 8-2. Output signal delays are not normally shown for combinational logic circuits, but one should realize that worst-case delays often cannot be ignored.

8-2-2   Sequential Logic Circuits

Up until this time we have concentrated on combinational logic circuits. Circuits which are not combinational fall into a class called sequential logic circuits as shown in Fig. 8-1. Sequential logic circuits are further divided into two major classes: synchronous or clock mode, and asynchronous (which includes fundamental mode and pulse mode). Sequential logic circuits, whether synchronous or asynchronous, have one common characteristic. The implementation of any sequential logic function results in a circuit whose present state output signals are determined by external input and feedback signals to the circuit as shown in the block diagram models in Fig. 8-3a and b. The external input signals are often referred to as the "primary input variables," while the feedback signals (those signals taken from the outputs of the circuit and used as inputs) are referred to as the "state or secondary variables." As a result, a sequential logic circuit has the

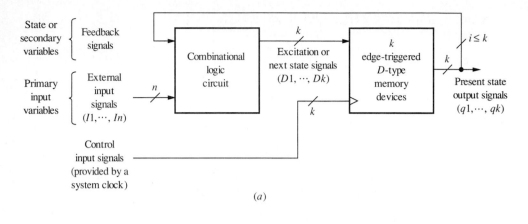

(a)

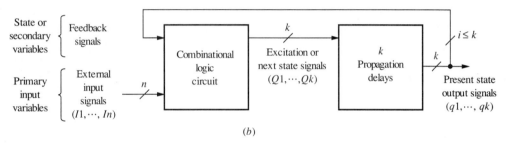

(b)

FIGURE 8-3

(a) Block diagram model for synchronous sequential logic circuits, (b) block diagram model for asynchronous sequential logic circuits.

property that its present state output signals (the output signals from the memory devices or propagation delays—where the propagation delays represent the worst-case delays through the longest delay path from the inputs to each output) are a function not only of its primary or external input signals but also of the previous values of the present state or feedback signals.

Because of the memory property of sequential logic circuits, they can be designed to execute various sequences of events that require a recorded history of past events. For example, a parallel adder circuit or a parallel comparator circuit requiring multiple copies of a simple circuit (iterations of the same circuit) can be designed as a sequential circuit (a serial adder or serial comparator) using less hardware at the expense of time (by using the same simple adder or simple comparator over and over). When speed is of the essence, hardware implementations that require more circuitry and less time are important; however, in many cases hardware implementations requiring less circuitry and more time are desirable. Sequential circuits thus allow a designer to trade off time for space. The block diagram organization of a serial adder and serial comparator is left as an exercise for the student.

8-2-3   Model for Synchronous Sequential Logic Circuits

Many useful synchronous sequential logic circuits use a combinational logic circuit that feeds signals to memory (storage) devices as shown in Fig. 8-3*a*. In this type of circuit, feedback signals are derived from the output lines of edge-triggered *D*(data)-type memory devices. It is not important to understand at this time what an edge-triggered *D*-type memory device is; however, it is important to understand that each memory device has a control input that is driven by a synchronizing signal provided by an astable device called a 'system clock'. Edge-triggered *D*-type memory devices, as well as other available memory devices, are presented later in this chapter. A block diagram of a system clock is illustrated in Fig. 8-4*a*.

A system clock (or clock) provides the idealized signal waveform CK illustrated in the timing diagram in Fig. 8-4*b*. The waveform shows that the output signal for the clock is never stable at one logic state for very long. A clock signal provides discrete timing events based on the period of the clock waveform [clock period = 1/(clock frequency)]. Many companies manufacture crystal-controlled clock oscillators. These astable circuits are predesigned and are available as off-the-shelf devices. Crystal-clock oscillators are available over a wide range of fixed frequencies and provide designers with a single package solution for implementing the clock in synchronous sequential logic designs.

The block diagram model in Fig. 8-3*a* illustrates one among many different models for synchronous sequential logic circuits. This model serves to show the types of logic elements or devices that are associated with synchronous circuits. For this model, there are *n* external input signals, *k* combinational logic output signals, *k* present state output signals, *i* feedback signals, and *k* control input signals (the control input signals to the memory devices are all supplied by the same system clock). The logic values that the *k* present state output signals will

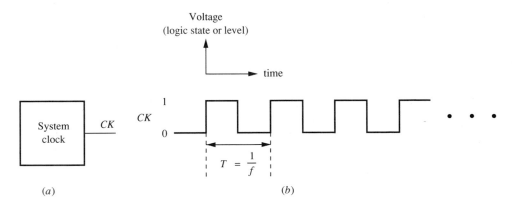

(a) *(b)*

FIGURE 8-4
(*a*) System clock block diagram, (*b*) clock waveform timing diagram.

assume next depend on the current values of the external input signals and the present state output signals i being fed back. The k combinational logic output signals are called excitation signals, and are obtained using the same methods presented earlier for combinational logic design. During the time period before the system clock ticks (before a new clock cycle), to initiate a new set of output values, the k present state output signals will be stable and can be observed. After the clock ticks, a new set of output values becomes the present state output signals and can be observed, again after a short delay (the propagation delay of the memory devices).

The present state output signals after each tick of the clock are functions of the excitation signals. Since the excitation signals determine the values that the present state output signals will assume next, the excitation signals either specify the next state signals or are the actual next state signals depending on the type of memory devices used. The model in Fig. 8-3a is for a synchronous design utilizing D-type memory devices, and $D1$ through Dk are the actual next state signals. These memory devices will be presented later in this chapter.

New values for the k present state output signals occur after each new clock cycle (after the clock ticks), thereby providing outputs that are synchronized with the system clock. Circuits that rely on a system clock are classified as synchronous sequential logic circuits. Synchronous sequential logic circuits are relatively easy to design utilizing the independent recurring discrete time samples or instants provided by each separate timing event (a timing event occurs when the clock ticks). The clock and the external input signal restrictions for synchronous or clock mode sequential circuit operation are illustrated in Fig. 8-5. Prior to each clock tick the circuit must be stable. This simply means that all the external input signals and excitation signals in the network have had time to reach a steady state value prior to the occurrence of the next clock cycle or timing event. The time between timing events represents the period of the clock waveform T and is often referred to as the 'state time'. In synchronous sequential logic circuits, the state time represents the minimum time a circuit can remain in the state specified by its state variables, that is, the minimum time the present state output signals can retain their current logic values.

After each clock timing event represented by the clock signal CK making a transition from 0 to 1 in Fig. 8-5, any number of the external input signals applied to the circuit are allowed to change their values. All the external input signals that change in value including the excitation signals must be allowed enough time to settle (reach a steady state value) before the next timing event occurs. Some memory devices require the excitation signals be stable prior to a 'memory setup time' (a brief period of time where the memory data input signals must be stable prior to each clock signal timing event). Certain memory devices require a 'memory hold time' (a brief period of time where the memory data input signals must be present after the clock signal timing event has passed). Excitation input signals are required to be held stable until after the memory hold time. If the excitation signals are not allowed to meet the setup and the hold time for

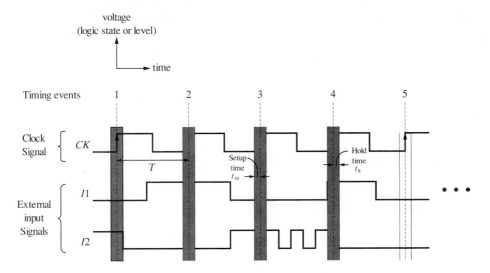

Comments (*a*) Circuit must be stable (*T* is sufficiently long), and all the external
input signals and excitation signals have time to settle prior to each
clock time event.

(*b*) Additional signal constraints can occur due to setup and hold time
requirements imposed by certain memory devices used in the circuit.
Since *I*1 and *I*2 are normally used to generate excitation inputs, *I*1
could violate the setup time requirement while *I*2 could violate the
hold time requirement of the circuit for timing event 4.

FIGURE 8-5
Clock and external input signal restrictions for synchronous or clock mode sequential circuit operation.

the memory devices being used, then one should expect erratic circuit operation;
however, setup and hold time requirements are quite easy to meet in synchronous
design.

The shaded area in Fig. 8-5 illustrates the possible setup and hold time
required by some memory devices. The time period beginning at the memory
setup time and ending at the memory hold time is the period in which the
excitation signals must remain stable. Keep in mind that the excitation signals
in a synchronous circuit are also the memory data input signals as shown in Fig.
8-3*a*. The remaining time period until the beginning of the next setup time is
the allowable transition period in which the excitation signals can be changed
by the external input signals. Using faster memory devices requires the use of
faster combinational logic circuits to meet the memory setup time. In this case
the two-level logic circuits we have been stressing are used rather than multilevel
logic circuits that are obtained by factoring functions (the circuit in Fig. E1-2*b*,
for example, is a three-level logic circuit). Circuits obtained by factoring have

longer delay paths and hence must be used with care in order to avoid violating memory setup time.

By far, the vast majority of sequential logic circuits today are designed using a system clock, and for this reason, synchronous sequential logic circuit design is where we will place our emphasis. Briefly, we can define a synchronous sequential logic circuit as a circuit whose memory output signals depend on its excitation signals at recurring discrete time instants provided by a system clock signal. The fundamentals of synchronous sequential logic circuit design will be presented in Chapter 9 when the design of synchronous frequency dividers, counters, shift registers, and controllers are presented.

8-2-4 Model for Asynchronous Sequential Logic Circuits

Sequential logic circuits that do not rely on a system clock for synchronization are called asynchronous sequential logic circuits. Asynchronous sequential logic circuits can be implemented using a combinational logic circuit with feedback as shown in Fig. 8-3b. The box labeled k propagation delays can also contain either controlled or uncontrolled memory devices. In either case the memory device control inputs are not driven by a system clock. The model in Fig. 8-3b has n external input signals, k excitation or next state signals, k present state output signals, and i feedback signals. Since asynchronous sequential logic circuits do not rely on a system clock to provide a new set of output values for the present state output signals, they must rely on the external input signals to provide a change in the state specified by the state variables. Any signal that is not synchronized can be called an asynchronous signal since its time of change (from 0 to 1 or 1 to 0) is not predictable. An asynchronous sequential logic circuit can be defined as a circuit whose present state output signals depend on its excitation signals at time instants determined by a logic change of one of its external input signals. An external input signal logic change can be a change from 0 to 1 or from 1 to 0. An external input signal logic change can also be a change from 0 to 1 back to 0 (a positive pulse), or it can be a change from 1 to 0 back to 1 (a negative pulse).

In order for asynchronous sequential circuits to operate properly, they must be restricted to operating either in fundamental mode or pulse mode. In fundamental mode each of the external input signals can only change one at a time from 0 to 1, or 1 to 0, and the circuit must be in a stable condition (all the signals in the circuit must be stable; that is, they must have reached a steady state value) when a change occurs. Each external input signal change from 0 to 1 or from 1 to 0 signifies a timing event as illustrated in Fig. 8-6a.

The external input signal restrictions for asynchronous sequential circuit operation in pulse mode are illustrated in Fig. 8-6b for positive pulses. Each positive pulse, a change from 0 to 1 back to 0 generated by an external input signal, signifies a timing event. The circuit must be in a stable condition when

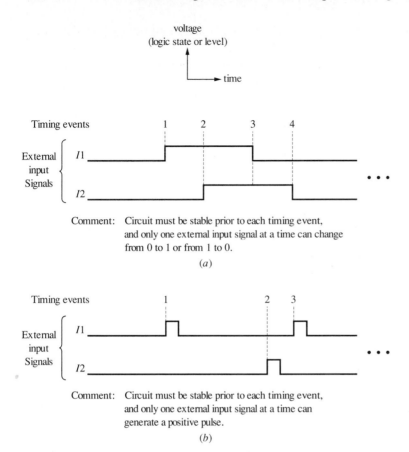

Comment: Circuit must be stable prior to each timing event, and only one external input signal at a time can change from 0 to 1 or from 1 to 0.

(a)

Comment: Circuit must be stable prior to each timing event, and only one external input signal at a time can generate a positive pulse.

(b)

FIGURE 8-6
External input signal restrictions for asynchronous sequential circuit operation *(a)* fundamental mode, *(b)* pulse mode.

each pulse occurs. A pulse mode circuit can also be designed to operate with negative pulses. If signals $I1$ and $I2$ are complemented in Fig. 8-6*b*, one can observe that three negative pulses are obtained.

Designing asynchronous sequential logic circuits is not an easy task compared to designing synchronous sequential logic circuits. To simplify the task somewhat, memory devices can be used in the box labeled propagation delays in the block diagram model shown in Fig. 8-3*b*. Even when this is done, however, asynchronous sequential logic circuits are generally designed as a last resort when the synchronization provided by a system clock cannot be used. This can be the case when a synchronous design operates too slowly because of the required frequency of the clock, or a clock signal may simply not be readily available for the particular application. Basic memory devices (such as latches or flip-flops) are asynchronous sequential logic circuits since they do not rely on a system clock. In the next section we will present basic memory devices along with various tools that are used to analyze their behavior. The fundamentals of asynchronous sequential logic circuit design will be presented in Chapter 10 for both fundamental mode and pulse mode circuits.

8-3   THE BASIC BISTABLE MEMORY DEVICES

Memory or storage devices are fundamental sequential components used in the design of both synchronous and asynchronous sequential logic circuits. A sequential logic device that has two and only two stable output states is called a bistable element or a flip-flop. In this section we will present the most basic forms of sequential logic circuits. The most basic forms of flip-flop circuits are two cross-coupled NOR gates or two cross-coupled NAND gates, and are commonly referred to as latches. To investigate the properties of these latch circuits we will utilize the following analysis tools: (*a*) 'circuit delay model', (*b*) 'characteristic equation', (*c*) 'characteristic or present state/next state table', (*d*) 'state diagram', (*e*) 'Karnaugh map', (*f*) 'ASM chart', (*g*) 'transition map', (*h*) 'flow map', and (*i*) 'timing diagram'. Many of these analysis tools can also be used in the design of sequential circuits as we shall see in Chapters 9 and 10.

8-3-1   Circuit Delay Model

An *S-R* latch (or Set-Reset latch) is the simplest controllable form of sequential circuit. This flip-flop sequential device can retain its output state of 0 or 1 only as long as power is applied to the circuit. The rectangular symbol for an *S-R* latch is shown in Fig. 8-7*a*. The *S* (set) and *R* (reset) inputs of the *S-R* latch allow the binary value of the output signal *Q* of the memory device to be changed. When *S* = 1 and *R* = 0, output signal *Q* changes to a 1, and output signal *Q* changes to a 0 when *S* = 0 and *R* = 1. In overly simplified terms, a latch is set when its *Q* output is 1, and this is accomplished by making *S* true or 1. A latch is reset or cleared when its *Q* output is 0, and this is accomplished by making *R* true or 1. A gate-level implementation using cross-coupled NOR gates is shown in Fig. 8-7*b*.

Figure 8-8*a* illustrates an alternate form of *S-R* latch. An implementation of this *S-R* latch using cross-coupled NAND gates is shown in Fig. 8-8*b*. This form

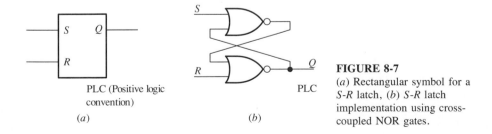

PLC (Positive logic convention)

(*a*)

PLC

(*b*)

FIGURE 8-7
(*a*) Rectangular symbol for a *S-R* latch, (*b*) *S-R* latch implementation using cross-coupled NOR gates.

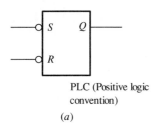

PLC (Positive logic
convention)

(*a*)

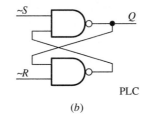

PLC

(*b*)

FIGURE 8-8
(*a*) Rectangular symbol for
alternate form of *S-R* latch,
(*b*) *S-R* latch implementation
using cross-coupled NAND
gates.

of latch is often referred to as an $\sim S$-$\sim R$ (or $\overline{S}$-$\overline{R}$) latch. By labeling the inputs $\sim S$ and $\sim R$, output Q changes to a 1 (the set state) when S is true or 1 (input $\sim S$ = 0), and output Q changes to a 0 (the reset or cleared state) when R is true or 1 (input $\sim R$ = 0). Both forms of *S-R* latches shown in Fig. 8-7a and Fig. 8-8a are bistable memory elements, but the output function or equation of each circuit is different. For this reason, each circuit has a slightly different characteristic; however, the output state represented by the signal name Q has two stable states so both circuits are bistable memory elements or flip-flops. To allow us to easily distinguish between these two basic cross-coupled *S-R* latch circuits, we will refer to the cross-coupled NOR gate latch as the *S-R* NOR latch and the cross-coupled NAND gate latch as the *S-R* NAND latch. Remember that the inputs of the *S-R* NAND latch are represented in complemented form, that is, $\sim S$ and $\sim R$. The operations of basic *S-R* latch circuits are very important. If one understands the operation of basic latch circuits, then more complex memory devices such as gated flip-flops and edge-triggered flip-flops will be easy to understand, since their operation is quite similar. For this reason, the analysis of *S-R* latch circuits will be presented in a very thorough manner. Circuits that are drawn in this and the following chapters will utilize the positive logic convention or direct polarity indication unless otherwise stated.

     The delay model shown in Fig. 8-9 will be used to analyze the operation of an *S-R* NOR latch. The input signal Q to the box labeled Δt (delay) is called the excitation input signal or the next state output signal. The delay output signal q is called the present state output signal. Recall that the output delay Δt in the delay box represents the worst-case delay from the inputs to the output (which

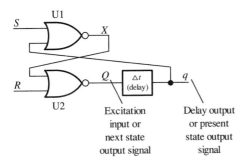

Excitation
input or
next state
output signal

Delay output
or present
state output
signal

FIGURE 8-9
Delay model for *S-R* NOR latch.

is $2t_{pd}$ in this case—see Chapter 5 subsection 5-4-2 for definition of t_{pd}). Delay input values are traditionally called next state values since these are the values that will shortly be present on the delay outputs. Delay output values are traditionally called present state values since these values record the history of the previous inputs. The delay output signal q in Fig. 8-9 becomes equal to the next state output signal Q after the delay time Δt has elapsed. This relationship can be represented by Eq. 8-1.

$$q = Q \text{ after } \Delta t \tag{8-1}$$

8-3-2 Characteristic Equation

The delay or present state output signal q is also fed back as an input to NOR gate U1 in Fig. 8-9. The excitation input signal Q, which is also the next state output signal, can be written in terms of input signals q, S, and R in the same manner as one writes a combinational logic function:

$$Q = \overline{S + q} + R = (S + q) \cdot \overline{R} = S \cdot \overline{R} + q \cdot \overline{R} \tag{8-2}$$

The Boolean function expressed by Eq. 8-2 is called the characteristic equation of the *S-R* NOR latch. The advantage of using the delay model in Fig. 8-9 is twofold. First, it illustrates conceptually why the present state output signal q is equal to Q after the delay time Δt. Second, it allows us to write the next state output signal Q in Eq. 8-2 as though we are dealing with a combinational logic circuit with input signals q, S, and R. Equation 8-2 is a sequential logic function since the variable Q (remember $q = Q$ after Δt) appears on both sides of the equation, illustrating the feedback nature of the *S-R* NOR latch circuit. The next state value Q is a function of the external input signals S, R, and the previous state value Q (called the present state value) represented as q.

Example 8-1

(*a*) Draw the *S-R* NAND latch circuit shown in Fig. 8-8*b* using an output delay box to represent the delay of the circuit. Use the signal name Q for the next state output signal, and the signal name q for the delay or present state output signal.

(*b*) Obtain the characteristic equation for the circuit in (*a*).

Solution

(*a*) Figure E8-1 shows the *S-R* NAND latch circuit with an output delay box. The next state output signal Q and the delay output signal q are placed on the input and output lines of the output delay box as shown in the figure.

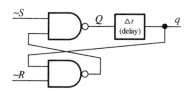

FIGURE E8-1

(*b*) The characteristic equation for the *S-R* NAND latch is the Boolean function for the next state output signal *Q*. The next state output signal *Q* is written in terms of the input signals *q*, ~*S*, and ~*R*.

$$Q = \overline{\sim S \cdot \overline{(q \cdot \sim R)}} = \overline{\sim S} + q \cdot \sim R \qquad (8\text{-}3)$$

Remember that the delay output signal *q* becomes equal to the value of signal *Q* after a delay of Δt; that is, $q = Q$ after Δt. Since *Q* appears on both sides of the equation, the equation is a sequential logic function.

8-3-3 Characteristic or Present State/ Next State Table

Equation 8-2 can be expressed in the form of a truth table called a present state/next state table (or PS/NS table) as illustrated in Table 8-1. All inputs are listed just like in a normal truth table starting with the present state output signal *q*, which is also an input signal, followed by the external input signals *S* and *R*. The next state output signal *Q* is treated as the only output signal in the table. Since the PS/NS table shows the characterization of a circuit, it is also commonly referred to as the characteristic table.

Table 8-1 shows all possible external input signal combinations for *S* and *R* for both the 0 and 1 conditions of the present state output signal *q*. If the next state output signal *Q* has the same value as the present state output signal *q*, the circuit is stable, since *q* does not have to change after the delay time Δt; otherwise, the circuit is unstable. In other words, if the present state output signal *q* changes from a 0 to a 1 or from a 1 to a 0 after the delay time Δt, then *q* is unstable. If the present state output signal *q* is a 0 or a 1 and remains the same value after the delay time Δt, then *q* is stable. In most cases, when the *S-R* NOR latch is used in a circuit application, the signal at point *X* in the circuit in Fig. 8-9 is available as an output signal. In these cases the input signal conditions *S R* = 11 are not

TABLE 8-1
Present state/next state or characteristic table for the *S-R* NOR latch circuit

Present state (PS) output signal *q*	External input signals *S R*	Next state (NS) output signal *Q*	Present state condition of *q*	Comment	$X = \overline{S+Q}$ $= \overline{S} \cdot \overline{Q}$ (after Δt)
0	0 0	0	Stable	Reset state ($Q = 0$)	1
0	0 1	0	Stable	Reset state ($Q = 0$)	1
0	1 0	1	Unstable	Set state ($Q = 1$)	0
0	1 1	0	Stable	$S R = 11$ not normally allowed	0
1	0 0	1	Stable	Set state ($Q = 1$)	0
1	0 1	0	Unstable	Reset state ($Q = 0$)	1
1	1 0	1	Stable	Set state ($Q = 1$)	0
1	1 1	0	Unstable	$S R = 11$ not normally allowed	0

normally allowed as indicated in Table 8-1. When both of the feedback signals for the *S-R* NOR latch circuit are used as present state output signals, the two output signals are complements except when $S\,R\,=\,11$ as shown in Table 8-1 (see column *X* in the table). This condition can easily be observed from the circuit diagram. When *S* and *R* are both 1; that is, both *S* and *R* inputs are trying to set and reset output *Q* at the same time, the output signals of both NOR gates are 0, thus preventing these output signals from being the complement of one another.

8-3-4   State Diagram

The characteristic equation and the characteristic table are two ways to describe the operation of the *S-R* NOR latch circuit. Using the information in the characteristic table, we can graphically represent the same information using a 'state diagram' as shown in Fig. 8-10*a*. The circles in a state diagram represent the output signal states of the circuit. State a represents the present state output signal when $q\,=\,0$ and state b represents the present state output signal when $q\,=\,1$. Directed line segments or arches indicate state transitions from a present state output value to the next state output value. Transitions from one present state output value to another are based on the external input values applied to the circuit. A state diagram provides a graphical representation of the characteristic of a circuit in the same manner as a present state/next state (PS/NS) table or characteristic table provides a tabular representation of the characteristic of a circuit.

8-3-5   ASM Chart

Another name for a sequential logic circuit is a 'state machine'. Since the storage capacity or number of bistable devices is not infinite in a sequential logic circuit, the term 'finite state machine' is also commonly used. An alternate graphical representation of the characteristic of a sequential logic circuit is provided by a diagram called an 'algorithmic state machine' (ASM) chart. This representation describes the operation of a circuit in much the same way as a programming flow chart. In an ASM chart there are three primary symbols, the state box (rectangular shape), the decision box (diamond shape), and the conditional output box (cylindrical shape). Only state boxes and decision boxes are shown in Fig. 8-10*b* since conditional output boxes are not needed. A state box can contain an output list of signal names that are generated when the circuit is in a particular state. The state assignment or binary code assigned to a state is shown at the upper right corner of the state box. A decision box describes the inputs to the circuit represented by the ASM chart. Each decision box has two exit paths, one to take if the input condition represented by the Boolean expression placed in the decision box is true (represented by a 1 on an exit path), and the other if the input condition is false (represented by a 0 on an exit path). When a conditional output box is used, it describes other outputs (those listed in the box) that are dependent on the state of the circuit in addition to one or more inputs.

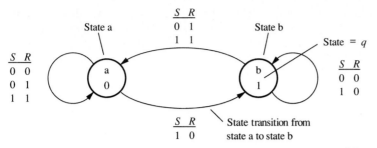

(a) State diagram for a S-R cross-coupled NOR gate latch.

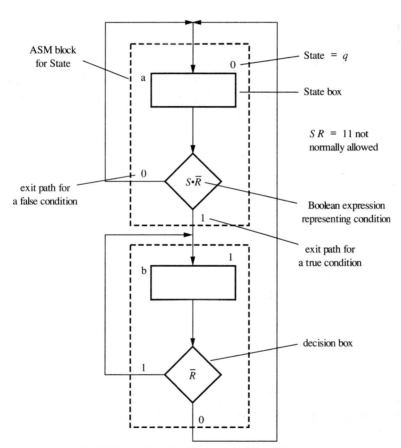

(b) ASM chart for a S-R cross-coupled NOR gate latch.

FIGURE 8-10
Graphical representations for a S-R cross-coupled NOR gate latch.

Notice the dotted lines in Fig. 8-10*b* that define the boundaries that make up each ASM block. In the simplest form, an ASM block consists of only a state box; however, a block can consist of several decision and conditional output boxes as well. An ASM block can have one entrance path to the state box and any number of exit paths. A path through an ASM block from its entrance point to an exit point is called a 'link path'. Each ASM block represents a state in the circuit as indicated by the state symbolic names *a* and *b*. When the circuit represented by the ASM chart is in a particular state, the output signals in the output list become true. The restriction that is placed on the formation of each ASM block is that for each state and stable set of input signals there must be only one next state. When a link path is followed through an ASM block and a conditional output box is encountered, the output signals in the output list become true.

When analyzing or designing sequential logic circuits, sometimes it may be easier to work with the format provided by a truth table (PS/NS table), while at other times the format provided by a state diagram or an ASM chart may be easier to work with. Comparing the state diagram and ASM chart in Fig. 8-10, one can observe that when the *S-R* NOR latch circuit is in the reset state ($q = 0$) and $S\,R = 10$ ($S \cdot \overline{R}$ is true or 1) then the circuit becomes set ($q = 1$). To keep the latch circuit set requires the $S\,R$ inputs be either 00 or 10 ($\overline{R}$ is true or 1); otherwise, the circuit becomes reset. Both the state diagram and the ASM chart in Fig. 8-10 accurately describe the operation of the *S-R* cross-coupled NOR gate latch.

Example 8-2

(*a*) Write the characteristic equation for the *S-R* NAND latch circuit in Fig. E8-1, and obtain its PS/NS table. Identify the external input conditions for the set state and the reset state, and also identify the external input signal conditions which are not normally used for the *S-R* NAND latch circuit; that is, observe the circuit and determine when the feedback output signals are not complements.

(*b*) Obtain the state diagram for the *S-R* NAND latch circuit. Let state 'a' represent the present state output signal when $q = 0$, and state 'b' represent the present state output signal when $q = 1$.

(*c*) Obtain the ASM chart for the *S-R* NAND latch circuit.

Solution

(*a*) By analyzing the circuit in Fig. E8-1, we can write the characteristic equation as follows.

$$Q = \overline{\sim S \cdot \overline{(q \cdot \sim R)}} = \sim S \; + \; q \cdot \sim R$$

The characteristic table or PS/NS table for the *S-R* NAND latch circuit can be determined from the product terms in the characteristic equation. The product terms provide the information to determine the input values for the next state output signal Q. Observing the circuit diagram for the *S-R* NAND latch in Fig. E8-1, one can see that when S and R are both 1 ($\sim S$ and $\sim R$ are both 0), the outputs of both NAND gates are 1. This is also the case where the output Q is trying to be set and reset at the same time (a contradiction in logic). External input signal conditions $\sim S \sim R = 00$ (or $S R = 11$) prevent the NAND gate output signals from being complements of one another and are therefore not normally allowed.

Present state (PS) output signal q	External input signals $\sim S \sim R$		Next state (NS) output signal Q	Comment
0	0	0	1	$S R = 11$ not normally allowed
0	0	1	1	set state
0	1	0	0	reset state
0	1	1	0	reset state
1	0	0	1	$S R = 11$ not normally allowed
1	0	1	1	set state
1	1	0	0	reset state
1	1	1	1	set state

(b) The state diagram for the *S-R* NAND latch circuit is shown in Fig. E8-2a. The PS/NS table is used to generate the diagram.

(c) The ASM chart for the *S-R* NAND latch circuit is shown in Fig. E8-26. The PS/NS table or the state diagram can be used to obtain the ASM chart. Notice that the Boolean expression in a decision box can be complemented if the true and false conditions on the decision box exit paths are complemented.

8-3-6 Karnaugh Map

The format provided by a Karnaugh map with the external input signals listed horizontally and the present state output signal listed vertically is also useful in representing the information contained in a PS/NS table, state diagram, or ASM chart. This form of Karnaugh map representation is useful in two ways. It is used as a reduction tool, and also as a tool to describe sequential logic circuit operation. The Karnaugh map representation for the *S-R* NOR latch circuit is shown in Fig. 8-11a. To be consistent with the state diagram and ASM chart representations,

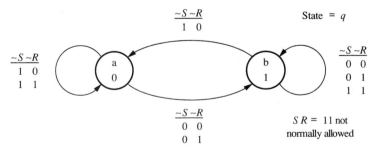

FIGURE E8-2a

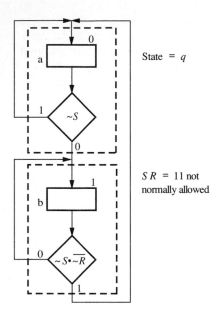

State = q

$SR = 11$ not
normally allowed

FIGURE E8-2*b*

state identifying symbols 'a' and 'b' can be written beside the present state values
for $q = 0$ and $q = 1$ on the left side of the map as illustrated in Fig. 8-11*a*. The
characteristic equation for the *S-R* NOR latch, Eq. 8-2, can be obtained directly
from the Karnaugh map in Fig. 8-11*a* or vice versa.

The Karnaugh map in Fig. 8-11*b* is used as an analysis tool. The 0s and 1s
shown in the cells of the map represent the next state output signal Q. When the
next state output signal Q has the same value as the present state output signal q,
the circuit is stable; otherwise, the circuit is unstable. The *S-R* NOR latch bistable
circuit can only exist in a stable state after the external input signals change and

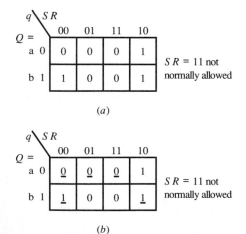

$SR = 11$ not
normally allowed

$SR = 11$ not
normally allowed

FIGURE 8-11
S-R NOR latch (*a*) Karnaugh map, (*b*) Karnaugh
map with stable states identified.

the present state output q is allowed time to settle ($q = Q$ after Δt). When the circuit is in state a; that is, present state output signal $q = 0$, the circuit is stable for the input conditions $S\,R = 00$, 01, or 11, since the next state output signal $Q = 0$. When the circuit is in state b; that is, present state output signal $q = 1$, the circuit is stable for the input conditions $S\,R = 00$ or 10, since the next state output signal $Q = 1$. Each cell in the Karnaugh map in Fig. 8-11b that contains an output signal condition that is stable is called a stable state. We have marked the stable states in the Karnaugh map using an underline symbol (some books use a circle instead of a underline symbol). These are the same states identified in Table 8-1 and in the state diagram in Fig. 8-10a. Stable states are identified in the state diagram by input conditions that show a directed line segment leaving a state and returning to the same state.

8-3-7   Transition Map

An alternate way of representing the circuit characteristic of the *S-R* NOR latch is with a transition map (or transition table) as shown in Fig. 8-12. The transition map is usually obtained from the Karnaugh map. A transition map gets its name from the Karnaugh map-like organization of the variables on the map; however, the variables need not be organized in the same fashion as a Karnaugh map. The same information, when presented in tabular format, is called a transition table. In the transition map for the *S-R* NOR latch shown in Fig. 8-12, each cell representing a stable state contains an underline symbol. The absence of an underline symbol in a cell in the transition map represents an unstable state. Notice when the external input conditions change in the transition map, the *S-R* NOR latch always ends up in a stable state. Beginning in a stable state, the transition map provides a method of describing how the present state output signal of the *S-R* NOR latch circuit responds to input changes. A horizontal movement in the transition map occurs during the time interval Δt when the external input conditions change and q does not change. When the external input conditions change and q changes, the following movements in the transition map take place during the time interval Δt: (a) a horizontal movement from the stable present state value to an unstable present state value occurs, and (b) a vertical movement from the unstable present state value to a stable present state value occurs. To show a vertical movement, an arrow is drawn in the transition map from each cell representing an unstable state to a cell representing a stable state where the movement will stop.

$S\,R = 11$ not normally allowed

FIGURE 8-12
Transition map of *S-R* NOR latch.

A transition map can be used to analyze the response of a circuit to its input changes in the following manner. Assume the circuit is initially in the stable state represented by $q\,S\,R = 001$. When the external input conditions change to $S\,R = 00$, the value of q remains unchanged, and the circuit responds by moving to stable state $q\,S\,R = 000$ during the time interval Δt (a horizontal movement in the transition map). Now assume that the external input conditions change to $S\,R = 10$. This requires q to change from 0 to 1 during the time interval Δt. This means that the circuit responds by moving to the unstable state $q\,S\,R = 010$ (a horizontal movement in the transition map), and then to stable state $q\,S\,R = 110$ (a vertical movement in the transition map).

8-3-8   Flow Map

The flow map (or flow table) is yet another way to represent the circuit characteristic of the *S-R* NOR latch as illustrated in Fig. 8-13*a*. Like the transition map, the flow map for a circuit is usually obtained from the Karnaugh map. (The same information represented in tabular format is called a flow table). State values are used in the cells of a Karnaugh map. A flow map can be obtained from a Karnaugh map by replacing these state values with state symbols as shown in Fig. 8-13*a*. To obtain the flow map for the *S-R* NOR latch using symbols a and b, each 0 in the Karnaugh map is replaced by the symbol a and each 1 is replaced by the symbol b. Each cell representing a stable state in the flow map contains an underline symbol (as mentioned earlier some books use circles instead of underline symbols). Cells representing unstable states do not contain an underline symbol. To illustrate how a flow map is used to analyze the operation of the *S-R* NOR latch, consider that the circuit is in a stable state and the external input conditions change such that q remains in the same state (remains in state a or state b) after Δt. In this case, during the time interval Δt, a horizontal movement occurs in the flow map to the new stable state specified by the external input conditions. When

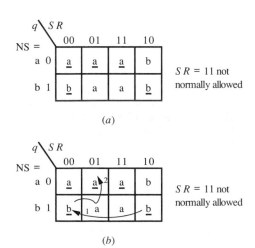

(a)

(b)

S R = 11 not normally allowed

FIGURE 8-13
S-R NOR latch (*a*) flow map, (*b*) sequence of events in a flow map.

the external input conditions cause q to change to a different state (from a to b or from b to a) a horizontal movement in the flow map is required, followed by a vertical movement to reach a stable state during the time interval Δt. Notice that the flow stops in a stable state; that is, in a cell containing an underline symbol.

Like a transition map, a flow map can also be used to analyze the response of a circuit to its input changes. We consider the flow map more useful since the sequence of events can be shown by arrows in the flow map. Using arrows to show a sequence of events in a transition map could be confusing unless the arrows depicting the sequence and the arrows showing unstable to stable movements are clearly delineated. Consider the following sequence beginning in stable state q $S\ R\ =\ 110$. When the external input conditions change to $S\ R\ =\ 00$, during the time interval Δt the circuit responds by moving from stable state b to stable state b at $q\ S\ R\ =\ 100$ (a horizontal movement in the flow map shown by arrow 1 in Fig. 8-13b). If the external input conditions are changed to $S\ R\ =\ 01$, the circuit responds during time interval Δt by moving to unstable state a (a horizontal movement, shown by arrow 2) and then on to stable state a (a vertical movement, also shown by arrow 2).

Example 8-3

(a) Obtain the transition map for the S-R NAND latch circuit shown in Fig. E8-1. Let state a represent the present state output signal $q\ =\ 0$, and state b represent the present state output signal $q\ =\ 1$. Identify the stable states in the transition map using an underline symbol.

(b) Obtain the flow map for the same circuit by representing the present state output signal $q\ =\ 0$ as state a, and present state output signal $q\ =\ 1$ as state b. Use arrows to show the sequence followed by the S-R NAND latch circuit for the following situations: (1) the circuit is stable in state q ~S ~$R\ =\ 010$ until the external input signal conditions change to ~S ~$R\ =\ 11$, (2) after the circuit becomes stable the external input signal conditions are changed to ~S ~$R\ =\ 01$, (3) after the circuit becomes stable the external input signal conditions are changed to ~S ~$R\ =\ 11$, and (4) after the time interval Δt, the external input signal conditions are changed back to ~S ~$R\ =\ 10$. Show the value of the present state output signal q for each of the external input signal conditions (1) through (4).

Solution

(a) Since the transition map is generated from the Karnaugh map, we need to obtain the Karnaugh map first. The Karnaugh map can be obtained from the characteristic equation, the PS/NS table (or the characteristic table), the state diagram, or the ASM chart for the S-R NAND latch. The Karnaugh map for the S-R NAND latch circuit is shown in Fig. E8-3a. Stable states in the map are identified by the cells that contain the underline symbol.

Using the Karnaugh map in Fig. E8-3a, we obtain the transition map shown in Fig. E8-3b. Each cell representing a stable state contains an underscore. An arrow is drawn from each unstable state to the corresponding stable state for the specified external input signal conditions. When external input signal conditions

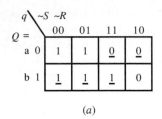

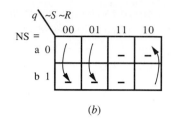

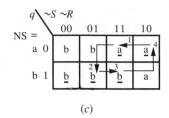

(a)   (b)   (c)

FIGURE E8-3

occur causing the latch circuit output signal to change to an unstable state, the arrow indicates the direction from that state to the next stable state. Notice that the S-R NAND latch circuit output signal is stable in the states $q \sim S \sim R = 011, 010, 100, 101$ and 111 (those states with underline symbols in the cells of the transition map); however, it is unstable in the states $q \sim S \sim R = 000, 001$, and 110 (those states without underline symbols in the cells of the transition map).

(b) The flow map for the S-R NAND latch circuit is shown in Fig. E8-3c. The flow map is obtained from the Karnaugh map by substituting the symbols a and b in the cells of the map for 0 and 1 respectively. Each cell representing a stable state contains an underline symbol. When external input signal conditions cause the present state output signal q of the circuit to change to an unstable state, a horizontal movement followed by a vertical movement occurs to put the circuit in a stable state. Arrow 1 in the flow map in Fig. E8-3c shows the sequence the circuit output signal follows from stable state $q \sim S \sim R = 010$ when the external input signal conditions change to $\sim S \sim R = 11$. The circuit stabilizes (the circuit output signal reaches a steady state value) after time interval Δt. Arrow 2 shows the sequence the circuit output signal follows when the external input signal conditions change to $\sim S \sim R = 01$. After the circuit stabilizes, the sequence the circuit output signal follows when the external input signal conditions change to $\sim S \sim R = 11$ is shown by arrow 3. After the appropriate delay Δt, arrow 4 shows the sequence the circuit present state output signal follows when $\sim S \sim R$ changes back to 10. By simply observing the flow map, we can see the sequence followed by the present state output signal q beginning in stable state $q \sim S \sim R = 010$ is 0 after Δt (for $\sim S \sim R = 11$), 1 after Δt (for $\sim S \sim R = 01$), 1 after Δt (for $\sim S \sim R = 11$), and 0 after Δt (for $\sim S \sim R = 10$).

In cases where the external input conditions $S R = 11$ for the S-R NOR latch circuit are allowed, these input signals (S and R) must not change to 0 simultaneously. If this happens, the next stable state of the circuit will be unknown because a 'critical race' condition exists between the two signals S and R. A 'critical race' occurs between two signals that are required to change at the same time when the next stable state is dependent on the delay paths in the circuit. Due to the different delay paths from S to q and R to q, the circuit responds as though the external input signals are sequenced in either one of the following

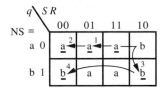

FIGURE 8-14

Illustrating a critical race condition: From stable state a, $S R = 11$ to 01 to 00 results in stable state a (arrows 1 and 2). From stable state a, $S R = 11$ to 10 to 00 results in stable state b (arrows 3 and 4).

ways: (1) $S R$ changes from 11 to 01 to 00, or (2) $S R$ changes from 11 to 10 to 00. The circuit response for (1) is shown in the flow map in Fig. 8-14 by arrows 1 and 2 and results in stable state a ($q = 0$); however, the circuit response for (2) is shown in the flow map in Fig. 8-14 by arrows 3 and 4, and results in stable state b ($q = 1$). Depending on the circuit delays, the next stable state will be unknown when the external input conditions S and R change from 1 to 0 at the same time. This situation is undesirable and should never be allowed to happen since the action of the circuit is unpredictable. Recall from Fig. 8-6a, one basic rule for asynchronous circuit operation is to allow only one external input signal to change at a time (this is commonly referred to as operating a circuit in fundamental mode).

When a race occurs between two signals that are required to change at the same time, and the next stable state is the same regardless of the delay paths in the circuit, then the race is 'noncritical'. Fortunately all other race conditions for the *S-R* NOR latch circuit are noncritical. As an example of a noncritical race with the circuit initially in stable state a at $q S R = 001$, observe the flow table in Fig. 8-15 for the following cases: (1) $S R$ changes from 01 to 10 passing through 00 (arrows 1 and 2), and (2) $S R$ changes form 01 to 10 passing through 11 (arrows 3 and 4). Notice that for either case the circuit response is the same and results in stable state b ($q = 1$). Noncritical race conditions such as this are not a problem since the action of the circuit is predictable.

8-3-9 Timing Diagram

The operation of the *S-R* NOR latch circuit in Fig. 8-9 can also be analyzed by observing its properties in the switching waveforms shown in the timing diagram in Fig. 8-16. Since waveforms for both q and X are shown in the figure, it is more accurate to consider each output as having a separate propagation delay of $1t_{pd}$ (where t_{pd} is one gate delay). Directed line segments pointing from an

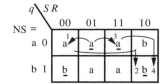

FIGURE 8-15

Illustrating a noncritical race condition: From stable state a $S R = 01$ to 00 to 10 results in stable state b (arrows 1 and 2). From stable state a, $S R = 01$ to 11 to 10 results in stable state b (arrows 3 and 4).

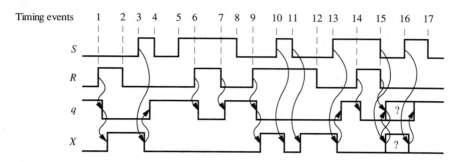

FIGURE 8-16
Timing diagram for a S-R NOR latch.

input transition to an output transition and from an output transition to another output transition indicate the sequence of events. When a single external input signal S or R changes, causing the output signals q and X to change, the sequence followed by the output signals is well defined—either $q = \overline{X}$ or $q = X = 0$ after Δt. When the external input signal conditions $S R = 11$ occur and both inputs are changed simultaneously to $S R = 00$, a critical race condition exists, and the state of output signals q and X are unknown until either S or R changes to a 1. If the external input signal conditions $S R = 11$ are not allowed to occur, then the critical race condition and the output signal conditions $q = X = 0$ cannot occur. By narrowing the normal operating range of the S-R NOR latch circuit to the external input conditions $S R = 00, 01,$ or 10, the action of the circuit is well defined and very useful.

Example 8-4

(a) Analyze the action of the sequential logic circuit shown in Fig. E8-4a. Draw the timing diagram waveforms for the signals $\sim S$, $\sim R$, and Q, showing their logic values before the push-button switch is pressed, while the switch is being pressed, and after the switch is released.

(b) Discuss the importance of the circuit and where this circuit might be used.

Solution

(a) A mechanical switch such as a switch on a computer keyboard, a light switch for a room in a house, or a switch like the one shown in Fig. E8-4a, all have the characteristic that is known as contact bounce when the switch is actuated. When the button of the switch shown in Fig. E8-4a is pushed, contact bounce occurs for a brief period of time (usually several milliseconds) before the contact comes to rest in a new position. The waveforms for input signals $\sim S$ and $\sim R$, and the output signal Q are shown in the timing diagram in Fig. E8-4b. The explanation in the timing diagram takes into consideration the nature of the contact bounce of the mechanical switch. Input signal timing events are numbered consecutively from 1 through 22. An input signal timing event occurs when either of the input signals $\sim S$ or $\sim R$ changes state either from a 0 to a 1 or a 1 to a 0, as shown in the timing diagram. We assume that the timing events 3 through 10 and 13 through 22 result from contact bounce.

   To illustrate the action of the *S-R* NAND latch circuit, the flow map for the circuit is shown in Fig. E8-4*c* with each input signal timing event indicated by corresponding numbered arrows in the map. The waveform for the output signal *Q* is obtained by observing the action of the *S-R* NAND latch output signal *Q* in the flow map based on the input signal conditions $\sim S$ and $\sim R$ in the timing diagram beginning with $\sim S \sim R = 01$. Figure E8-4*d* illustrates the sequence the output signal *Q* follows. Notice that contact bounce does not affect the state of the output signal *Q*; therefore, when the push-button switch is pressed and released, a single negative pulse is provided at the *Q* output of the *S-R* NAND latch circuit.

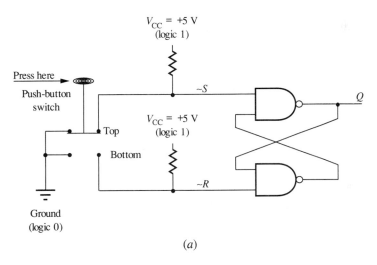

(*a*)

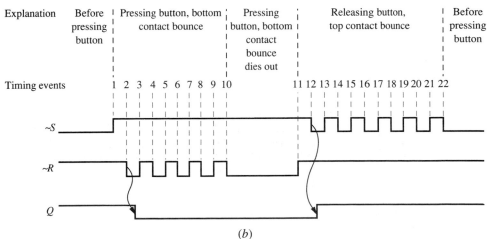

(*b*)

FIGURE E8-4

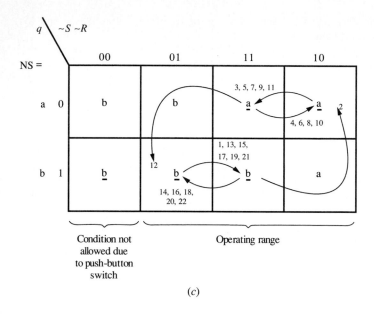

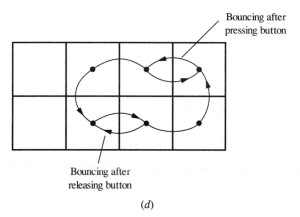

FIGURE E8-4

(*b*) The circuit is important when it is necessary to provide a single transition from 1 to 0 or 0 to 1 for a binary signal produced by the actuation of a mechanical switch. This type of hardware debounce circuit can be used in the laboratory when testing digital circuits, or in a computer application when it is necessary to provide a bounce free signal at an input to the computer circuit. In many computer keyboard applications, the mechanical switches are debounced using a software routine that accounts for the longest possible delay time that contact bounce is expected to occur.

Figure 8-17 provides a summary of the different tools or techniques that we have used to characterize the *S-R* cross-coupled NOR gate-latch circuit. Observe

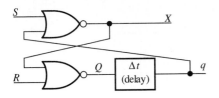

(a) S-R NOR latch circuit delay model

$$Q = S \cdot \bar{R} + q \cdot \bar{R}$$

(b) Characteristic equation

PS	External inputs		NS	
q	S	R	Q	Comment
0	0	0	0	Reset state
0	0	1	0	Reset state
0	1	0	1	Set state
0	1	1	0	$SR = 11$ not normally allowed
1	0	0	1	Set state
1	0	1	0	Reset state
1	1	0	1	Set state
1	1	1	0	$SR = 11$ not normally allowed

(c) Characteristic or present state/next state table

(d) State diagram

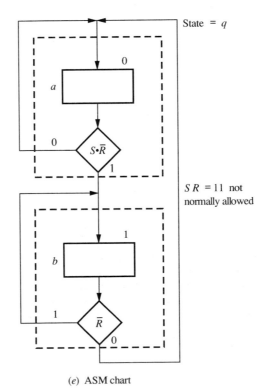

(e) ASM chart

FIGURE 8-17
Summary of sequential circuit analysis tools applied to a *S-R* NOR latch cirucit.

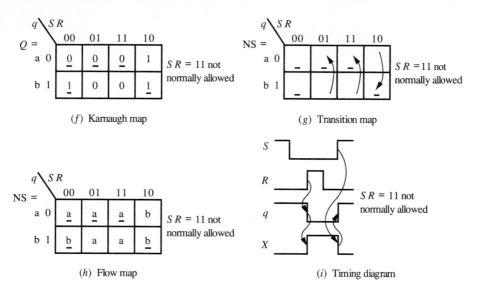

(f) Karnaugh map   (g) Transition map

(h) Flow map   (i) Timing diagram

FIGURE 8-17
(*continued*)

that the propagation delays associated with the signals q and X are not shown in the timing diagram to keep the timing diagram as simple as possible. Other sequential circuits can be analyzed using one or more of the techniques in Fig. 8-17. The design of sequential circuits also involves the use of many of these same techniques.

We have used the lower case symbol q to represent the dynamic nature of the present state output signal, and the upper case symbol Q to represent the static nature of the next state output signal. Table 8-2 illustrates the wide variety of

TABLE 8-2
Different symbols for present state and next state output signals

Symbol for present state output signal	Symbol for next state output signal
q	Q
Q	Q^+
Q	Q
Q_0	Q
Qn	$Q(n + 1)$
Qt	$Q(t + 1)$
Q	$Q(t + 1)$
y	Y
Y	Y^+

symbols that are commonly used by different data and text books to represent the present state output signal and the next state output signal for sequential circuits. One should know the meaning of the terms 'present state' and 'next state' and how these terms apply to sequential circuits instead of memorizing the symbology.

8-4 ADDITIONAL BISTABLE MEMORY DEVICES

In order to choose the proper memory devices for synchronous and asynchronous sequential circuits, one must understand the different groups of memory devices that are available. In general, bistable devices or flip-flops can be divided into three major groups: latches (either basic or gated), master-slave (pulse-triggered) flip-flops, and edge-triggered flip-flops as illustrated in Fig. 8-18. Memory devices in each of these major groups use the basic S-R latch circuits as building blocks. In this section we present the memory devices that make up these three major groups and discuss their properties. Within each group, an S-R, D, J-K or T-type latch or flip-flop can be constructed. Each of these types are also presented.

8-4-1 Gated S-R Latches

As we mentioned earlier, the asynchronous sequential S-R latch circuits shown in Figs. 8-19a and b are the simplest types of memory devices. By using AND gates or NAND gates with the basic latch circuits as shown in Figs. 8-19c and d, a control or enable input is added to provide additional capability. When the control or enable input signal C is 0, signals applied to the S or R inputs can not effect the cross-coupled NOR or NAND latch circuit; however, when the control input signal C is 1, signals applied to the S or R inputs can effect the cross-coupled NOR or NAND latch circuit. A circuit with a control input applied in the manner shown in Figs. 8-19c or d is often called a gated input circuit or simply

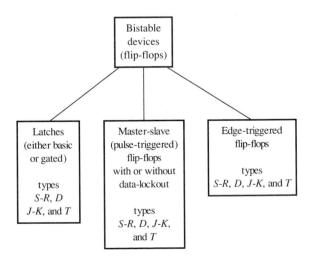

FIGURE 8-18
Three major groups of bistable devices.

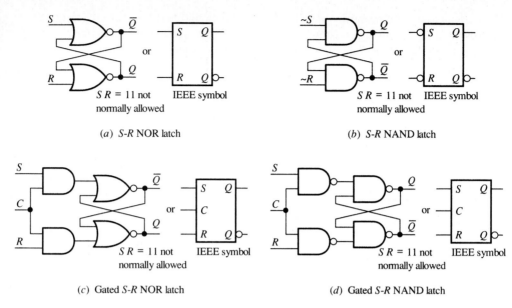

(a) S-R NOR latch

(b) S-R NAND latch

(c) Gated S-R NOR latch

(d) Gated S-R NAND latch

FIGURE 8-19
Basic and gated *S-R* latch circuits.

a gated circuit. The rectangular symbols shown beside each circuit are the IEEE recommended symbols for the respective latches. The outputs labeled Q are not part of the IEEE symbol.

Example 8-5

(a) Write the characteristic equation for the gated *S-R* NOR latch circuit shown in Fig. 8-19c.

(b) Obtain the transition map and the state diagram for the gated *S-R* NOR latch circuit.

Solution

(a) To obtain the characteristic equation, the gated *S-R* NOR latch circuit is drawn as shown in Fig. E8-5a. It is not necessary to draw the delay model, but it helps to emphasize the difference between the present state output q and the next state output. The characteristic equation is the equation that expresses the next state output Q of the latch circuit in terms of the external inputs C, S, and R and the

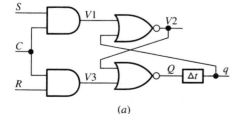

(a)

FIGURE E8-5

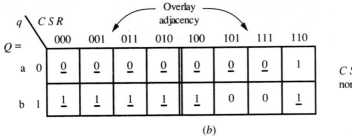

(b)

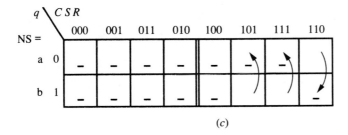

(c)

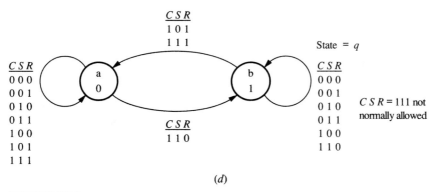

(d)

FIGURE E8-5

present state output q. The signals labeled $V1$ through $V3$ are simply used as temporary variables.

From Fig. E8-5a we can write

$$V1 = C \cdot S$$

$$V2 = \overline{V1 + q} = \overline{C \cdot S + q}$$

$$V3 = C \cdot R$$

$$Q = \overline{V2 + V3}$$

$$= \overline{\overline{C \cdot S + q} + C \cdot R}$$

$$= (C \cdot S + q) \cdot \overline{(C \cdot R)}$$

$$= (C \cdot S + q) \cdot (\overline{C} + \overline{R})$$

$$= C \cdot S \cdot \overline{R} + q \cdot \overline{C} + q \cdot \overline{R} \tag{8-4}$$

(b) To obtain the transition map, we first draw the Karnaugh map using the characteristic equation. The Karnaugh map for the gated *S-R* NOR gate latch is shown in Fig. E8-5b. By choice, the external input variables listed across the top of the map divide the map into two distinct parts such that the right half overlays the left half of the map.

The transition map is shown in Fig. E8-5c. An underline symbol is used in each cell representing a stable state. Directed arrows show the transition from an unstable state to a stable state. Notice that the left side of the transition map has all stable states, so that when the output q is 0 or 1 and $C = 0$ the circuit stays in the same output state even if S and R change. A transition to a different state can only occur when the control input $C = 1$. If the circuit is in stable state $q\ C\ S\ R = 0100$ (sometime referred to as the total present state 0100), and the S input changes to a 1, then the circuit goes from stable state $q\ C\ S$ $R = 0100$ to unstable state 0110 and on to stable state 1110 where $q = 1$, as expected (the set state). The input signal conditions $S\ R = 11$ should not be allowed to happen when $C = 1$, because of the possibility of both $S\ R$ input signals changing back to 00 simultaneously and creating a critical race. $S\ R = 11$ also forces Q and $\bar{Q}$ to contradict their definition as complements.

The state diagram for the gated *S-R* NOR latch can also be obtained from the Karnaugh map as is shown in Fig. E8-5d. Both the transition map and the state diagram depict the operation of the gated *S-R* NOR latch circuit.

8-4-2   Gated *D* Latch

Since latches are more useful with a control input, that is, a gated input circuit, we will concentrate on the *D*, *J-K*, and *T* latch circuits with gated inputs (each of these configurations can also exist without a control input). A gated *S-R* latch can be converted to a gated *D* latch by the simple addition of an Inverter as shown in Fig. 8-20. The rectangular symbol shown to the right of the gated *D* latch circuit is the recommended IEEE symbol. As was mentioned earlier, the Qs labeled in the IEEE symbol are not part of the symbol, and are used simply to provide continuity with the material presented in the text.

The Inverter in the gated *D* latch serves two different purposes. The Inverter provides a latch circuit with only one input (not including the control input) called the data input. The Inverter also insures that the $S\ R$ latch inputs are complemented so that the signal conditions $S\ R = 11$ cannot occur. To set the gated *D* latch circuit requires input signal conditions $C\ D = 11$, and to reset the circuit requires

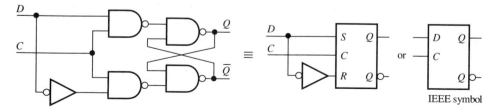

FIGURE 8-20
Gated *D* latch.

$CD = 10$. All latch circuits whether they be *S-R*, *D*, *J-K*, or *T*, have the property of transparency. The transparency property simply means that the outputs respond to an input signal change when the control input signal is 1. For a gated *D* latch, the data input signal *D* is transferred to the *Q* output when the control (enable) input signal is 1, and the *Q* output signal follows the data input signal *D* as long as the control input remains a logic 1. The gated *D* latch is often called a transparent latch for this reason.

Using the characteristic equation for the *S-R* NAND gate latch, Eq. 8-3, we can write the characteristic equation for the gated *D* latch circuit in Fig. 8-20 as follows:

$$Q = \sim\! S + q \bullet \sim\! R$$
$$= C \bullet D + q \bullet \overline{(C \bullet \overline{D})}$$
$$= C \bullet D + q \bullet (\overline{C} + D)$$
$$= C \bullet D + q \bullet \overline{C} + q \bullet D \tag{8-5}$$

Using the characteristic equation, we obtain the Karnaugh map, the flow map, and the state diagram for the gated *D* latch as shown in Fig. 8-21*a* through

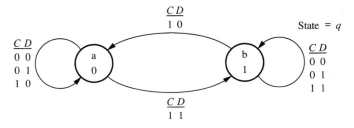

(*a*) Karnaugh map for gated *D* latch

(*b*) Flow map for gated *D* latch

(*c*) State diagram for gated *D* latch

FIGURE 8-21
Different analysis tools used to show the properties of a gated *D* latch.

c. Each of these analysis tools can be used to analyze the action of the Q output in response to a change of the D input signal. In the flow map for example, when the circuit is in stable state $q\,C\,D = 010$ and the D input changes to 1, the circuit response is shown by arrow 1. Arrow 2 shows the response of the output Q when the D input changes back to 0 while C remains unchanged. This illustrates the transparent property of the gated D latch.

Data present on the D input, that is, either a logic 1 or 0, is stored or latched into the gated D latch when the control input changes from a 1 to a 0. For proper operation of the latch, the data must meet the data setup time, t_{su}, and data hold time, t_h, for the memory device being used. Data that arrives after the data setup time requirement or changes prior to the data hold time requirement can cause the device to operate erratically. It is the designer's responsibility to investigate the data book for the device being used, obtain the data setup and data hold time information, and insure that the device is being operated within its designed capabilities.

Figure 8-22 shows a partial circuit diagram for a set of eight transparent latch circuits in a single package. The input labeled $\overline{OC}$ is the output control input for the 3-state Inverter. Pulling $\overline{OC}$ to a logic 0 enables the Q outputs of the latches, while pulling $\overline{OC}$ to a logic 1, places the outputs in a high impedance state and thus allows a bus-organized system to utilize the bus. Like $\overline{OC}$ which controls all the 3-state output Inverters, the C input controls all the individual latch control inputs. Each individual input control, labeled $C1$, controls its local D input, labeled $1D$. This notation is called dependency notation. For more information concerning dependency notation refer to *Overview of IEEE Standard 91-1984,* in Appendix A, Section 4.0. The property of transparency of the gated D latch circuit is clearly illustrated in the timing diagram shown in Fig. 8-22*b*. Notice that when C is 1 output, $Q1$ follows input $D1$. Propagation delay is shown as t_{pd} and its value is dependent on which input, C or $D1$, causes the output $Q1$ to change.

8-4-3   Gated *J-K* Latch

A gate-level diagram of a gated *J-K* latch circuit is shown in Fig. 8-23. The recommended IEEE symbol (with Qs added) is also shown in the figure. Like the gated D latch which utilizes an Inverter to insure the basic latch circuit S and R inputs remain complemented, the gated *J-K* latch uses gating to steer either the J input signal or K input signal, and thus keep the S and R inputs complemented. If the J and K inputs are a logic 1 at the same time the control input C is 1, then the latch circuit should toggle the output. If the Q output is a logic 1, toggling the output changes it to a 0, and if the Q output is a logic 0, toggling the output changes it to a 1. Unfortunately, the input signal conditions $C\,J\,K = 111$ provide an unstable circuit as illustrated in the following example; otherwise, the gated *J-K* latch circuit has the same properties as the gated *S-R* latch circuit. Steering is achieved when the control input C is 1 by allowing only the J input to be enabled when the Q output is 0, and allowing only the K input to be enabled when the Q output is 1. Thus when the Q output is 0 and the J in-

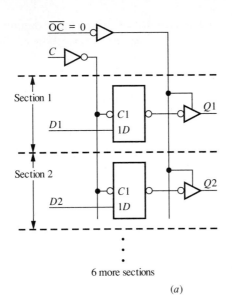

Partial circuit diagram
of a 74ACT11373,
Octal *D*-type transparent
latches with 3-state outputs.

t_w = 5 ns min, Pulse duration
      (enable, *C* high)

t_{su} = 3.5 ns min, Setup time
      (data before *C*↓)

t_h = 3.5 ns min, Hold time
      (data after *C*↓)

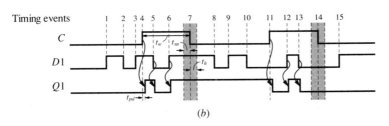

FIGURE 8-22
(*a*) Partial circuit diagram of a 74ACT11373, (*b*) timing diagram illustrating setup and hold time requirements and transparent property of latch.

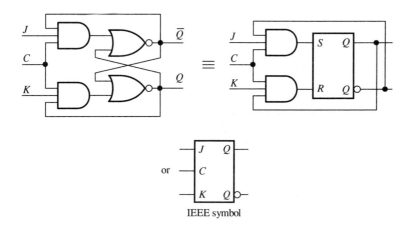

FIGURE 8-23
Gated *J-K* latch.

put is a logic 1, the J input is enabled so the latch can be set. When the Q output is 1 and the K input is a logic 1, the K input is enabled so the latch can be reset or cleared.

Example 8-6

(a) Write the characteristic equation for the gated J-K latch circuit shown in Fig. 8-23.

(b) Obtain the Karnaugh map and the state diagram for the gated J-K latch circuit. Discuss any unusual properties associated with the circuit as observed from its Karnaugh map or state diagram.

(c) Obtain the ASM chart for the gated J-K latch circuit.

Solution

(a) From Fig. 8-23 we can write the characteristic equation for the S-R latch circuit as

$$Q = S \cdot \overline{R} + q \cdot \overline{R}$$

where,

$$S = \overline{q} \cdot C \cdot J$$

and,

$$R = q \cdot C \cdot K$$

so,

$$Q = (\overline{q} \cdot C \cdot J) \cdot \overline{(q \cdot C \cdot K)} + q \cdot \overline{(q \cdot C \cdot K)}$$
$$= (\overline{q} \cdot C \cdot J) \cdot (\overline{q} + \overline{C} + \overline{K}) + q \cdot (\overline{q} + \overline{C} + \overline{K})$$
$$= \overline{q} \cdot C \cdot J + \overline{q} \cdot C \cdot J \cdot \overline{K} + q \cdot \overline{C} + q \cdot \overline{K}$$
$$= \overline{q} \cdot C \cdot J \cdot (1 + \overline{K}) + q \cdot \overline{C} + q \cdot \overline{K}$$
$$= \overline{q} \cdot C \cdot J + q \cdot \overline{C} + q \cdot \overline{K} \tag{8-6}$$

(b) To obtain the flow map we first draw the Karnaugh map using the characteristic equation. The Karnaugh map for the gated J-K latch is shown in Fig. E8-6a. Again by choice, the external input variables listed across the top of the map divide the map into two distinct parts such that the right half overlays the left half of the map.

Because the J input is qualified by $\overline{q} \cdot C$ and the K input is qualified by the $q \cdot C$, as shown in the circuit diagram, the circuit can only be set when it is currently reset or cleared, and can only be reset or cleared when it is currently

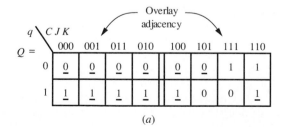

(a)   **FIGURE E8-6**

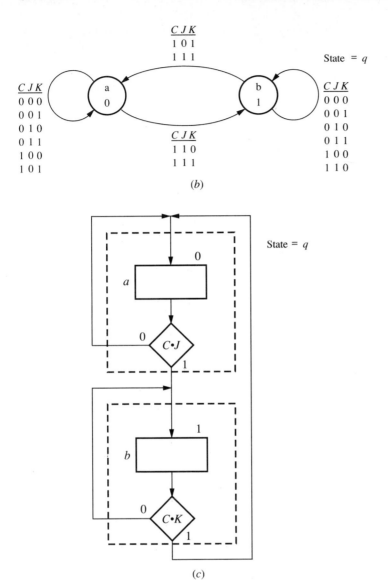

(b)

(c)

FIGURE E8-6

set. When J and K are both 1 while C is 1, the circuit does not have a stable state (there are no stable states in column $C\,J\,K = 111$ in the Karnaugh map), so the circuit oscillates at a frequency determined by the propagation delays through the circuit. By 'oscillates' we mean that the output repeatedly changes from logic 1 to 0, back to 1, back to 0, and so on for as long as the external input conditions $C\,J\,K = 111$ exist. This is not a desirable property, and renders the gated J-K latch circuit as practically useless. One possible way to make the circuit perform the toggle function in a normal manner is to use extremely short positive pulses to trigger the circuit at the control input. The problem that occurs is a result of having a control input signal with a pulse width that is too long. The duration of the positive pulse must be shorter than the propagation delay through the circuit. The circuit should toggle each time a positive pulse is applied to the control input. It is obvious that the gated *J-K* latch circuit does not do this without deliberately controlling the pulse width of the signal applied to the control input—an undesirable solution in most cases.

The state diagram for the gated *J-K* latch is shown in Fig. E8-6*b*. The state diagram illustrates the oscillatory nature of the circuit for the input conditions $C\,J\,K = 111$. While the external input conditions $C\,J\,K = 111$ exist, the state diagram illustrates that the circuit output signal *q* changes logic state from 1 to 0 and back repeatedly.

(*c*) The ASM chart is shown in Fig. E8-6*c*. The oscillatory nature can also be observed for the input conditions $C\,J\,K = 111$ in the ASM chart.

8-4-4 Gated *T* Latch

The gated *T* latch circuits shown in Fig. 8-24 suffer the same erratic oscillating property as the gated *J-K* latch circuit. Each implementation races from $q = 0$ to $q = 1$ back to $q = 0$, and so on endlessly, while the input conditions $C\,T = 11$ exist.

The characteristic equation for the gated *T* latch is listed as follows:

$$Q = \bar{q}\cdot C\cdot T + q\cdot\bar{C} + q\cdot\bar{T} \qquad (8\text{-}7)$$

The Karnaugh map in Fig. 8-25 shows that the gated *T* latch is unstable for the input conditions $C\,T = 11$. This instability can also be observed from the state diagram shown in Fig. 8-26.

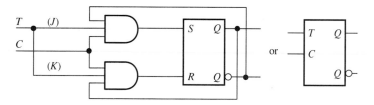

(*a*) Converting a gated *J-K* latch to a gated *T* latch

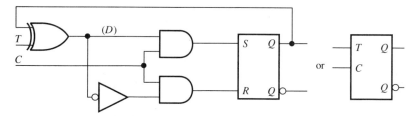

(*b*) Converting a gated *D* latch to a gated *T* latch

FIGURE 8-24
Gated *T* latches.

$q \backslash C\,T$

$Q =$

	00	01	11	10
0	$\underline{0}$	$\underline{0}$	1	$\underline{0}$
1	$\underline{1}$	$\underline{1}$	0	$\underline{1}$

FIGURE 8-25
Karnaugh map for gated T latch.

Gated *J-K* and *T* latch circuits are not considered useful circuit configurations because of their instability when $C\,J\,K = 111$ and $C\,T = 11$; however, with sufficiently narrow positive pulses applied to the control input, a gated *J-K* or *T* latch circuit can be made to perform normal circuit operation. Normal circuit operation means that the circuit output q will change from a logic 0 to a logic 1 upon the arrival of a narrow control pulse, and remain in a logic 1 until the arrival of the next narrow control pulse, which causes the output q to change to a logic 0. In other words, only one state change will occur for each positive control pulse. Narrow positive pulses can be generated using a pulse-narrowing circuit as shown in Fig. 8-27a.

The timing diagram in Fig. 8-27b illustrates the operation of the pulse-narrowing circuit. When a positive pulse is applied to the pulse-narrowing circuit in Fig. 8-27a, The Inverter delay Δt_1 results in a narrow pulse of width Δt_1 being supplied at the output of the AND element. Adding an odd number of Inverters increases the delay Δt_1; however, if the pulse width of the output is too wide, a race condition exists as discussed earlier. If the pulse width is too narrow, the control input of the gated latch will simply not have time to respond. A gated latch circuit that utilizes a pulse-narrowing circuit at its control input can be made to operate normally, but this type of circuit is not considered as reliable as other alternatives such as master-slave (pulse-triggered) and edge-triggered flip-flops. Master-slave and edge-triggered flip-flops do not require narrow pulses for proper circuit operation.

8-4-5 Master-Slave Flip-Flops

The master-slave (pulse-triggered) flip-flop evolved as a result of trying to solve the problem associated with control input signal pulses that are too long. If very

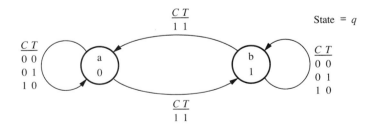

$$\begin{array}{c} C\,T \\ 1\ 1 \end{array}$$

State $= q$

$$\begin{array}{c} C\,T \\ 0\ 0 \\ 0\ 1 \\ 1\ 0 \end{array}$$

a
0

b
1

$$\begin{array}{c} C\,T \\ 0\ 0 \\ 0\ 1 \\ 1\ 0 \end{array}$$

$$\begin{array}{c} C\,T \\ 1\ 1 \end{array}$$

FIGURE 8-26
State diagram for gated T latch.

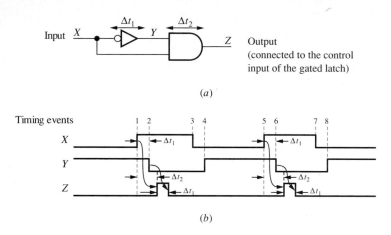

(a)

(b)

FIGURE 8-27
(*a*) Pulse-narrowing circuit, (*b*) timing diagram illustrating the operation of the pulse-narrowing circuit.

narrow control input signal pulses are generated, then the present state outputs that are fed back to the inputs of memory devices will only cause one transition for each pulse applied to the control input. This is the desired result. If the present state output could also be fed back to combinational logic inputs in a circuit, without causing a race condition or oscillating, then that memory device could also be used to build larger sequential circuits. Gated-latch circuits cannot be used in feedback applications in synchronous designs simply because of their transparent property. As soon as the control input signal goes to its logic 1 state, as illustrated in the timing diagram in Fig. 8-22, the output $Q1$ follows the input $D1$. If the $Q1$ output were fed back to a combinational logic circuit to contribute to the next state value of $Q1$, a race condition could occur during the time the latch circuit is transparent. This race condition could cause the circuit to oscillate with the result that the next state value, that would be stored or latched at the time the control input signal goes to a logic 0, would be indeterminate or unpredictable — a totally unacceptable condition.

A master-slave flip-flop is designed to interrupt the logic connection between the inputs and the outputs during the time the input control signal is a logic 1. Removing the logic connection between the input and output signals afforded by the master-slave design provides the following benefits: (*a*) it removes the transparency property and hence the race condition, and (*b*) it provides a memory device, which can be used in synchronous sequential designs (designs where the present state outputs can be used to determine the next state outputs, synchronized by a system clock).

To understand how the transparency property is removed by interrupting the signal path from the inputs to the outputs, observe the circuit diagram of the master-slave *J-K* flip-flop shown in Fig. 8-28.

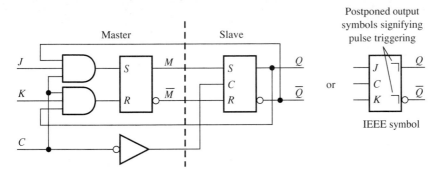

FIGURE 8-28
Master-slave (pulse-triggered) *J-K* flip-flop.

The master-slave *J-K* flip-flop is simply two gated *S-R* latches connected in cascade such that the master (the first gated *S-R* latch) drives the slave (the second gated *S-R* latch). The present state outputs of the slave are fed back to the inputs of the master such that the *J* input is qualified by $\bar{q} \cdot C$ while *K* is qualified by $q \cdot C$, providing steering and thus allowing the toggle property. When the external control input signal is a logic 1, the master is transparent from its inputs to its outputs. Because of the Inverter, during this same time period the slave is being disabled (its control input signal is a logic 0), and thus is not transparent. By disabling the slave, the present state output signals from the slave remain stable and cannot cause a race condition through the slave via the outputs of the master. In effect this interrupts the logic connection between the inputs and the outputs of the device during the time the input control signal is a logic 1. When the external control input signal makes a transition from a logic 1 to a logic 0, the data present at the *J* and *K* inputs are captured, the master is disabled, and the slave is enabled. Changes at the master's inputs are of no consequence at this time since it is now disabled and its outputs are stable. The slave is now transparent, but the master is supplying it with stable inputs. After a short delay time of Δt, the slave's outputs also become stable. Since the master is disabled, feedback around the entire circuit is again interrupted. The race condition is eliminated by this cascaded or double rank approach, allowing the master-slave flip-flop to toggle only once for each external control signal pulse.

Because the master-slave principle utilizes both edges of the external signal applied to the control input, this type of flip-flop is a pulse-triggered bistable. The recommended IEEE symbol for a standard master-slave (pulse-triggered) flip-flop is represented in Fig. 8-28. The postponed output symbols shown inside the rectangular symbol of the master-slave flip-flop indicate that the internal state of the output, that is, the next state value of the output, is postponed until the external control signal returns to its 0 logic state.

Observing the timing diagram in Fig. 8-29, one can see that the master-slave *J-K* flip-flop would ideally operate on the major timing events 1, 2, 3, 5, 6, 8,

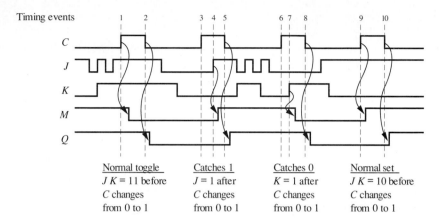

Normal toggle
$JK = 11$ before
C changes
from 0 to 1

Catches 1
$J = 1$ after
C changes
from 0 to 1

Catches 0
$K = 1$ after
C changes
from 0 to 1

Normal set
$JK = 10$ before
C changes
from 0 to 1

In all cases the slave output signal Q follows the
master output signal M when C changes from 1 to 0

FIGURE 8-29
Timing diagram for a master-slave (pulse-triggered) *J-K* flip-flop.

9, and 10. When the external control signal changes from a 0 to 1, the master becomes transparent and the slave is disabled. When the external control signal changes from 1 to 0 to make the master disabled, the slave becomes transparent. However, due to the fact that the master is transparent during the time the external control signal is a logic 1, when the slave is reset or cleared, the *J* input can change to 1 after the rising edge of *C*. The master catches this set condition. This property of the master-slave *J-K* flip-flop is called 1s catching since the set or 1 condition was caught by the master after the 0 to 1 timing event of the external control signal. The 1s catching property is illustrated in Fig. 8-29 by timing event 4. The set condition is then passed on to the slave on the following 1 to 0 transition of the external control signal at timing event 5.

In a similar manner, when the slave is set, the *K* input can change to a 1 after the rising edge of *C*, and the master catches this reset or cleared condition. Since the reset or cleared condition is caught, this property of the master-slave *J-K* flip-flop is called 0s catching. The 0s catching property is illustrated in Fig. 8-29 by timing event 7. The reset or cleared condition is passed on to the slave on the next falling edge of the external control signal at timing event 8. The 1s catching and 0s catching property exists for all master-slave (pulse-triggered) flip-flops that do not contain data-lockout (the data-lockout property will be discussed below). A master-slave flip-flop can be constructed as a *J-K*, *R-S*, *D* or *T* type flip-flop. The 1s and 0s catching property of master-slave flip-flops needs to be fully understood by the designer using this type of flip-flop. From the timing diagram it should be noted that the output of the slave in a master-slave (pulse-triggered) flip-flop always follows the logic level of the master when the external control signal *C* changes from 1 to 0; that is, it follows the master on the falling edge of the external control signal. The 1s catching or 0s catching property of master-slave (pulse-triggered) flip-flops can result in improper circuit operation if this type of flip-flop is used in a synchronous design. Improper circuit operation occurs when the external input signals have not become stable (reached a steady state condition) prior to the 0 to 1 transition of the control input signal.

In order to improve the master-slave (pulse-triggered) *J-K* flip-flop, a data-lockout feature can be added to prevent 1s and 0s catching. With data-lockout, the master-slave (pulse-triggered) *J-K* flip-flop has a setup time requirement for the *J* and *K* inputs as well as a hold time requirement with reference to the 'rising edge' (the 0 to 1 transition) of the signal applied to the control input. During the setup and hold time, the *J* and *K* inputs must be stable; however, after the hold time, the *J* and *K* inputs can be changed without affecting the state of the master. If the *J* and *K* inputs change after the hold time, while the control input is at a logic 1, 1s and 0s catching will not occur. The recommended IEEE symbol for a master-slave *J-K* flip-flop with data-lockout is shown in Fig. 8-30. It should be noted that the symbol has both a dynamic input qualifying symbol at the control input and a postponed output qualifying symbol at each output. The dynamic input symbol shown in Fig. 8-30 simply means that an external transition from 0 to 1 at the control input produces a 'transitory internal 1 state', and the internal logic state of the control input is 0 at all other times. The postponed output symbol means that the output can only change after the completion of the positive pulse at the control input (when the pulse changes from 1 back to 0).

A master-slave *S-R* flip-flop is constructed in the same manner as a master-slave *J-K* flip-flop, only it does not provide toggling capability since steering is removed. An Inverter added between the *S* and *R* inputs, or the *J* and *K* inputs of a master-slave *S-R* or *J-K* flip-flop, results in a master-slave *D* flip-flop. A master-slave *T* flip-flop is simply a master-slave *J-K* flip-flop with both inputs tied together and renamed *T*.

Example 8-7

(a) Draw a circuit diagram for a master-slave *S-R* flip-flop using two-gated *S-R* NOR latches and an Inverter. Also draw and label a rectangular logic symbol for the master-slave *S-R* flip-flop to identify the pulse-triggered nature of the circuit.

(b) Complete the waveforms for the outputs *M*, *Q*, *MDL* and *QDL* shown in Fig. E8-7*a*. Assume that the outputs *M* and *Q* are the master and slave output signals for a master-slave *S-R* flip-flop without data-lockout. Assume that this flip-flop

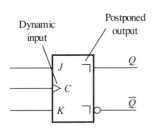

FIGURE 8-30
IEEE symbol for a data-lockout master-slave *J-K* flip-flop.

has 0 setup and hold time where the setup time is relative to the C input signal change from 0 to 1, and the hold time is relative to the C input signal change from 1 to 0. The outputs MDL and QDL are the master and slave output signals for a master-slave S-R flip-flop with data-lockout. Assume that this flip-flop also has a 0 setup and hold time. In this case, the setup time is relative to the C input change from 0 to 1; however, the hold time is also relative to the C input signal change from 0 to 1. If the waveforms for Q and QDL look different, explain why. Which device (the master-slave or the master-slave with data-lockout) has more desirable setup and hold time characteristics?

Solution

(a) The logic diagram for a master-slave S-R flip-flop is shown in Fig. E8-7b, and the rectangular-shape logic symbol for the same circuit is shown in Fig. E8-7c. A master-slave S-R flip-flop has the same restrictions as a gated S-R latch or a simple S-R NOR latch in that the input signal conditions $S\,R = 11$ are not normally allowed when the control input C is 1.

(b) The waveforms for the two different types of master-slave S-R flip-flops are shown in Fig. E8-7d. A slight propagation delay was used when drawing the waveforms. The master-slave S-R flip-flop (when data-lockout is not mentioned, the lockout feature is not normally available) is a 1s and 0s catcher; it catches the 0 at timing event 4 and the 1 at timing event 11, which happen to violate the

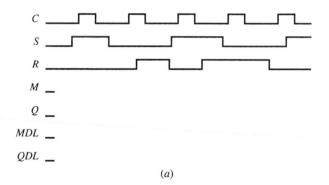

(a)

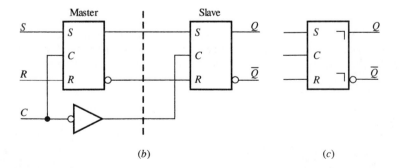

(b)   (c)

FIGURE E8-7

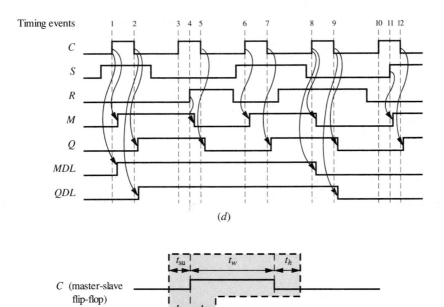

Timing events

(d)

In each case, the shaded areas represent the time when the S, R, D, J, K,or T
inputs must be stable, depending on the type of flip-flop.

(e)

FIGURE E8-7

setup time of the device. The master-slave S-R flip-flop with data-lockout, on
the other hand, does not pay attention to the S or the R inputs after its hold
time, so the R signal change at timing event 4 and the S signal change at timing
event 11 do not violate the hold time of the device. The master-slave flip-flop
with data-lockout only triggers at timing events 1, 3, 6, 8, and 10 which occur
on a 0 to 1 transition of C.

  Figure E8-7e illustrates the general data setup and hold time characteristics
for a master-slave flip-flop with and without data-lockout. Notice that for proper
circuit operation the S and R inputs must be stable for a longer period of time for
the master-slave S-R flip-flop without data-lockout for proper circuit operation.
Circuit operation is more predictable for the master-slave S-R flip-flop with data-
lockout because the S and R inputs only have to be stable for a brief period of
time slightly before the 0 to 1 transition of C, during the 0 to 1 transition of C,
and slightly after the 0 to 1 transition of C. Compare this time to the master-
slave S-R flip-flop without data-lockout which requires the S and R inputs to be
stable slightly before the 0 to 1 transition of C, during the 0 to 1 transition of
C, during the entire time C is 1 (this time is fixed for the device and is given
as a minimum pulse width t_w(MIN)), and slightly after the 1 to 0 transition of

C as shown in Fig. E8-7*e*. The master-slave *S-R* flip-flop with data-lockout has more desirable setup and hold time characteristics.

Because of the popularity of edge-triggered flip-flops, master-slave flip-flops are seldom used in new designs. To understand existing designs, however, it is still important to have a good working knowledge of all types of latches and flip-flops.

8-4-6   Edge-Triggered Flip-Flops

Master-slave flip-flops load a master latch on the rising edge of the control input signal while the slave latch is disabled so that the feedback path is broken. The content of the master latch is then transferred to the slave latch on the falling edge of the control input signal while the master latch is disabled. The time difference required to trigger the master and then trigger the slave to effect the transfer provides a delayed output and can be a disadvantage to high speed operation. To overcome this disadvantage and achieve higher speed flip-flop devices, edge-triggered flip-flops were developed.

An edge-triggered flip-flop is a memory storage device that is designed to respond to a positive or negative going signal applied to its control input. The edge-triggered *D* flip-flop is the type of flip-flop used in the majority of designs (this is the device referred to as the edge-triggered *D*-type memory device in Fig. 8-3*a*). The recommended IEEE symbol for a positive edge-triggered *D* flip-flop is shown in Fig. 8-31*a*. Notice that the logic symbol contains the dynamic input qualifying symbol at the control input like a master-slave flip-flop with data-lockout, but it does not contain a postponed output qualifying symbol. The logic symbol for the negative edge-triggered *D* flip-flop contains a negation symbol outside the symbol outline. When the negation symbol and the dynamic input symbol are combined as shown in Fig. 8-31*b*, it means that an external transition from 1 to 0 at the control input produces a 'transitory internal 1 state', and that the internal logic state of the control input is 0 at all other times.

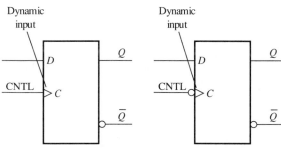

(*a*) Positive edge-triggered
D flop-flop (triggers
on a 0 to 1 transition
at the control input)

(*b*) Negative edge-triggered
D flop-flop (triggers
on a 1 to 0 transition
at the control input)

FIGURE 8-31
IEEE recommended logic
symbols for edge-triggered *D*
flip-flops.

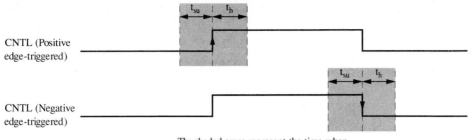

The shaded areas represent the time when
the external input(s) must be stable

FIGURE 8-32
General setup and hold time requirements, for positive and negative edge-triggered flip-flops.

The setup and hold time requirements for a positive and negative edge-triggered flip-flop are shown in Fig. 8-32. Notice that the positive edge-triggered flip-flop and data-lockout master-slave flip-flop (see Fig. E8-7e) have the same relative setup and hold time requirements; that is, setup and hold times are relative to a 0 to 1 transition of the control input signal. The output signal for an edge-triggered flip-flop occurs immediately after it is triggered (only delayed by the propagation delays of the gates making up the device). The output signal for a master-slave flip-flop with data-lockout (also called variable-skew) is delayed until the slave is transparent, that is, until the trailing edge of the control pulse occurs (plus the slave's propagation delay). This is the primary timing difference between these two types of flip-flops.

The circuit shown in Fig. 8-33a is an example of a positive edge-triggered *D* flip-flop. Notice that this particular circuit is quite easy to understand. When *C* is 0 (prior to the edge of the positive pulse), output *Q1* of the first latch circuit (the master) is transparent and follows the input signal *D*. As soon as *C* changes from 0 to 1, the value of *Q1* stabilizes, since the master is disabled. Output *Q2* of the second latch circuit (the slave) is transparent and follows the value of *Q1*.

The properties of this circuit can also be observed from its flow table or its state diagram. First, the circuit is drawn with gates as shown in Fig. 8-33b. In feedback circuits, a delay box is required to form a timing break in each feedback loop so that the next state equations can be written using combinational logic techniques in terms of the present state values. Observe where the delay boxes for Δt_1 and Δt_2 are placed in Fig. 8-33b. Since the master (or input) latch acts transparent when the control input signal is 0, output *y1* has time to settle to the value of *Y1* prior to the control input signal making a transition from 0 to 1, provided that the external input signal D meets the setup time of $3t_{pd}$ (assuming t_{pd} is an average gate or Inverter delay). By placing a timing break as shown by the delay box Δt_1, *Y1* can be written as a function of *y1*, *D*, and *C*. The second output, or slave latch, acts transparent when the control input signal is 1. Since the master latch is disabled as soon as the 0 to 1 transition occurs at input *C*, the slave (or output latch) is either set or reset depending on the

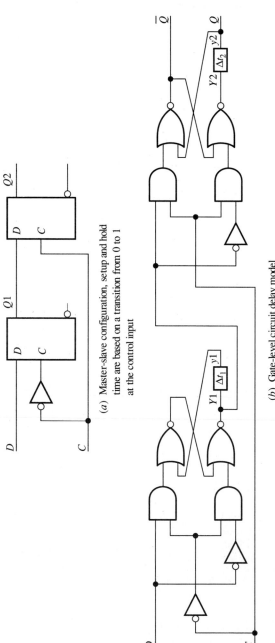

(a) Master-slave configuration, setup and hold time are based on a transition from 0 to 1 at the control input

(b) Gate-level circuit delay model

FIGURE 8-33
Positive edge-triggered *D* flip-flop.

value of $Y1$ when the control input reaches 1. The timing break represented by delay box Δt_2 allows $Y2$ to be written as a function of $y1$, $y2$, D, and C. Obtaining the next state equations for $Y1$ and $Y2$ facilitates the construction of the Karnaugh maps for the circuit. The Karnaugh maps are next combined to obtain a composite Karnaugh map as discussed below. The flow map and state diagram for the circuit can be easily obtained from the composite Karnaugh map. The other analysis tools such as the transition map and timing diagram can also be obtained, if desired, to help analyze the operation of the circuit.

The characteristic equations for $Y1$ and $Y2$ for the positive edge-triggered D flip-flop circuit in Fig. 8-33b can be written as follows.

$$Y1 = \overline{(y1 \; + \; \overline{C} \cdot D)} + \overline{C} \cdot D$$

$$= (y1 \; + \; \overline{C} \cdot D) \cdot (C + D)$$

$$= y1 \cdot C \; + \; y1 \cdot D \; + \; \overline{C} \cdot D \tag{8-8}$$

$$Y2 = \overline{(Y1 \cdot C \; + \; y2)} + \overline{Y1} \cdot C$$

$$= (Y1 \cdot C \; + \; y2) \cdot (Y1 + \overline{C})$$

$$= Y1 \cdot C \; + \; y2 \cdot Y1 \; + \; y2 \cdot \overline{C}$$

$$= Y1 \cdot (C + y2) \; + \; y2 \cdot \overline{C}$$

$$= (y1 \cdot C \; + \; y1 \cdot D \; + \; \overline{C} \cdot D) \cdot (C + y2) \; + \; y2 \cdot \overline{C}$$

$$= y1 \cdot C \; + \; y1 \cdot y2 \cdot C \; + \; y1 \cdot C \cdot D \; + \; y1 \cdot y2 \cdot D$$

$$\quad + \; y2 \cdot \overline{C} \cdot D \; + \; y2 \cdot \overline{C}$$

$$= y1 \cdot C \; + \; y1 \cdot y2 \cdot D \; + \; y2 \cdot \overline{C} \tag{8-9}$$

Using the characteristic equations for $Y1$ and $Y2$, the Karnaugh maps are drawn as illustrated in Fig. 8-34a and b. A composite Karnaugh map for $Y1$ $Y2$ can be generated as shown in Fig. 8-34c using the separate maps. With a little practice, the composite Karnaugh map can be drawn directly from the characteristic equations. First, fill in the map for the 1s for $Y1$, leaving room for the 1s for $Y2$. Prior to listing the 1s for $Y2$, fill in the 0s for $Y1$. Then fill in the map for the 1s for $Y2$ and the 0s for $Y2$. The order of filling in the composite map for the next state $Y1$ $Y2$ can be reversed if desired. Using a composite Karnaugh map saves time and effort since only one map is filled in. An underline symbol is placed in each cell in the composite Karnaugh map in Fig. 8-34c representing a stable state; that is, on the same line, the binary value for $Y1$ $Y2$ is the same as the binary value for $y1$ $y2$. The flow map for the circuit can be obtained from the composite Karnaugh map as shown in Fig. 8-34d.

To illustrate the operation of the positive edge-triggered D flip-flop, arrows have been drawn in the flow map shown in Fig. 8-35 to assist in the following discussion. Keep in mind that prior to the 0 to 1 transition of the control input signal, the D input signal must be stable before the setup time for reliable flip-flop operation. Assume that the circuit is initially in stable state $y1$ $y2$ C D =

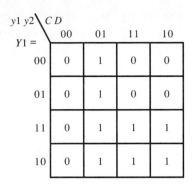

(a) Karnaugh map for Y1

yl y2 \ CD

	00	01	11	10
Y1 =				
00	0	1	0	0
01	0	1	0	0
11	0	1	1	1
10	0	1	1	1

(b) Karnaugh map for Y2

yl y2 \ CD

	00	01	11	10
Y2 =				
00	0	0	0	0
01	1	1	0	0
11	1	1	1	1
10	0	0	1	1

(c) Composite Karnaugh map for Y1 Y2

yl y2 \ CD

	00	01	11	10
Y1 Y2 =				
00	00	10	00	00
01	01	11	00	00
11	01	11	11	11
10	00	10	11	11

(d) Flow map

yl y2 \ CD

		00	01	11	10
NS =					
a	00	a	d	a	a
b	01	b	c	a	a
c	11	b	c	c	c
d	10	a	d	c	c

FIGURE 8-34

Separate Karnaugh maps, composite Karnaugh map, and flow table for positive edge-triggered D flip-flop in Figure 8-33.

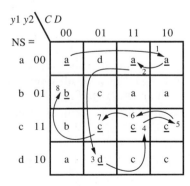

yl y2 \ CD

		00	01	11	10
NS =					
a	00	a	d	a	a
b	01	b	c	a	a
c	11	b	c	c	c
d	10	a	d	c	c

FIGURE 8-35

Flow map analysis for the positive edge-triggered D flip-flop in Figure 8-33.

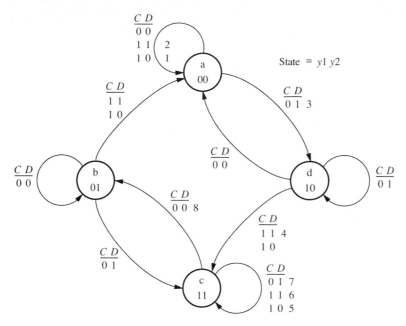

FIGURE 8-36
State diagram analsysis for the positive edge-triggered *D* flip-flop in Figure 8-33.

0000. When *C D* changes to 10 (the control input changes from 0 to 1), the operation of the circuit is shown by arrow 1, and the output *y2* remains 0 (the flip-flop remains reset). If *C D* changes to 11 (*D* changes after the control input is high), *y2* does not change as shown by arrow 2. When *C D* changes to 01, arrow 3 shows the flow of the circuit. Notice that *y2* does not change. If *C D* changes to 11, (the control input changes from 0 to 1), then *y2* changes to 1 (the flip-flop gets set) as shown by arrow 4. As before, if *D* changes after the control input is high by changing *C D* to 10 and then back to 11, *y2* remains set as indicated by arrows 5 and 6. When *C D* changes to 01 and finally back to 00, arrows 7 and 8 show that *y2* remains set. Figure 8-35 provides a complete description of the operation of the positive edge-triggered *D* flip-flop circuit shown in Fig. 8-33. Additional paths can be followed through the flow map to verify the edge-triggering property of the device for various input combinations.

The state diagram for the positive edge-triggered *D* flip-flop is illustrated in Fig. 8-36. The state diagram was drawn using the flow map and therefore contains the same information as the flow map and the composite Karnaugh map. Decimal numbers are placed beside the *C D* binary input conditions corresponding to the numbered arrows in Fig. 8-35. Either tool can be used to analyze the operation of the positive edge-triggered *D* flip-flop. Some designers prefer to use one type of analysis tool, while others prefer a different representation. One should have a working knowledge of all the different analysis tools, since these same tools are also used to design sequential circuits.

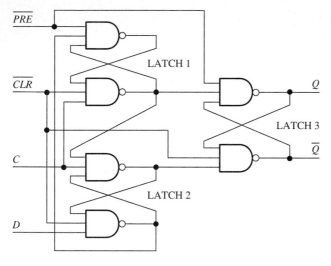

(*a*) Gate-level circuit (74LS74A)

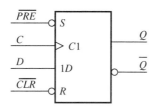

(*b*) IEEE logic symbol with dependency notation

FIGURE 8-37
Positive edge-triggered *D* flip-flop with direct preset and direct clear.

Another positive edge-triggered *D* flip-flop circuit is illustrated in Fig. 8-37*a*. Notice that this circuit has two additional inputs called $\overline{PRE}$ (PRE is an abbreviation for preset and means the same as set) and $\overline{CLR}$ (CLR is an abbreviation for clear and means the same as reset). Flip-flops often have either one or both of these inputs so that a circuit utilizing these devices can be placed in an initial state. At power up (when the power is turned on), the initial state of the flip-flops in a circuit is unknown due to racing conditions as power is being applied to the circuit. After the power switch is turned on and power to the circuit stabilizes (which usually takes several ms), the flip-flops in a circuit can be set by $\overline{PRE}\ \overline{CLR} = 01$ or cleared by $\overline{PRE}\ \overline{CLR} = 10$. The $\overline{PRE}$ and $\overline{CLR}$ inputs are sometimes referred to as direct preset (or set) and direct clear (or reset). These inputs normally control the output latch of a flip-flop and act the same as $\sim S$ and $\sim R$ for the S-R NAND latch, and for this reason the conditions $PRE\ CLR = 11$ are not normally allowed. For the control input to gain control, the $\overline{PRE}$ and $\overline{CLR}$ inputs are normally left in a binary 11 condition ($PRE\ CLR = 00$) until they are used to preset or clear the device, and then their values are again returned to 11.

The IEEE logic symbol for the positive edge-triggered D flip-flop circuit is shown in Fig. 8-37b. The input named $C1$ inside the symbol outline controls the input named $1D$. The inputs named S and R are not controlled by the $C1$ input. In synchronous applications $C1$ is driven by the clock. Since $C1$ controls the $1D$ input, $1D$ is a synchronous input. The S and R inputs (which are not controlled by the $C1$ input) are asynchronous inputs.

Example 8-8. Draw a circuit for the positive edge-triggered D flip-flop shown in Fig. 8-37a using delay boxes to form timing breaks for each feedback path. Obtain the characteristic equations and the composite Karnaugh map for the circuit. Obtain the flow map, and indicate by arrows the circuit operation for the following circuit conditions: assume $y1\ y2\ y3\ C\ D\ =\ 11000$ and $C\ D$ changes in the following sequence (1) 01, (2) 11, (3) 10, (4) 00, and (5) 10. At the end of the sequence, what state is the flip-flop left in, the set state or the reset state?

Solution The circuit is redrawn in Fig. E8-8a using delay boxes to form timing breaks for each feedback path. The $\overline{PRE}$ and $\overline{CLR}$ inputs are each pulled to a logic 1, which is their normal state. These inputs are only used to preset or clear the device independently of the C and D inputs.

The next state equations $Y1$, $Y2$, and $Y3$ can be written from the circuit as follows. Notice that the equations are written beginning at the output and working backwards. The reverse technique could also be used.

$$Y1 = \overline{(1 \cdot \overline{(y2 \cdot 1 \cdot D)} \cdot y1) \cdot 1 \cdot C}$$
$$= \overline{(y2 \cdot D) \cdot y1}\ +\ \overline{C}$$
$$= (\overline{y2} + \overline{D}) \cdot y1\ +\ \overline{C}$$
$$= y1 \cdot \overline{y2}\ +\ y1 \cdot \overline{D}\ +\ \overline{C} \qquad\qquad (8\text{-}10)$$

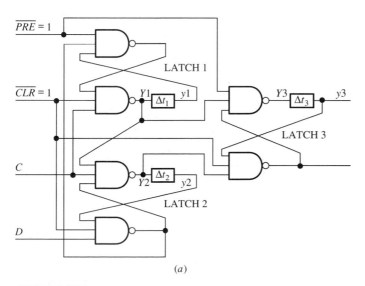

(a)

FIGURE E8-8

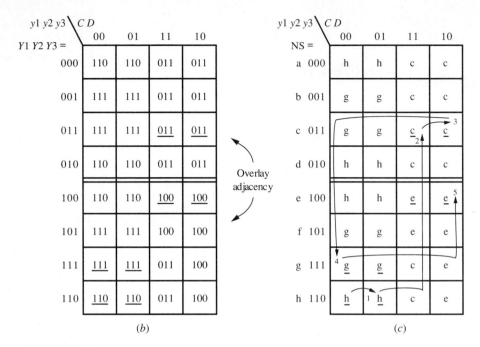

FIGURE E8-8

$$Y2 = \overline{Y1 \bullet C \bullet \overline{(y2 \bullet 1 \bullet D)}}$$

$$= \overline{(y1 \bullet \overline{y2} + y1 \bullet \overline{D} + \overline{C}) \bullet C \bullet \overline{(y2 \bullet 1 \bullet D)}}$$

$$= \overline{(y1 \bullet \overline{y2} + y1 \bullet \overline{D} + \overline{C})} + \overline{C} + y2 \bullet D$$

$$= (\overline{y1} + y2) \bullet (\overline{y1} + D) \bullet C + \overline{C} + y2 \bullet D$$

$$= (\overline{y1} + \overline{y1} \bullet D + \overline{y1} \bullet y2 + y2 \bullet D) \bullet C + \overline{C} + y2 \bullet D$$

$$= \overline{y1} \bullet C + \overline{y1} \bullet C \bullet D + \overline{y1} \bullet y2 \bullet C + y2 \bullet C \bullet D + \overline{C} + y2 \bullet D$$

$$= \overline{y1} \bullet C + \overline{C} + y2 \bullet D \tag{8-11}$$

$$Y3 = \overline{1 \bullet Y1 \bullet \overline{(y3 \bullet 1 \bullet Y2)}}$$

$$= \overline{1 \bullet (y1 \bullet \overline{y2} + y1 \bullet \overline{D} + \overline{C}) \bullet \overline{(y3 \bullet 1 \bullet (\overline{y1} \bullet C + \overline{C} + y2 \bullet D))}}$$

$$= \overline{(y1 \bullet \overline{y2} + y1 \bullet \overline{D} + \overline{C})} + y3 \bullet (\overline{y1} \bullet C + \overline{C} + y2 \bullet D)$$

$$= (\overline{y1} + y2) \bullet (\overline{y1} + D) \bullet C + \overline{y1} \bullet y3 \bullet C + y3 \bullet \overline{C} + y2 \bullet y3 \bullet D$$

$$= (\overline{y1} + \overline{y1} \bullet D + \overline{y1} \bullet y2 + y2 \bullet D) \bullet C + \overline{y1} \bullet y3 \bullet C$$

$$\quad + y3 \bullet \overline{C} + y2 \bullet y3 \bullet D$$

$$= \overline{y1} \bullet C + \overline{y1} \bullet C \bullet D + \overline{y1} \bullet y2 \bullet C + y2 \bullet C \bullet D + \overline{y1} \bullet y3 \bullet C$$

$$\quad + y3 \bullet \overline{C} + y2 \bullet y3 \bullet D$$

$$= \overline{y1} \bullet C + y2 \bullet C \bullet D + y3 \bullet \overline{C} + y2 \bullet y3 \bullet D \tag{8-12}$$

Using the characteristic equations, the composite Karnaugh map and flow map are drawn as shown in Figs. E8-8b and c.

Following the sequence of input events for the positive edge-triggered D flip-flop in the flow map results in the flip-flop being left in the present state of $y1$ $y2$ $y3 = 100$. At the end of the sequence the present state output variable $y3 = 0$; therefore, the flip-flop is in the reset state.

An example of a negative edge-triggered J-K flip-flop is illustrated in gate form in Fig. 8-38a. The IEEE logic symbol for a negative edge-triggered J-K flip-flop is shown in Fig. 8-38b. The synchronous inputs are $1J$ and $1K$ since these are the inputs controlled by the control input $C1$. S and R are asynchronous inputs.

To trigger a negative edge-triggered flip-flop requires that the control signal make a 1 to 0 transition; however, if an Inverter is placed in series with the control input, the flip-flop can be triggered by a signal that makes a 0 to 1 transition. This technique allows either a positive or negative edge-triggered flip-flop to be used as the opposite-type triggered flip-flop as illustrated in Figs. 8-39a and b.

Edge-triggered S-R and T flip-flops are not normally available as off-the-shelf devices. A J-K flip-flop can be used as a S-R flip-flop if one uses the J as the S input and K as the R input. The function of a S-R flip-flop is the same as the function for a J-K flip-flop except when $S = R = 1$ (remember this condition

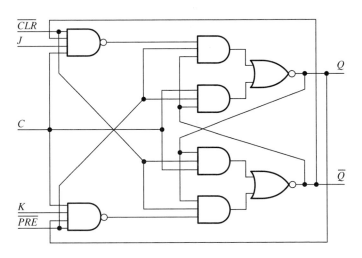

(*a*) Gate-level circuit (74LS76A)

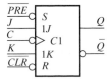

(*b*) IEEE logic symbol with dependency notation

FIGURE 8-38
Negative edge-triggered J-K flip-flop with direct preset and direct clear.

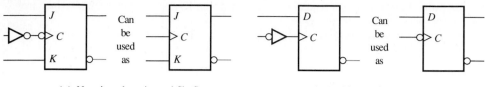

(a) Negative edge-triggered flip-flop
with an Inverter used as a positive
edge-triggered flip-flop.

(b) Positive edge-triggered flip-flop
with an Inverter used as a negative
edge-triggered flip-flop.

FIGURE 8-39
Using an edge-triggered flip-flop and an Inverter to trigger on the opposite edge.

is not normally allowed). By connecting the *J* and *K* inputs together and using this as the *T* input, a *J-K* flip-flop is converted to a *T* flip-flop. Figures 8-40*a* and *b* illustrate these configurations in graphical form.

8-5 REDUCED CHARACTERISTIC AND EXCITATION TABLES FOR BISTABLE DEVICES

To easily distinguish between the four types of bistable devices in the last section, it is customary to use a reduced or shortened characteristic table (PS/NS table). In a reduced characteristic table, the signal for the control input may either be indicated in the table or implied. To keep the table easy to remember, the signal for the control input is usually implied. If the control input signal is implied, the reduced characteristic table specifies the next state output of each type of device in terms of its external input(s) and its present state output, whether it is a latch, master-slave flip-flop, or edge-triggered flip-flop. The reduced characteristic tables for bistable types *S-R*, *D*, *J-K*, and *T* are shown in Fig. 8-41.

Using the reduced characteristic table for a bistable allows the designer to quickly recall the important input-output characteristics for a particular type device, that is, to analyze the operation of the device. Recall that upper case

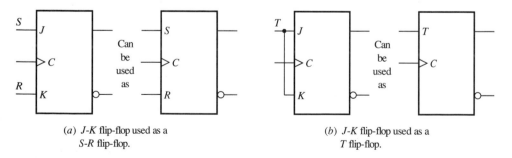

(a) *J-K* flip-flop used as a
S-R flip-flop.

(b) *J-K* flip-flop used as a
T flip-flop.

FIGURE 8-40
Using a *J-K* flip-flop as a *S-R* flip-flop or a *T* flip-flop.

S	R	Q	Operation
0	0	q	no change
0	1	0	reset
1	0	1	set
1	1	X	not normally allowed

(a) S-R type bistable

D	Q	Operation
0	0	reset
1	1	set

(b) D-type bistable

J	K	Q	Operation
0	0	q	no change
0	1	0	reset
1	0	1	set
1	1	$\bar{q}$	toggle

(c) J-K type bistable

T	Q	Operation
0	q	no change
1	$\bar{q}$	toggle

(d) T-type bistable

FIGURE 8-41
Reduced characteristic tables for latches, master-slave flip-flops, and edge-triggered flip-flops.

Q represents the next state output, and lower case q represents the present state output. A 0 in the Q column is a reset operation, while a 1 in the Q column is a set operation. In the reduced characteristic tables for bistable types S-R, J-K, and T, a q in the Q column (for $SR = 00$, $JK = 00$ and $T = 0$) means that the next state of the bistable is the same as the present state (a hold operation); however, a $\bar{q}$ in the Q column (for $JK = 11$ and $T = 1$) means that the next state of the bistable is the complement of the present state, that is, a toggle operation.

The D-type bistable has the simplest reduced characteristic table, and its bistable equation can be written by inspection as $Q = D$. The bistable equation for any other bistable can also be obtained from its reduced characteristic table. Bistable equations are used in the analysis of synchronous sequential circuits. The analysis process will be presented in Chapter 9.

To analyze circuits that contain latches and flip-flops, the reduced characteristic tables for the bistable types S-R, D, J-K, and T should be committed to memory.

Example 8-9. Obtain a minimum bistable equation for the S-R, J-K, and T type bistables using their reduced characteristic tables.

Solution Using the minterm designators and the characteristic numbers for the next state output Q for the S-R type bistable, one can write the following standard sum of products equation.

$$Q(S, R) = f_0 \cdot m_0 + f_1 \cdot m_1 + f_2 \cdot m_2 + f_3 \cdot m_3$$
$$= q \cdot \bar{S} \cdot \bar{R} + 0 \cdot \bar{S} \cdot R + 1 \cdot S \cdot \bar{R} + - \cdot S \cdot R$$
$$= q \cdot \bar{S} \cdot \bar{R} + S \cdot \bar{R} + - \cdot S \cdot R$$

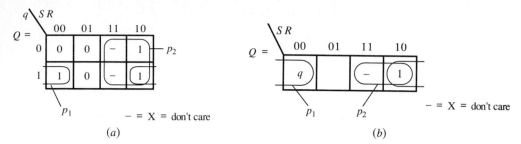

FIGURE E8-9

Using this equation, the Karnaugh map shown in Fig. E8-9a can be plotted and read to obtain the following minimum bistable equation for the *S-R* type bistable.

$$Q = p_1 + p_2$$
$$= q \cdot \overline{R} + S \qquad (8\text{-}13)$$

As an alternate method, one can plot the Karnaugh map shown in Fig. E8-9b. This Karnaugh map is a reduced map with q as a map-entered variable. Reading the map using the map-entered variable technique presented in Chapter 4 Section 4-6, we can write the minimum bistable equation for the *S-R* type bistable as follows.

$$Q = p_1 + p_2$$
$$= q \cdot \overline{R} + S$$

Obtaining a minimum bistable equation for the *J-K* and *T* type bistables is left as an exercise for the student.

To design synchronous sequential logic circuits using bistables, a different type of table called an excitation table is utilized. To design a circuit, a designer must first determine the output sequence the circuit must follow. Knowing the output sequence and the type of bistable devices to be used in the design, the designer must then determine the bistable input condition(s) to accomplish the required output sequence. The excitation table specifies the required external input condition(s) for each possible bistable output transition from its present state q to its next state Q. The excitation tables for bistable types *S-R*, *D*, *J-K*, and *T* are shown in Fig. 8-42.

The don't cares in the excitation tables for the *S-R* and *J-K* type bistables mean that the input value can be either a 0 or a 1 and cause the required present state to next state transition. For example, if the present state $q = 0$ and it is desired to keep the next state $Q = 0$, then $S R = 01$ will cause the bistable to reset; however, since it is already reset, $S R = 00$ will not cause any action, and the bistable will remain reset. In this case, R can be either a 0 or a 1 and the result is the same, so R is listed in the excitation table as a don't care by the symbol X. The input S is a don't care for the transition $q Q = 11$, since $S R$ = 10 or 00 will cause the output of the bistable to remain set. The same reasoning

q	Q	S	R
0	0	0	X
0	1	1	0
1	0	0	1
1	1	X	0

(*a*) *S-R* type bistable

q	Q	D
0	0	0
0	1	1
1	0	0
1	1	1

(*b*) *D*-type bistable

q	Q	J	K
0	0	0	X
0	1	1	X
1	0	X	1
1	1	X	0

(*c*) *J-K* type bistable

q	Q	T
0	0	0
0	1	1
1	0	1
1	1	0

(*d*) *T*-type bistable

FIGURE 8-42
Excitation tables for latches, master-slave flip-flops, and edge-triggered flip-flops.

holds for *J-K* type bistables remaining in the reset condition or the set condition. *K* is a don't care for the transition $q\,Q = 01$ because the *K* input is qualified with q, so when $q = 0$, $K \cdot q = 0$. It doesn't matter whether *K* is a 0 or a 1. The *J* input is qualified with $\bar{q}$, so when $q = 1$, $J \cdot \bar{q} = 0$; in this case, it doesn't matter whether *J* is a 0 or a 1. The *D* type bistable has the simplest excitation equation since the *D* input is equal to the next state output regardless of the present state value; that is, $D = Q$. The excitation equation for the *T* type bistable can be seen to be an Exclusive OR function of the present state and next state variables; that is, $T = q \oplus Q$, and toggling occurs when $T = 1$. It does not toggle otherwise.

To design circuits that contain latches and flip-flops, the excitation tables for the bistable types *S-R*, *D*, *J-K*, and *T* should be committed to memory. These tables are utilized in designing synchronous sequential circuits in Chapter 9.

8-6 METASTABILITY AND SYNCHRONIZATION USING BISTABLES

Bistables are normally assumed to have only two stable states. It is possible for a bistable under certain conditions, however, to have a third state called a metastable state. This state occurs, for example, when a *S-R* NOR or NAND latch circuit's Q and $\bar{Q}$ output signals have the same voltage value at a logic level between a logic 0 and a logic 1. In this condition the output signals of both NOR or NAND gates that make up the latch circuit are perfectly balanced at an undefined logic level between 0 and 1. A metastable state most often occurs in a bistable when either the input data setup or hold time specification, t_{su} or t_h, is violated. Metastability is a transitory condition that can last many times longer than the latches propagation delay time, t_{pd}. If a bistable enters a metastable state, the time interval that it remains in that state before it eventually resolves into a stable state is quite unpredictable.

To reduce the possibility of metastability in synchronous sequential circuits and provide usable signals, an edge-triggered *D* flip-flop is often used to synchronize an asynchronous external input signal with the system clock. This type of synchronizer circuit is shown in Fig. 8-43*a*. The timing diagram in Figure 8-43*b* shows the clock signal *CK*, the asynchronous input signal *ASYNIN* (the asynchronous signal that needs to be synchronized), and the synchronous output signal *SYNOUT* at the output of the synchronizer (the asynchronous signal after it is synchronized). Immediately after each clock pulse timing event, the logic value of the asynchronous input signal *ASYNIN* meeting the data setup and hold time is synchronized with the system clock signal *CK* and is available as the synchronized output signal *SYNOUT* at the output of the *D* flip-flop. If the asynchronous input signal does not meet the data setup or hold time of the edge-triggered *D* flip-flop, a transitory metastable state can occur at the output of the *D* flip-flop and cause erratic operation, since the output would then be synchronized to an 'unknown state' for an unknown period of time. A method of completely eliminating metastability is not known; however, synchronizing asynchronous input signals is a commonly used solution. Usually the best that a designer can do is investigate memory devices that resolve metastable states faster.

To catch narrow pulses occuring just before or just after a clock timing event, like the second positive pulse shown for the asynchronous input signal *ASYNIN* in Fig. 8-43*b*, either the system clock frequency needs to be sufficiently increased, or a multiple-stage synchronizer must be used. A multiple-stage synchronizer utilizing an input latch to drive an edge-triggered flip-flop is left as an exercise for the student.

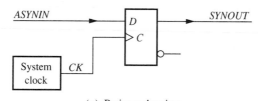

(*a*) Basic synchronizer.

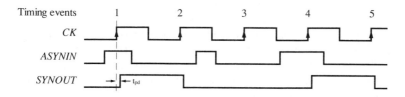

(*b*) Basic synchronizer timing diagram.

FIGURE 8-43
Synchronizer.

In the general case, metastability can result when two or more signals in a sequential circuit (synchronous or asynchronous) change at or about the same time. Remember, each timing event occuring in a circuit due to a signal changing state must be given enough time to settle or become stable prior to the next timing event, or a sequential circuit cannot be expected to operate predictably. In a synchronous circuit all the external input signals (except the asynchronous input signals, such as the direct preset and direct clear input signals) can change state at any time, provided that the output signals of the combinational logic circuit they are driving, all become stable prior to the setup and hold time requirements of the memory devices being used in the circuit (see Fig. 8-3a).

REFERENCES

1. Roth, Jr., C. H., *Fundamentals of Logic Design,* 3d ed., West Publishing, St. Paul, Minnesota, 1985.
2. Hill, J. H., and Peterson, G. R., *Introduction to Switching Theory & Logic Design,* 3d ed., John Wiley & Sons, New York, 1981.
3. Mano, M. M., *Digital Design,* Prentice-Hall, Englewood Cliffs, New Jersey, 1984.
4. Prosser, F. P., and Winkel, D. E., *The Art of Digital Design, An Introduction to Top-Down Design,* Prentice-Hall, Englewood Cliffs, New Jersey, 1987.
5. Breeding, K. J., *Digital Design Fundamentals,* Prentice-Hall, Englewood Cliffs, New Jersey, 1989.
6. McCluskey, E. J., *Logic Design Principles,* Prentice-Hall, Englewood Cliffs, New Jersey, 1986.
7. Kohavi, Z., *Switching and Finite Automata Theory,* 2d ed., McGraw-Hill, New York, 1978.
8. Fletcher, W. I., *An Engineering Approach to Digital Design,* Prentice-Hall, Englewood Cliffs, New Jersey, 1980.
9. Clare, C. R., *Designing Logic Systems Using State Machines,* McGraw-Hill, New York, 1973.
10. Rhyne, Jr., V. T., *Fundamentals of Digital Systems Design,* Prentice-Hall, Englewood Cliffs, New Jersey, 1973.
11. Johnson, E. L., and Karim, M. A., *Digital Design, A Pragmatic Approach,* Prindle, Weber & Schmidt, Boston, Massachusetts, 1987.
12. *The TTL Data Book Volume 2,* Texas Instruments, Dallas, Texas, 1985.
13. *Advanced CMOS Logic Data Book,* Texas Instruments, Dallas, Texas, 1988.

PROBLEMS

Section 8-2   Combinational Logic Circuits
Versus Sequential Logic Circuits

8-1. Describe how the lumped delays represented by the Δt's in the combinational logic circuit model in Fig. 8-2 can be calculated in an actual circuit implementation.

8-2. Draw a block diagram for a serial adder circuit that uses only one full adder and a block representing a set of memory devices to store the sum and carry output bits.

8-3. Draw a block diagram for a serial comparator circuit that uses only one full comparator and a block representing a set of memory devices to store the less than, equal to, and greater than output bits.

8-4. A synchronous sequential circuit has one, but an asynchronous sequential circuit does not. What is it? Sketch its output waveform. What do synchronous and asynchronous sequential circuits have in common that combinational circuits do not have?

8-5. Draw a clock signal *CK* with four 1-to-0 timing events, a nonzero setup time, t_{su}, and a nonzero hold time, t_h. Also draw one excitation input signal *EI* that follows the sequence 0101 and meets the setup and hold time requirements (bit 3 in the sequence occurs in time for the first clock timing events, bit 2 occurs in time for the second clock timing event, and so on).

8-6. Discuss what marks a timing event in each of the following types of sequential circuits.
  (*a*) a synchronous or clock mode circuit
  (*b*) a fundamental mode circuit
  (*c*) a pulse mode circuit

Section 8-3   The Basic Bistable Memory Devices

8-7. Draw rectangular symbols for the following latch circuits.
  (*a*) S-R latch for cross-coupled NOR gates using the positive logic convention
  (*b*) S-R latch for cross-coupled NOR gates using direct polarity indication
  (*c*) S-R latch for cross-coupled NAND gates using the positive logic convention
  (*d*) S-R latch for cross-coupled NAND gates using direct polarity indication

8-8. Analyze the latch circuit shown in Fig. P8-8 by obtaining the following sequential circuit analysis tools.
  (*a*) the delay model for the circuit using the positive logic convention
  (*b*) the characteristic equation
  (*c*) the characteristic or PS/NS table
  (*d*) the state diagram
  (*e*) the ASM chart
  (*f*) the Karnaugh map
  (*g*) the transition map
  (*h*) the flow map
  (*i*) the timing diagram

8-9. Repeat Prob. 8-8 for the latch circuit shown in Fig. P8-9.

8-10. Repeat Prob. 8-8 for the latch circuit shown in Fig. P8-10.

8-11. Use the flow map of the cross-coupled NAND latch and show that the circuit has a critical race when *S R* = 11 (the circuit is forced to set and reset at the same time), and then *S* and *R* go to 0 simultaneously.

8-12. Analyze the action of the sequential logic circuit in Fig. P8-12 considering the effect of contact bounce. Draw timing diagram waveforms of the signals *S*, *R*, and *Q* to show the following.

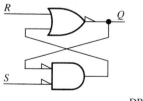

DPI

FIGURE P8-8

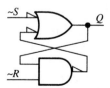

DPI

FIGURE P8-9

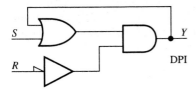

FIGURE P8-10

(a) the logic values before the push-button switch is pressed
(b) the logic values while the switch is being pressed
(c) the logic values after the switch is released

Section 8-4   Additional Bistable Memory Devices

8-13. List the three major groups of bistable devices.

8-14. Draw a gate level circuit diagram for a gated *S-R* latch using NAND gates. Draw the IEEE rectangular logic symbol for the circuit.

8-15. Obtain the following tools and answer the following question for a gated *S-R* latch circuit using NAND gates.
(a) the delay model for the circuit for the Q output
(b) the characteristic equation for the Q output
(c) the transition map
(d) the flow map
(e) the state diagram
(f) the ASM chart
(g) what input conditions are not normally allowed (use the flow map to justify your answer)

8-16. Draw a gate level circuit diagram for a gated *D* latch using an Inverter, AND gates, and NOR gates. Draw the IEEE rectangular logic symbol for the circuit.

8-17. Obtain the following tools and answer the following question for a gated *D* latch circuit using an Inverter, AND gates, and NOR gates.
(a) the delay model for the circuit for the Q output
(b) the characteristic equation for the Q output
(c) the transition map

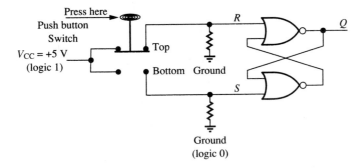

FIGURE P8-12

(*d*) the flow map

(*e*) the state diagram

(*f*) the ASM chart

(*g*) what happens when $C\,D = 11$ and both C and D are changed to 0 simultaneously (use the flow map to justify your answer)

8-18. Draw a gate level circuit diagram for a gated *J-K* latch using NAND gates. Draw the IEEE rectangular logic symbol for the circuit.

8-19. Obtain the following tools and answer the following questions for a gated *J-K* latch circuit using NAND gates.

(*a*) the delay model for the circuit for the Q output

(*b*) the characteristic equation for the Q output

(*c*) the flow map

(*d*) when is the circuit unstable

(*e*) how can the circuit be made to operate properly

8-20. Draw a gate-level circuit diagram for a gated *T* latch using AND and NOR gates. Draw the IEEE rectangular logic symbol for the circuit.

8-21. Obtain the following tools and answer the following question for a gated *T* latch circuit using AND and NOR gates.

(*a*) the delay model for the circuit for the Q output

(*b*) the characteristic equation for the Q output (use the characteristic equation for the *S-R* NOR latch)

(*c*) the flow map

(*d*) is the circuit always stable (if not, how can the circuit be made to operate properly)

8-22. Draw a gate-level circuit diagram for a master-slave *J-K* flip-flop using an Inverter and NAND gates. Provide an asynchronous preset $(\overline{PRE})$ and clear $(\overline{CLR})$ input on the slave latch. Now draw the same functional flip-flop with $\overline{PRE}$ and $\overline{CLR}$ using Inverters, AND gates, and NOR gates. Draw the IEEE rectangular logic symbol for a master-slave *J-K* flip-flop.

8-23. How can a master-slave (pulse-triggered) *J-K* flip-flop function properly without a pulse narrowing circuit while a gated *J-K* latch cannot?

8-24. What groups of flip-flops do not possess the property of 1s and 0s catching?

8-25. Complete the waveforms for M and Q shown in Fig.P8-25 for a master-slave (pulse-triggered) *D* flip-flop where M is the output of the master and Q is the output of the slave. Number the timing events beginning with 1. Describe what happens at timing events 6 and 11.

8-26. What must a designer do, concerning the synchronous inputs, to insure proper operation of latches and flip-flops?

8-27. Complete the waveforms for M and Q shown in Fig.P8-27 for a master-slave (pulse-triggered) *T* flip-flop where M is the output of the master and Q is the output of

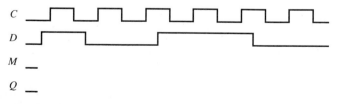

FIGURE P8-25

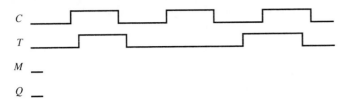

FIGURE P8-27

the slave. Number the timing events beginning with 1. Describe what happens at timing events 2 and 7. Does the T input signal violate the data setup and hold time for a pulse-triggered flip-flop? Explain.

8-28. Draw the IEEE rectangular logic symbol for the master-slave (pulse-triggered) *J-K* flip-flop without data-lockout and then with data-lockout.

8-29. Complete the waveforms for M and Q (master and slave outputs of a pulse-triggered flip-flop without data-lockout) and *MDL* and *QDL* (master and slave outputs of a pulse-triggered flip-flop with data-lockout) shown in Fig. P8-29. Do Q and *QDL* result in the same state after the last 1 to 0 transition of C shown in the diagram? If not, explain why.

8-30. Show a sketch comparing the setup and hold time for a pulse-triggered and a positive edge-triggered flip-flop.

8-31. Draw IEEE rectangular logic symbols for positive edge-triggered *J-K*, *S-R*, and *D* flip-flops. Now draw the same negative edge-triggered flip-flop types.

8-32. Draw timing waveforms to show how the master-slave configuration shown in Fig. 8-33 operates as an edge-triggered *D* flip-flop. Let D be a logic 1 at the first timing event, and a 0 at the second timing event. Show waveforms for the master edge-triggered output *MET* and the slave edge-triggered output *SET*. Assume that the flip-flop is initially in the reset state.

8-33. Use the flow map shown in Fig. 8-35 and numbered arrows to indicate the operation of the positive edge-triggered *D* flip-flop circuit for the following sequence of inputs.

$$y1 \ y2 \ C \ D = 0000 \quad \text{initial condition}$$
$$C \ D = \quad 01$$
$$C \ D = \quad 11 \quad \text{output } y2 \text{ gets set}$$

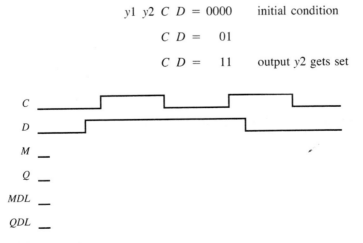

FIGURE P8-29

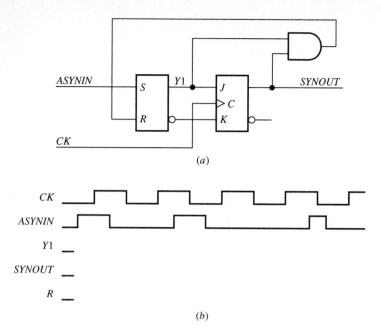

(a)

(b)

FIGURE P8-41

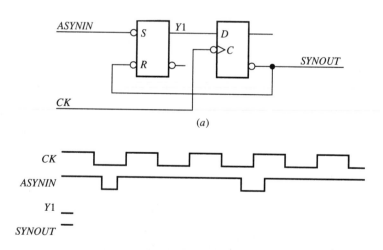

(a)

FIGURE P8-42

$$C \ D = \quad 10$$

$$C \ D = \quad 00$$

$$C \ D = \quad 10 \qquad \text{output } y2 \text{ gets reset}$$

$$C \ D = \quad 00$$

8-34. Show how a negative edge-triggered D flip-flop and an Inverter can be used as a positive edge-triggered D flip-flop.

8-35. Show how a negative edge-triggered J-K flip-flop and an Inverter can be used as a positive edge-triggered J-K flip-flop.

8-36. Show how a negative edge-triggered J-K flip-flop can be used either as a negative edge-triggered T flip-flop or as a negative edge-triggered S-R flip-flop.

Section 8-5   Reduced Characteristic and Excitation Tables for Bistable Devices

8-37. Obtain a minimum bistable equation for the J-K type bistable using its reduced characteristic table.
  (*a*) beginning with its standard sum of products expression
  (*b*) by utilizing the map-entered variable technique

8-38. Obtain a minimum bistable equation for the T type bistable using its reduced characteristic table.
  (*a*) beginning with its standard sum of products expression
  (*b*) by utilizing the map-entered variable technique

8-39. Write the reduced characteristic tables for the S-R, D, J-K, and T type bistables from memory.

8-40. Write the excitation tables for the S-R, D, J-K, and T type bistables by reorganizing their respective reduced characteristic tables.

Section 8-6   Metastability and Synchronization Using Bistables

8-41. Analyze the multiple-stage synchronizer circuit shown in Fig. P8-41*a* and answer the following questions.
  (*a*) Is this a positive pulse or negative pulse synchronizer?
  (*b*) What is the advantage of using a multiple-stage synchronizer circuit like the one in Fig. P8-41*a*, compared to a single-stage synchronizer circuit for the same clock frequency?
  (*c*) Describe the purpose of the AND gate in the circuit.
  (*d*) Complete the waveforms for $Y1$, *SYNOUT*, and R shown in Fig. P8-41*b*.

8-42. Analyze the multiple-stage synchronizer circuit shown in Fig. P8-42*a* and answer the following questions.
  (*a*) Is this a positive pulse or negative pulse synchronizer?
  (*b*) What does a multiple-stage synchronizer circuit like the one in Fig. P8-42*a* do that a signal stage synchronizer circuit cannot do for the same clock frequency?
  (*c*) Complete the waveforms for $Y1$ and *SYNOUT* shown in Fig. P8-42*b*.

CHAPTER
9

SYNCHRONOUS SEQUENTIAL LOGIC CIRCUIT DESIGN

9-1 INTRODUCTION AND INSTRUCTIONAL GOALS

This chapter is perhaps one of the most important chapters in this text. The various tools that were introduced in Chapter 8 are used in this chapter to design synchronous sequential circuits. The topics include the synchronous design and analysis process, Moore and Mealy state machine design, timing diagram descriptions, state reduction, state assignment, D, T, and J-K flip-flop design procedures, counter design, and state machine design using programmable devices. The emphasis is placed on presenting various design descriptions, techniques, and tools that can be applied by a designer to a broad class of synchronous design problems.

In Section 9-2, Moore, Mealy, and Mixed Type Synchronous State Machines, the process for synchronous design and analysis is presented. General steps in the design process are discussed, and a Moore state machine is designed using a D flip-flop to illustrate some of the techniques. Synchronous circuits using various edge-triggered flip-flop types are presented from the point of view of design simplicity. Hazards and critical races are discussed in relationship to synchronous sequential circuits. General steps are provided for analyzing synchronous sequential circuits. Examples illustrate the techniques commonly used to analyze a synchronous sequential state machine design.

Synchronous Sequential Design of Moore and Mealy Machines in Section 9-3 provides various methods that can be used to obtain either a Moore state machine design or a Mealy state machine design. Various logic descriptions including the state diagram, ASM chart, and timing diagram are presented. State reduction is presented using an implication table to investigate redundant states, and the problem of state assignment is discussed. Separate procedures are presented for designing synchronous state machines using *D*, *T*, and *J-K* flip-flops.

We present various counter designs in Section 9-4, Synchronous Counter Design. Both synchronous and asynchronous counter characteristics are presented. Different types of synchronous counter designs are discussed in detail. Design procedures are discussed for counters that can follow any user-prescribed state sequence, and for shift register counters that can only follow a specified state sequence.

This chapter would not be complete without Section 9-5, Designing Synchronous Sequential State Machines using Programmable Devices. For small synchronous circuits, programmable devices can be used to provide only a single package in most cases. Thus, state machines using programmable devices with built in flip-flops also shorten design time, allow design changes to be easily made, and are generally more reliable. The software package we discuss (PLDesigner by Minc) allows the designer to separate the design phase from the device selection phase. Examples show state machine designs using each of the following entry methods: logic equation, truth table (PS/NS table), state machine, and waveform (timing diagram).

This chapter should prepare you to

1. Draw block diagrams for Moore type, Mealy type, and mixed type state machines, and explain their differences.
2. List the general steps involved in synchronous sequential design, and discuss the significance of each step.
3. Discuss the advantages and disadvantages of designing synchronous state machines using *D*, *T*, and *J-K* edge-triggered flip-flops.
4. Describe the 'set or hold method' of obtaining the next state equations from a timing diagram, a state diagram, or an ASM chart.
5. Apply the 'set or hold method' to obtain the *D* excitation input equations for a synchronous sequential design.
6. List the general steps involved in synchronous sequential analysis, and discuss the significance of each step.
7. Use the synchronous sequential analysis process to analyze synchronous sequential designs.

8. Utilize a timing diagram to describe a synchronous sequential design, and draw a state diagram or an ASM chart from the timing diagram description.

9. Apply state reduction using an implication table.

10. Determine the number of flip-flops for a given number of states in a synchronous design.

11. Calculate the number of unique state assignment codes in a synchronous state machine given the number of rows in its flow table.

12. Design synchronous state machines using D, T, and J-K flip-flops, utilizing their excitation tables.

13. Explain the differences in the operation of synchronous and asynchronous counters.

14. Design synchronous counters to follow any prescribed output sequence.

15. Design synchronous shift register counters of the following types: ring counter, twisted ring counter, and maximum length shift counter.

16. Explain the term 'illegal state recovery' and describe how illegal state recovery applies to synchronous sequential state machine design.

17. Use the 'design phase' of the software package PLDesigner to create synchronous sequential logic designs using the following entry methods: logic equation, truth table, state machine and waveform.

9-2   MOORE, MEALY, AND MIXED TYPE SYNCHRONOUS STATE MACHINES

Three types of circuit models for synchronous state machines are illustrated in Fig. 9-1. Each model is different because of its output method. The circuit model in Fig. 9-1*a* is called a synchronous state machine with Moore type outputs in recognition of its founder E. F. Moore. Notice that Moore type external outputs are dependent on only the present state of the circuit, and are independent of the external inputs to the circuit. A state machine with only Moore type outputs is often referred to as a Moore type machine or simply a Moore machine. The external outputs in the synchronous state machine circuit model shown in Fig. 9-1*b* are dependent on both the external inputs and the present state of the circuit. A Mealy state machine is one that provides only these types of outputs, called Mealy type outputs in recognition of the founder G. H. Mealy. The circuit model illustrated in Fig. 9-1*c* is a synchronous state machine with both Moore and Mealy type external outputs and is called a mixed type machine.

9-2-1   Synchronous Design Procedure

There are no set rules on where a synchronous sequential design should start, since this decision is often made by the designer as a result of the design specification or the designer's preference. A design can even originate from a word description; if this is the case, it is often organized using one or more of the tools presented

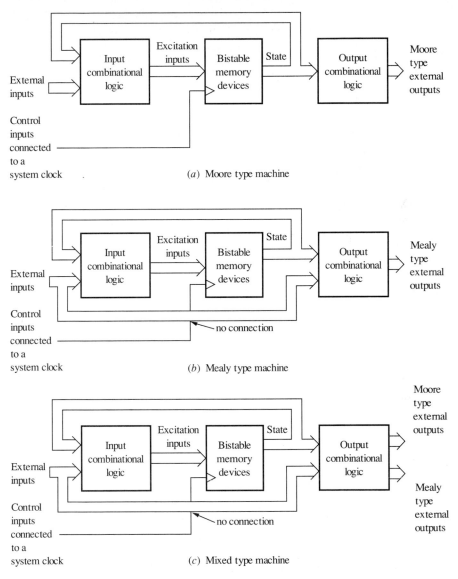

FIGURE 9-1
Circuit models of synchronous state machines.

in Chapter 8. Like combinational logic design, the goal is to obtain realizable sequential logic circuits that work, using as few devices as possible. An outline of the sequential design process is provided in Table 9-1.

In synchronous design the designer will often generate a PS/NS table, state diagram, ASM chart, flow map, or timing diagram that shows the specification for a proposed circuit. The object is to select the minimum number of flip-flops

TABLE 9-1
Synchronous design process

	Design steps	Comments
Step 1	Organize design specification into a PS/NS table, state diagram, ASM chart, flow map, or timing diagram	
Step 2	Perform state reduction	Can result in fewer flip-flops
Step 3	Determine minimum number of flip-flops and assign state variables (one variable for each flip-flop), then assign a unique code to each state	Binary code selection affects combinational logic complexity
Step 4	Select type of flip-flop, then determine excitation input equations and Moore and/or Mealy output equations	Excitation input equations are easier to obtain for D-type flip-flops and harder to obtain for T, J-K, S-R, and J-$\overline{K}$ flip-flops
Step 5	Draw synchronous sequential logic circuit diagram	The system clock is connected to the control inputs of all flip-flops so that they are synchronously controlled

to satisfy the input-output conditions for the circuit. A set of variables such as $y1$, $y2$, $y3$, . . . , yn called state variables are assigned to the outputs of the bistable memory devices or flip-flops in the circuit, that is, one variable to each flip-flop. For n state variables, there are 2^n possible binary combinations or memory states for the circuit. A state of the circuit can therefore be referred to as the binary value of the state variables representing the flip-flops in the circuit (see Fig. 8-3a). Each state is assigned a unique binary code. When added to the description of the circuit, the state assignment codes show the logic transition of each flip-flop output from one state to the next state, that is, present state, next state information. The excitation input equations for the flip-flops in the circuit and the Moore and/or Mealy output equations are determined using the present state/next state information along with any input requirement. Using the excitation input and Moore and/or Mealy output equations, a sequential circuit diagram can be drawn. Synchronous circuits always utilize a system clock to drive the control inputs of the flip-flops in the circuit simultaneously.

State reduction is used to obtain the minimum number of states. After the minimum number of states has been determined, the number of flip-flops can be determined. Assigning a unique binary code to each state is called state assignment. Different assignments affect the complexity of the combinational logic used in the circuit. State reduction and state assignment will be presented later in this chapter.

9-2-2 Designing with Edge-Triggered Devices

As we discussed in Chapter 8, synchronous sequential circuit designs today pre-dominately incorporate edge-triggered D-type flip-flops; however, edge-triggered T, J-K, S-R, and J-$\overline{K}$ types can also be used. Edge-triggered D, J-K, and J-$\overline{K}$ flip-flops are available as off-the-shelf devices; however, edge-triggered T and S-R flip-flops are not usually available as off-the-shelf devices. The excita-tion table for each is listed in the combined excitation table shown in Fig. 9-2.

Any new edge-triggered type flip-flop can be added to the table in Fig. 9-2 if the excitation inputs are known. Reduced characteristic tables are supplied for available flip-flops in manufacturer's data books. The excitation input require-ments for any flip-flop can be evaluated from the reduced characteristic table of the flip-flop by simply reorganizing the table. Edge-triggered D flip-flops are more often used because they (*a*) are readily available as off-the-shelf devices, (*b*) have a single D input, and (*c*) have a simple completely specified excitation input function ($D = Q$ as observed in Fig. 9-2). D flip-flops are considered the easiest to design with.

The following example illustrates a simple and efficient way of designing with edge-triggered D flip-flops. An alternate design procedure that can be applied to any edge-triggered flip-flop type will be presented later in this chapter.

> **Example 9-1.** Design a Moore type synchronous state machine with three external inputs $X1$, $X2$, and $X3$ and one output Z. The output Z goes to 1 when $X1 \cdot X2 \cdot X3 = 1$ at the next system clock timing event. The output Z stays at 1 as long as $X3 = 0$; otherwise, the output goes to 0.
>
> (*a*) Draw an ASM chart to represent the design.
> (*b*) Use a positive edge-triggered D flip-flop in the design.
>
> *Solution*
>
> (*a*) An ASM chart that describes the operation of the circuit is shown in Fig. E9-1a. Only two states are necessary to provide the values 0 and 1 for the output Z. The state variable y is assigned to the output of the flip-flop used in the circuit. A binary 0 is assigned to state variable y for state a, and a binary 1 is assigned to state variable y for state b. A transition from one state to the next only occurs at each system clock timing event. The system clock signal is

State transitions		Excitation inputs				
PS	NS					
q	Q	D	T	$J\ K$	$S\ R$	$J\ \overline{K}$
0	0	0	0	0 X	0 X	0 X
0	1	1	1	1 X	1 0	1 X
1	0	0	1	X 1	0 1	X 0
1	1	1	0	X 0	X 0	X 1

FIGURE 9-2
Combined excitation table for flip-flops.

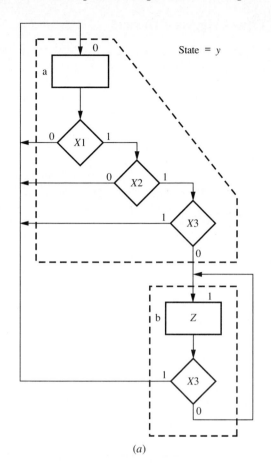

State = y

(a)

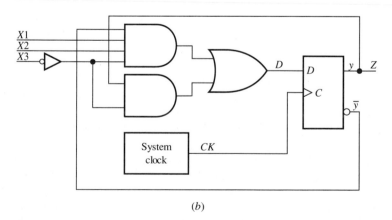

(b)

FIGURE E9-1

implied; therefore, it is not shown in the ASM chart describing the synchronous design.

(b) Notice in the ASM chart that Z is 1 when the flip-flop is set. Like combinational logic circuit design, one can either write the state equation to set the flip-flop or write the state equation to reset or clear the flip-flop. It is the same concept as writing a Boolean equation for the 1s or the 0s for a combinational logic function, only now we write the Boolean equation for the 1s or the 0s for the sequential logic function next state output Y. Let's choose to write the equation for the 1s of the next state output Y. The equation is obtained by writing every link path Boolean expression for each state that results in a next state output $Y = 1$, and OR them all together. The resulting next state output equation for Y is simply applied to the D input of the D-type flip-flop, since $D = Y$ (see Fig. 9-2).

    Following the link paths from state a to b ($y = 0$ to $y = 1$) and from state b to b ($y = 1$ to $y = 1$) that result in a next state output of 1, we obtain the following next state output equation. For documentation purposes, an explanation is provided just to the right and on the same line as each Boolean expression in the equation. This method of generating Boolean equations for next state outputs can be called the 'set or hold method' and only applies when designing with D-type flip flops. To obtain the 0s for the sequential logic function, the method can be called the 'clear or hold method'.

$$Y = \bar{y} \cdot X1 \cdot X2 \cdot \overline{X3} \qquad Y = 1 \text{ when } y = 0 \text{ AND } X1 \cdot X2 \cdot \overline{X3} = 1$$
$$\text{(set operation from state a to b)}$$

$$+ \, y \cdot \overline{X3} \qquad Y = 1 \text{ when } y = 1 \text{ AND } X3 = 0$$
$$\text{(hold operation from state b to b)}$$

Since the Moore output Z is dependent only on the state of the machine, Z is obtained for the 1s of the function by writing every Boolean expression for each state that results in an output $Z = 1$, and OR them all together.

$$Z = y \qquad\qquad Z = 1 \text{ when } y = 1$$

    Using the relationship $D = Y$ allows us to draw the combinational logic circuit for the D input to the positive edge-triggered D flip-flop as shown in Fig. E9-1b. The system clock is represented by a block diagram.

    Output transitions can occur only on the positive rising edge of the CK input signal (these transitions mark the timing events for the circuit).

    We are assuming that inputs $X1$, $X2$, and $X3$ have previously been synchronized with the system clock. If this is not the case, these inputs need to be synchronized using a D-type flip-flop (see Section 8-6 in Chapter 8).

    The edge-triggered T flip-flop is also fairly easy to design with, since its excitation input function is completely specified and can be represented by the relationship ($T = q \oplus Q$). The T flip-flop is not usually available as an off-the-shelf device, perhaps because a J-K flip-flop can easily be used as a T flip-flop.

    An edge-triggered J-K flip-flop (or J-$\bar{K}$ flip-flop) has twice as many inputs as a D or T flip-flop; therefore, it requires twice as many excitation input equations. One can see by observing the J and K entries in the combined excitation table in Fig. 9-2 that these excitation inputs are incompletely specified; that is, there are don't cares in the *J* and *K* columns. Because of these don't cares, *J-K* flip-flops are also a little harder to use; that is, the design procedure takes longer. However, the don't cares are an advantage in some designs, since they often allow functions to be obtained with less combinational logic than is required for *D* and *T* flip-flops.

Designing with edge-triggered *S-R* flip-flops is practically the same as designing with *J-K* flip-flops. Because the *S-R* flip-flop has one less don't care entry in both the *S* and *R* columns of its excitation table compared to *J-K* flip-flops, more combinational logic can be required than is necessary for *J-K* flip-flops. Like the *T* flip-flop, the *S-R* flip-flop is not typically available as an off-the-shelf device (a *J-K* flip-flop can be used as an *S-R* flip-flop if desired). The *S-R* flip-flop is not normally used in synchronous sequential design; therefore, it will not be discussed further.

The input combinational logic block in Fig. 9-1 does not have to be hazard-free for synchronous circuits, since all the excitation signals (combinational signals) are required to be stable (all hazards have died out) prior to the next system clock timing event. Logic hazards in combinational logic circuits are therefore not eliminated, but may simply be ignored by either (*a*) choosing only edge-triggered memory devices for the design, (*b*) choosing faster combinational logic devices, (*c*) choosing two-level combinational logic circuits (which are faster than circuits with multiple levels), or (*d*) choosing a slower system clock frequency (slowing down the frequency of the clock). Critical races due to feedback are normally eliminated in synchronous sequential logic circuits by utilizing edge-triggered flip-flops. In temporary storage type applications requiring no feedback, either latches or edge-triggered flip-flops can be used.

9-2-3 Synchronous Analysis Procedure

If you will recall, in Chapter 8 we analyzed a number of different asynchronous sequential bistable circuits using the following procedure. We wrote the next state output equation(s) from the circuit diagram. From the next state output equation(s) we used one or more of the tools presented in Chapter 8 to show the input-output behavior of the circuit. An outline of the synchronous analysis process is provided in Table 9-2.

After the next state equations are obtained, the general procedure for analyzing a synchronous circuit is the same as the procedure used to analyze an asynchronous circuit. To obtain the next state equations, a state variable such as $y1, y2, \ldots, yn$ is assigned to each of the flip-flop outputs. Next, the excitation input equations for the flip-flops are written, and the external (Moore and/or Mealy) output equations are written from the circuit diagram. The excitation input equation(s) for each flip-flop are then substituted into the bistable equation for the particular flip-flop type to obtain the next state output equation. A minimum bistable equation for a flip-flop can be obtained from its reduced characteristic table as illustrated in Example 8-9. The next state and external output

TABLE 9-2
Synchronous analysis process

	Analysis steps	Comments
Step 1	Assign a state variable to each flip-flop in the synchronous sequential logic circuit	
Step 2	Write the excitation input equations for each flip-flop and also write the Moore and/or Mealy output equations	
Step 3	Substitute the excitation input equations into the bistable equations for the flip-flops to obtain the next state output equations	The bistable equation for a flip-flop can be obtained from its reduced characteristic table
Step 4	Obtain a composite Karnaugh map using the next state and external (Moore and/or Mealy) output equations	Separate Karnaugh maps can be used if desired
Step 5	Use the composite Karnaugh map to obtain a PS/NS table, state diagram, ASM chart, flow map, or timing diagram to show the behavior of the circuit	

equations are used to obtain a composite Karnaugh map to show the state and output behavior of the circuit. From the composite Karnaugh map for the next state and external outputs, a PS/NS table, state diagram, ASM chart, flow map, or timing diagram can also be drawn to show the behavior of the circuit.

The following examples are used to illustrate the analysis procedure for synchronous sequential logic circuits.

Example 9-2. Analyze the synchronous Mealy machine in Fig. E9-2a to obtain its state diagram.

Solution Since the flip-flops have already been assigned state variables $y1$ and $y2$, the next step is to write the Boolean equations for the circuit. There are two flip-flop excitation input equations and one Mealy type output equation.

$$D1 = \overline{y1} \cdot y2 \cdot \overline{X}$$

$$D2 = X + \overline{y1} \cdot y2$$

$$Z = y1 \cdot y2 \cdot X$$

The bistable equation for a D flip-flop is $Q = D$ (see the reduced characteristic

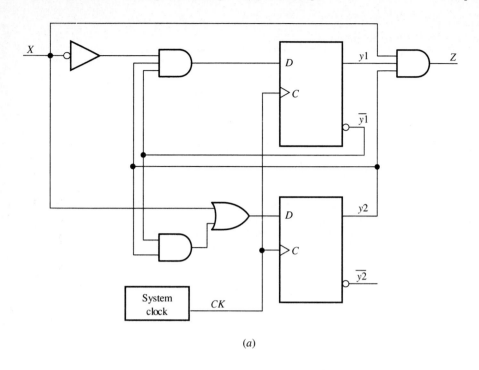

(a)

$Y1\ Y2,\ PS\ Z =$

y1 y2 \ X	0	1
00	00, 0	01, 0
01	11, 0	01, 0
11	00, 0	01, 1
10	00, 0	01, 0

(b)

NS, PS Z =

y1 y2 \ X	0	1
a 00	a, 0	b, 0
b 01	c, 0	b, 0
c 11	a, 0	b, 1
d 10	a, 0	b, 0

(c)

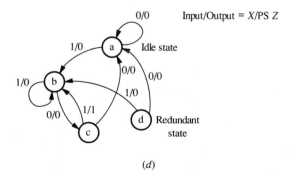

(d)

FIGURE E9-2

table for the *D*-type bistable in Fig. 8-41*b*). The next state output equations *Y*1 and *Y*2 are simply obtained by substituting *D*1 and *D*2 into the bistable equations *Y*1 = *D*1, and *Y*2 = *D*2. This results in the following next state output equations.

$$Y1 = \overline{y1} \cdot y2 \cdot \overline{X}$$

$$Y2 = X + \overline{y1} \cdot y2$$

Using the next state output equations *Y*1 and *Y*2 and the Mealy output equation *Z*, we can obtain the composite Karnaugh map in Fig. E9-2*b*. In a synchronous design every square in the Karnaugh map represents a stable state. The system clock signal is implied; therefore, it is not shown in the composite Karnaugh map. From the composite Karnaugh map we can obtain the flow map shown in Fig. E9-2*c*.

A Mealy state diagram can be drawn as shown in Fig. E9-2*d* using the flow map. The Mealy state diagram clearly shows the behavior of the Mealy synchronous circuit or machine. The circuit is called a sequence detector. By observing the state diagram, it can be seen that the circuit detects the *X* input sequence 101 and provides an output *Z* = 1 each time that sequence is repeated. In the state diagram, moving to state b, then c, then back to b results in a Mealy output of 1 on the return to b. Of course, the input sequence must be synchronized with three successive clock timing events to cause the output. Mealy outputs are a function of both the state and the external inputs, and are shown in the state diagram on the directed line segments connecting the states.

State a is the idle state in the Mealy state diagram for the sequence detector. State d is identical to state a, since it has the same next states and outputs. Identical states are discussed later when state reduction is presented.

Example 9-3. Analyze the synchronous Moore machine shown in Fig. E9-3*a* to obtain its state diagram.

Solution Using the assigned variables *y*1 and *y*2 for the two *J-K* flip-flops, we can write the four excitation input equations and the single Moore output equation as follows.

$$J1 = y2 \cdot X$$

$$K1 = \overline{y2}$$

$$J2 = X$$

$$K2 = \overline{X}$$

$$Z = y1 \cdot \overline{y2}$$

The bistable equation for a *J-K* flip-flop is $Q = q \cdot \overline{K} + \overline{q} \cdot J$ (solving for the function *Q* in the reduced characteristic table for the *J-K* type bistable in Fig. 8-41*c*). Writing the equations in terms of the present state variable *y* and the next state variable *Y*, we obtain $Y1 = y1 \cdot \overline{K1} + \overline{y1} \cdot J1$ and $Y2 = y2 \cdot \overline{K2} + \overline{y2} \cdot J2$. Substituting the excitation input equations obtained from the circuit into these equations, we obtain the next state output equations for flip-flops 1 and 2 as follows.

$$Y1 = y1 \cdot \overline{(y2)} + \overline{y1} \cdot y2 \cdot X$$
$$= y1 \cdot y2 + \overline{y1} \cdot y2 \cdot X$$

$$Y2 = y2 \cdot \overline{(X)} + \overline{y2} \cdot X$$
$$= y2 \cdot X + \overline{y2} \cdot X$$
$$= X$$

Using the next state and external output equations $Y1$, $Y2$, and Z, one can obtain the composite Karnaugh map in Fig. E9-3b. If one desires, a separate Karnaugh

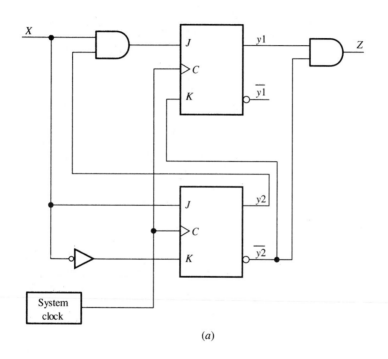

(a)

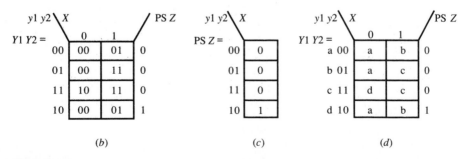

(b) $\qquad$ (c) $\qquad$ (d)

FIGURE E9-3

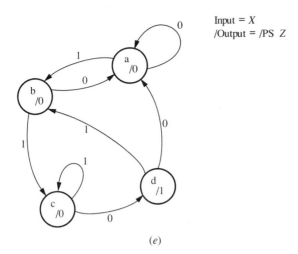

Input = X
/Output = /PS Z

(e)

FIGURE E9-3

map can be drawn as shown in Fig. E9-3c for the Moore output equation. The system clock signal is implied in the composite Karnaugh map. From the composite Karnaugh map we can obtain the flow map in Fig. E9-3d.

A Moore state diagram can be drawn as shown in Fig. E9-3e using the flow map. The Moore output Z is placed in each state circle, since this output is only a function of the state and not the external input X. The Moore state diagram clearly shows the behavior of the Moore synchronous circuit. Observing the state diagram, one can see that the circuit provides a Z output of 1 if the external input X changes with the sequence 110 for three successive clock timing events.

9-3   SYNCHRONOUS SEQUENTIAL DESIGN OF MOORE AND MEALY MACHINES

In this section we present a more formal design procedure for Moore and Mealy machines using D, T and J-K flip-flops. The process is introduced using a simple but realistic synchronous machine design example which is first checked for state reduction using an implication table. The state assignment problem is also discussed, providing the student with a brief insight of the magnitude of the problem of assigning state codes to state variables. It should be noted that each design procedure for a particular flip-flop type is presented independent of the design procedures for the other types. A single simple design example is used throughout our discussion so that a comparison can be drawn between the various types.

9-3-1   Timing Diagram Description

Suppose we desire to obtain a circuit to provide an output signal frequency f_Z that is $\frac{1}{4}$ the frequency f_{CK} of the system clock signal supplied to the circuit or $f_Z = \frac{1}{4} f_{CK}$. The timing diagram in Fig. 9-3 shows the system clock signal CK

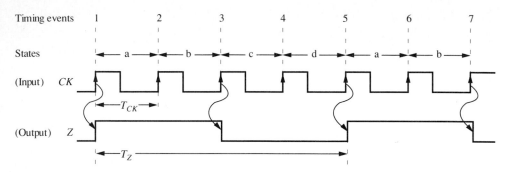

FIGURE 9-3
Timing diagram for synchronous machine M1.

and an output signal Z with a frequency $\frac{1}{4}$ the frequency of the system clock signal; that is, the period of Z is 4 times the period of CK or $T_Z = 4T_{CK}$, where $T = 1/f$. We will refer to the synchronous machine represented by the timing diagram in Fig. 9-3 as M1 (machine 1). Since the timing events are shown as 0 to 1 transitions, we are specifying that positive edge-triggered flip-flops will be used in the circuit implementation. To keep the timing diagram simple, the output propagation delays for output Z at timing events 1, 3, 5, etc. are not shown. When circuit timing is critical or when a circuit fails to work as expected (which can occur for fast clock frequencies, slow logic devices, and/or circuits with multiple levels of logic), separate timing diagrams are usually drawn to check specifically for critical timing paths. This is especially true for larger systems. This problem is not within the scope of this text.

From the timing diagram for synchronous machine M1, we can draw the Moore state diagram as shown in Fig. 9-4. It is customary to draw a state diagram for a synchronous circuit with the system clock signal implied. Since positive edge-triggered flip-flops will be used in the design, a transition from one state to the next state occurs when the clock signal changes from a logic 0 to a logic 1 (or from a low-level voltage to a high-level voltage for positive logic). The Moore

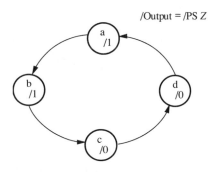

M1 (Moore machine)

FIGURE 9-4
Moore state diagram without state assignment codes.

state diagram shows the logic output value of Z for each of the four required states of the machine a through d. The Z output value is placed in the circle for each state, since Moore outputs are only a function of the state of the machine.

Example 9-4. Figure E9-4a shows a synchronous Moore machine we will call M2 (machine 2), and Fig. E9-4b shows a synchronous Mealy machine we will call M3. Both M2 and M3 have one external input X. When the external input X is 1, each provides an output signal Z that is $\frac{1}{4}$ the frequency of the system clock signal, just like M1 in the timing diagram in Fig. 9-3. When X is 0 on the next clock timing event, both M2 and M3 go to state d (an idle or rest state). Both M2 and M3 stay in the idle state until X is 1 on the next clock timing event, allowing both machines to move to state a.

(a) Draw an equivalent ASM chart for a Moore and Mealy machine for M2 and M3.

(b) Draw a timing diagram for both M2 and M3 for the input sequence 00111101100010110 where the MSB in the binary number represents the first value of X in the sequence. Show the changes in the asynchronous input X occurring approximately $\frac{1}{4}$ of a cycle prior to the next clock timing event (to illustrate adequate setup time). Assume that output Z is initially 0 and the machine is in the idle or reset state. Ignore output delays when drawing the output waveform.

Solution

(a) A Moore output is a function of only the state of the machine. In the Moore ASM chart for M2 shown in Fig. E9-4c, the Moore output Z is placed in the state box in state a and also in the state box in state b. This means that the Moore output Z is 1 only when the machine M2 is in states a and b, and the Moore output Z is 0 in each of the other states. The Mealy ASM chart for M3 is shown in Fig. E9-4d. Since a Mealy output is a function of both the state of the machine and the external input, the Mealy output Z is shown listed in a

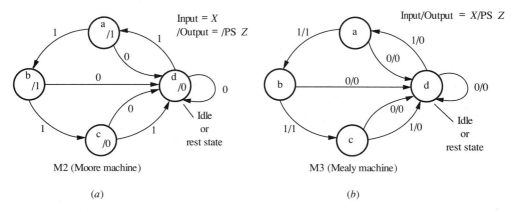

M2 (Moore machine)   M3 (Mealy machine)

(a)   (b)

FIGURE E9-4

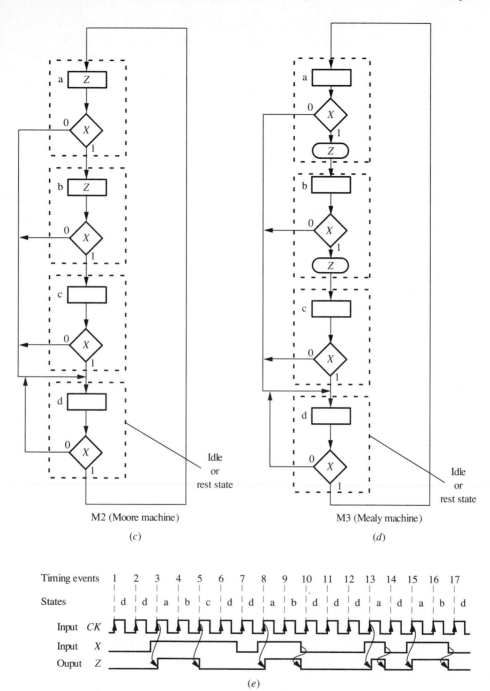

M2 (Moore machine)

(c)

M3 (Mealy machine)

(d)

Idle
or
rest state

Idle
or
rest state

FIGURE E9-4

conditional output box in state a, and a separated conditional output box in state b. The Mealy output Z is 1 when the machine M3 is in state d and changes to state a $(X = 1)$, and when the machine is in state a and changes to state b $(X = 1)$; otherwise, the Mealy output Z is 0.

Some designers prefer to work with state diagrams, while others prefer to work with ASM charts. It helps to know both representations when communicating with other designers.

(b) The timing diagram for M3 is shown in Fig. E9-4e. Notice that the Mealy output can change either after a clock timing event or immediately after an external input change. A Moore output, however, can only change after a clock timing event. The timing diagram for M2 is left as an exercise for the student.

9-3-2 State Reduction by Implication Table

Sometimes in the design of synchronous machines, the design specification can lead to a PS/NS table and eventually to a circuit that contains more states, and hence more flip-flops, than are actually necessary to do the job. For s states the smallest number of flip-flops n required to represent all the states is found by using the relationship

$$2^n \geq s \tag{9-1}$$

To investigate whether or not a particular machine has been proposed with more states than are necessary, an 'implication table' ('pair chart' or 'merger table') can be used. An implication table is basically a state reduction tool or technique that allows a designer to check for redundant states. If it is determined that the number of states for a circuit can be reduced while maintaining the same external input-output requirements, it may be possible to reduce the number of flip-flops in the circuit; however, even if the number of states can be reduced, the number of flip-flops may not necessarily be reduced as indicated by Eq. 9-1.

The present state/next state (PS/NS) table for the synchronous machine M1 can be written as follows.

PS	NS	PS output Z
a	b	1
b	c	1
c	d	0
d	a	0

Two state pairs that have identical outputs and also change to the same next states, as a result of the clock signal and/or external inputs, are defined as 'identical states'. Identical states are redundant. One of the two states can be eliminated, reducing the total number of states in the design. Identical states can often be spotted in a PS/NS table (flow table or flow chart) by simply inspecting the table or the chart (see Fig. E9-2d which contains identical states: state a and state d). Using an implication table, a logic designer can systematically compare each state in the PS/NS table with every other state to determine pairs of equivalent states. Equivalent states are also redundant. One of the two state pairs can be eliminated to reduce the total number of states in the design. Two state pairs, such as m-n, are implied to be equivalent when they have identical outputs, but their next states are dependent on the implied equivalence of two other states such as r-s. If state pairs r-s prove to be equivalent states, then state pairs m-n are also equivalent states. 'Sets of equivalent states are determined indirectly by determining all the state pairs that are not equivalent (their outputs are not identical)'. In the PS/NS table for M1 (which is also repeated in Fig. 9-5a for reference), we compare state a with b, c, and d, then state b with c, and d, and finally state c with d to determine the status of each pair of states. A pair of states can be identical, implied equivalent, or nonequivalent states. The implication table in Fig. 9-5b allows the status of these comparisons to be listed in an organized manner. For large designs, multiple passes through the implication table are typically required to determine all the nonequivalent states. Implied equivalent states that remain after an exhaustive search to eliminate all nonequivalent states are equivalent by elimination.

Notice that the implication table is constructed in the following manner. Along the x axis all the states are listed in order from left to right except the last state. Along the y axis all the states are listed in order from top to bottom except the first state. This organization provides an individual square for listing the status of all possible combinations of two states in the PS/NS table.

To fill in the implication table, we make row comparisons, two at a time, in the PS/NS table. In effect, we are testing each state listed on the x axis in the implication table with a corresponding state listed on the y axis. If two states are identical then their next states and outputs are identical, and this is recorded in the appropriate square in the implication table by a check mark. If two states are nonequivalent, then their outputs are not identical. This is recorded

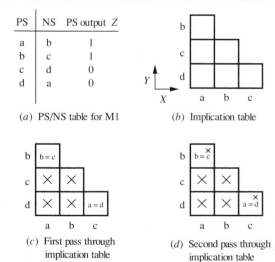

PS	NS	PS output Z
a	b	1
b	c	1
c	d	0
d	a	0

(a) PS/NS table for M1

(b) Implication table

(c) First pass through implication table

(d) Second pass through implication table

FIGURE 9-5
Implication table for M1 to check for state reduction.

in the appropriate square in the implication table by a X mark. If two states are implied equivalent then their outputs are identical but their next states are not, and the output states that need to be checked for equivalence are recorded in the appropriate square in the form m = n. The first pass through the implication table for M1 is shown in Fig. 9-5c. The square at the intersection of states a and b implies equivalent states that depend on the equivalence of states b and c; consequently, b = c is entered in the square. The square at the intersection of states c and d implies equivalent states that depend on the equivalence of states a and d; consequently, a = d is entered in the square. All other states are nonequivalent. The equivalence of states b and c, and of states a and d, are then checked on the second pass through the implication table to see what they are dependent on. Both states b and c, and a and d are dependent on nonequivalent states (the squares at the intersection of states b and c, and a and d each contain an X); therefore, the squares containing b = c and a = d are marked by an X as shown in Fig. 9-5d. Since there are no remaining implied equivalent states to check after the second pass we can say that there are no identical or equivalent states in M1. Thus a minimum of four states must be used in the design. When identical or equivalent states are found, such as m equivalent to n, we can simply replace n and m everywhere in the PS/NS state table with the same state name to reduce the number of states of the machine.

As we have demonstrated, the input-output requirements of a machine must be specified in order to apply this state reduction procedure. The reduction procedure has been demonstrated for a machine with Moore type outputs, but the same procedure applies equally well for machines with Mealy type outputs.

Example 9-5

(a) Obtain the PS/NS tables for the synchronous machines M2 and M3 in Example 9-4.

(b) Apply the state reduction procedure to obtain the implication table for the Mealy machine M3.

Solution

(a) The PS/NS table for the Moore machine M2 can be written in the following compact form.

	NS		
PS	$X=0$	$X=1$	PS output Z
a	d	b	1
b	d	c	1
c	d	d	0
d	d	a	0

The PS/NS table for Mealy machine M3 is often written as follows.

	NS, PS output Z	
PS	X = 0	X = 1
a	d, 0	b, 1
b	d, 0	c, 1
c	d, 0	d, 0
d	d, 0	a, 0

(*b*) The implication table is shown in Fig. E9-5. The b = c and a = d entries are implied equivalent states which are found on the second pass not to be equivalent. Since the number of states for the Mealy machine M3 cannot be reduced, two flip-flops are still required to carry out the design.

9-3-3   State Assignment

A minimum of two flip-flops are required in the design for M1, since there are four states shown in the state diagram. If we had found from the implication table in Fig. 9-5 that M1 required only three states (a pair of states were discovered to be identical or equivalent), Eq. 9-1 shows that we would still need two flip-flops to implement M1. If we had found from the implication table that M1 required only two states (two pairs of states were discovered to be identical or equivalent), Eq. 9-1 shows that we would only need one flip-flop to implement M1. Since two flip-flops are required, there are two state variables (one for each flip-flop) which we will name $y1$ and $y2$. These bistable outputs provide four different states a through d which we can assign the state values (codes) $y1\ y2 = 00\ 01\ 10$ and 11 in any order.

The sequence or order of assigning the state values 00 through 11 to the state names a through d is totally arbitrary in synchronous design, but the choice of the state assignment codes can greatly affect the gate complexity of both the excitation input equations that determine the next state of each flip-flop, as well as the output equations of the circuit. By choosing certain assignments, the excitation and output equations in a synchronous design can contain fewer literals and require fewer gates. In counter designs, a specified sequence is followed as a result of Moore type outputs being taken directly from the flip-flop outputs in the circuit. In this case, the state assignment codes are dictated by the counting sequence and cannot be altered. Many have researched the state assignment process to try

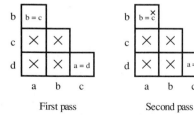

First pass   Second pass   **FIGURE E9-5**

to obtain a general procedure that will lead to an optimum state assignment; that is, the process of selecting state assignment codes that always result in a minimum number of combinational logic devices to realize any design. This problem is quite complex and generally requires a lengthy computer search.

The number of possible state assignment codes C is given by the following relationship.

$$C = 2^v(2^v - 1) \cdots (2^v - r + 1) = \frac{2^v!}{(2^v - r)!} \tag{9-2}$$

where v is the number of state variables and r is the number of rows in the PS/NS table or flow table.

Equation 9-2 results from the fact that 2^v codes can be chosen for the first row, $2^v - 1$ codes can be chosen for the second row, and so forth, on down to the last row where only $2^v - r + 1$ codes can be chosen.

The relationship for C in Eq. 9-2 does not provide distinct codes. A distinct code is sometimes defined as a code that cannot be obtained from another code either by permutation and/or by complementation of the state variables. Notice that v state variables can be permuted in $v!$ ways and complemented in 2^v ways.

If one considers that the $v!$ permuted state variables in the above relationship do not provide distinct state assignments, then C can be divided by $v!$. It has been suggested that $C/v!$ represents the number of distinct state assignment codes when using D flip-flops; however, when using J-K flip-flops, $C/(v!2^v)$ represents the number of distinct state assignment codes. In the case of D flip-flops, the $v!$ permuted state variables are considered redundant state assignment codes; however, for J-K flip-flops both the $v!$ permuted and 2^v complemented state variables are considered redundant state assignment codes.

Example 9-6. Determine the number of distinct state assignment codes for a synchronous state machine with five rows in its flow table that is designed with

(a) D flip-flops
(b) J-K flip-flops.

Solution Since $2^n \geq s$ where $v = n$ and $r = s$, for $r = 5$, $v = 3$.

(a) For D flip-flops we can calculate the distinct state assignment codes using the relationship $C/v!$ where C is given by Eq. 9-2.

$$\text{Distinct state assignment codes} = \frac{C}{v!} = \frac{2^v!}{(2^v - r)!v!}$$

$$= \frac{2^3!}{(2^3 - 5)!3!}$$

$$= \frac{8!}{3!3!}$$

$$= 1120$$

(*b*) For *J-K* flip-flops we can calculate the distinct state assignment codes using the relationship $C/(v!2^v)$ where C is given by Eq. 9-2.

$$\text{Distinct state assignment codes} = \frac{C}{v!2^v} = \frac{2^v!}{(2^v - r)!v!2^v}$$

$$= \frac{(2^v - 1)!}{(2^v - r)!v!}$$

$$= \frac{(2^3 - 1)!}{(2^3 - 5)!3!}$$

$$= \frac{(8 - 1)!}{3!3!}$$

$$= 140$$

Using all 1120 or 140 distinct state assignment codes for a flow table with 5 rows and 3 state variables to determine an optimum state assignment quickly becomes a monumental task if attempted manually, so this is generally not done. With this many state assignment codes, a designer will usually just pick a code for each state, either randomly or using some simplified rules, and proceed with the design.

Today with the extremely high integration capabilities and resulting low cost of logic functions, state assignment techniques, although important, are a little less important than they once were. Some researchers have even demonstrated that an optimum state assignment when using one type of bistable device is not necessarily optimum when using a different type of bistable device. A state assignment using the gray code for M1 is shown in Fig. 9-6. The gray code (also called a unit distance code) has the advantage that a minimum number of bits change (only one bit) as machine transitions occur at the clock timing events. The state assignment codes for M1 are $y1\ y2 = 00$ for a, 01 for b, 11 for c, and 10 for d—the counting sequence for gray code (see Table 2-3 in Chapter 2).

The most important thing to remember about assigning codes to synchronous state machines is simply that each state must be assigned a unique code; that

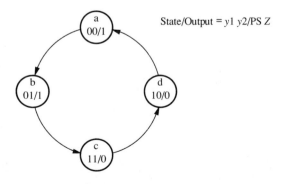

State/Output = $y1\ y2$/PS Z

M1 (Moore machine)

FIGURE 9-6
Moore state diagram for M1 with state assignment codes.

is, two states cannot be assigned the same code. Also keep in mind that when designing synchronous circuits with a large number of rows in the flow table, one can use many different state assignment codes to obtain a workable design. For a small machine with four rows or less in the flow table, it is possible to do an exhaustive manual search for an optimum state assignment. In the case of four rows there are only 12 distinct state assignment codes when using D flip-flops, and 3 distinct state assignment codes when using J-K flip-flops. A good rule of thumb is to assign state codes such that a minimum number of bits change (for each state change), if possible. The general state assignment problem is beyond the scope of this text. For information on different state assignment methods that can often lead to excitation input equations containing fewer literals, refer to McCluskey, Dietmeyer, and Rhyne in the references at the end of this chapter.

9-3-4   *D* Flip-Flop Design Procedure

After defining state variables and assigning codes to each state, a composite Karnaugh map (or separate Karnaugh maps if desired) is obtained for the next state and external outputs. As an example of the D flip-flop design procedure, we show the composite Karnaugh map for the next state and external outputs for M1 in Fig. 9-7a. Separate Karnaugh maps for each of the next state outputs and the Moore output can be drawn, if one prefers, as shown in Figs. 9-7b and c.

   These maps can be obtained from either the state diagram or the PS/NS table for M1. A set of reduced equations is obtained by reading either the composite Karnaugh map or the separate Karnaugh maps. This procedure is fine for D-type flip-flops, since the D excitation inputs are equal to the next state outputs. This is why D flip-flops are the most popular choice. A designer only has to obtain the reduced next state output equations and substitute these into the relationship $Di = Yi$. The reduced next state output equations can be written from the composite Karnaugh map for M1 as follows.

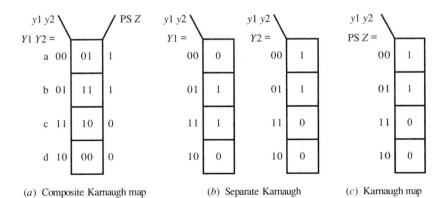

(*a*) Composite Karnaugh map for the next state and external outputs for M1.

(*b*) Separate Karnaugh maps for the next state ouputs for M1.

(*c*) Karnaugh map for the Moore output for M1.

FIGURE 9-7
Karnaugh maps for M1 when using D flip-flops.

$$Y1 = y2$$

$$Y2 = \overline{y1}$$

$$Z = \overline{y1}$$

For D-type flip-flops, $D1 = Y1 = y2$ and $D2 = Y2 = \overline{y1}$. Using the excitation input equations $D1$ and $D2$ and the Moore output equation Z, the synchronous circuit diagram for M1 can be drawn as shown in Fig. 9-8 using discrete devices (nonprogrammable devices).

The system clock is connected to the control input of both D flip-flops to insure that the flip-flops are driven synchronously by the same clock timing events. After the circuit is drawn, the excitation input signals $D1$ and $D2$ and the state variables $y1$, $\overline{y1}$, $y2$, and $\overline{y2}$ are usually removed from the final circuit diagram (the detailed logic diagram), since they are only used to obtain the circuit.

When working with programmable devices, the equations can be written by inspection from a state diagram or ASM chart using the 'set or hold method' (as illustrated in Example 9-1), and the software package can be instructed to handle the reduction process. This is the procedure most often followed for large problems, so that there is no need for awkwardly large, time consuming Karnaugh maps. If you will remember in Chapter 7 when we discussed PALs, registered types such as the PAL16R4 have four positive edge-triggered D flip-flops available inside the package, in addition to the programmable logic (see the circuit for the PAL16R4 in Fig. 7-25). Other registered PALs such as the PAL16RP8 (see circuit in Appendix C) have eight positive edge-triggered D flip-flops available with programmable polarity outputs, and the popular PALC22V10 (see Fig. 7-26) has ten positive edge-triggered D flip-flops available, also with programmable polarity outputs.

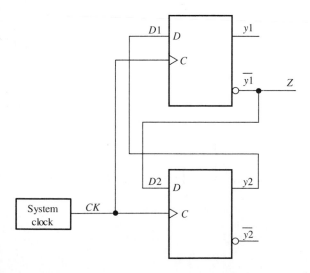

FIGURE 9-8
Synchronous circuit diagram for M1 using positive edge-triggered D flip-flops.

Example 9-7. Obtain the synchronous circuit equations for M3 using the state diagram in Fig. E9-4*b* or the ASM chart in Fig. E9-4*d*. Use the following state assignments: $y1\ y2 = 00$ for state d (the idle state), 01 for state a, 11 for state b, and 10 for state c.

(*a*) Obtain the reduced next state and external output equations using a composite Karnaugh map or separate Karnaugh maps, if desired.

(*b*) Obtain the next state and external output equations by inspection from the state diagram or ASM chart (use the 'set or hold method' for the next state outputs and the 1s for the Z output).

(*c*) Use the reduced set of equations to obtain a circuit diagram for M3 using positive edge-triggered *D* flip-flops.

Solution

(*a*) The flow map for M3 can be obtained as shown in Fig. E9-7*a*. The flow map is then used to obtain the composite Karnaugh map for M3 as shown in Fig. E9-7*b*. Separate maps can be drawn, if desired, or the composite map can be read in turn for each next state output Y1 and Y2 and the Mealy output Z.

Reading the composite map, we obtain the following next state output and Mealy output equations for the 1s of the functions.

$$Y1 = y2 \cdot X$$

$$Y2 = \overline{y1} \cdot X$$

$$Z = y2 \cdot X$$

(*b*) Since we have already discussed how to read the next state output equations from an ASM chart in Example 9-1, we will read the equations for M3 from the state diagram. The state diagram is shown in Fig. E9-7*c* with the state assignment for each state included.

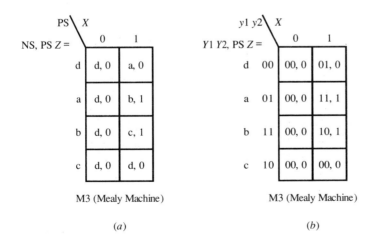

NS, PS Z =	PS \ X	0	1
	d	d, 0	a, 0
	a	d, 0	b, 1
	b	d, 0	c, 1
	c	d, 0	d, 0

M3 (Mealy Machine)

(*a*)

Y1 Y2, PS Z =		y1 y2 \ X	0	1
	d	00	00, 0	01, 0
	a	01	00, 0	11, 1
	b	11	00, 0	10, 1
	c	10	00, 0	00, 0

M3 (Mealy Machine)

(*b*)

FIGURE E9-7

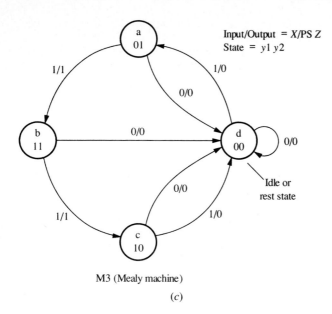

Input/Output $= X/PS\ Z$
State $= y1\ y2$

M3 (Mealy machine)

(c)

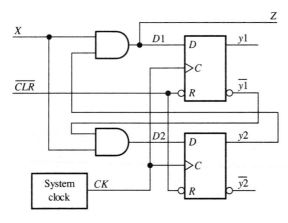

M3 (Mealy machine)

(d)   **FIGURE E9-7**

To provide a comparison, the output equations for the 1s of the functions $Y1$ and $Y2$, are obtained using the 'set or hold method'. For example, for $Y1$ we simply write every Boolean expression for each state that will result in a next state output $Y1$ of 1, and OR them all together. The explanation for each Boolean expression is provided to the right of the expression for documentation purposes.

$$Y1 = \overline{y1} \cdot y2 \cdot X \qquad\qquad Y1 = 1 \text{ when } \overline{y1} \cdot y2 = 1 \text{ AND } X = 1$$
$$\text{(set operation from state a to b)}$$

$$+ \ y1 \cdot y2 \cdot X \qquad\qquad Y1 = 1 \text{ when } y1 \cdot y2 = 1 \text{ AND } X = 1$$
$$\text{(hold operation from state b to c)}$$

$$Y2 = \overline{y1} \cdot \overline{y2} \cdot X \qquad Y2 = 1 \text{ when } \overline{y1} \cdot \overline{y2} = 1 \text{ AND } X = 1$$
(set operation from state d to a)

$$+ \ \overline{y1} \cdot y2 \cdot X \qquad Y2 = 1 \text{ when } \overline{y1} \cdot y2 = 1 \text{ AND } X = 1$$
(hold operation from state a to b)

Recall that the Mealy output Z is dependent on both the state of the machine and the external input X. To write the Mealy output Z for the 1s of the function simply involves locating each directed line segment that contains a $Z = 1$; write every Boolean expression for each state that will result in a PS output Z of 1; and OR them all together as shown below.

$$Z = \overline{y1} \cdot y2 \cdot X \qquad Z = 1 \text{ when } \overline{y1} \cdot y2 = 1 \text{ AND } X = 1$$

$$+ \ y1 \cdot y2 \cdot X \qquad Z = 1 \text{ when } y1 \cdot y2 = 1 \text{ AND } X = 1$$

Observe that these Boolean functions when reduced are the same as those obtained in (a) above.

(c) The circuit diagram for M3 is shown in Fig. E9-7d using discrete devices. Remember that $D1 = Y1$, and $D2 = Y2$ when using D flip-flops. Also notice that $Z = Y1 = D1$.

Notice that the state assignment was made so that M3 can be placed in the idle or rest state on the next timing event by changing X to 0. If the flip-flops have an asynchronously reset or clear input as shown in Fig. E9-7d, this input can also be used to place the machine in the idle state. The asynchronous signal $\overline{CLR}$ (NOT clear) is applied to the reset inputs. This signal is normally 1. When $\overline{CLR}$ is changed to 0; that is, $CLR = 1$ (also $R = 1$), the flip-flops are reset or cleared and remain reset until $\overline{CLR}$ is changed back to 1.

9-3-5 *T* Flip-Flop Design Procedure

After defining state variables and assigning codes to each state, a composite Karnaugh map (or separate Karnaugh maps if desired) is obtained for the next state and external outputs. From this map, a composite Karnaugh map (or separate Karnaugh maps if desired) is obtained for the T excitation inputs.

As an example for T flip-flops, the design procedure will be carried out for M1 (see the state diagram in Fig. 9-6). In Fig. 9-9 we show the composite Karnaugh map for the next state and external outputs for M1, and the composite Karnaugh map for the T excitation inputs.

The composite Karnaugh map for the next state and external outputs can be obtained from either the state diagram or the PS/NS table for M1. The composite Karnaugh map for the T excitation inputs is obtained using the composite Karnaugh map for the next state and external outputs and the relationship $Ti = yi \oplus Yi$. As observed in Fig. 9-2, the excitation input for a T flip-flop is the Exclusive OR of its present state and next state outputs; that is, $T = q \oplus Q$. The Exclusive OR relationship must be applied to the present state/next state transition for each state variable in the composite Karnaugh map for the next state and external outputs. For example, looking at the second row of the composite Karnaugh map for the next state and external outputs, we observe that $y1 \ y2 = 01$

$$Ti = yi \oplus Yi$$

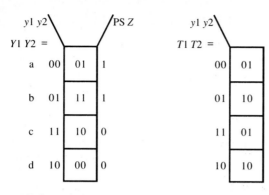

(a) Composite Karnaugh map for the next state and external outputs for M1.

(b) Composite Karnaugh map for the *T* excitation inputs for M1.

FIGURE 9-9
Composite Karnaugh maps for M1 when using *T* flip-flops.

and $Y1\ Y2 = 11$. This results in $T1\ T2 = 1(0 \oplus 1)\ 0(1 \oplus 1)$ in the second row of the composite Karnaugh map for the *T* excitation inputs. All other *T* excitation input values are determined in the same manner.

A set of reduced equations can now be obtained by reading the composite Karnaugh map (or separate maps if desired) for the *T* excitation inputs and the Karnaugh map for the next state and external outputs for the PS out put Z.

$$T1 = y1 \oplus y2$$
$$T2 = y1 \odot y2 = \overline{y1} \oplus y2 \qquad (P10: \text{see Section 7-6})$$
$$Z = \overline{y1}$$

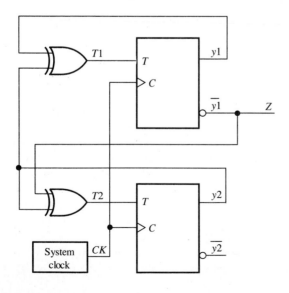

FIGURE 9-10
Synchronous circuit diagram for M1 using positive edge-triggered *T* flip-flops.

Using the excitation input equations $T1$ and $T2$ and the Moore output equation Z, the synchronous circuit diagram for M1 can be drawn as shown in Fig. 9-10 using positive edge-triggered T flip-flops (most often J-K flip-flops are used as T flip-flops).

The system clock is connected to the control input of both T flip-flops to insure that the flip-flops are driven synchronously by the same clock timing events.

Example 9-8. Design a synchronous sequential logic circuit that responds to the timing events of the system clock and produces the repetitive outputs listed in the following input-output table. Use positive edge-triggered T flip-flops in the design.

Timing events	Outputs
1	000
2	001
3	010
4	011
5	100
6	101
7	110
8	111
9	000
10	001
⋮	

(a) Obtain a Moore state diagram for the circuit.

(b) Choose a state assignment that requires minimum combinational logic for the Moore outputs.

(c) Show each step in the design process.

Solution

(a) The Moore state diagram is shown in Fig. E9-8a.

(b) Eight states are required to provide the eight different combinations of Moore outputs listed in the input-output table. For eight states, three flip-flops are required. State variables $y1$, $y2$, and $y3$ are used to represent the output states of the three bistable memory devices. A state assignment that results in minimum combinational logic is one in which the state assignment codes match the Moore binary output combinations. The Moore state diagram in Fig. E9-8b shows the state assignment such that $y1 = Z1$, $y2 = Z2$, and $y3 = Z3$.

(c) Using the Moore state diagram in Fig. E9-8b, the composite Karnaugh map for the next state outputs can be obtained as shown in Fig. E9-8c (the external outputs are not shown since they are equal to the next state outputs). The relationship $T = q \oplus Q$ is used to obtain the composite Karnaugh map for the T excitation inputs shown in Fig. E9-8d.

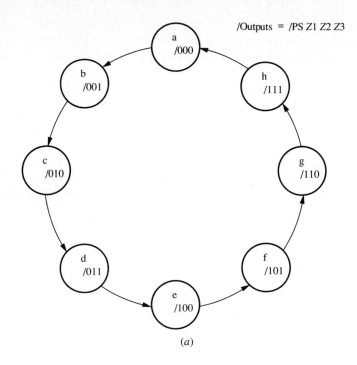

/Outputs = /PS Z1 Z2 Z3

(a)

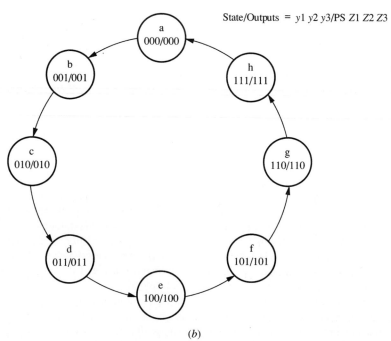

State/Outputs = y1 y2 y3/PS Z1 Z2 Z3

(b)

FIGURE E9-8

$$Ti = yi \oplus Yi$$

$Y1\ Y2\ Y3\ =$

y1 \ y2 y3	00	01	11	10
0	001	010	100	011
1	101	110	000	111

(c)

$T1\ T2\ T3\ =$

y1 \ y2 y3	00	01	11	10
0	001	011	111	001
1	001	011	111	001

(d)

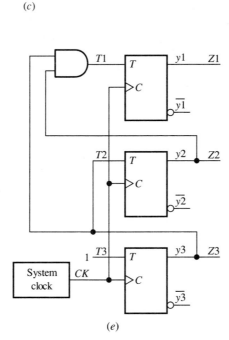

(e)

FIGURE E9-8

The T excitation input equations can be written as follows.

$$T1 = y2 \cdot y3$$

$$T2 = y3$$

$$T3 = 1$$

A synchronous circuit using positive edge-triggered T flip-flops to provide the correct input-output response is shown in Fig. E9-8e.

Since the circuit in Fig. E9-8e responds to the clock signal in binary counting sequence, the circuit is called a 3-bit binary counter. By drawing a timing diagram, one can show that the signal provided by Z1 is $\frac{1}{8}$ of the frequency of the clock signal, Z2 is $\frac{1}{4}$ of the frequency of the clock signal, and Z3 is $\frac{1}{2}$ of the frequency of the clock signal. This is left as an exercise for the student. Circuits with fixed input and output frequency ratios are also called frequency dividers or scalers.

As illustrated in Example 9-8, *T* flip-flops can also be used to design synchronous sequential circuits with a little more work. The state assignment chosen for a design can often be selected to reduce the input and/or output combinational logic to a manageable level using a particular type of flip-flop. The *D* flip-flop implementation for a 3-bit counter, even with the same state assignment, requires more combinational logic gates for the excitation inputs. The *D* flip-flop design is left as an exercise for the student.

9-3-6 *J-K* Flip-Flop Design Procedure

After defining state variables and assigning codes to each state, a composite Karnaugh map (or separate Karnaugh maps if desired) is obtained for the next state and external outputs. From this map, a composite Karnaugh map for the *J* excitation inputs and one for the *K* excitation inputs (or separate maps if desired) are obtained.

As an example for *J-K* flip-flops, the design procedure will be carried out for M1 (see the state diagram in Fig. 9-6). In Fig. 9-11 we show the composite Karnaugh map for the next state and external outputs for M1, and the composite Karnaugh maps for the *J* and *K* excitation inputs.

The composite Karnaugh map for the next state and external outputs can be obtained from either the state diagram or the PS/NS table for M1. The composite Karnaugh maps for the *J* and *K* excitation inputs are obtained using the composite Karnaugh map for the next state and external outputs, and the excitation table for

y_i	Y_i	J_i	K_i
0	0	0	X
0	1	1	X
1	0	X	1
1	1	X	0

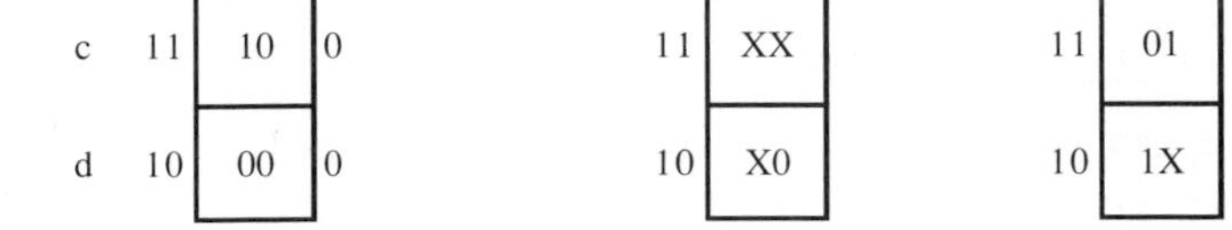

(*a*) Composite Karnaugh map for the next state and external outputs for M1.

(*b*) Composite Karnaugh maps for the *J* and *K* excitation inputs for M1.

FIGURE 9-11
Composite Karnaugh maps for M1 when using *J-K* flip flops.

the *J-K* flip-flop. It is helpful to first write the excitation table for the *J-K* flip-flop, as shown in Fig. 9-11*b*. Then use it as a reference. For example, looking at the first row of the composite Karnaugh map for the next state and external outputs, we observe that $y1\ y2 = 00$ and $Y1\ Y2 = 01$. This results in $J1\ K1 = 0X$ ($y1 = 0$ to $Y1 = 0$) and $J2\ K2 = 1X$ ($y2 = 0$ to $Y2 = 1$) by observing the excitation table for the *J-K* flip-flop. These values are placed in the appropriate *J* and *K* columns of the separate composite Karnaugh maps for the excitation inputs. All other *J-K* excitation input values are determined in the same manner.

A set of reduced equations can now be obtained by reading the composite Karnaugh maps (or separate maps if desired) for the *J* and *K* excitation inputs, and the Karnaugh map for the next state and external outputs for the PS output Z.

$$J1 = y2$$
$$J2 = \overline{y1}$$
$$K1 = \overline{y2}$$
$$K2 = y1$$
$$Z = \overline{y1}$$

Using the excitation input equations for $J1$, $J2$, $K1$, $K2$, and the Moore output equation for Z, the synchronous circuit diagram for M1 can be drawn as shown in Fig. 9-12 using positive edge-triggered *J-K* flip-flops.

The system clock is connected to the control input of both *J-K* flip-flops to insure that the flip-flops are driven synchronously by the same clock timing events.

Example 9-9. A circuit that is used to supervise the operation of other circuits is often referred to as a controller. The bus arbiter circuit like the one shown in Fig. E9-9*a* is an example of a simple controller. The bus arbiter circuit in Fig. E9-9*a* has two input signals *RA* and *RB* (which represent bus request *A* and request *B*) and two output signals *EA* and *EB* (which represent bus enable *A* and enable *B*). The

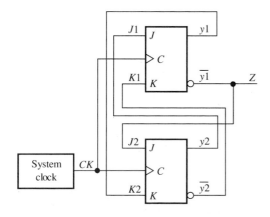

FIGURE 9-12
Synchronous circuit diagram for M1 using positive edge-triggered *J-K* flip-flops.

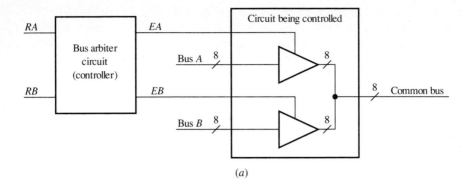

(a)

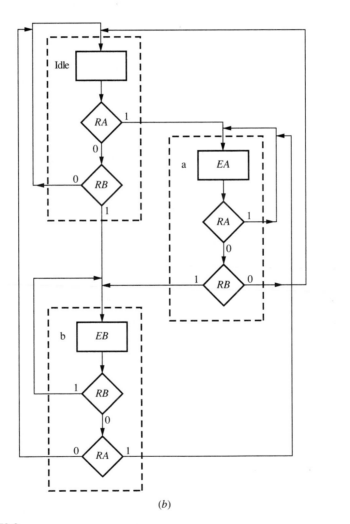

(b)

FIGURE E9-9

PS	NS Inputs *RA RB*				PS outputs *EA EB*
	00	01	11	10	
Idle	Idle	b	a	a	00
a	Idle	b	a	a	10
b	Idle	b	b	a	01
c	Idle	-	-	-	- -

- = don't care

(c)

y_i	Y_i	J_i	K_i
0	0	0	X
0	1	1	X
1	0	X	1
1	1	X	0

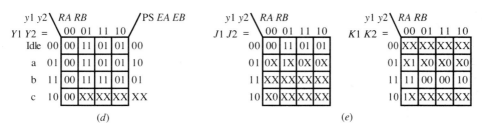

$Y_1\ Y_2 =$

$y_1\ y_2$ \ *RA RB*	00	01	11	10	PS *EA EB*
Idle 00	00	11	01	01	00
a 01	00	11	01	01	10
b 11	00	11	11	01	01
c 10	00	XX	XX	XX	XX

(d)

$J_1\ J_2 =$

$y_1\ y_2$ \ *RA RB*	00	01	11	10
00	00	11	01	01
01	0X	1X	0X	0X
11	XX	XX	XX	XX
10	X0	XX	XX	XX

$K_1\ K_2 =$

$y_1\ y_2$ \ *RA RB*	00	01	11	10
00	XX	XX	XX	XX
01	X1	X0	X0	X0
11	11	00	00	10
10	1X	XX	XX	XX

(e)

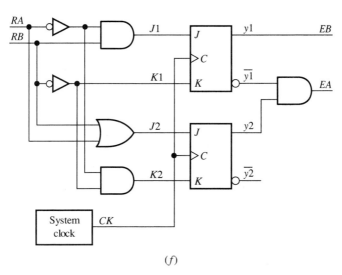

(f)

FIGURE E9-9

bus enable output signals *EA* and *EB* are used to control the three state inputs of three state buffers in the circuit being controlled. Only one bus at a time can be enabled when it is requested; otherwise, a bus conflict results when bus *A* and bus *B* try to share the common bus at the same time.

Design the bus arbiter circuit so that it behaves in the following manner. For inputs *RA RB* = 00 the circuit either goes to the idle state (outputs *EA EB* = 00 disabling both bus *A* and bus *B* buffers in the circuit being controlled), or remains in the idle state. When the inputs *RA RB* change to 10 or 11 the circuit goes to state a (outputs *EA EB* = 10 enabling the bus *A* buffers and disabling the bus *B* buffers). When the inputs *RA RB* change to 01 while in the idle state, the circuit goes to state b (outputs *EA EB* = 01 disabling bus *A* buffers and enabling bus *B* buffers). To go from state a to state b requires inputs *RA RB* = 01, and to go from state b to state a requires inputs *RA RB* = 10.

(*a*) Obtain an ASM chart for the bus arbiter circuit design.

(*b*) Obtain a PS/NS table including the Moore outputs for the bus arbiter circuit design.

(*c*) Obtain the composite Karnaugh map for the next state and external outputs using the ASM chart or the PS/NS table and the following state assignments: *y*1 *y*2 = 00 for the idle state, 01 for state a, 11 for state b.

(*d*) Obtain the composite Karnaugh maps for the *J* and *K* excitation inputs. Write the excitation input and external output equations for the design.

(*e*) Draw the circuit implementation for the bus arbiter using positive edge-triggered *J-K* flip-flops.

Solution

(*a*) An ASM chart for the bus arbiter circuit is shown in Fig. E9-9*b*. Notice that the two output signals *EA* and *EB* are only a function of the states and are therefore Moore type outputs.

(*b*) The PS/NS table is shown in Fig. E9-9*c*. For the bus arbiter circuit to be self-starting (or self-correcting), it must be able to begin in any state and return to the intended range of operating states. The intended range of operating states for the bus arbiter includes the idle state, state a, and state b. Since three states are required in the design, two flip-flops must be used. For two flip-flops there are four states. If state c (the fourth state) cannot change to one of the states in the intended operating range, the machine can hang up and not function properly. For inputs *RA RB* = 00 we show the next state of c as the idle state so that the machine will surely be self-starting. All the rest of the next state values for state c, including the output, can be listed as don't cares as shown.

(*c*) The composite Karnaugh map for the next state and external outputs is shown in Fig. E9-9*d*. The fourth state, c, has been given the state assignment code 10.

(*d*) The composite Karnaugh maps for the *J* and *K* excitation inputs are shown in Fig. E9-9*e*. Reading the composite Karnaugh maps, we obtain the following excitation input and external output equations.

$$J1 = \overline{RA} \cdot RB$$

$$J2 = RA + RB$$

$$K1 = \overline{RB}$$

$$K2 = \overline{RA} \cdot \overline{RB}$$

$$EA = \overline{y1} \cdot y2$$

$$EB = y1$$

(e) The circuit implementation for the bus arbiter using positive edge-triggered *J-K* flip-flops is shown in Fig. E9-9f.

9-4 SYNCHRONOUS COUNTER DESIGN

Practically all digital logic systems utilize counters in their design. Digital instruments, for example, use counters to control the precise measurement of the number of events per unit time. Digital computers, on the other hand, use counters to control both the sequence and execution of programming steps. The synchronous sequential circuit M1 designed in Section 9-3 is a form of synchronous counter used to perform frequency scaling. The synchronous binary counter circuit illustrated in Fig. E9-8a has the most common counting sequence—straight binary; however, a counter can be designed to have any counting sequence. Counter circuits are classified as either synchronous or asynchronous as illustrated in Fig. 9-13.

9-4-1 Synchronous Counters Versus Asynchronous Counters

In synchronous circuits the system clock drives the control inputs to all the flip-flops in the design. As a result, all the flip-flops in a synchronous counter design change state at the same time. An example of an asynchronous 3-bit binary counter is shown in Fig. 9-14a. Notice that the control inputs are not driven by the same signal source, but by each succeeding flip-flop output. In general, synchronous counters have superior frequency performance (operate at higher frequencies) compared to asynchronous counters.

Asynchronous binary counters are easy to design, provided that a straight binary sequence is followed; otherwise, the design procedure is highly dependent on the creativity and ingenuity of the designer. Since each flip-flop in the design is triggered in a delayed ripple fashion, as illustrated in the timing diagram in Fig. 9-14b, the next state of the counter is delayed until all the flip-flops reach a stable state. For this reason, asynchronous counters are also called ripple counters. The ripple effect causes false or momentary states that can cause output decoding

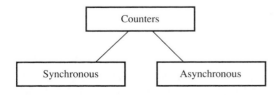

FIGURE 9-13
The classification of counter circuits.

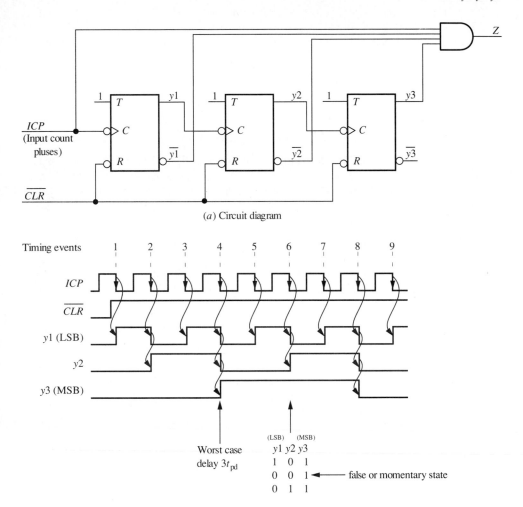

$$Z = ICP \cdot \overline{y1} \cdot \overline{y2} \cdot y3$$

(*a*) Circuit diagram

(*b*) Timing diagram for asynchronous 3-bit binary counter showing worst case delay (settling time) at timing event 4, and false or momentary state $y1\ y2\ y3$ = 001 at timing event 6.

FIGURE 9-14
Asynchronous 3-bit binary counter.

glitches. The output decoding circuit (see output Z in Fig. 9-14*a*) must include a qualifying signal, such as the signal *ICP* (where ICP is an abbreviation for input count pulses), that only enables the output decoding circuit after all the flip-flops have reached a steady state value; otherwise, a logic 1 glitch (a momentary positive pulse) could occur at timing event 6 (see Fig. 9-14*b*). If a decoded output without such a qualifying signal is used to drive the edge-sensitive control input of another device, a hazard or glitch (momentary erroneous pulse) can occur in the decoded output signal. This will most likely cause false triggering of the device being driven by this signal. Since problems of this nature do not normally occur in synchronous designs, asynchronous counter designs are not utilized as often as synchronous counter designs. If a designer uses off-the-shelf MSI asynchronous counters, the designer must be aware of the existence of false or momentary states and consider ways of dealing with decoding glitches.

9-4-2 Designing Synchronous Counters

A basic synchronous binary counter stores a binary code or number and increments the binary number with each clock timing event. A counter is sometimes described by the number of bits (or flip-flops) it contains, such as a 3-bit counter. A counter can be described by the number of states in the primary or main counting sequence, such as a modulo 5 counter (also called a divide by 5 counter). Some shift register counters have specific names that identify the counter, such as a ring counter, a twisted ring counter (switch-tail, Johnson, or Moebius counter), and a maximum length shift counter.

Like most synchronous sequential circuits, it is important that counters be self-correcting or self-starting; otherwise, if the counter for some unexpected reason finds itself in a state that does not have a path to the primary counting sequence (perhaps due to a temporary power glitch), the counter can hang up and fail to operate. When all the states of a counter are being used in the primary counting sequence, a counter will always be self-correcting. An n-bit counter with 2^n states that counts in straight binary is called a binary counter. A binary counter is self-correcting because all the states appear in the primary counting sequence. To guarantee proper circuit operation, an n-bit counter with less than 2^n states in its primary counting sequence must be designed to be self-correcting. This is accomplished by ensuring that any unused or illegal states have a path back to the primary counting sequence. This is sometimes referred to as illegal state recovery. Designing counters to be self-correcting (or have illegal state recovery) is particularly important when using T and J-K flip-flops. Since the output of a D flip-flop goes to the reset condition ($Q = 0$) when the D input is 0, D flip-flops are generally better suited than T and J-K flip-flops for illegal state recovery. Illegal state recovery can be aided by providing an all zeros state in the primary counting sequence, thus allowing the D flip-flops in the design to return to this state.

An example of a decade counter is shown in Fig. 9-15. This counter counts from 0 to 9 in binary, and can also be called a BCD counter. Since four bits are required to achieve a binary 9 (1001), there are a total of 16 states, 10 states in the primary counting sequence and 6 states unused. The counter has an external input X that allows it to count up when X is 1 and down when X is 0.

Synchronous counters can be designed to follow any desired output sequence. Once the output sequence is specified, state reduction cannot usually be performed, since the number of states is fixed by the output sequence. Counters are most often designed as Moore machines to minimize the output circuit complexity. Reduced circuit complexity is achieved by equating the state of the circuit to the Moore output variables. In Fig. 9-15 we can make the following

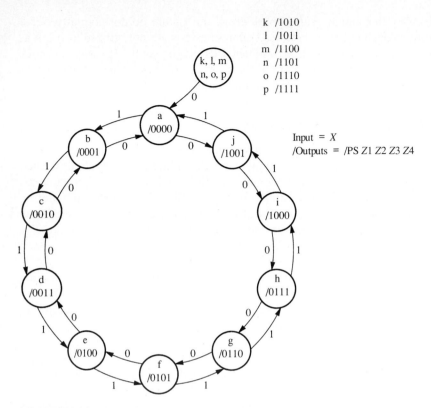

FIGURE 9-15
State diagram for synchronous BCD up-down counter.

state assignment to simplify the design $y1 = Z1$, $y2 = Z2$, $y3 = Z3$, $y4 = Z4$. Using the state diagram, one can obtain the composite Karnaugh map for the next state outputs and the composite Karnaugh map for the excitation inputs for T flip-flops as shown in Fig. 9-16.

Using the composite Karnaugh map in Fig. 9-16b we can write the following T excitation input equations for the BCD up-down counter.

$$T1 = \overline{X} \cdot \overline{y2} \cdot \overline{y3} \cdot \overline{y4} + y1 \cdot y2 + y1 \cdot y3 + X \cdot y2 \cdot y3 \cdot y4 + X \cdot y1 \cdot y4$$

$$T2 = \overline{X} \cdot y2 \cdot \overline{y3} \cdot \overline{y4} + \overline{X} \cdot y1 \cdot \overline{y3} \cdot y4 + y1 \cdot y2 + X \cdot y3 \cdot y4$$

$$T3 = \overline{X} \cdot \overline{y1} \cdot y2 \cdot \overline{y4} + \overline{X} \cdot y1 \cdot \overline{y2} \cdot y4 + \overline{X} \cdot y3 \cdot \overline{y4} + y1 \cdot y3 + X \cdot \overline{y1} \cdot y4$$

$$T4 = \overline{y1} + y4 + \overline{y2} \cdot \overline{y3}$$

A synchronous circuit diagram for the BCD up-down counter can now be drawn according to the T excitation input equations using edge-triggered T flip-flops (or J-K flip-flops used as T flip-flops). This exercise is left for the student.

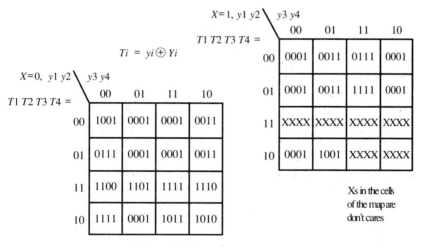

(a) Composite Karnaugh map for the
next state outputs.

$$Ti = yi \oplus Yi$$

(b) Composite Karnaugh map for the
T excitation inputs.

FIGURE 9-16
Composite Karnaugh maps for synchronous BCD up-down counter.

To save design time, it is advised that data books be consulted for the logic family one intends to use to see if the MSI counter needed already exists. Many single-package MSI designs exist in both synchronous and asynchronous types that may be available as an off-the-shelf device to solve the designer's problem. See Appendix B for a list of available Digital Logic Products from Texas Instruments. Several counters are listed in this reference section.

Example 9-10. Design a 2-bit synchronous counter with a binary output sequence. The counter is to count forward or up when the external input signal X is 0, and backward or down when the external input signal $X = 1$. The PS/NS table is listed as follows.

PS	NS X=0	X=1	PS outputs Z1 Z1
a	b	d	00
b	c	a	01
c	d	b	10
d	a	c	11

(a) Show both an ASM chart and a state diagram for the design.

(b) Design the counter using both D and T flip-flops. Use the 'set or hold method' to obtain the Boolean equations for the D flip-flop design.

(c) Obtain the circuit diagram for the implementation with the least complexity.

Solution

(a) An ASM chart for the counter is shown in Fig. E9-10a. An equivalent state diagram is shown in Fig. E9-10b. The state diagram has been modified slightly, indicating state transitions for $X = 0$ with the expression $\overline{X}$ and state transitions for $X = 1$ with the expression X. Some designers prefer to show Boolean expressions for the external input conditions in this manner on state diagrams rather than list the external inputs and separately show their corresponding 0 and 1 values. A state change occurs in Fig. E9-10b from state a to state b when the input condition $\overline{X}$ is true ($X = 0$).

The state assignment is made such that $y1 = Z1$, $y2 = Z2$ which results in minimum output decoding (none).

(b) The excitation input equations for the D flip-flop implementation can be written directly from either the ASM chart or the state diagram using the 'set or hold method' as follows.

$$D1 = Y1 = \overline{y1} \cdot \overline{y2} \cdot X \qquad \text{set operation from state a to d}$$
$$+ \ \overline{y1} \cdot y2 \cdot \overline{X} \qquad \text{set operation from state b to c}$$
$$+ \ y1 \cdot \overline{y2} \cdot \overline{X} \qquad \text{hold operation from state c to d}$$
$$+ \ y1 \cdot y2 \cdot X \qquad \text{hold operation from state d to c}$$
$$D2 = Y2 = \overline{y1} \cdot \overline{y2} \cdot \overline{X} \qquad \text{set operation from state a to b}$$
$$+ \ \overline{y1} \cdot \overline{y2} \cdot X \qquad \text{set operation from state a to d}$$
$$+ \ y1 \cdot \overline{y2} \cdot X \qquad \text{set operation from state c to b}$$
$$+ \ y1 \cdot \overline{y2} \cdot \overline{X} \qquad \text{set operation from state c to d}$$

These equations can be reduced by Boolean algebra as follows.

$$D1 = \overline{(y1 \oplus y2)} \cdot X \ + \ y1 \oplus y2) \cdot \overline{X}$$

$$= y1 \oplus y2 \oplus X$$

$$D2 = \overline{y2} \cdot \overline{X} \ + \ \overline{y2} \cdot X$$

$$= \overline{y2}$$

The composite Karnaugh map for the next state outputs is shown in Fig. E9-10c (the external outputs are the same as the present state outputs and need

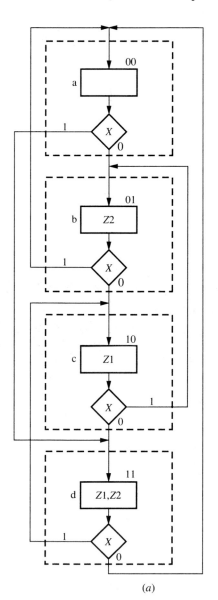

(a)

FIGURE E9-10

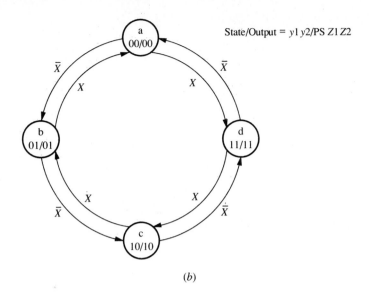

State/Output = $y1\,y2/PS\ Z1\ Z2$

(b)

$$Ti = yi \oplus Yi$$

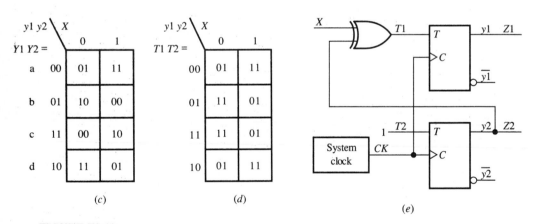

(c) (d) (e)

FIGURE E9-10

not be shown). The composite Karnaugh map for the T excitation inputs is shown in Fig. E9-10d.

Reading the composite Karnaugh map for the T excitation inputs, we can write the T excitation input equations as follows.

$$T1 = y2 \cdot \overline{X} + \overline{y2} \cdot X = y2 \oplus X$$

$$T2 = 1$$

(c) By observing the excitation equations, we can see that the T implementation has the least complexity. The circuit implementation using positive edge-triggered T flip-flops is shown in Fig. E9-10e.

9-4-3 Shift Register Counters

A single edge-triggered bistable device or flip-flop is the simplest form of register. Combinations of several flip-flops and associated control logic are used in the design of multiple bit registers. Simple storage registers consist of several flip-flops (usually 8) often controlled by a single control signal. These types of registers allow temporary data storage for binary words or addresses. Shift registers are another type of register that consist of several flip-flops (4, 5, 8, and 16 are common) with control logic to allow not only data storage but data manipulation operations such as shift right or shift left. Data can be loaded into the flip-flops of the shift register in the following ways depending on the design: parallel-in parallel-out (called broadside), parallel-in serial-out, serial-in parallel-out, and serial-in serial-out. In normal operation a shift register shifts data from flip-flop to flip-flop in the register, synchronized with the clock timing events. The flip-flops used in shift registers are normally D-type or S-R type used as D-type.

There are three basic types of shift register counters: (*a*) ring counters, (*b*) twisted ring counters (also called switch-tail ring counters, Johnson counters, or Moebius counters), and (*c*) maximum length shift counters. Ring counters are designed with a shift register connecting the output signal of the last flip-flop in the shift register back to the input of the first flip-flop in the register. If the output signal of the last flip-flop is first inverted and then fed back, the counter is a twisted ring counter. If signals from the last two flip-flop outputs are fed back to the input of the first flip-flop via an Exclusive OR or Exclusive NOR gate, then the counter is called a maximum length shift counter. Different counting sequences can be obtained with shift register counters depending on the number of bits used in the counter and the type of feedback used. All three types of shift register counters mentioned above are not self-correcting, and thus require ingenuity to determine illegal state recovery circuitry. This is the main problem with shift register counters in general. Any design using a shift register counter should include illegal state recovery circuitry.

State assignment codes for ring counters and twisted ring counters are fixed by the circuit configuration. The codes that result in the primary counting sequence for ring counters and twisted ring counters are balanced or symmetrical. As a result, these counters are well-suited for generating timing pulses. No output decoding circuitry is required for a ring counter, and minimum output decoding is required for the twisted ring counter.

9-4-4 Ring Counter Design

A basic 3-bit ring counter design is shown in Fig. 9-17*a*. This circuit has the counting sequences illustrated by the state diagram in Fig. 9-17*b*. For n flip-flops (or 3 flip-flops), the number of states in the primary counting sequence is n (or 3). This leaves $2^n - n$ (or 5) unused or illegal states. The basic counter is not self-correcting and, at power up, may even find itself in an illegal state and never initiate the primary counting sequence. The primary counting sequence is the one

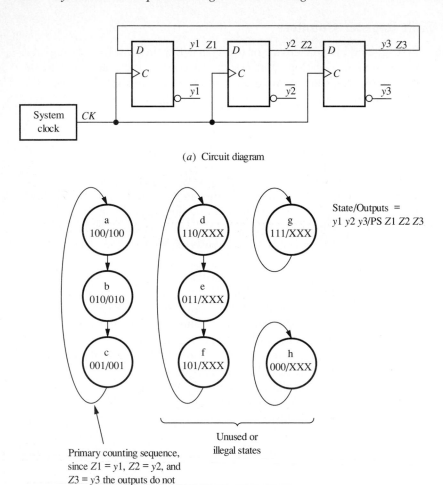

(*a*) Circuit diagram

State/Outputs =
y1 y2 y3/PS Z1 Z2 Z3

Primary counting sequence,
since Z1 = y1, Z2 = y2, and
Z3 = y3 the outputs do not
need to be decoded

Unused or
illegal states

(*b*) Counting sequences

FIGURE 9-17
Basic 3-bit ring counter.

that does not need output decoding circuitry if each Moore output is a flip-flop outputs; that is, $Z1 = y1$, $Z2 = y2$, $Z3 = y3$. To provide illegal state recovery, the circuit can be changed so that $D1 = \overline{y1} \cdot \overline{y2}$. With this simple addition, the circuit will always find the primary counting sequence should it end up in an unused or illegal state. The recovery circuit works as follows. The feedback input $D1$ will be 0 until $y1\, y2 = 00$ at which time $D1$ will be 1. This logic will force all the illegal states to systematically return to the primary counting sequence. For example, if state 011 is entered, the next state would normally be 101 without the recovery logic; however, with the recovery logic, the feedback signal will force the next state to be 001, which is in the primary counting sequence.

Obtaining the state diagram with the illegal state recovery circuitry added is left as an exercise for the student.

9-4-5   Twisted Ring Counter Design

A basic 3-bit twisted ring counter is shown in Fig. 9-18*a*. For n flip-flops (or 3 flip-flops) the number of states in the primary counting sequence is $2n$ (or 6). This leaves $2^n - 2n$ (or 2) unused or illegal states. The state diagram for this circuit is shown in Fig. 9-18*b*. To provide a separate output for any state, the state needs to be decoded; however, only a 2-input AND gate is required to decode any state as shown in Fig. 9-19 for Moore output Z1 for state e. To provide illegal state recovery, the basic 3-bit twisted ring counter can be changed so that $D1 = \overline{y3} + y1 \cdot \overline{y2} \cdot y3$. With this addition, the circuit will always recover to the primary counting sequence if the circuit should find itself in states $y1\ y2\ y3$ $= 010$ or 101. In the equation for $D1$, $\overline{y3}$ is the complemented feedback signal that makes the counter a twisted ring counter, as opposed to just a ring counter. The expression $y1 \cdot \overline{y2} \cdot y3$ is the recovery part of the equation for $D1$. Analysis of the circuit with illegal state recovery is left as an exercise for the student.

9-4-6   Maximum Length Shift Counter Design

An example of a maximum length shift counter is illustrated in Fig. 9-20. For n flip-flops (or 4 flip-flops), the number of states in the primary counting sequence is $2^n - 1$ (or 15). This leaves 1 unused or illegal state. When an Exclusive OR function is used in the feedback circuit as illustrated in Fig. 9-20, a 0 is fed back to the input if the circuit finds itself in the 0000 state, and the circuit cannot recover. One method allowing illegal state recovery includes decoding this state and using the decoded output to parallel load 1111. Another method is to decode the 0000 state and AND the decoded output with the CK signal to synchronously *set* all the flip-flops on the next clock timing event. When an Exclusive NOR function is used in the feedback circuit, a 1 is fed back to the input if the circuit finds itself in the 1111 state, and the circuit cannot recover. In this case, the correction circuit must decode the illegal state of 1111 and either use the decoded output to parallel load 0000 or decode the 1111 state and AND the decoded output with the CK signal to synchronously *reset* all the flip-flops on the next clock timing event. The state diagrams for the Exclusive OR and Exclusive NOR forms of maximum length shift counters are left as exercises for the student.

> **Example 9-11.** Use a shift register counter to design a circuit to provide the output signals illustrated by the timing diagram in Fig. E9-11*a*. Output signals Z1 through Z4 shown in the timing diagram are used to control another circuit in a digital system. The circuit must operate as error-free as possible; illegal state recovery is a required feature of the circuit.
>
> **Solution** A state diagram showing the required output sequence is drawn in Fig. E9-11*b*. The Moore output sequence is the same as the state sequence provided by

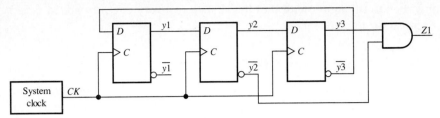

(*a*) Circuit diagram

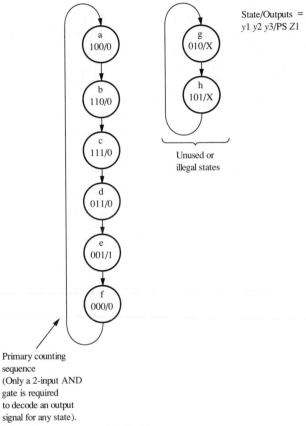

State/Outputs =
y1 y2 y3/PS Z1

Unused or
illegal states

Primary counting
sequence
(Only a 2-input AND
gate is required
to decode an output
signal for any state).

(*b*) Counting sequences

FIGURE 9-18
Basic 3-bit twisted ring counter.

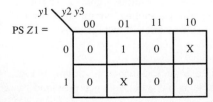

y1 \ y2 y3				
PS Z1 =	00	01	11	10
0	0	1	0	X
1	0	X	0	0

FIGURE 9-19
Karnaugh map for Moore output Z1 for state e.

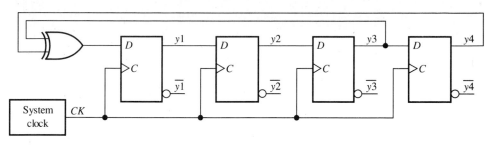

FIGURE 9-20
Exclusive OR form of maximum length shift counter.

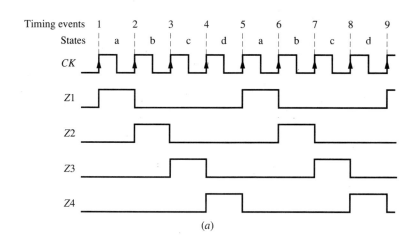

(*a*)

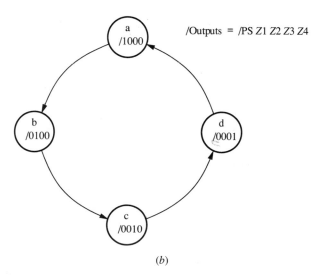

(*b*)

FIGURE E9-11

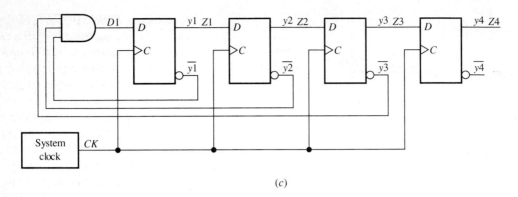

(c)

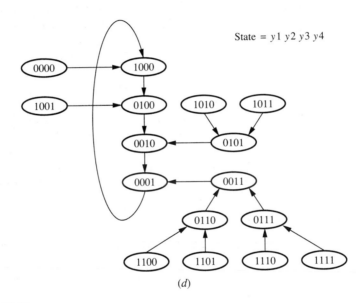

State = $y1\ y2\ y3\ y4$

(d)

FIGURE E9-11

a 4-bit ring counter if each Moore output is equated to one of the state variables so that $Z1 = y1$, $Z2 = y2$, $Z3 = y3$, and $Z4 = y4$; therefore, with illegal state recovery circuitry, a 4-bit ring counter can be used to generate the required output signals.

A 4-bit ring counter has four flip-flops, four states in its primary counting sequence, and $2^n - n$ (or 12) illegal states. Providing the basic 4-bit ring counter with a feedback signal $D1 = \overline{y1 \cdot y2 \cdot y3}$ supplied to the first flip-flop will cause all illegal states to return to the main counting sequence. A 4-bit ring counter implementation for the circuit is shown in Fig. E9-11c.

A simplified state diagram is shown for the circuit in Fig. E9-11d. The state diagram can be obtained by analyzing the circuit in Fig. E9-11c to verify that all illegal states have a path to the primary counting sequence.

Notice that when the circuit is first turned on, it takes at most three timing events (three clock cycles) to place the circuit in its primary counting sequence.

9-5 DESIGNING SYNCHRONOUS STATE MACHINES USING PROGRAMMABLE DEVICES

In this section we introduce synchronous sequential logic circuit design using the software package PLDesigner by Minc. A system flow chart for PLDesigner is shown in Fig. 9-21. PLDesigner illustrates the convenience and versatility of available software designed to increase the productivity of the logic designer. The PLDesigner software package can be used for both combinational and sequential designs (see problems at the end of this chapter), but in this section we will only present examples for synchronous sequential designs.

Recall in Section 7-8 that it is necessary when using PALASM and most other software packages to select the device prior to starting the design. It is the designer's responsibility to have a good working knowledge of each type of programmable logic device to insure that the reduced equations fit into the chosen device. If the reduced equations do not fit either logically or physically, then the designer must partition the equations so that they will fit into multiple devices. PLDesigner is one of the few software packages available that separates the 'design phase' from the 'device-selection phase'. Various software tools are used to implement a design using PLDesigner. Using the 'CONFIGURATION'

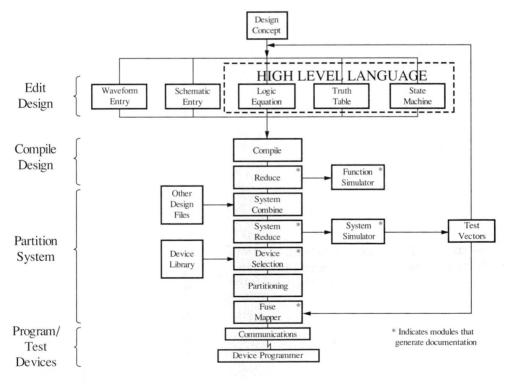

FIGURE 9-21
PLDesigner System Flow chart (Courtesy Mine Incorporated)

tool, the following four equation reduction levels are provided: Sum-of-Products, Espresso, Espresso/Exact, and Quine-McCluskey. Espresso is a trademark of the University of California at Berkeley. The design phase is quite flexible and allows the designer to edit or create a design with the 'EDIT DESIGN' tool (see Edit Design in Fig. 9-21) using the following entry methods: waveform, schematic, logic equation, truth table, and state machine. Once the 'design file' is obtained using one of these methods, it is compiled using the 'COMPILE DESIGN' tool to obtain the reduced equations (see Compile Design in Fig. 9-21). Functional simulation is also provided by PLDesigner as shown in the system flowchart in Fig. 9-21, but this will not be discussed. Device selection is not considered until after the design phase.

After the design phase, the system is partitioned using the 'PARTITION SYSTEM' tool to select possible solutions (see Partition System in Fig. 9-21) based on the 'available file' and on 'constraints' specified by the designer. The available file contains those devices the designer prefers to use, while the constraints consist of such categories as manufacturer, logic family, number of devices, package type, temperature, maximum current usage, and maximum propagation delay, just to name a few. Devices that do not meet the constraints are not selected for possible implementation. The constraints are prioritized by the designer with a 1 representing a low priority and a 10 representing a high priority. A device is then selected by the designer from among the possible ordered solutions presented by PLDesigner after partitioning. Once the device is selected, the fuse map and the pinout diagram is generated and appended to the documentation [.DOC] file for the design. The fuse map is either Internal, JEDEC, ASCII, or INTEL format, and the desired format is specified with the 'CONFIGURATION' tool prior to using the 'PARTITION SYSTEM' tool. The fuse map can then be used to program the device using the 'PROGRAM/TEST DEVICES' tool (see Program/Test Devices in Fig. 9-21). PLDesigner Version 1.2 supports the Data I/O 29A and 29B device programmers. A generic interface is also provided to support device programmers from other manufacturers.

A designer does not have to understand each and every PLD in order to design with an advanced software package such as PLDesigner. This allows a designer to concentrate on the design phase, and lets PLDesigner provide the transition from the design phase to actual PLD circuit implementation.

9-5-1 Design Examples Using PLDesigner

The following examples will illustrate the procedure for creating synchronous design files using the logic equation, truth table, state machine, and waveform entry methods. This is the design phase of PLDesigner. After the design file is obtained, we will show the section of the documentation file containing the design equations created by the language compiler. In each of the following examples we used PLDesigner V1.2, and the Quine-McCluskey equation reduction level was selected. In the following examples we treat each entry method as a separate entity. The fact that different design files for text entry and waveform entry can be combined to create one file that can then be partitioned into one or more devices as a single design is one of the outstanding features of PLDesigner.

9-5-2   Logic Equation Entry Method

Example 9-12. Figure E9-12a shows the state diagram for synchronous machine M1 presented earlier in Fig. 9-6. Recall that this is a Moore machine that has an output Z that is $\frac{1}{4}$ the frequency of the system clock signal.

Obtain a logic equation design file for the synchronous machine M1 shown in Fig. E9-12a using the text editor provided by the PLDesigner software package. Also, obtain the documentation file containing the reduced equations for M1.

Solution A logic equation design file called M1_LE.SRC can be created using the text editor as shown in Fig. E9-12b (the extension .SRC stands for the SouRCe code of the design file). To access the text editor to create the design file, the 'EDIT DESIGN' tool is used. Either the Equation [.SRC], State Machine [.SRC], or Truth Table [.SRC] option can be selected, since the same editor is used for all forms of text entry.

The format for a design file using text entry is quite similar to the Pascal programming language. Each entry statement is terminated by a semicolon, and comments are preceded by a double quote. The "set or hold method" is used to write the logic equations for the next state variables Y1 and Y2 directly from the state diagram for M1. The Moore output Z is written in terms of the present state variables y1 and y2 for the 1s of the function.

After obtaining the logic equation design file, M1_LE.SRC, the next step is to compile the file using the 'COMPILE DESIGN' tool. After the reduced equations are generated by the compiler, they are placed in the documentation file M1_LE (the M1_LE.DOC file). The documentation file is accessed via the EDIT DESIGN

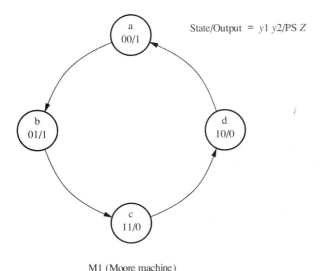

State/Output = y1 y2/PS Z

M1 (Moore machine)

(a)

FIGURE E9-12

```
┌══════════ PLDesigner V1.2 - MINC Copyright 1987-88 - Text Editor ══════════┐
│                                             Press <ESC> for command line    │
│   Window:1 File:C:\MINCWORK\M1_LE.SRC            Line:1    Col:1    AI INS   │
└═════════════════════════════════════════════════════════════════════════════┘
```

```
"Design file
TITLE               State Machine M1;
COMMENT             Logic Equation entry format;

FUNCTION M1_LE;

INPUT CK;

OUTPUT y1, y2 CLOCKED_BY CK;
OUTPUT Z;

        Y1 = /y1*y2                "set
           +  y1*y2;               "hold
        Y2 = /y1*/y2               "set
           + /y1*y2;               "hold
         Z = /y1*/y2 + /y1*y2;

END M1_LE;
```

(*b*)

```
┌══════════ PLDesigner V1.2 - MINC Copyright 1987-88 - Text Editor ══════════┐
│                                             Press <ESC> for command line    │
│   Window:1 File:C:\MINCWORK\M1_LE.DOC            Line:55   Col:1    AI INS   │
└═════════════════════════════════════════════════════════════════════════════┘
```

```
EQUATIONS FOR FUNCTION M1_LE

INPUT SIGNALS:
        HIGH_TRUE   CK

OUTPUT SIGNALS:
        HIGH_TRUE   Y1
        HIGH_TRUE   Y2
        HIGH_TRUE   Z

REDUCED EQUATIONS:

Y1.CLK          = CK ;
  .D            = Y2 ;

Y2.CLK          = CK ;
  .D            = /Y1 ;

Z.EQN           = /Y1 ;
```

(*c*)

FIGURE E9-12

tool. Compiling M1_LE.SRC using the COMPILE DESIGN tool results in the M1_LE.DOC file shown in Fig. E9-12c.

The reduced equations for M1_LE show that two D flip-flops are required. The control inputs of both the $y1$ and $y2$ flip-flops are driven by the system clock. The D input of the $y1$ flip-flop is driven by the signal $y2$, while the D input of the $y2$ flip-flop is driven by the signal $\overline{y1}$. The Moore output Z is $\overline{y1}$.

Notice that the reduced equations obtained by PLDesigner shown in the M1_LE.DOC file agree with the reduced equations obtained manually for M1 in subsection 9-3-4, and can be represented by the circuit shown in Fig. 9-8. Any input or output signal name specified in the design file without the keyword LOW_TRUE preceding the signal name is assumed to be HIGH_TRUE. In addition to the reduced equations, the .DOC file also provides the signal list for the equations, that is, the list of each input and output signal specified in the design file.

Example 9-13. Obtain a logic equation design file for the synchronous machine M3 shown in Fig. E9-13a using the text editor provided by the PLDesigner software package. Also obtain the documentation file containing the reduced equations for M3. Notice that M3 is a Mealy machine.

Solution A logic equation design file called M3_LE.SRC can be created using the text editor as shown in Fig. E9-13b. The logic equations for the next state variables $Y1$ and $Y2$ can be written directly from the state diagram for M3 using the "set or hold method." The Mealy output Z is written in terms of the present state variables $y1$ and $y2$ and the input variable X (for the 1s of the function).

       Using the COMPILE DESIGN tool to compile M3_LE.SRC results in the M3_LE.DOC file shown in Fig. E9-13c. The reduced equations for M3 in the M3_LE.DOC file can be manually verified as in Example 9-7 (although this is not necessary), or verified by using either the truth table entry method or the state machine entry method.

9-5-3 Truth Table Entry Method

Example 9-14. Obtain a truth table design file for the synchronous machine M1 shown in Fig. E9-12a using the text editor provided by the PLDesigner software package. Also obtain the documentation file containing the reduced equations for M1.

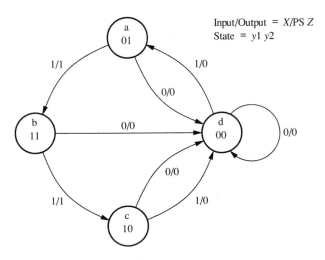

Input/Output = X/PS Z
State = $y1$ $y2$

M3 (Mealy machine)

(a)

FIGURE E9-13

```
============ PLDesigner V1.2 - MINC Copyright 1987-88 - Text Editor ==========
                                        Press <ESC> for command line
   Window:1 File:C:\MINCWORK\M3_LE.SRC       Line:3    Col:23    AI INS
```

```
"Design file
TITLE               State Machine M3;
COMMENT             Logic Equation entry format;

FUNCTION M3_LE;

INPUT CK, X;

OUTPUT y1, y2 CLOCKED_BY CK;
OUTPUT Z;

    Y1 = /y1*y2*X          "set
       +  y1*y2*X;         "hold
    Y2 = /y1*/y2*X         "set
       + /y1*y2*X;         "hold
     Z = /y1*y2*X + y1*y2*X;

END M3_LE;
```

(*b*)

```
============ PLDesigner V1.2 - MINC Copyright 1987-88 - Text Editor ==========
                                        Press <ESC> for command line
   Window:1 File:C:\MINCWORK\M3_LE.DOC       Line:56   Col:1    AI INS
```

```
EQUATIONS FOR FUNCTION M3_LE

INPUT SIGNALS:
        HIGH_TRUE   CK
        HIGH_TRUE   X

OUTPUT SIGNALS:
        HIGH_TRUE   Y1
        HIGH_TRUE   Y2
        HIGH_TRUE   Z

REDUCED EQUATIONS:

Y1.CLK            = CK ;
  .D              = X*Y2 ;

Y2.CLK            = CK ;
  .D              = X*/Y1 ;

Z.EQN             = X*Y2 ;
```

(*c*)

FIGURE E9-13

Solution A truth table design file called M1_TT.SRC can be created using the text editor as shown in Fig. E9-14*a*. The format required for the truth table is shown in Fig. E9-14*a*. Notice that the input and output variables and values in the table must be separated by a double colon, while the individual input and output variables and values must be separated by a comma. For each row in the table, the present state values are on the left, while the next state and *the present state output values* are on the right; that is, double colons mark the middle of the table. The truth table shown in Fig. E9-14*a* is actually the PS/NS table for M1.

Compiling M1_TT.SRC, using the COMPILE DESIGN tool (the extension .SRC stands for the SouRCe code for the design file), results in the M1_TT.DOC file shown in Fig. E9-14*b*.

```
╔══════════ PLDesigner V1.2 - MINC Copyright 1987-88 - Text Editor ══════════╗
║                                            Press <ESC> for command line     ║
║  Window:1 File:C:\MINCWORK\M1_TT.SRC       Line:1     Col:1      AI INS      ║
╚═════════════════════════════════════════════════════════════════════════════╝
```

```
"Design file
TITLE              State Machine M1;
COMMENT            Truth Table or State Table entry format;

FUNCTION M1_TT;

INPUT CK;
OUTPUT y1, y2 CLOCKED_BY CK;
OUTPUT Z;
     TRUTH_TABLE

   y1,y2::Y1,Y2,Z;

      0,0::0,1,1;
      0,1::1,1,1;
      1,1::1,0,0;
      1,0::0,0,0;

   END;"Truth table

END M1_TT;
```

(*a*)

```
╔══════════ PLDesigner V1.2 - MINC Copyright 1987-88 - Text Editor ══════════╗
║                                            Press <ESC> for command line     ║
║  Window:1 File:C:\MINCWORK\M1_TT.DOC       Line:45    Col:1      AI INS      ║
╚═════════════════════════════════════════════════════════════════════════════╝
```

```
EQUATIONS FOR FUNCTION M1_TT

INPUT SIGNALS:
        HIGH_TRUE   CK

OUTPUT SIGNALS:
        HIGH_TRUE   Y1
        HIGH_TRUE   Y2
        HIGH_TRUE   Z

REDUCED EQUATIONS:

Y1.CLK             = CK ;
  .D               = Y2 ;

Y2.CLK             = CK ;
  .D               = /Y1 ;

Z.EQN              = /Y1 ;
```

(*b*)

FIGURE E9-14

Notice that the reduced equations obtained by PLDesigner shown in the M1_TT.DOC file agree with the reduced equations obtained for M1 in the M1_LE.DOC file in Fig. E9-12*c*.

Example 9-15. Obtain a truth table design file for the synchronous machine M3 shown in Fig. E9-13*a* using the text editor provided by the PLDesigner software package. Also obtain the documentation file containing the reduced equations for M3.

```
┌─══════════ PLDesigner V1.2 - MINC Copyright 1987-88 - Text Editor ═══════┐
│                                          Press <ESC> for command line    │
│  Window:1 File:C:\MINCWORK\M3_TT.SRC              Line:1    Col:1    AI INS │
└──────────────────────────────────────────────────────────────────────────┘
```

```
"Design file
TITLE               State Machine M3;
COMMENT             Truth Table or State Table entry format;
FUNCTION M3_TT;
INPUT CK, X;
OUTPUT y1, y2 CLOCKED_BY CK;
OUTPUT Z;
    TRUTH_TABLE
    y1,y2,X::Y1,Y2,Z;

    0,0,0::0,0,0;
    0,0,1::0,1,0;
    0,1,1::1,1,1;
    0,1,0::0,0,0;
    1,1,1::1,0,1;
    1,1,0::0,0,0;
    1,0,1::0,0,0;
    1,0,0::0,0,0;

    END;"Truth table
END M3_TT;
```

(a)

```
┌─══════════ PLDesigner V1.2 - MINC Copyright 1987-88 - Text Editor ═══════┐
│                                          Press <ESC> for command line    │
│  Window:1 File:C:\MINCWORK\M3_TT.DOC              Line:56   Col:1    AI INS │
└──────────────────────────────────────────────────────────────────────────┘
```

```
EQUATIONS FOR FUNCTION M3_TT

INPUT SIGNALS:
        HIGH_TRUE   CK
        HIGH_TRUE   X

OUTPUT SIGNALS:
        HIGH_TRUE   Y1
        HIGH_TRUE   Y2
        HIGH_TRUE   Z

REDUCED EQUATIONS:

Y1.CLK          = CK ;
  .D            = X*Y2 ;

Y2.CLK          = CK ;
  .D            = X*/Y1 ;

Z.EQN           = X*Y2 ;
```

(b)

FIGURE E9-15

Solution A truth table design file called M3_TT.SRC can be created using the PLDesigner text editor as shown in Fig. E9-15a. The present state and input variables are shown on the left of the double colons in the truth table or PS/NS table, and the next state and Mealy present state output variables are shown on the right.

Using the COMPILE DESIGN tool to compile M3_TT.SRC results in the M3_TT.DOC file shown in Fig. E9-15b. As expected, PLDesigner produces the same reduced equations as those obtained for M3 in the M3_LE.DOC file in Fig. E9-13*c*.

9-5-4 State Machine Entry Method

Example 9-16. Obtain a state machine design file for the synchronous machine M1 shown in Fig. E9-12*a* using the text editor provided by the PLDesigner software package. Also obtain the documentation file containing the reduced equations for M1.

Solution A state machine design file called M1_SM.SRC can be created using the PLDesigner text editor as shown in Fig. E9-16*a*. The format required for state machine language is shown in Fig. E9-16*a*. Each state consists of the word STATE, a state name (a, b, c, or d), the code that is assigned to each particular state name (enclosed in brackets), and a colon. After the colon the Moore output value is specified, followed by the path (a GOTO *state name*) that the machine must follow at the next clock timing event. If the state bits are not identified and a state assignment code is not provided in the design file for each state, then PLDesigner will assign a code to each state. This is one advantage of using the state machine entry method compared to the logic equation and truth table entry methods.

Compiling M1_SM.SRC using the COMPILE DESIGN tool results in the M1_SM.DOC file shown in Fig. E9-16*b*. The reduced equations are the same as those obtained using the logic equation and truth table entry methods.

Example 9-17. Obtain a state machine design file for the synchronous machine M3 shown in Fig. E9-13*a* using the text editor provided by the PLDesigner software

```
========= PLDesigner V1.2 - MINC Copyright 1987-88 - Text Editor =========
                                     Press <ESC> for command line
  Window:1 File:C:\MINCWORK\M1_SM.SRC      Line:21   Col:10    AI INS
```

```
"Design file
TITLE              State Machine M1;
COMMENT            State Machine entry format;

FUNCTION M1_SM;

INPUT CK;
OUTPUT y1, y2 CLOCKED_BY CK;
OUTPUT Z;
     STATE_MACHINE M1;

     CLOCKED_BY CK;
     STATE_BITS[y1,y2];
          STATE a[0]: Z = 1; GOTO b;
          STATE b[1]: Z = 1; GOTO c;
          STATE c[3]: Z = 0; GOTO d;
          STATE d[2]: Z = 0; GOTO a;

     END M1;"State machine

END M1_SM;
```

(*a*)

FIGURE E9-16

```
============ PLDesigner V1.2 - MINC Copyright 1987-88 - Text Editor ============
                                        Press <ESC> for command line
    Window:1 File:C:\MINCWORK\M1_SM.DOC    Line:55    Col:1      AI INS
```

EQUATIONS FOR FUNCTION M1_SM

INPUT SIGNALS:
      HIGH_TRUE   CK

OUTPUT SIGNALS:
      HIGH_TRUE   Y1
      HIGH_TRUE   Y2
      HIGH_TRUE   Z

REDUCED EQUATIONS:

```
Y1.CLK          = CK ;
  .D            = Y2 ;

Y2.CLK          = CK ;
  .D            = /Y1 ;

Z.EQN           = /Y1 ;
```

(b)

FIGURE E9-16

package. Also obtain the documentation file containing the reduced equations for M3.

Solution A state machine design file called M3_SM.SRC can be created using the PLDesigner text editor as shown in Fig. E9-17*a*. In each state an IF THEN ELSE statement is used with the appropriate value of the input variable *X* to specify the path that the machine must follow at the next clock timing event. The keywords

```
============ PLDesigner V1.2 - MINC Copyright 1987-88 - Text Editor ============
                                        Press <ESC> for command line
    Window:1 File:C:\MINCWORK\M3_SM.SRC    Line:1    Col:1      AI INS
```

```
"Design file
TITLE               State Machine M3;
COMMENT             State Machine entry format;

FUNCTION M3_SM;
INPUT CK, X;
OUTPUT y1, y2 CLOCKED_BY CK;
OUTPUT Z;
      STATE_MACHINE M3;
      CLOCKED_BY CK;
      STATE_BITS[y1,y2];
          STATE d[0]: IF X = 1 THEN BEGIN  Z = 0; GOTO a; END
                              ELSE BEGIN  Z = 0; GOTO d; END;
          STATE a[1]: IF X = 1 THEN BEGIN  Z = 1; GOTO b; END
                              ELSE BEGIN  Z = 0; GOTO d; END;
          STATE b[3]: IF X = 1 THEN BEGIN  Z = 1; GOTO c; END
                              ELSE BEGIN  Z = 0; GOTO d; END;
          STATE c[2]: IF X = 1 THEN BEGIN  Z = 0; GOTO d; END
                              ELSE BEGIN  Z = 0; GOTO d; END;
      END M3;"State machine
END M3_SM;
```

(a)

FIGURE E9-17

```
╔══════════ PLDesigner V1.2 - MINC Copyright 1987-88 - Text Editor ══════════╗
║                                              Press <ESC> for command line  ║
║  Window:1 File:C:\MINCWORK\M3_SM.DOC            Line:56    Col:1     AI INS ║
╚════════════════════════════════════════════════════════════════════════════╝
```

```
EQUATIONS FOR FUNCTION M3_SM

INPUT SIGNALS:
        HIGH_TRUE   CK
        HIGH_TRUE   X

OUTPUT SIGNALS:
        HIGH_TRUE   Y1
        HIGH_TRUE   Y2
        HIGH_TRUE   Z

REDUCED EQUATIONS:

Y1.CLK              = CK ;
  .D                = X*Y2 ;

Y2.CLK              = CK ;
  .D                = X*/Y1 ;

Z.EQN               = X*Y2 ;
```

(*b*)

FIGURE E9-17

BEGIN and END are used in the same manner as in the Pascal language to bracket two or more equations or commands.

Using the COMPILE DESIGN tool to compile M3_SM.SRC results in the M3_SM.DOC file shown in Fig. E9-17*b*. As expected, the reduced equations are the same as those obtained using the logic equation and truth table entry methods.

9-5-5 Waveform Entry Method

In the following waveform examples, the waveform editor only allows waveforms to be drawn with zero setup and hold time. This means that when the rising edge of the system clock occurs, all inputs are assumed to be at the correct value. It is the responsibility of the designer to meet minimum setup and hold times in the final circuit implementation.

Example 9-18. Obtain a waveform design file for the synchronous machine M1 shown in Fig. E9-12*a* using the waveform editor provided by the PLDesigner software package. Also obtain the documentation file for M1 containing the reduced equations.

Solution Using the PLDesigner waveform editor, a waveform design file can be created for M1, called M1_W.WAV, as shown in Fig. E9-18*a* (the extension .WAV stands for the WAVe code for the design file). The timing events and states shown above the waveform editor display in Fig. E9-18*a* are used to emphasize the relationship of each waveform to the corresponding signal in the state diagram in Fig. E9-12*a*. The waveform editor assumes that each specified output (*y*1 and *y*2 in this case) is provided by a *D* flip-flop that is driven by the system clock. Since the *Z* output is a Moore output, it is not shown as one of the output waveforms. The Sec/Div indicator at the bottom right hand side of Fig. E9-18*a* simply provides

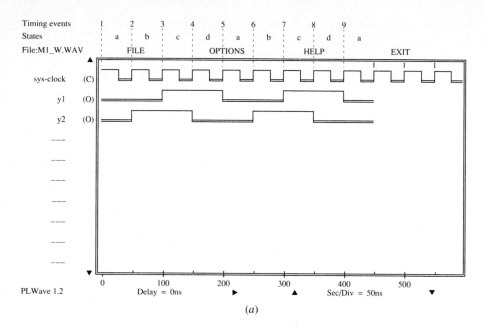

(a)

```
================ PLDesigner V1.2 - MINC Copyright 1987-88 - Text Editor ================
                                           Press <ESC> for command line
  Window:1 File:C:\MINCWORK\M1_W.DOC        Line:44   Col:1      AI INS
```

EQUATIONS FOR FUNCTION M1_W

OUTPUT SIGNALS:

        HIGH_TRUE   y1
        HIGH_TRUE   y2

REDUCED EQUATIONS:

```
y1.CLK              = sys_clock ;
  .D                = y2 ;

y2.CLK              = sys_clock ;
  .D                = /y1 ;
```

(b)

FIGURE E9-18

a scale factor for our example, since the problem does not specify the frequency of the system clock.

  After obtaining the waveform design file, M1_W.WAV, the next step is to compile the file using the 'COMPILE DESIGN' tool. Nine timing events (just over two complete cycles for M1) are provided in the design file for M1_W.WAV, since this happens to be the minimum number required by the compiler before it can generate the reduced equations. After the reduced equations are generated by the compiler, they are placed in the documentation file for M1_W (the M1_W.DOC file). The documentation file M1_W.DOC is shown in Fig. E9-18*b*.

The reduced equations for $y1$ and $y2$ in the M1_W.DOC file agree with the equations in the M1_LE.DOC file in Example 9-12. The Moore output Z (or Mealy output, if we were obtaining a design that contained a Mealy output) can be added to a waveform design file by selecting OPTIONS, and adding a FBINCLUDE file. This file simply lists $y1$ and $y2$ as inputs, Z as an output, and the Moore output equation $Z = \overline{y1} \cdot \overline{y2} + \overline{y1} \cdot y2$ for the 1s of the function (PLDesigner can be used to handle the reduction process). The FBINCLUDE file is obtained using the logic equation entry method that was presented in Examples 9-12 and 9-13. When the M1_W.WAV file is compiled after the FBINCLUDE file containing the Moore output equation, the documentation file M1_W.DOC then contains not only the reduced equations for $y1$ and $y2$ but also the reduced equation for the Moore output Z. Creating a documentation file that contains the reduced equations for a waveform design file and a text design file is left as an exercise for the student.

Example 9-19. Obtain a waveform design file for the synchronous machine M2 shown in Fig. E9-19a using the waveform editor provided by the PLDesigner software package. Also obtain the documentation file containing the reduced equations for M2.

Solution Using the PLDesigner waveform editor, two slightly different waveform design files can be created for M2, called M2_W1.WAV shown in Fig. E9-19b and M2_W2.WAV shown in Fig. E9-19c. Both of these waveform design files were created using the state diagram for M2.

     Both of these waveform design files also represent an accurate design description for M2. The only difference is that M2_W1.WAV begins with state d, while M2_W2.WAV begins with state c. Compiling these files results in the documentation files M2_W1.DOC and M2_W2.DOC in Figs. E9-19d and e.

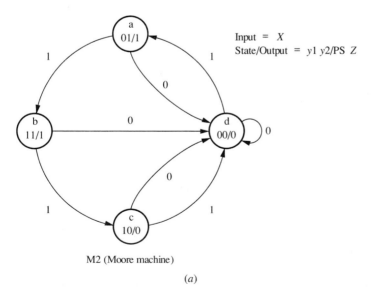

Input = X
State/Output = y1 y2/PS Z

M2 (Moore machine)

(a)

FIGURE E9-19

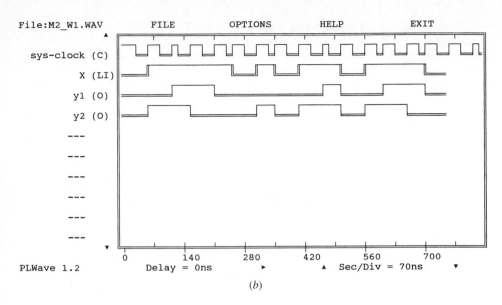

(*b*)

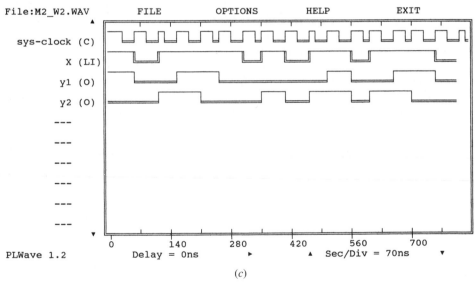

(*c*)

FIGURE E9-19

In the documentation file for M2_W1, signals X, $y1$, and $y2$ are all HIGH_TRUE; however, in the documentation file for M2_W2, signal X is HIGH_TRUE, signal $y1$ is LOW_TRUE, and signal $y2$ is HIGH_TRUE. The reduced equations in the M2_W1.DOC and M2_W2.DOC files are equivalent, as seen by writing the equations using either all HIGH_TRUE signals or all LOW_TRUE signals.

The signal list and reduced equations for M2_W2 can be written as follows using signal names with level indication.

```
╔══════════ PLDesigner V1.2 - MINC Copyright 1987-88 - Text Editor ══════════╗
║                                               Press <ESC> for command line  ║
║   Window:1 File:C:\MINCWORK\M2_W1.DOC              Line:64    Col:1    AI INS║
╚═════════════════════════════════════════════════════════════════════════════╝
```

EQUATIONS FOR FUNCTION M2_W1

INPUT SIGNALS:

      HIGH_TRUE   X

OUTPUT SIGNALS:

      HIGH_TRUE   y1
      HIGH_TRUE   y2

REDUCED EQUATIONS:

```
y1.CLK              = sys_clock ;
  .D                = X*y2 ;

y2.CLK              = sys_clock ;
  .D                = X*/y1 ;
```

(d)

```
╔══════════ PLDesigner V1.2 - MINC Copyright 1987-88 - Text Editor ══════════╗
║                                               Press <ESC> for command line  ║
║   Window:1 File:C:\MINCWORK\M2_W2.DOC              Line:64    Col:1    AI INS║
╚═════════════════════════════════════════════════════════════════════════════╝
```

EQUATIONS FOR FUNCTION M2_W2

INPUT SIGNALS:

      HIGH_TRUE   X

OUTPUT SIGNALS:

      LOW_TRUE   y1
      HIGH_TRUE   y2

REDUCED EQUATIONS:

```
y1.CLK              = sys_clock ;
  .D                = /y2 + /X ;

y2.CLK              = sys_clock ;
  .D                = X*y1 ;
```

(e)

FIGURE E9-19

Signal list: $X(H)$, $y1(L)$, $y2(H)$

$y1.D(L) = \overline{y2}(H) + \overline{X}(H)$

$y2.D(H) = X(H) \cdot y1(L)$

The signal list and reduced equations for M2_W2 can be rewritten using all HIGH_TRUE signals by converting the signal $y1(L)$ which has a low level indicator, to the equivalent form $\overline{y1}(H)$, which has a high level indicator as follows.

Signal list: $X(H)$, $\overline{y1}(H)$, $y2(H)$

$$\overline{y1}.D(\text{H}) = \overline{y2}(\text{H}) + \overline{X}(\text{H})$$

$$y2.D(\text{H}) = X(\text{H}) \cdot \overline{y1}(\text{H})$$

so, $y1.D = y2 \cdot X$ (by DeMorgan's theorem)

$$y2.D = X \cdot \overline{y1}$$

The last two equations are the same equations provided in the M2_W1.DOC file. This demonstrates that different, but equivalent, equations are possible for different, but equivalent, waveform design files.

9-5-6   Schematic Entry Method

Sometimes a designer may prefer to obtain a design without using the tools provided by a software design package. At a later time, the designer or someone else may decide that the circuit should be implemented by a PLD. Assuming that the schematic or circuit diagram is available, it would be nice if the schematic could be used to obtain a design file without redesigning the circuit. In some cases, it is possible to utilize PLDesigner with the OrCAD, P-CAD, or FutureNet schematic capture packages to create a design file. P-CAD is a trademark of Personal CAD Systems Incorporated. The procedure used to create a schematic and translate it into a design file usable by PLDesigner is beyond the scope of this text.

REFERENCES

1. Mano, M. M., *Digital Design,* Prentice-Hall, Englewood Cliffs, New Jersey, 1984.
2. Prosser, F. P., and Winkel, D. E., *The Art of Digital Design, An Introduction to Top-Down Design,* Prentice-Hall, Englewood Cliffs, New Jersey, 1987.
3. Breeding, K. J., *Digital Design Fundamentals,* Prentice-Hall, Englewood Cliffs, New Jersey, 1989.
4. Kohavi, Z., *Switching and Finite Automata Theory,* 2d ed., McGraw-Hill, New York, 1978.
5. Fletcher, W. I., *An Engineering Approach to Digital Design,* Prentice-Hall, Englewood Cliffs, New Jersey, 1980.
6. Clare, C. R., *Designing Logic Systems Using State Machines,* McGraw-Hill, New York, 1973.
7. Roth, Jr., C. H., *Fundamentals of Logic Design,* 3d ed., West Publishing, St. Paul, Minnesota, 1985.
8. Rhyne, Jr., V. T., *Fundamentals of Digital Systems Design,* Prentice-Hall, Englewood Cliffs, New Jersey, 1973.
9. Dietmeyer, D. L., *Logic Design of Digital Systems,* 3d ed., Allyn and Bacon, Boston, Mass., 1988.
10. Edwards, H. E., *The Principles of Switching Circuits,* The M. I. T. Press., Cambridge, Mass., 1973.
11. Johnson, E. L., and Karim, M. A., *Digital Design, A Pragmatic Approach,* Prindle, Weber & Schmidt Publishers, Boston, Mass., 1987.
12. McCluskey, E. J., *Logic Design Principles,* Prentice-Hall, Englewood Cliffs, New Jersey, 1986.
13. Hill, J. H., and Peterson, G. R., *Introduction to Switching Theory & Logic Design,* 3d ed., John Wiley & Sons, New York, 1981.
14. Story, J. R., Harrison, H. J., and Reinhard, E. A., "Optimum State Assignment for Synchronous Sequential Circuits," *IEEE Transactions on Computers,* December, 1972, pp. 1365–1373.

15. Noe, P. S., "Remarks on the SHR-Optimal State Assignment Procedure," *IEEE Transactions on Computers,* Volume C-22, Number 9, September, 1973, pp. 873–875.
16. Noe, P. S., and Rhyne, V. T., "A Modification to the SHR-Optimal State Assignment Procedure," *IEEE Transactions on Computers,* Volume C-23, Number 3, March, 1974, pp. 327–329.
17. *PLDesigner User's Manual, Programmable Logic Device Design Program,* Version 1.2, Minc, Colorado Springs, Colorado, 1988.
18. *Schematic Design Tools, OrCAD/SDT III,* OrCAD Systems, Hillsboro, Oregon, 1987.
19. *Programmable Logic, Data Book,* Texas Instruments, Dallas, Texas, 1988.
20. *Programmable Logic Devices Databook and Design Guide,* National Semiconductor Corporation, Santa Clara, CA, 1989.
21. *PAL Device Handbook,* Advanced Micro Devices, Sunnyvale, CA, 1988.
22. *PAL Device Data Book,* Advanced Micro Devices, Sunnyvale, CA, 1988.

PROBLEMS

Section 9-2   Moore, Mealy, and Mixed Type Synchronous State Machines

9-1. Given input variables $X1$ and $X2$, present state variables $y1$, $y2$, and $y3$, and output variables $Z1$ through $Z8$, identify the Moore type and Mealy type outputs in the following list.
 (a) $Z1 = y1 \cdot \overline{y2} \cdot X1$
 (b) $Z2 = \overline{y1} \cdot y3$
 (c) $Z3 = y1 \cdot \overline{y2} \cdot \overline{y3}$
 (d) $Z4 = y3 \cdot X1 \cdot \overline{X2}$
 (e) $Z5 = \overline{y2}$
 (f) $Z6 = \overline{y1} \cdot \overline{y3} \cdot \overline{X1}$
 (g) $Z7 = y1 + y2$
 (h) $Z8 = \overline{y2} + X2$

9-2. Discuss the difference between Moore and Mealy outputs.

9-3. Design a Moore type synchronous state machine with two external inputs $X1$ and $X2$ and one output Z. When $\overline{X1} \cdot X2 = 1$ at the next clock timing event, output Z goes to 1. Output Z then goes to 0 unless $X2 = 1$ causing the output to stay at 1.
 (a) Draw an ASM chart to represent the design.
 (b) Use the 'set or hold method' to obtain the next state output equation for a negative edge-triggered D flip-flop. Also obtain the equation for output Z.
 (c) Draw the logic circuit.

9-4. Modify the design specification in Prob. 9-3 such that $\overline{X1} \cdot X2 = 0$ causes the output Z to stay at 1. $\overline{X1} = 1$ causes the output Z to go to 0; otherwise, output Z stays at 1.
 (a) Draw a state diagram to represent the design.
 (b) Use the 'set or hold method' to obtain the next state output equation for a positive edge-triggered D flip-flop. Also obtain the equation for output Z.
 (c) Draw the logic circuit.

9-5. Design a Mealy type synchronous state machine with one external input X and two external outputs $Z1$ and $Z2$. When X is 1 at the next tick of the clock, the machine changes from state a to state b and $Z1$ $Z2$ = 01. The machine stays in state b as long as X is 1 and provides outputs $Z1$ $Z2$ = 10. When X is 0, the machine

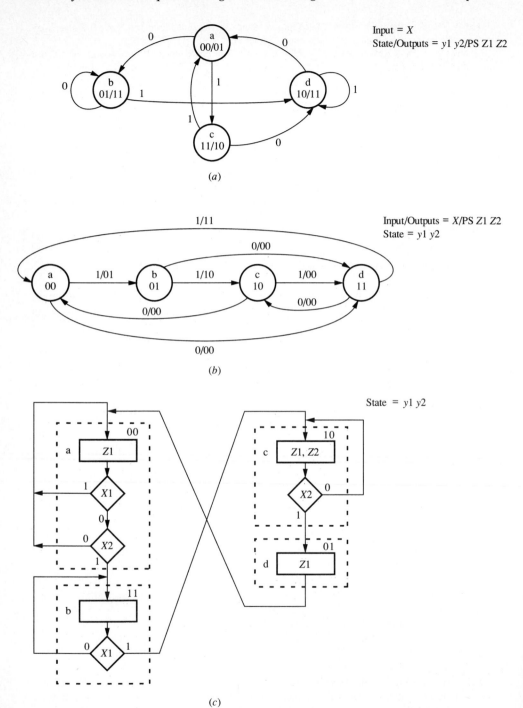

(a)

(b)

(c)

FIGURE P9-6

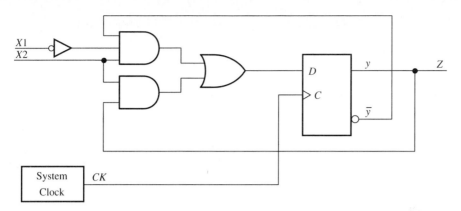

FIGURE P9-7

changes back to state a with $Z1\ Z2 = 00$, and output $Z1\ Z2$ remains 00 as long as the machine stays in state a.

(*a*) Draw an ASM chart to represent the design.

(*b*) Draw a state diagram to represent the design.

(*c*) Use either (*a*) or (*b*) to obtain the next state and external output equations for a positive edge-triggered D flip-flop.

(*d*) Minimize the equations and draw the circuit diagram.

9-6. For the synchronous machine in Fig. P9-6a through *c*, obtain the next state and external output equations for D flip-flops. Use the 'set or hold method' to write the next state output equations. Do not minimize the equations.

9-7. Analyze the state machine in Fig. P9-7 and obtain the following items:

(*a*) the excitation input and external output equations

(*b*) the flow map

(*c*) the state diagram

9-8. Analyze the state machine in Fig. P9-8 and obtain the following items:

(*a*) the excitation input and external output equations

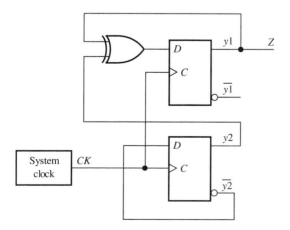

FIGURE P9-8

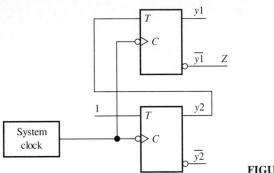

FIGURE P9-9

    (*b*) the flow map
    (*c*) the state diagram

9-9. Analyze the state machine in Fig. P9-9 and obtain the following items:
    (*a*) the excitation input and external output equations
    (*b*) the flow map
    (*c*) the ASM chart

9-10. Analyze the state machine in Fig. P9-10 and obtain the following items:
    (*a*) the excitation input and external output equations
    (*b*) the flow map
    (*c*) the ASM chart

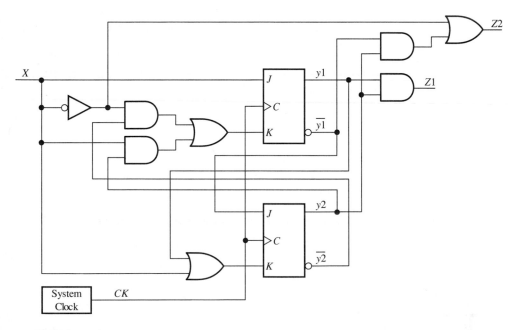

FIGURE P9-10

Section 9-3 Synchronous Sequential Design of Moore and Mealy Machines

9-11. Design a synchronous circuit using positive edge-triggered D flip-flops that provides an output signal Z, that is a subdivision of the clock frequency CK, as illustrated in the timing diagram in Fig. P9-11.

  (*a*) What is the frequency of the output signal Z compared to the clock frequency?

  (*b*) Obtain the PS/NS table for the design.

  (*c*) Obtain the excitation input and external output equations for the design using a composite Karnaugh map. Use the state assignments $y1\ y2 = 00$ (state a), 01 (state b), and 10 (state c). Force all unused states to go to 00 (state a).

  (*d*) Draw the circuit diagram.

9-12. Design a synchronous circuit using negative edge triggered D flip-flops that provides an output signal $Z\ \frac{1}{5}$ the frequency of the system clock CK.

  (*a*) Draw the timing diagram showing the system clock signal CK and the output signal Z.

  (*b*) Obtain the PS/NS table.

  (*c*) Obtain the composite Karnaugh map for the next state and external output signals. Use a straight binary counting sequence for the state assignments beginning with $a = 0$, $b = 1$, $c = 2$, etc., and force all unused states to go to 0.

  (*d*) Obtain the excitation input and external output equations.

  (*e*) Draw the circuit diagram.

9-13. Draw the timing diagram for M2 shown in Fig. E9-4a for the input sequence 00111101100010110 where the MSB in the binary number represents the first value of X in the sequence. Show the change in the asynchronous input occurring approximately $\frac{1}{4}$ of a cycle prior to the next clock timing event. Assume that output Z is initially 0, and that the machine is initially in the idle or rest state. Ignore output delays when drawing the output waveform. Discuss when the Moore output Z can change, compared to when a Mealy output can change.

9-14. Apply the state reduction procedure for the Mealy machine in Fig. P9-14.

  (*a*) Obtain the PS/NS table for the machine.

  (*b*) Use an implication table to check for redundant states. Can the number of flip-flops be reduced for this machine?

9-15. Consider the Moore machine in Fig. P9-15.

  (*a*) Obtain its PS/NS table.

  (*b*) Check for redundant states using an implication table. What is the minimum number of flip-flops required for this machine before reduction? After reduction?

9-16. How many ways can m state variables be permuted? Be complemented?

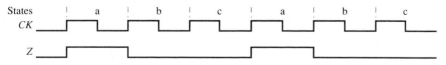

FIGURE P9-11

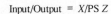

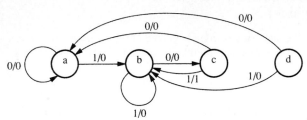

FIGURE P9-14

9-17. Use the relationships $C_{DFF} = 2^v!/(2^v - r)!v!$ and $C_{JKFF} = 2^v!/(2^v - r)!v!2^v$ (where C_{DFF} and C_{JKFF} represent the number of distinct state assignment codes for D and J-K type flip-flops) to complete the following table.

r (rows)	v (state variables)	C_{DFF}	C_{JKFF}
2	1		
3	2		
4	2		
5	3		
6	3		
7	3		
8	3		
9	4		

9-18. Design a synchronous circuit using positive edge-triggered D flip-flops, with one external input signal X and one Mealy output signal Z. When X remains 1 for three clock timing events, the Mealy output goes to 1 on the third clock timing event. All other combinations of X and clock timing events result in a Mealy output of 0, and the machine returns to the 0 state. Part of the timing diagram is illustrated in Fig. P9-18.
(*a*) Draw the Mealy state machine diagram.

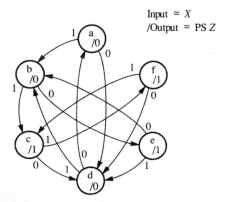

FIGURE P9-15

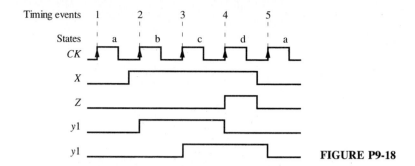

FIGURE P9-18

(*b*) Obtain the PS/NS table for the machine.

(*c*) Obtain minimized excitation input and external output equations for the machine using a composite Karnaugh map.

(*d*) Draw the circuit diagram.

9-19. Design a synchronous circuit using positive edge-triggered T flip-flops for the state diagram in Fig. P9-6*a*.

(*a*) Obtain the PS/NS table for the machine.

(*b*) Obtain reduced excitation input and external output equations.

(*c*) Draw the circuit diagram.

9-20. Design a synchronous circuit using negative edge-triggered T flip-flops to provide an output signal Z $\frac{1}{6}$ the frequency of system clock CK.

(*a*) Draw the timing diagram showing the system clock signal CK and the output signal Z.

(*b*) Obtain the PS/NS table.

(*c*) Obtain minimized excitation input and external output equations. Use a gray code counting sequence for the state assignments beginning with state 0, and force all unused states to go to 0.

(*d*) Draw the circuit diagram.

9-21. Obtain a positive edge-triggered T flip-flop design for the following PS/NS table. Let the state variables be $y1$ $y2$, and the state assignments for states a through d follow the binary counting sequence.

(*a*) Obtain the composite Karnaugh map.

(*b*) Write minimized excitation and output equations.

(*c*) Draw the circuit diagram.

	NS, PS Z	
PS	$X=0$	$X=1$
a	b,0	a,0
b	c,0	b,1
c	d,0	c,0
d	a,0	d,1

9-22. Draw the timing diagram for the circuit in Fig. E9-8*e*. Show that $f_{Z1} = \frac{1}{8} f_{CK}$, $f_{Z2} = \frac{1}{4} f_{CK}$, and $f_{Z3} = \frac{1}{2} f_{CK}$.

$y1\ y2$ X $Y1\ Y2,$ PS $Z2 =$		0	1	PS $Z1$
a	00	01, 1	11, 0	0
b	01	01, 1	10, 1	0
c	11	10, 1	00, 0	1
d	10	00, 1	10, 0	0

FIGURE P9-24

9-23. Design a synchronous circuit using positive edge-triggered *J-K* flip-flops to provide the timing diagram in Fig. P9-11.
  (*a*) Obtain the flow map.
  (*b*) Obtain the excitation input and external output equations for the design using a composite Karnaugh map with state assignments $y1\ y2 = 00$ (state a), 01 (state b), 11 (state c), and 10 (state d). Force all unused states to go to state a.
  (*c*) Draw the circuit diagram.
9-24. Design a synchronous circuit using negative edge-triggered *J-K* flip-flops for the state machine represented by the composite Karnaugh map in Fig. P9-24.
  (*a*) Obtain the composite excitation maps.
  (*b*) Obtain reduced excitation input and Moore and Mealy output equations.
  (*c*) Draw the logic diagram for the synchronous mixed-type machine.
9-25. Design a synchronous circuit using positive edge-triggered *J-K* flip-flops for the state machine in Fig. P9-25.
  (*a*) Obtain the flow map.
  (*b*) Write minimized excitation input and output equations.
  (*c*) Draw the circuit diagram.
9-26. Briefly describe the function of a controller.

Section 9-4 Synchronous Counter Design

9-27. What type of problem can occur when an asynchronous counter is used?
9-28. How can you tell when a counter is synchronous or asynchronous from its circuit diagram?

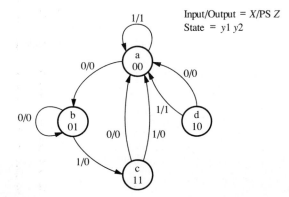

FIGURE P9-25

9-29. Why must a counter be self-correcting or self-starting?

9-30. How can one be sure that a counter is self-correcting or self-starting? Which type of flip-flop is generally better suited for illegal state recovery, when 0 is one of the state assignments in the primary counting sequence?

9-31. For practice, design a synchronous counter to provide a BCD Moore output sequence.

(a) Considering the state assignments for the circuit and the Moore outputs, what technique is generally used for counters to reduce the complexity of the circuit? Explain. Use this technique in the design.

(b) Carry out the design using positive edge-triggered *D* flip-flops. Draw the circuit diagram.

(c) Carry out the design using positive edge-triggered *J-K* flip-flops. Draw the circuit diagram.

9-32. Design a synchronous counter that has a Moore output decimal sequence of 0, 1, 3, 5, 7, and then repeats this same sequence over and over. To insure illegal state recovery, force all unused or illegal states to go to 0.

(a) Draw the state diagram.

(b) Carry out the design using positive edge-triggered *D* flip-flops. Draw the circuit diagram.

(c) Carry out the design using negative edge-triggered *T* flip-flops. Draw the circuit diagram.

(d) Carry out the design using positive edge-triggered *J-K* flip-flops. Draw the circuit diagram.

9-33. For each of the following Moore output sequences, design a synchronous counter that uses positive edge-triggered *D* flip-flops. The sequence should repeat indefinitely. Provide illegal state recovery by directing all unused or illegal states to the 0 state. Draw the circuit diagram for each counter.

(a) 0, 1, 3, 4, 7

(b) 0, 6, 4, 3, 2, 5

(c) 0, 2, 4, 6, 7, 5, 3

9-34. Repeat Prob. 9-33 for positive edge-triggered *T* flip-flops.

9-35. Repeat Prob. 9-33 for negative edge-triggered *J-K* flip-flops.

9-36. Design a synchronous 2-bit ring counter with illegal state recovery circuitry. Analyze the design and draw a simplified state diagram showing all possible states for the machine.

9-37. Repeat Prob. 9-36 for a synchronous 3-bit ring counter.

9-38. Repeat Prob. 9-36 for a synchronous 5-bit ring counter.

9-39. Design a synchronous 2-bit twisted ring counter with illegal state recovery circuitry. Analyze the design and draw a simplified state diagram showing all possible states for the machine.

9-40. Repeat Prob. 9-39 for a synchronous 3-bit twisted ring counter.

9-41. Repeat Prob. 9-39 for a synchronous 4-bit twisted ring counter.

9-42. Design a synchronous 2-bit maximum length shift counter with an Exclusive OR function that has illegal state recovery to decode the illegal state and use the decoded output to set all the flip-flops on the next clock timing event. Analyze the design and draw a simplified state diagram showing all the possible states for the machine.

9-43. Repeat Prob. 9-42 for a 3-bit maximum length shift counter.

9-44. Repeat Prob. 9-42 for a 4-bit maximum length shift counter.

9-45. Design a synchronous 2-bit maximum length shift counter with an Exclusive NOR function that has illegal state recovery to decode the illegal state and use the decoded output to reset all the flip-flops on the next clock timing event. Analyze the design and draw a simplified state diagram showing all the possible states for the machine.

9-46. Repeat Prob. 9-45 for a 3-bit maximum length shift counter.

9-47. Repeat Prob. 9-45 for a 4-bit maximum length shift counter.

Section 9-5   Designing Synchronous State Machines Using Programmable Devices

(Problems 9-48 through 9-52 are combinational logic problems to be worked using the PLDesigner software package.)

9-48. Obtain minimized expressions for the Boolean functions in the following problems in Chapter 3.
  (*a*) Problem 3-17
  (*b*) Problem 3-18
  (*c*) Problem 3-19

9-49. Obtain minimized expressions for the Boolean functions in the following problems in Chapter 3.
  (*a*) Problem 3-23
  (*b*) Problem 3-26
  (*c*) Problem 3-27
  (*d*) Problem 3-42

9-50. Obtain minimized expressions for the functions in the truth tables in the following problems in Chapter 4.
  (*a*) Problem 4-18
  (*b*) Problem 4-20

9-51. Obtain minimized equations for the functions in the full-comparator truth table in Example 5-7 in Chapter 5.

9-52. Obtain minimized equations for the seven-segment outputs in the truth table in Fig. E5-17*a* in Chapter 5:
  (*a*) For a common cathode display.
  (*b*) For a common anode display.

(Problems 9-53 through 9-59 are synchronous sequential logic problems to be worked using the PLDesigner software package.)

9-53. Obtain minimized equations for the synchronous state machine represented by the ASM chart in Fig. E9-1*a*.
  (*a*) Use the 'set or hold method' and the logic equation entry method.
  (*b*) Use the state machine entry method.

9-54. Obtain minimized equations for the synchronous state machine represented by the composite Karnaugh map in Fig. E9-2*b*.
  (*a*) Use the truth table entry method.
  (*b*) Use the state machine entry method.

9-55. Obtain minimized equations for the synchronous state machine represented by the composite Karnaugh map in Fig. E9-3*b*.
  (*a*) Use the truth table entry method.
  (*b*) Use the state machine entry method.

9-56. Obtain minimized equations for the synchronous state machine represented by the state diagram shown in Fig. 9-15. Let the state variables be $y1$ $y2$ $y3$ $y4$ such that $y1 = Z1$, $y2 = Z2$, $y3 = Z3$, and $y4 = Z4$. Use the state machine entry method.

9-57. Obtain minimized equations for the synchronous state machine represented by the state diagram in Fig. 9-18 after modifying the diagram so that unused or illegal state h goes to state b to provide illegal state recovery.

9-58. Obtain minimized equations for the synchronous state machine represented by the timing diagram in Fig. E9-4e. Let the state variables be $y1$ $y2$ with the following state assignments: $y1$ $y2$ = 00 (state d), 01 (state a), 10 (state c), and 11 (state b). Add FBINCLUDE file under OPTIONS to generate the Mealy output as part of the waveform design file.

9-59. Obtain minimized equations for the synchronous state machine represented by the timing diagram in Fig. P9-11. Let the state variables be $y1$ $y2$ with the following state assignments: $y1$ $y2$ = 00 (state a), 01 (state b), and 10 (state c).

(*a*) Use the waveform entry method to allow the Moore output equation to be included as part of the waveform design file.

(*b*) Check the solution to insure that the equations for the state machine generated by PLDesigner have illegal state recovery for state d.

CHAPTER
10

ASYNCHRONOUS SEQUENTIAL LOGIC CIRCUIT DESIGN

10-1 INTRODUCTION AND INSTRUCTIONAL GOALS

Various tools and techniques for analyzing fundamental mode asynchronous sequential logic circuits were presented in Chapter 8. The analysis process was carried out as we investigated basic memory circuits ranging from basic latches to edge-triggered D flip-flops. Each of these circuits are asynchronous sequential circuits and can be designed by the techniques presented in this chapter. It is important to have a good foundation including both synchronous and asynchronous design methodology, so both types of sequential circuit design can be understood and mastered.

In Section 10-2, Asynchronous Design Fundamentals, differences in synchronous and asynchronous designs are compared. Input signals for clock mode, fundamental mode, and pulse mode circuits are discussed with an emphasis placed on the input signal restrictions for fundamental mode circuits.

Asynchronous Sequential Design of Fundamental Mode Circuits in Section 10-3 provides a design methodology for fundamental mode circuits. Beginning with a timing diagram, each of the following design tools are discussed in detail: timing tables, primitive flow maps, implication tables, merger diagrams, reduced flow maps, transition diagrams, composite Karnaugh maps for the next state and

584

external outputs, and composite Karnaugh maps for the excitation inputs to *S-R* latch circuits. The design methodology presented stresses that fundamental mode sequential circuits should be designed to have illegal state recovery, a race-free state assignment, logic hazard-free equations, and essential hazard removal.

Asynchronous sequential logic design would be incomplete without Section 10-4, Asynchronous Sequential Design of Pulse Mode Circuits. Circuits presented in this section have input signals which are asynchronous pulses that cannot overlap, but the design methodology is very similar to synchronous design with a few exceptions (there is no master clock). The maximum pulse width of each individual input pulse can cause problems in pulse mode circuits unless double-rank circuits are utilized as discussed in this section.

This chapter should prepare you to

1. List the design differences between synchronous and asynchronous circuits, and discuss the merits of each type of design.
2. Draw waveforms that illustrate the timing events for clock mode, fundamental mode, and pulse mode sequential circuits.
3. List the input signal restrictions that are placed on fundamental mode asynchronous circuits, and discuss the significance of each restriction.
4. Describe additional types of hazards and races that can occur in fundamental mode asynchronous circuits, and illustrate how each type is treated.
5. List the general steps involved in asynchronous fundamental mode design, and discuss the significance of each step.
6. Utilize a timing diagram to describe a fundamental mode design, and obtain a timing chart from the timing diagram description.
7. Obtain a primitive flow map from a timing diagram or state diagram.
8. Utilize an implication table to identify all pairs of equivalent or compatible states in a primitive flow map.
9. Draw a merger diagram and obtain groups of mergeable states from the diagram.
10. Obtain a reduced flow map from a primitive flow map and mergeable set of states.
11. Make a race-free assignment by using a transition diagram.
12. Obtain the composite Karnaugh map for the next state and external outputs using the state assignments in a transition diagram.
13. Obtain composite Karnaugh maps for the excitation inputs to *S-R* NOR latches.
14. Write logic hazard-free equations for fundamental mode asynchronous sequential circuits, and draw the circuit diagrams.
15. Determine when an essential hazard can occur, and be able to recommend possible solutions to prevent the essential hazard from occurring.
16. Design fundamental mode asynchronous circuits.

17. List the input signal restrictions that are placed on pulse mode asynchronous circuits, and discuss the significance of each restriction.

18. Analyze and design double-rank pulse mode asynchronous sequential circuits.

10-2   ASYNCHRONOUS DESIGN FUNDAMENTALS

In Chapter 9 we presented synchronous sequential logic design and demonstrated the relative ease of the synchronous design methodology. The design methodology for asynchronous sequential logic circuits is generally much more involved.

10-2-1   Synchronous and Asynchronous Design Differences

Asynchronous sequential logic circuits are not dependent on a system clock, and this is their main advantage. In some applications, sequential designs that do not incorporate a system clock are desirable. For example, when two independent systems such as a computer and a printer must communicate without a common system clock, an asynchronous design is often used. In other cases where speed is important, an asynchronous design can be required. Asynchronous circuits can operate faster than synchronous circuits because a state change can take place, once the circuit reaches a stable state, immediately after an external input signal changes. In comparison, a state change can only take place in a synchronous or clock mode circuit at discrete time intervals. The time interval for a synchronous circuit is the period of the system clock. In this case, the external inputs to the circuit must meet the set up and hold time requirements for the edge-triggered flip-flop being used in the design. Time between successive timing events must be chosen to be slightly longer than the longest delay path in the circuit. From the standpoint of speed, this is a disadvantage; however, from the standpoint of ease of design, it is an advantage. Since sufficient time is allowed between timing events for a synchronous circuit to become stable, hazards are given time to disappear prior to the next timing event.

A summary of the three different types of sequential logic circuits is shown in Fig. 10-1. The timing diagram illustrated below each sequential circuit type depicts the kinds of input signals that are used to trigger a change in state.

Since each input signal change triggers a state transition in a fundamental mode asynchronous sequential circuit, as illustrated in Fig. 10-1, restrictions must be placed on the manner in which the input signals can be applied to the circuit. Fundamental mode asynchronous sequential circuit input signal restrictions can be stated by the following two rules.

Rule 1: Only one input signal is allowed to change at one time.

Rule 2: Before the next input signal is allowed to change, the circuit must be given time to reach a new stable state.

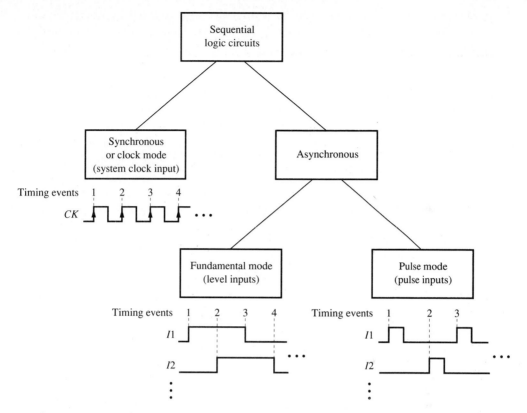

FIGURE 10-1
Summary of sequential logic circuit types and the corresponding external input signals required to provide a change in state.

The rules are closely tied together and must not be violated or the fundamental mode circuit will probably not function as desired. Rule 1 is commonly referred to as operating a circuit in fundamental mode. Rule 2 provides some insight into the speed of a fundamental mode asynchronous circuit, since the longest delay path from the input to the output of an asynchronous circuit is a measure of the speed of the circuit.

Hazards must also be considered when designing asynchronous circuits. Consider the case of the output of one circuit (the first circuit) driving the inputs of another circuit (the second circuit). The second circuit may perform unpredictably if the first circuit is not designed hazard-free. To insure that glitches or hazards (spurious signals) do not inadvertently act as timing events to other circuits, asynchronous circuits are generally designed logic hazard-free. As we discussed in Section 7-9, logic hazards can be removed by adding appropriate redundant terms to cover each occurrence of adjacent 1s (0s) not already contained in the same p-subcubes (r-subcubes) in the minimum covering of the function; that is, cross cover the minimum prime implicants in the SOP or POS Boolean equations for the functions. Function hazards which can occur when more than one input signal is changed at one time must be restricted from occurring by operating an asynchronous circuit in fundamental mode; that is, Rule 1 must be followed. Otherwise, function hazards can cause critical races that result in unpredictable circuit behavior.

In the design of fundamental mode asynchronous circuits, a circuit can start in a stable state and cycle through one or more unstable states before reaching the next stable state (see Fig. 8-13b). If a race condition exists, an asynchronous circuit can end up in the wrong state if the race is critical. State assignment codes are chosen in fundamental mode designs primarily to prevent critical races between stable state transitions rather than to reduce the logic complexity of the circuit. The process is called choosing a race-free state assignment. Critical races can be caused by assigning binary codes to the state variables in such a manner that two or more bits must change when a state transition occurs. This problem can be solved by assigning adjacent state assignment codes as discussed below.

10-3 ASYNCHRONOUS SEQUENTIAL DESIGN OF FUNDAMENTAL MODE CIRCUITS

An asynchronous design description can be specified in a number of different ways. The biggest problem the designer faces is simply obtaining a rigorous enough description to satisfy the requirements of the design. An asynchronous design can be described using a word statement, PS/NS table, Karnaugh map, state diagram, ASM chart, flow map, or a timing diagram, provided that each of these tools includes all the required information.

Table 10-1 is a summary of a set of design steps to outline a general design methodology for the asynchronous sequential design of fundamental mode circuits, that is, asynchronous circuits with level inputs.

10-3-1 Timing Diagram Specification

The starting point for an asynchronous fundamental mode design is often provided in the form of a word description, timing diagram, state diagram, or ASM chart. Consider the timing diagram shown in Fig. 10-2a for fundamental mode machine one (FM1). We will now present the asynchronous fundamental mode design process enumerated in Table 10-1 using FM1. Each level input change for $X1$ and $X2$ in Fig. 10-2a is marked as a timing event in Fig. 10-2b. The time between two successive timing events represents a new state, and the states are labeled in alphabetical order beginning with state a. The timing diagram is terminated only after all possible circuit operations have been represented at least once. The last state shown can therefore be represented by one or more previous states.

10-3-2 Using a Timing Table

Before starting the asynchronous design process, it is useful to check if the timing diagram can be implemented as a combinational logic circuit, that is, if $Z =$

TABLE 10-1
Asynchronous fundamental mode design process

	Design steps	Comments
Step 1	Organize design specification into a timing diagram, state diagram, or ASM chart	
Step 2	Obtain a primitive flow map from the design specification	
Step 3	Use an implication table to obtain all pairs of equivalent or compatible states. Next, use a merger diagram to choose sets of states that can be merged to obtain a minimum number of states	Usually results in a reduction in the number of required states
Step 4	Obtain a reduced flow map using the set of mergeable states chosen in step 3	
Step 5	Determine the number of state variables, assign state variable names, choose a race-free state assignment for the states in the reduced flow map, and obtain the composite Karnaugh map for the next state and external outputs.	If a race-free assignment cannot be obtained, additional states can be added to ensure a race-free design
Step 6	Obtain hazard-free equations for the next state output functions or the excitation input equations for S-R latches; draw the circuit diagram	Implementations using S-R latches are inherently logic hazard-free if the excitation input equations are logic hazard-free.

$Z(X1,X2)$. This can be done by either closely observing the timing diagram or moving this information to a timing table where comparisons can perhaps be seen more easily. In a timing table, each state is listed in the same sequence it occurs in the timing diagram on a separate row with the corresponding input/output conditions. The timing table for FM1 is shown on the following page and, as one can observe, it is simply a tabular form of the timing diagram for FM1.

Notice that state f occurs when $X1\ X2 = 00$ with an output of $Z = 1$ while state h occurs when $X1\ X2 = 00$ with an output of $Z = 0$. The Z output is not just dependent on the inputs $X1$ and $X2$, but also on the previous state of the circuit. The circuit required to generate the Z output must therefore be some form of sequential circuit with feedback and memory capability.

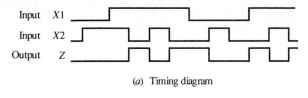

(a) Timing diagram

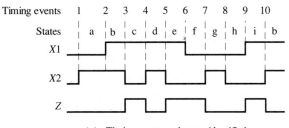

(c) Timing events and states identified

FIGURE 10-2
Fundamental mode machine one (FM1).

Timing table for FM1

State	X1	X2	Z	
a	0	1	0	
b	1	1	0	
c	1	0	1	
d	1	1	0	
e	1	0	1	
f	0	0	1	←
g	0	1	0	
h	0	0	0	← Note that h ≠ f
i	1	0	1	
b	1	1	0	

10-3-3 Obtaining a Primitive Flow Map

The next step in the design procedure is to use the timing diagram or timing table to obtain either a primitive flow table (PS/NS table, or flow chart) or a primitive flow map. We will use the map format rather than the table format. A primitive flow map is a map with one stable state listed in each row of the map. When a flow map is written with more than one stable state in each row, it is often referred to as a reduced flow map (see Fig. E8-8c). An unfilled primitive flow map for FM1 is shown in Fig. 10-3a. Each present state is shown as a separate row and listed on the left side of the map just like the timing table. The external input signals are represented as columns on the top of the map. Next state transitions are represented inside each square corresponding to the external input signals that cause the transition (this will be filled in shortly). The present state output signal Z (Moore output) is listed for each state down the right side of the map.

Once the flow map is set up as shown in Fig. 10-3a, it must be filled in. The flow map can be filled in by transferring the information about FM1 from

PS \ X1 X2	00	01	11	10	PS Z
NS = a					0
b					0
c					1
d					0
e					1
f					1
g					0
h					0
i					1

(*a*) Primitive flow map for FM1 showing each present state as a separate row and the corresponding output.

PS \ X1 X2	00	01	11	10	PS Z
NS = a	–	<u>a</u>	b	–	0
b	–	–	<u>b</u>	c	0
c	–	–	d	<u>c</u>	1
d	–	–	<u>d</u>	e	0
e	f	–	–	<u>e</u>	1
f	<u>f</u>	g	–	–	1
g	h	<u>g</u>	–	–	0
h	<u>h</u>	–	–	i	0
i	–	–	b	<u>i</u>	1

– = don't care

(*b*) Completed primitive flow map for FM1.

FIGURE 10-3
Primitive flow map for FM1.

either its timing diagram or its timing table. Beginning in state a (or row a), FM1 stays in state a for input signals $X1\ X2 = 01$; consequently, state a is listed under the column $X1\ X2 = 01$ and underlined as shown in Fig. 10-3b. When the next state is the same state as the present state, the state is underlined to show that it is a stable state. In state a, FM1 moves to state b for input signals $X1\ X2 = 11$; consequently, state b is listed in column $X1\ X2 = 11$ as shown in Fig. 10-3b. State b is not underlined, since it is an unstable state in state a. The input signals $X1\ X2 = 00$ and 10 do not occur while in state a; consequently, don't cares are entered in columns $X1\ X2 = 00$ and 10 in state a. This completes all entries in the primitive flow map for state a. Following the same procedure for each of the other states results in the completed primitive flow map for FM1 in Fig. 10-3b.

10-3-4 Using an Implication Table

The next step in the design process is to obtain an implication table (also called a pair chart or a merger table). An implication table can be used to identify all pairs of equivalent or compatible states. By equivalent or compatible states we mean states whose next state and output entries are not conflicting and therefore can be merged. The procedure for obtaining the implication table is basically the same procedure that was presented in Chapter 9, subsection 9-3-2, for synchronous sequential design. When comparing pairs of states to determine if they are equivalent or compatible, implied equivalent or implied compatible, or nonequivalent or noncompatible states, the stability or nonstability of the states being compared is unimportant. Multiple passes through the implication table are generally required to determine all nonequivalent or noncompatible states, with the remaining states being either equivalent or compatible.

The first pass through the implication table for FM1 is shown in Fig. 10-4a. Additional passes can be made through the implication table as shown in Fig. 10-4b to determine all nonequivalent or noncompatible states (nonmergeable states). The squares in an implication table that are left showing implied equivalent or compatible states, after multiple passes are made to determine all nonequivalent or noncompatible states, are equivalent or compatible states, that is, mergeable states, by elimination.

It is not necessary to draw the implication table for each pass, since the mergeable states can be determined by making multiple passes using the same implication table. The implication table for FM1 indicates that the following pairs of states can be merged.

a,b a,d b,d a,g a,h c,e b,g b,h c,i c,f d,g d,h e,i e,f f,i g,h

10-3-5 Merger Diagrams

It is possible to obtain larger groups of equivalent or compatible states among the pairs of states listed above. This can be done in an organized manner for the entire primitive flow map by using a merger diagram. A merger diagram is a visual tool that allows the designer to observe patterns to obtain groups of mergeable states. To draw a merger diagram, each state in a primitive flow map is listed and then

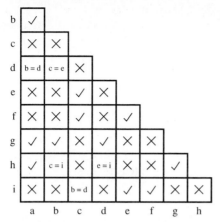

(a) First pass through the implication table for FM1.

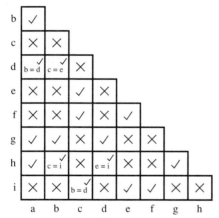

(b) Implication table for FM1 after all nonequivalent and/or noncompatible states have been determined.

FIGURE 10-4
Implication table for FM1.

circled, and equivalent or compatible states are connected by line segments. Two examples of merger diagrams are shown in Fig. 10-5.

The merger diagram in Fig. 10-5a is for a primitive flow map with four states, while the merger diagram in Fig. 10-5b is for a different primitive flow map that contains seven states. Interpreting the merger diagram in Fig. 10-5a, one can obtain the following sets of mergeable states.

	a b,c,d
or	a,b c,d
or	a b,c d
or	a c b,d

Each of these four sets will result in a circuit with a slightly different complexity; therefore, only one set needs to be chosen. The first and second sets

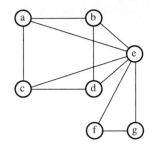

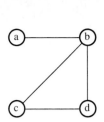

(*a*) A merger diagram for a primitive flow map with four states.

(*b*) A merger diagram for a primitive flow map with seven states.

FIGURE 10-5 Merger diagrams.

of mergeable states result in two states, but the third and fourth sets results in three states, and should not be chosen. In general, a design with a larger number of states is more complex and hence the reason for seeking to reduce the number of states required for a design. The procedure for merging states will be presented after the merger diagram for FM1 is obtained.

The merger diagram in Fig. 10-5*b* provides the following sets of mergeable states, and each set results in two states. Again, only one of these sets needs to be chosen.

$$a,b,c,d,e \quad f,g$$

or
$$a,b,c,d \quad e,f,g$$

The primitive flow map for FM1 contains nine different states and its merger diagram is shown in Fig. 10-6.

The merger diagram for FM1 has two groups of strongly connected states. By merging these two groups of states, a reduced flow map containing only two states instead of nine states can be obtained. The two states are obtained by merging the following groups of states:

$$a,b,d,g,h \quad c,e,f,i$$

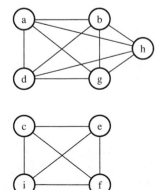

FIGURE 10-6 Merger diagram for FM1.

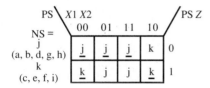

(*a*) Merging states in the
first and second rows.

PS \ *X*1 *X*2   PS Z
NS =
      00 01 11 10

| NS = j (a, b, d, g, h) | $\underline{j}$ | $\underline{j}$ | $\underline{j}$ | k | 0 |
| k (c, e, f, i) | $\underline{k}$ | j | j | $\underline{k}$ | 1 |

(*b*) Changing to a single state
name for each row.

FIGURE 10-7
Reduced flow map for FM1.

10-3-6 Obtaining a Reduced Flow Map

After choosing a set of mergeable states, the next step in the asynchronous design process is to use those groups to obtain a reduced flow map. This process is initiated by listing the states to be merged in the first and second rows of the reduced flow map of FM1, as shown in Fig. 10-7*a*. For the first group of mergeable states (states a,b,d,g,h listed in the first row of the reduced flow map), one begins by transferring the next state and output information from the primitive flow map to the first row of the reduced flow map.

To merge each of the other states in the first row, states b,d,g, and h, their next state information is added in turn to the information already present in the first row. Do not change the names of any of the states at this time, since this will be done later. During the merging process, stable states always have precedence over unstable states. For example, when the next state information for state b is added to the first row, state b is underlined to indicate that it is stable, and when state h is finally added, it is underlined to indicate it is a stable state.

After merging the states listed in the first row, the states listed in the second row must be merged in the same way. Figure 10-7*a* shows the result of merging the states in the primitive flow map to obtain a reduced flow map for FM1. Since there are only two rows in the reduced flow map, we can now provide a single state name for each row in the map as shown in Fig. 10-7*b*.

10-3-7 Making Race-Free State Assignments Using a Transition Diagram

The number of state variables n is now determined by the relationship $2^n \geq s$ where s represents the number of states or rows in the reduced flow map. State variable names can now be assigned. If more than two state variables are required, state assignment codes should be chosen so that no more than one bit changes between each stable state transition to prevent the possibility of critical races. This is called making a race-free state assignment. A race is caused by different delay paths in the circuit. For a noncritical race, the final stable state that the circuit reaches is the same regardless of which state variable changes first and wins the race; however, for a critical race, the final stable state that the circuit reaches can be unpredictable. By choosing a race-free state assignment, asynchronous circuit operation will be predictable, since all races are eliminated. To achieve a race-free assignment for stable state transitions for an asynchronous machine, it is necessary to assign state codes that are adjacent; that is, their binary values must be different by only one bit.

Example 10-1. Obtain a race-free state assignment for the Mealy asynchronous sequential machine represented by the state diagram shown in Fig. E10-1a.

Solution A flow map for the Mealy machine can be drawn from the state diagram as shown in Fig. E10-1b. Notice that the Mealy output is a function of both the state and the input and therefore is included beside each next state entry in the flow map.
To obtain a race-free assignment, a transition diagram can be drawn as illustrated in Fig. E10-1c. External input signals and Mealy or Moore output signals need not be considered when constructing a transition diagram, since only the state transitions are of importance.
Once the transition diagram is drawn, a race-free state assignment can be chosen. Since there are four states in the transition diagram, two state variables

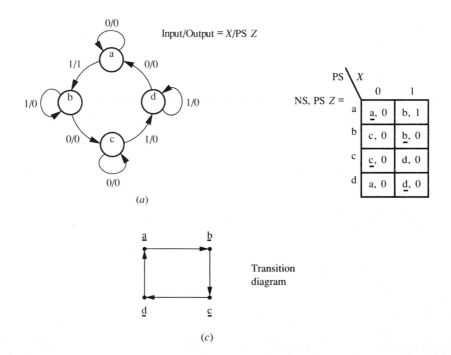

(a)

(c)

Transition diagram

FIGURE E10-1

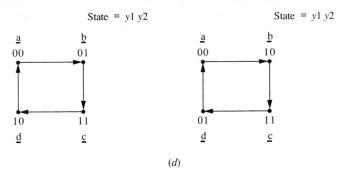

(d)

FIGURE E10-1
(*continued*)

must be chosen; we chose $y1$ and $y2$. In this case several different race-free state assignments are possible, and two different race-free state assignments are shown in Fig. E10-1d.

There are cases when adjacent state codes cannot be chosen between stable state transitions because of an insufficient number of states in the reduced flow map. The procedure in this situation is to add one or more states to allow the assignment of adjacent state codes.

Example 10-2. The transition diagrams shown in Fig. E10-2a and b each represent a different asynchronous machine, FM2 and FM3. Obtain a race-free state assignment for each machine.

Solution Notice that a race-free state assignment cannot be made for a machine, such as FM2, with only three states. By increasing the number of states to four, FM2 can be made race-free. To allow the assignment of adjacent state codes, and hence obtain a race-free state assignment, state d is added to the transition diagram as illustrated in Fig. E10-2c. The state assignment code map in Fig. E10-2d illustrates that a and b have adjacent codes as well as a and c, but that b and c do not have adjacent codes. A stable state transition from state b to state c or from state c to state b must pass through unstable or transitory state d, since the state assignment code for d has an adjacent code to both b and c.

Because of the transition from state a to c in Fig. E10-2b, a minimum of three state variables are required to obtain a race-free state assignment for FM3. One possible race-free state assignment is shown in the transition diagram in Fig. E10-2e. Notice that three unstable or transitory states (e, f, and g) must be used to allow the stable state transitions to be race-free. The state assignment code map for FM3 shown in Fig. E10-2f helps to identify adjacent assignments. A universal race-free state assignment for an asynchronous machine with four stable states is an assignment that is race-free for any asynchronous fundamental mode machine with four stable states. The state assignment for FM3 is not universal, since stable states b and d do not have adjacent codes. Obtaining a universal race-free state assignment code for any asynchronous machine that contains four stable states is left as an exercise for the student.

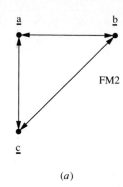

FM2

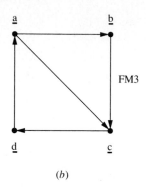

FM3

(a)   (b)

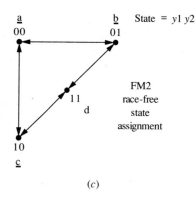

State = y1 y2

FM2
race-free
state
assignment

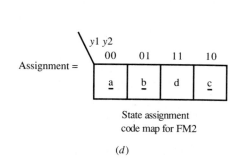

State assignment
code map for FM2

(c)   (d)

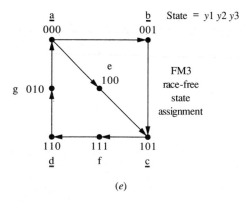

State = y1 y2 y3

FM3
race-free
state
assignment

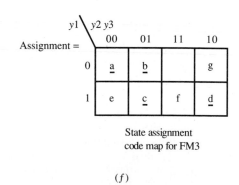

State assignment
code map for FM3

(e)   (f)

FIGURE E10-2

Ideally, a race-free state assignment is chosen for an asynchronous design; however, any state assignment that is critical race-free can be used.

10-3-8 Composite Karnaugh Map for the Next State and External Outputs

The transition diagram for FM1 is shown in Fig. 10-8. A single state variable y is assigned. State code 0 is assigned to state j, and state code 1 is assigned to state k. Notice that there cannot be a critical race due to the state assignment for FM1, since it contains only one state variable.

The composite Karnaugh map for the next state and external outputs for FM1 is shown in Fig. 10-9. This map is obtained from the reduced flow map in Fig. 10-7*b* and the state assignments made in the transition diagram in Fig. 10-8.

Reading the composite Karnaugh map for the present state output Z and the next state output Y, we can obtain the following Boolean equations. Notice that we elected to write the equation for Y in terms of the sum of products for the 0s of the function Y.

$$Z = y$$

$$\overline{Y} = \overline{y} \cdot \overline{X1} \ + \ X2$$

or $$Y = (y + X1) \cdot \overline{X2}$$

Notice that the equation for the next state output Y is hazard free, since the prime implicants in the SOP equation overlap in the Karnaugh map (they are cross covered). Remember that it is important to obtain hazard free equations when designing asynchronous circuits to prevent spurious signals or glitches from occurring in the circuit.

Notice that the next state output Y for FM1 is a product of sums equation for the 0s of the function. The equation Y is also expressed in OR/AND form and can be expressed in NOR/NOR form by applying double complementation and DeMorgan's theorem as follows (see Chapter 3, Section 3-7):

$$Y = \overline{\overline{(y + X1) \cdot \overline{X2}}}$$

$$= \overline{\overline{y + X1} + X2} \qquad \text{NOR/NOR form}$$

Using the NOR/NOR form of the equation for the next state output Y, the fundamental mode asynchronous sequential logic circuit for FM1 can be drawn as shown in Fig. 10-10. The delay box is only used to assist in drawing the circuit and can be removed once the circuit is drawn. It should now be obvious that the design we have carried out for FM1 is the design of a cross-coupled NOR gate-latch circuit. Using the tools and procedures introduced for this design serves to illustrate how other fundamental mode circuits can be designed.

j   k   State $= y$

0   1

FIGURE 10-8
Transition diagram and state assignment for FM1.

y \ $X1\,X2$	00	01	11	10	$PS\ Z$

$Y =$

		00	01	11	10	
j	0	0	0	0	1	0
k	1	1	0	0	1	1

FIGURE 10-9

Composite Karnaugh map for the next state and external outputs for FM1.

Example 10-3. Design an asynchronous circuit with two inputs and one output. Input $X1$ is driven by the system clock. The second input $X2$ is an asynchronous input. If the asynchronous input $X2$ is a logic 1 when input $X1$ changes from 1 to 0, output Z follows the waveform of input $X1$ for one cycle, that is, until $X1$ changes again from 1 to 0. If the asynchronous input $X2$ is a logic 0 when input $X1$ changes from 1 to 0, output Z is a logic zero until $X1$ changes again from 1 to 0. The circuit should be designed so that it is logic hazard-free, has a race-free state assignment, and can recover from unused or illegal states.

(*a*) Obtain a timing diagram from the word description or specification of the circuit.

(*b*) Use the timing diagram to obtain a primitive flow map.

(*c*) Find all the pairs of equivalent or compatible states by utilizing an implication table.

(*d*) Use the pairs of equivalent or compatible states to draw a merger diagram to determine a set of mergeable states that can provide a minimum number of states for the design.

(*e*) Obtain a reduced flow map using the set of mergeable states chosen in (*d*), and assign a single state name to each group of states that is merged.

(*f*) Draw a transition diagram for the reduced flow map in (*e*) and then modify it, if necessary, to obtain a race-free state assignment.

(*g*) Obtain a composite Karnaugh map for the next state and external outputs without worrying about a race-free state assignment or illegal state recovery. Modify the composite Karnaugh map to obtain a map that will insure that (1) a race-free state assignment exists; that is, critical races cannot occur; (2) illegal state recovery exists; that is, the circuit cannot hangup in an illegal state.

(*h*) Obtain logic hazard-free equations for the next state and external outputs and draw a gate-level asynchronous sequential circuit diagram for the design.

(*i*) Use the composite Karnaugh map obtained in (*g*) for a race-free state assignment and illegal state recovery to obtain composite Karnaugh maps for the excitation inputs of S-R NOR latches.

(*j*) Obtain logic hazard-free equations for the S and R excitation inputs to S-R NOR latches. Then draw an asynchronous sequential circuit diagram for the design. Also draw the circuit using S-R NAND latches.

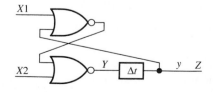

FIGURE 10-10

Fundamental mode asynchronous sequential circuit for FM1.

Solution

(a) Using the word description or specification of the circuit, one can obtain the timing diagram illustrated in Fig. E10-3a. The timing events and the states are named as shown in the figure.

(b) The primitive flow map can be obtained directly from the timing diagram as shown in Fig. E10-3b. There are seven rows, or one row for each state, and only one state is stable per row.

(c) The implication table for the design is shown in Fig. E10-3c.

(d) A merger diagram for the design is illustrated in Fig. E10-3d. One set of mergeable states is listed below the diagram in Fig. E10-3d.

(e) Figure E10-3e shows the reduced flow map for the set of mergeable states a,b,f,g c,d e. This set provides a state machine with just three states named h, i and j as shown in the reduced flow map in Fig. E10-3f.

(f) A transition diagram for the circuit is shown in Fig. E10-3g. Two state variables are required for three states, and these variables are chosen as y1 and y2.

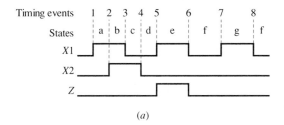

(a)

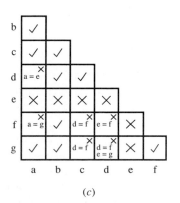

$-$ = don't care

(b)

(c)

FIGURE E10-3

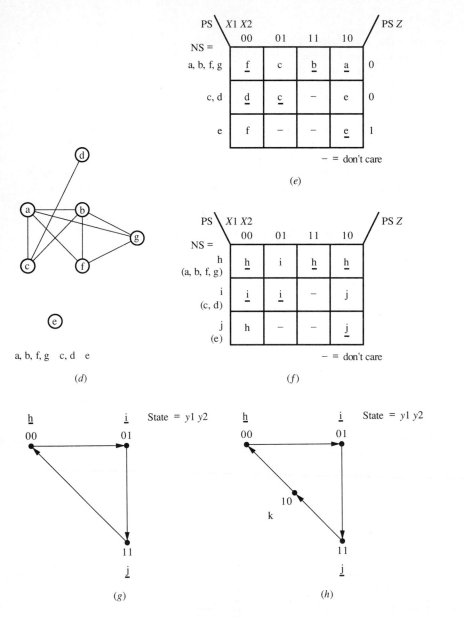

FIGURE E10-3
(*continued*)

Each state is assigned a binary code as shown in the figure. Since a race-free state assignment cannot be made for a state machine with only three stable states, a fourth transitory state, k, has been added to the modified transition diagram in Fig. E10-3*h* to provide a race-free state assignment.

(g) Figure E10-3*i* shows a composite Karnaugh map for the next state and external outputs. Don't care conditions are listed in the row representing the next state

$y1\,y2$ \ $X1\,X2$ — PS Z

$Y1\,Y1 =$

		00	01	11	10	
h	00	00	01	00	00	0
i	01	01	01	–	11	0
j	11	00	–	–	11	1
k	10	–	–	–	–	–

– = don't care

(i)

$y1\,y2$ \ $X1\,X2$ — PS Z

$Y1\,Y2 =$

		00	01	11	10	
h	00	00	01	00	00	0
i	01	01	01	–	11	0
j	11	10	–	–	11	1
k	10	00	01	00	00	–

– = don't care

(j)

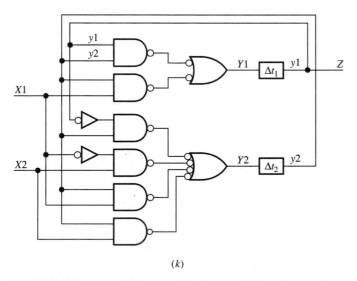

(k)

FIGURE E10-3
(*continued*)

yi Yi	Si Ri
0 0	0 X
0 1	1 0
1 0	0 1
1 1	X 0

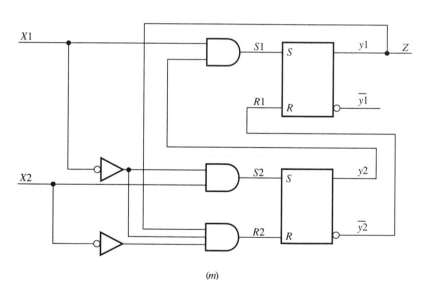

$$y1\ y2\ \backslash\ X1\ X2$$

S1 S2 =	00	01	11	10
00	00	01	00	00
01	0X	0X	−	1X
11	X0	−	−	XX
10	00	01	00	00

$$y1\ y2\ \backslash\ X1\ X2$$

R1 R2 =	00	01	11	10
00	XX	X0	XX	XX
01	X0	X0	−	00
11	01	−	−	00
10	1X	10	1X	1X

− = X = don't care

(*l*)

(*m*)

FIGURE E10-3 (*continued*)

values for state k. This is bad design practice. One should insure that the next state values for state k will place the circuit in a reasonable stable state, that is, a stable state in the primary sequence. The race conditions in the first column are also ignored. One should also insure that race conditions do not occur that cause a circuit failure by choosing a race-free state assignment.

The composite Karnaugh map in Fig. E10-3*j* has two improvements. There is a race condition in Fig. E10-3*i* due to state assignment code 11 the

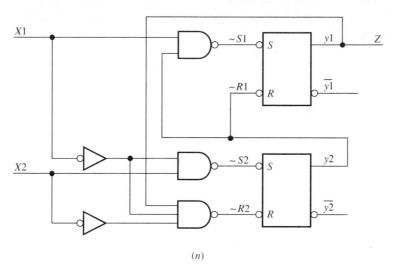

(n)

FIGURE E10-3   (continued)

fourth column of state j changing to state assignment code 00 in the first column
of state j. Since the first column has two stable states, the circuit can end
up in either stable state h (which it should) or stable state i (which it should
not). By changing state assignment code 00 in the first column of state j to
state assignment code 10 as shown in the transition diagram in Fig. E10-3h,
and then providing a state assignment code of 00 in the first column of state k,
the race condition is eliminated. This change in the state assignment codes in the
composite Karnaugh map for the next state and external outputs will force the
circuit to follow the state transitions shown in the transition diagram in Fig. E10-
3h. To insure that illegal state recovery occurs, the state assignments codes for
the don't cares in the second, third, and fourth columns of state k in Fig. E10-
3i are chosen such that the circuit changes to one of the stable states in the
other rows of the map. If the don't cares in the second, third, and fourth
columns of state k were all chosen as stable states, and one of these states
should accidentally occur, then the circuit could hangup. To provide a race-
free state assignment and illegal state recovery, the changes described above
have been made in the composite Karnaugh map shown in Fig. E10-3j.

(h) Reading the composite Karnaugh map for the next state and external outputs,
the following logic hazard-free equations can be written.

$$Y1 = y1 \cdot y2 \ + \ y2 \cdot X1$$

$$Y2 = \overline{y1} \cdot y2 \ + \ \overline{X1} \cdot X2 \ + \ y2 \cdot X1 \ + \ y2 \cdot X2$$

$$Z = y1$$

The term $y2 \cdot X2$ is necessary to insure that $Y2$ is logic hazard-free. Using these
equations, one can draw the fundamental mode asynchronous circuit diagram
as shown in Fig. E10-3k.

(i) S-R NOR latches can also be used to implement asynchronous circuits. As
in synchronous design, one must obtain the composite Karnaugh maps for the

excitation inputs. These maps are obtained from the composite Karnaugh map for the next state and external outputs, using the excitation table for the *S-R* type bistable. The excitation table for the *S-R* type bistable and the resulting composite Karnaugh maps for the *S* and *R* excitation inputs are shown in Fig. E10-3*l*.

(*j*) Since cross-coupled NOR and NAND latches can be demonstrated to be logic hazard-free, one simply has to write logic hazard-free excitation input equations to drive the inputs of these devices to obtain a logic hazard-free circuit. The following logic hazard-free *S* and *R* excitation input equations can be written from the composite Karnaugh maps in Fig. E10-3*l*.

$$S1 = y2 \cdot X1$$

$$S2 = \overline{X1} \cdot X2$$

$$R1 = \overline{y2}$$

$$R2 = y1 \cdot \overline{X1} \cdot \overline{X2}$$

Using these excitation input equations and the output equation $Z = y1$, one can draw the asynchronous sequential circuit in Fig. E10-3*m* using *S-R* NOR latches. Complementing both sides of the *S* and *R* excitation input equations obtained above allows one to draw the asynchronous sequential circuit in Fig. E10-3*n* using ~*S*-~*R* latches, that is, *S-R* NAND latches.

When a design is specified using a state diagram or ASM chart, it may be the case that many of the design tools presented in this section do not need to be used if state reduction is not required. The following example illustrates the design of a fundamental mode circuit beginning with a state diagram description.

Example 10-4. Design a fundamental mode asynchronous sequential circuit with the following behavior. The circuit has two external inputs *D* and *C* and one output *Z*. The circuit, which we will refer to as FM4, must perform as indicated by the state diagram shown in Fig. 10-4*a*. The circuit should be designed so that it is logic hazard-free.

Solution For this design one does not need to obtain a primitive flow map, implication table, merger diagram, reduced flow map, or transition diagram since FM4

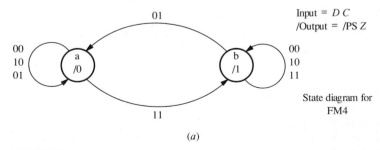

(*a*)

FIGURE E10-4

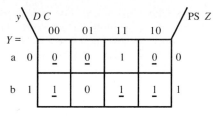

Composite Karnaugh
map for FM4

(b)

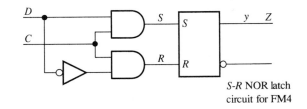

Gate level circuit
for FM4

(c)

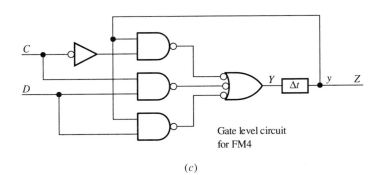

y	Y	S	R
0	0	0	X
0	1	1	0
1	0	0	1
1	1	X	0

S-R NOR latch
circuit for FM4

(e)

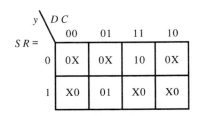

Composite Karnaugh
map for the S and R
excitation inputs for FM4

(d)

S-R NAND latch
circuit for FM4

(f)

FIGURE E10-4
(*continued*)

cannot be reduced to less than two states. For two states, one state variable is required which will be called y. Assigning the state code 0 to state a and state code 1 to state b, one can write the Moore output as $Z = y$. The composite Karnaugh map for the next state and external outputs can be drawn using the state diagram as shown in Fig. E10-4b.

Reading the composite Karnaugh map for FM4, a logic hazard-free Boolean equation for the next state output Y can be obtained for the 1s of the function as follows:

$$Y = y \cdot \overline{C} + D \cdot C + y \cdot D$$

The term $y \cdot D$ is included in the equation to insure that the circuit is logic hazard-free. A circuit implementation for FM4 using gate-level logic is shown in Fig. E10-4c.

Using the composite Karnaugh map for the next state output Y and the excitation table for the S-R type bistable, we can obtain the composite Karnaugh map for the S and R excitation inputs as shown in Fig. E10-4d.

Reading the composite Karnaugh map for the S and R excitation inputs, the following logic hazard-free Boolean equations can be written for the S and R excitation inputs for FM4.

$$S = D \cdot C$$

$$R = \overline{D} \cdot C$$

The excitation input equations obtained above allow us to draw a circuit for FM4 as shown in Fig. E10-4e using an S-R NOR latch. Complementing both sides of the S and R excitation input equations provides us with the equations to obtain the circuit illustrated in Fig. E10-4f using an S-R NAND latch.

10-3-9   Essential Hazards

There is one additional type of hazard we have not discussed in the design of fundamental mode asynchronous sequential circuits. This hazard is called an 'essential hazard' and occurs only in asynchronous sequential circuits as a result of certain specified next state sequences. Essential hazards, like logic hazards, are due to unequal signal delay paths that can exist in a circuit; however, essential hazards cannot be fixed by simply adding additional terms in the next state output or excitation input equations. To fix an essential hazard, one must first determine that such a hazard can exist. If such a hazard can exist, it is necessary to insure that the feedback delays for the state variables in the asynchronous circuit implementation are adequate, so that the essential hazard does not occur.

To see how an essential hazard can occur, observe the flow map presented in Fig. 10-11. Assume that the circuit is stable in state a and the external input signal X changes from 0 to 1 as shown by arrow 1. The circuit ends up in stable state b, which is the expected state for proper circuit operation. It has been proven (see Unger in the references at the end of this chapter) that an essential hazard can occur in a fundamental mode asynchronous sequential circuit if the circuit can end up in a different state (than the expected state) as a result of three consecutive changes in a single external input signal compared to only one change in the same external input signal. Now consider what can happen to the circuit represented by the flow map in Fig. 10-11 when the external input signal X changes from 0

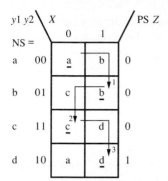

FIGURE 10-11
Reduced flow map for a circuit containing an essential hazard.

to 1 to 0 and back to 1 (three consecutive changes). Notice that the circuit can end up in state d due to three consecutive signal changes as shown by arrows 1, 2, and 3, and not the expected state b due to only a single signal change. Using Unger's test, an essential hazard can occur in the circuit represented by the flow map in Fig. 10-11 since the circuit can end up in state d when, in fact, only one signal change occurs which should place the circuit in state b.

A composite Karnaugh map for the next state and external outputs for the reduced flow map in Fig. 10-11 is shown in Fig. 10-12. Two state variables are required for four states, and $y1$ and $y2$ are chosen. As usual, a race-free state assignment is also chosen. The essential hazard described above can occur when the external input signal X changes from 0 to 1 and this change is not seen by the next state output signals $Y1$ and $Y2$ at the same time as the present state output signals $y1$ and $y2$, due to different propagation delay paths in the circuit. In other words, whether or not an essential hazard can occur in a circuit is dependent on a race between the change in an external input variable (X in this case) and the change in the state variables ($y1$ and $y2$ in this case). An essential hazard cannot occur in a circuit that contains only one state variable.

From the composite Karnaugh map shown in Fig. 10-12, we see that the essential hazard can occur for the following conditions. First, assume that the circuit is in state a with $y1 = y2 = 0$. Second, assume that the external input signal X changes from 0 to 1. The circuit generating the next state output $Y2$ now sees an input signal $X = 1$, and the two present state output signals $y1 = 0$ and

```
y1 y2 \ X              PS Z
            0     1
Y1 Y2 =
   a   00  00    01    0

   b   01  11    01    0

   c   11  11    10    0

   d   10  00    10    1
```

FIGURE 10-12
Composite Karnaugh map for the next state and external outputs for a circuit containing an essential hazard.

$y2 = 0$. These inputs cause $Y2$ to change to 1 and place the circuit in stable state $Y1\ Y2 = 01$. Third, now suppose the circuit generating the next state output $Y1$ sees an input signal $X = 0$ (this assumes that the signal is being delayed to the $Y1$ circuit). With $X = 1$, present state output signal $y1 = 0$, and the newly generated present state output $y2 = 1$, the next state output $Y1$ changes to 1 ($Y1$ should remain at 0 for proper circuit operation, but because of the delay in X to the $Y1$ circuit it doesn't). The circuit now moves to stable state $Y1\ Y2 = 11$ instead of staying in stable state 01. Fourth, now assume both the $Y1$ and $Y2$ circuits finally see the input signal $X = 1$. With the circuit in stable state 11 ($Y1 = Y2 = 1$) the circuit ends up in stable state 10 instead of the intended stable state 01. This problem can be fixed; that is, the essential hazard can be removed, by adding the proper amount of delay in the feedback path of $Y2$ so that both next state outputs $Y1$ and $Y2$ see the X input signal change and the present state output signals $y1$ and $y2$ all in the same time frame.

Full understanding of the operation of a circuit with an essential hazard, as discussed above, usually requires recording the delays in each part of the asynchronous circuit implementation and deriving a timing diagram for the input and state variable signals in the circuit. Logic hazard-free equations for the next state outputs $Y1$ and $Y2$, and the resulting circuit diagram for the circuit we are discussing are shown in Fig. 10-13.

The following discussion provides additional insight into the operation of a circuit with an essential hazard. This presentation parallels the discussion above and assumes that the circuit in Fig. 10-13 is initially in state $y1\ y2 = 00$. The

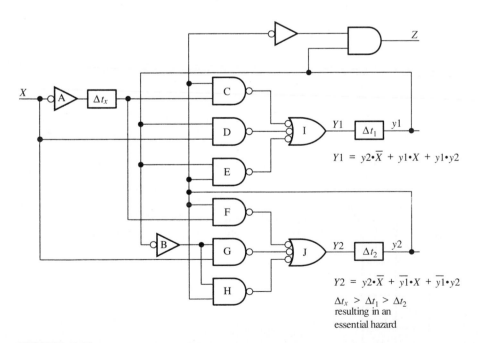

$$Y1 = y2 \cdot \overline{X} + y1 \cdot X + y1 \cdot y2$$

$$Y2 = y2 \cdot \overline{X} + \overline{y1} \cdot X + \overline{y1} \cdot y2$$

$\Delta t_x > \Delta t_1 > \Delta t_2$
resulting in an
essential hazard

FIGURE 10-13
Circuit diagram containing an essential hazard.

lumped delays Δt_x, Δt_1, and Δt_2 illustrate how the circuit can contain an essential hazard for the conditions $\Delta t_x > \Delta t_1 > \Delta t_2$. Inverter B and gate G are responsible for $Y2$ changing from 0 to 1 when the X input signal changes from 0 to 1. Because of the short delay provided by Δt_2, $y2$ becomes 1 ($y2 = Y2$ after Δt_2). Assuming that the output at the Δt_x delay is still 1 (it hasn't made the change to 0 just yet), gate C with inputs 1 and $y2 = 1$ causes output $Y1$ to go to 1. Due to the short delay provided by Δt_1, $y1$ becomes 1 ($y1 = Y1$ after Δt_1). Now assume that the output of the Δt_x delay finally goes to 0; gate F is responsible for $Y2$ changing from 1 to 0. With $Y1 = 1$ and $Y2 = 0$, the circuit ends up in state 10. Prevention of the essential hazards often requires a close analysis of the problem, as we have demonstrated by using the circuit diagram. One can basically understand that to remove the essential hazard required either changing the delay time for Δt_x such that $\Delta t_2 > \Delta t_x > \Delta t_1$ or adding a delay in the feedback path for $y2$ such that $\Delta t_2 + \Delta t_{add} > \Delta t_x > \Delta t_1$.

To insure the proper operation of a fundamental mode asynchronous circuit, a designer must check for essential hazards in the circuit and fix the delay paths so that essential hazards will not occur. In order for fundamental mode asynchronous sequential circuits to operate properly, without hanging up and generating spurious signals, a designer must strive to design circuits that have

1. illegal state recovery,
2. a race-free or critical race-free state assignment,
3. logic hazard-free equations, and
4. essential hazard removal.

10-4 ASYNCHRONOUS SEQUENTIAL DESIGN OF PULSE MODE CIRCUITS

Looking back in Fig. 10-1, one can observe that the input signals to pulse mode circuits are simply the presence or absence of pulses (the positive pulses being used represent a signal transition from 0 to 1 back to 0). As in fundamental mode, this places certain restrictions on the manner in which the input signals, the pulses, can be applied to the circuit. Pulse mode asynchronous sequential circuit input signal restrictions can be stated by the following rules.

Rule 1: Only one input signal pulse is allowed to occur at one time.

Rule 2: Before the next input pulse is allowed to occur, the circuit must be given time to reach a new stable state via a single state change.

Rule 3: Each applied input pulse has a minimum pulse width that is determined by the time it takes to change the slowest flip-flop in the circuit to a new stable state.

Rule 4: The maximum pulse width of an applied input pulse must be sufficiently narrow so that it is no longer present when the new present state output signals become available.

Notice that there are four input signal restrictions for pulse mode circuits compared to only two input signal restrictions for fundamental mode circuits. If any of these rules are violated the pulse mode circuit will most likely not function as desired. Rule 1 restricts the input signal pulses from occurring at the same time; that is, the input signal pulses must be mutually exclusive. A pulse mode asynchronous circuit that is operated such that the input signal pulses are mutually exclusive is often referred to as operating in fundamental mode. If you will recall, in Chapter 8, subsections 8-4-3 and 8-4-4, the signal applied to the control input of a gated *J-K* latch and a gated *T* latch had the restriction imposed by Rule 4. Precisely controlling the maximum pulse width of each pulse in a stream of pulses, as well as the propagation delays for the circuits they are driving, makes the design of pulse mode circuits somewhat undesirable, compared to both clock mode synchronous sequential circuits and fundamental mode asynchronous sequential circuits. If the maximum pulse width is not controlled or a double-rank circuit is not used, each input pulse can cause more than one state change (which is not allowed in pulse mode design). After we present the basic design fundamentals of pulse mode circuits, the double-rank circuit principle will be introduced to show how pulse mode circuits can be designed to function more reliably.

10-4-1   Principles of Pulse Mode Circuits

Pulse mode asynchronous circuits must utilize bistable memory devices, either latches or flip-flops, to keep the circuit in a stable state when no input pulses are being applied to the circuit. Since only one input signal pulse can occur at one time (Rule 1), a pulse mode circuit must respond to each input pulse by causing a single state change (Rule 2) from one stable state to the next stable state. Unstable states do not occur in pulse mode circuits, thus removing the task of selecting a race-free or critical race-free state assignment. To represent the presence of an input signal pulse when designing pulse mode circuits, each input signal pulse is written in uncomplemented form, that is, $X1, X2, \ldots$.

A pulse mode circuit is to be designed that has two inputs $X1$ and $X2$, one Mealy output $Z1$, and one Moore output $Z2$. The Mealy output is coincident with the third consecutive pulse on input $X2$, and the Moore output occurs after the third consecutive pulse on input $X2$, or after any pulse that occurs on input $X1$. The state diagram for the circuit based on the word description is shown in Fig. 10-14 and will be referred to as pulse mode machine one (PM1). Notice that PM1 is a mixed type machine since it contains both a Mealy and a Moore type output.

The flow map for PM1 can be obtained from the state diagram as shown in Fig. 10-15. Notice that the input conditions $X1\ X2 = 00\ 01\ 11$ and 10 that are normally shown in a flow map for either a synchronous or fundamental mode asynchronous design do not appear in a pulse mode design. Remember, in a pulse mode circuit the presence of a pulse at an input causes the circuit to change its state; however, the absence of a pulse causes no change in the circuit and

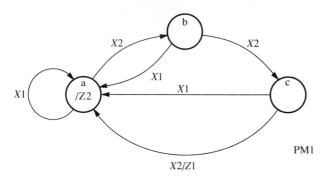

FIGURE 10-14
State diagram for pulse mode
machine one (PM1).

hence represents unimportant information. To represent the presence of a pulse in a pulse mode circuit, only the name of the pulse is listed. In the flow map for PM1, $X1$ represents input condition 10, and $X2$ represents input condition 01. Notice that $X1$ and $X2$ do not have adjacent codes. Input conditions 00 is not represented in the table because this is unimportant information, and input condition 11 is not represented in the table because this violates the restriction imposed by Rule 1.

A composite Karnaugh map for the next state and external outputs for PM1 is shown in Fig. 10-16a. Two state variables are required, and $y1$ and $y2$ are chosen. To insure illegal state recovery, PM1 is directed to go to state a from unused or illegal state d. The state assignment is arbitrary since a race-free state assignment is not a design requirement for pulse mode circuits. The Mealy and Moore outputs are also chosen to be 0 in state d. A designer could use don't cares in state d, but this is not good design practice because if something can go wrong, it probably will, so one should design the circuit so that it can recover.

For two state variables, two memory storage devices are required. *J-K* flip-flops can be used as *T* flip-flops for the memory storage devices. The composite Karnaugh map for the *T* excitation inputs is shown in Fig. 10-16b.

Since only one input pulse either $X1$ or $X2$ can occur at one time in a pulse mode circuit (also recall $X1$ and $X2$ do not have adjacent codes), the $T1$ and $T2$ excitation input and Mealy output equations must be obtained by considering each column separately in the composite maps. Logic hazard-free equations for $T1$, $T2$, $Z1$ and $Z2$ can be obtained as follows.

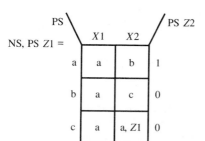

FIGURE 10-15
Flow map for PM1.

$$T_i = y_i \oplus Y_i$$

$Y_1 Y_2, PS Z_1 =$	$y_1 y_2$	X_1	X_2	$PS Z_2$
a	00	00, 0	01, 0	1
b	01	00, 0	11, 0	0
c	11	00, 0	00, 1	0
d	10	00, 0	00, 0	0

$T_1 T_2 =$	$y_1 y_2$	X_1	X_2
	00	00	01
	01	01	10
	11	11	11
	10	10	10

(*a*) Composite Karnaugh map for the next state and external outputs for PM1.

(*b*) Composite Karnaugh map for the *T* excitation inputs for PM1.

FIGURE 10-16
Composite Karnaugh maps for PM1 when using *T* flip-flops.

$$T_1 = y_1 \cdot X_1 \; + \; y_1 \cdot X_2 \; + \; y_2 \cdot X_2$$

$$T_2 = y_2 \cdot X_1 \; + \; \overline{y_1} \cdot \overline{y_2} \cdot X_2 \; + \; y_1 \cdot y_2 \cdot X_2$$

$$Z_1 = y_1 \cdot y_2 \cdot X_2$$

$$Z_2 = \overline{y_1} \cdot \overline{y_2}$$

One should observe from these equations that each product term in the excitation input and Mealy output equations must contain an input pulse signal (*X1* or *X2*), and that no input pulse signal appears in complemented form. An implementation for PM1 using positive edge-triggered *T* flip-flops (*J-K*s used as *T*s) is shown in Fig. 10-17.

With a little more effort, one could also obtain the composite Karnaugh maps for the *S* and *R* excitation inputs and use either *S-R* or ~*S*-~*R* latches in the design of pulse mode asynchronous circuits.

10-4-2   Using Double-Rank Pulse Mode Circuits

To provide a pulse mode design that can function more reliably, a double-rank circuit can be utilized that operates by using the same principle as a master-slave flip-flop. Figure 10-18 illustrates a double-rank circuit using an *S-R* latch as the master rank, and a *D* flip-flop as the slave rank. Each input pulse changes only the state of the master rank. The slave rank changes between input pulses. By utilizing a pulse mode circuit that consists of a master rank followed by a slave rank, the maximum pulse width input signal restriction required by Rule 4 is met without providing a sufficiently narrow pulse. The slave rank prevents the present

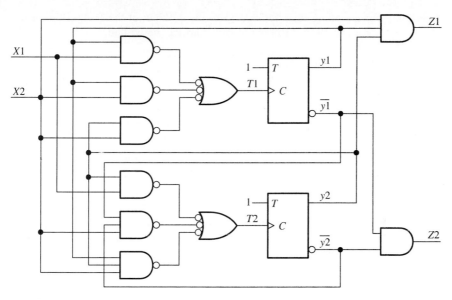

FIGURE 10-17
Pulse mode asynchronous sequential circuit for PM1 using positive edge-triggered T flip-flops.

state output signals, which are fed back to the inputs of the circuit, from changing until after the input pulse is gone.

As one can observe, the design (and also the analysis) of pulse mode circuits is quite similar to the design of synchronous sequential circuits with a few special considerations; that is, a master clock is not used and the presence but not the absence of a pulse is considered. Adding a slave rank allows the implementation of useful and reliable pulse mode circuits.

Example 10-5. Analyze the double-rank pulse mode circuit shown in Fig. 10-18 by obtaining its state diagram.

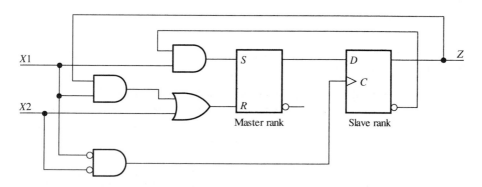

FIGURE 10-18
Double-rank pulse mode asynchronous sequential circuit.

Solution After assigning state variable $y1$ to the q output of the D flip-flop for the slave rank in Fig. 10-18, the following equations can be written.

$$S = \overline{y1} \cdot X1$$

$$R = y1 \cdot X1 \ + \ X2$$

$$Z = y1$$

The next state output equation for the circuit can be found by substituting the S and R excitation input equations into the bistable equation for the S-R latch.

$$Q = S \ + \ q \cdot \overline{R} \qquad S\text{-}R \text{ latch bistable equation (Eq. 8-13)}$$

so, $$Y1 = S \ + \ y1 \cdot \overline{R}$$

and, $$Y1 = \overline{y1} \cdot X1 \ + \ y1 \cdot \overline{(y1 \cdot X1 + X2)}$$

$$= \overline{y1} \cdot X1 \ + \ y1 \cdot (\overline{y1} + \overline{X1}) \cdot \overline{X2}$$

$$= \overline{y1} \cdot X1 \ + \ y1 \cdot \overline{X1} \cdot \overline{X2}$$

for pulse mode, $$= \overline{y1} \cdot X1$$

Complemented input signals are not included; that is, they are removed from the pulse mode design equation since these signals do not convey any information. In the expression $y1 \cdot \overline{X1} \cdot \overline{X2}$, $\overline{X1}$ and $\overline{X2}$ are removed leaving $y1$, which must also be removed, since $y1$ alone cannot cause the circuit to change states; recall that in a pulse mode circuit, every next state product term needs to contain an input pulse signal that can change the state of the circuit or it conveys no information. If the expression were $y1 \cdot X1 \cdot \overline{X2}$, then only $\overline{X2}$ would be removed, leaving the expression $y1 \cdot X1$, which contains an input pulse signal that can cause the ciruit to change states.

Using the next state output equation $Y1$ and the Moore output equation Z, we can obtain the composite Karnaugh map for the next state and external outputs as

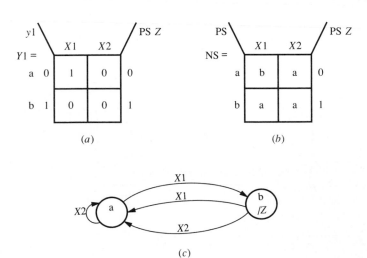

(a)

(b)

(c)

FIGURE E10-5

shown in Fig. E10-5a. Assigning 0 to state a and 1 to state b in Fig. E10-5a allows one to obtain the flow map in Fig. E10-5b.

Using the flow map, we can draw the state diagram for the circuit as shown in Fig. E10-5c.

The following example illustrates an application for a pulse mode type design. The method used to provide the input pulses meets the first three input signal restrictions imposed on pulse mode design, and the last input signal restriction can be met by using a double-rank circuit implementation.

Example 10-6. Design a combination lock that has three inputs $X1$, $X2$, and $X3$ and one Moore output Z that actuates the lock mechanism of a door. The door provides access to a secure laboratory containing expensive electronic and computing equipment. Each person that has authorization to use the laboratory is provided with an optical key the size of a credit card. The laboratory door opens when the key is inserted into an optical reader that is mounted next to the laboratory entrance. Design the combinational lock circuit as a pulse mode asynchronous sequential logic circuit. The Moore output is to occur only after the following sequence of input pulses are applied to the inputs of the circuit: $X2$ $X1$ $X2$ $X3$. Two consecutive $X1$ input pulses will return the circuit to an idle state and should be initiated prior to applying the sequence of pulses required to open the door. The circuit must be designed to recover if it enters an unused or illegal state.

(a) Assume that the optical reader can read three rows of slots on the optical key. Show the sequence of slots for the optical key that will meet the design requirements for the circuit. Row 1 provides the pulses for $X1$, row 2 provides the pulses for $X2$, and row 3 provides the pulses for $X3$.

(b) Draw a state diagram to generate the required combination lock sequence. Pulses that occur outside the intended sequence should return the circuit to the idle state.

(c) Obtain the flow map for the design.

(d) Use an implication table to check if the flow map can be reduced.

(e) Obtain the composite Karnaugh map for the next state and external outputs.

(f) Obtain the composite Karnaugh map for the T excitation inputs for the master rank.

(g) Read the composite Karnaugh maps and obtain the design equations for the combination lock.

(h) Draw the double-rank pulse mode asynchronous circuit implementation for the combination lock.

Solution

(a) A sketch of the optical card reader and the optical key with the required slots is shown in Fig. E10-6a.

(b) Using the word description for the circuit, a partial state diagram can be drawn to follow the required sequence from state a through state e as illustrated in Fig. E10-6b. This state diagram only provides the necessary sequence to open the lock. A more complete state diagram for the circuit is shown in Fig. E10-6c.

(c) The flow map for the design is shown in Fig. E10-6d.

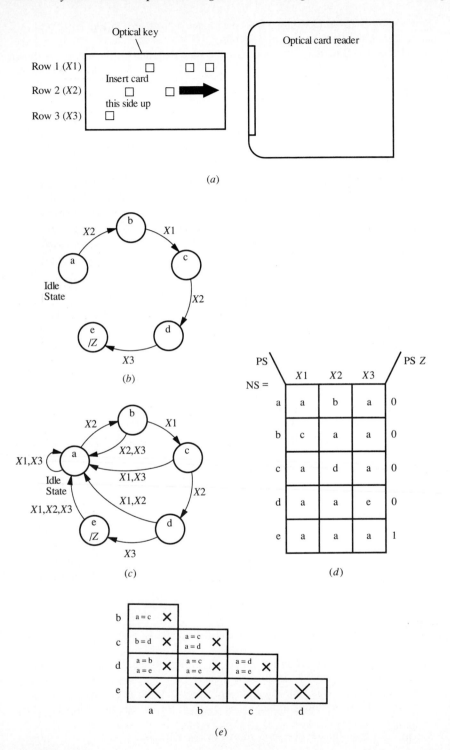

FIGURE E10-6

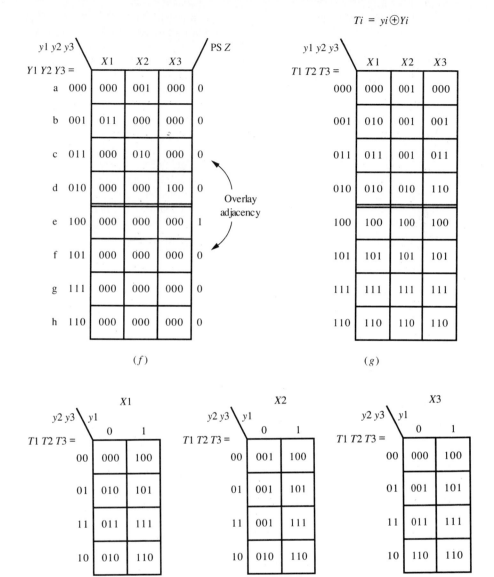

$$Ti = yi \oplus Yi$$

$y1\ y2\ y3$		PS Z		$y1\ y2\ y3$			

(f)

(g)

(h)

FIGURE E10-6
(*continued*)

(*d*) Using the flow map, the implication table is filled in as shown in Fig. E10-6*e* and indicates that state reduction is not possible.

(*e*) The composite Karnaugh map for the next state and external outputs is shown in Fig. E10-6*f*. The state variables were chosen as $y1$, $y2$, and $y3$. Since a pulse mode circuit is stable between pulses, critical races cannot occur; therefore, the state assignment codes can be assigned in an arbitrary manner, except that each state must be given a unique state code as shown in Fig. E10-6f. The unused or illegal states, f, g, and h, each return the circuit to the idle state to insure illegal state recovery.

(f) Using the composite Karnaugh map for the next state and external outputs and the T excitation equation $Ti = yi \oplus Yi$, the composite Karnaugh map for the T excitation inputs can be obtained as shown in Fig. E10-6g. Figure E10-6h shows an alternate way of presenting the composite Karnaugh map for the T excitation inputs. This multiple map method can be used if one keeps in mind that the input pulse shown above each map must be ANDed with the prime implicants obtained for that specific map (just like a map-entered variable). Also notice that the $y1$ variable is shown along the vertical columns in each of the smaller maps so that the information in the larger map in Fig. E10-6g can be quickly written in the squares in each of the smaller maps.

(g) The excitation input and Moore output design equations for the combination lock can be written as follows.

$$T1 = y1 \cdot X1 \; + \; y1 \cdot X2 \; + \; y1 \cdot X3 \; + \; y2 \cdot \overline{y3} \cdot X3$$

$$T2 = \overline{y1} \cdot y3 \cdot X1 \; + \; y2 \cdot X1 \; + \; y2 \cdot \overline{y3} \cdot X2 \; + \; y1 \cdot y2 \cdot X2 \; + \; y2 \cdot X3$$

$$T3 = y2 \cdot y3 \cdot X1 \; + \; y1 \cdot y3 \cdot X1 \; + \; \overline{y1} \cdot \overline{y2} \cdot X2 \; + \; y3 \cdot X2 \; + \; y3 \cdot X3$$

$$Z = y1 \cdot \overline{y2} \cdot \overline{y3}$$

(h) The circuit implementation for the combination lock is left as an exercise for the student.

REFERENCES

1. Unger, S. H., *Asynchronous Sequential Switching Circuits,* John Wiley & Sons, New York, 1969.
2. Kohavi, Z., *Switching and Finite Automata Theory,* 2d ed., McGraw-Hill, New York, 1978.
3. Lind, L. F., and Nelson, J. C. C., *Analysis and Design of Sequential Digital Systems,* Halsted Press, a Division of John Wiley & Sons, New York, 1977.
4. Hill, J. H., and Peterson, G. R., *Introduction to Switching Theory & Logic Design,* 3d ed., John Wiley & Sons, New York, 1981.
5. Johnson, E. L., and Karim, M. A., *Digital Design, A Pragmatic Approach,* Prindle, Weber & Schmidt, Boston, Mass., 1987.
6. Breeding, K. J., *Digital Design Fundamentals,* Prentice-Hall, Englewood Cliffs, New Jersey, 1989.
7. Mano, M. M., *Digital Design,* Prentice-Hall, Englewood Cliffs, New Jersey, 1984.
8. Fletcher, W. I., *An Engineering Approach to Digital Design,* Prentice-Hall, Englewood Cliffs, New Jersey, 1980.
9. McCluskey, E. J., *Logic Design Principles,* Prentice-Hall, Englewood Cliffs, New Jersey, 1986.

PROBLEMS

Section 10-2 Asynchronous Design Fundamentals

10-1. What are the two advantages of asynchronous sequential circuits compared to synchronous sequential circuits?

10-2. What types of input signals provide the trigger to cause a state transition for each of the following types of sequential circuits?

    (*a*) synchronous

    (*b*) asynchronous (Fundamental mode)

    (*c*) asynchronous (Pulse mode)

10-3. List the restrictions that must be placed on the manner in which the input signals are applied to a fundamental mode asynchronous circuit.

10-4. Are the inputs $X1$ $X2$ of a fundamental mode asynchronous circuit allowed to change from 01 to 10 simultaneously? Explain.

10-5. What determines the speed of the following types of sequential circuits?

    (*a*) an asynchronous circuit

    (*b*) a synchronous circuit

10-6. List two types of problems that must be considered when designing asynchronous sequential circuits that do not have to be considered when designing synchronous sequential circuits.

Section 10-3 Asynchronous Sequential Design of Fundamental Mode Circuits

10-7. Number each of the timing events in the timing diagram shown in Fig. P10-7. Label each state.

10-8. Obtain the timing table for the timing diagram shown in Fig. P10-7. Show that the circuit represented by the timing table is sequential rather than combinational in nature.

10-9. Number each of the timing events in the timing diagram shown in Fig. P10-9. Label each state.

10-10. Obtain the timing table for the timing diagram shown in Fig. P10-9. Show that the circuit represented by the timing table is sequential in nature and not combinational.

10-11. Complete the design for the fundamental mode circuit represented by the timing diagram in Fig. P10-7.

    (*a*) Obtain the primitive flow map.

    (*b*) Find all pairs of equivalent or compatible states using an implication chart.

    (*c*) Determine the minimum number of states for the design using a merger diagram.

    (*d*) Obtain a reduced flow map using a minimum set of mergeable states, and assign a single state name to each group of merged states.

    (*e*) Draw a transition diagram for the reduced flow map, and, if necessary, modify it to obtain a race-free state assignment.

    (*f*) Obtain a composite Karnaugh map with a race-free assignment and illegal state recovery.

    (*g*) Obtain logic hazard-free equations for the circuit, and draw a gate-level circuit implementation.

    (*h*) Obtain a logic hazard-free circuit implementation using *S-R* NOR latches.

    (*i*) Obtain a logic hazard-free circuit implementation using *S-R* NAND latches.

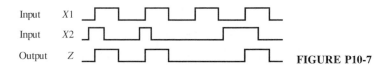

Input $X1$

Input $X2$

Output Z **FIGURE P10-7**

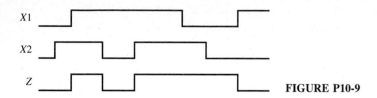

FIGURE P10-9

10-12. Obtain a universal race-free state assignment code for any asynchronous fundamental mode machine with four stable states. Show that the proposed state assignment is universal (all stable states have adjacent codes) by drawing the state assignment code map.

10-13. Repeat Prob. 10-11 for the timing diagram shown in Fig. P10-9.

10-14. Design a fundamental mode asynchronous sequential circuit with the following behavior. The circuit has two external inputs A and B, one Mealy output *SUM*, and one Moore output *CARRYOUT*. The circuit operation is that of a half adder as shown in the state diagram in Fig. P10-14. Design the circuit so that it is logic hazard-free. Choose a race-free state assignment.

   (*a*) Obtain the composite Karnaugh map for the next state and external outputs.

   (*b*) Obtain logic hazard-free equations for the circuit.

   (*c*) Draw a gate-level circuit implementation for the half adder circuit.

   (*d*) Obtain the composite Karnaugh map for the S and R excitation inputs.

   (*e*) Obtain logic hazard-free S and R excitation input equations, and draw the half adder circuit using S-R NOR latches.

   (*f*) Draw the half adder circuit using S-R NAND latches.

10-15. Repeat Prob. 10-14 for the asynchronous sequential comparator circuit represented by the state diagram in Fig. P10-15.

10-16. Is the state assignment in Fig. 10-12 race-free? Explain.

10-17. Briefly explain what an essential hazard is, what causes it, and how it can be observed from a flow map. Draw a flow map with one external input and two state variables that has an essential hazard that is different from the flow map illustrated in Fig. 10-11. Indicate by arrows the essential hazard in your flow map. Can an essential hazard occur in a flow map that contains an external input and a single state variable? Explain.

10-18. Is the circuit diagram shown in Fig. 10-13 logic hazard-free? Explain.

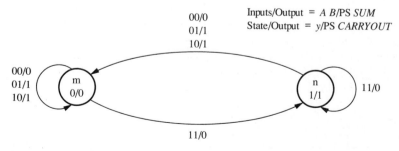

FIGURE P10-14

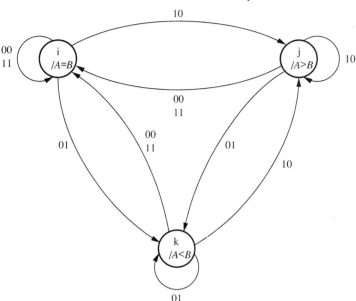

Inputs = AB
/Outputs = /PS $A=B$ $A>B$ $A<B$

FIGURE P10-15

10-19. Indicate by arrows each essential hazard in the flow maps in Fig. P10-19. Explain the technique for observing an essential hazard in a flow map.

10-20. Describe how an essential hazard can be fixed once it is determined that such a hazard exists in a circuit.

10-21. List four important items a designer must carefully consider when designing fundamental mode asynchronous circuits. Which of these items are also important when designing synchronous circuits?

$y_1\,y_2$ \ X		0	1
$Y_1\,Y_1 =$			
a	00	a	$\underline{a}$
b	01	$\underline{b}$	a
c	11	b	$\underline{c}$
d	10	$\underline{d}$	c

(a)

$y_1\,y_2$ \ X1 X2		00	01	11	10
$Y_1\,Y_1 =$					
m	00	$\underline{m}$	$\underline{m}$	r	r
n	01	$\underline{n}$	$\underline{n}$	$\underline{n}$	s
r	11	n	s	$\underline{r}$	$\underline{r}$
s	10	$\underline{s}$	$\underline{s}$	n	$\underline{s}$

(b)

FIGURE P10-19

Section 10-4 Asynchronous Sequential Design of Pulse Mode Circuits

10-22. List the restrictions that must be placed on the manner in which the input signals are applied to a pulse mode asynchronous circuit.

10-23. Describe what the term mutually exclusive means in reference to a pulse mode asynchronous circuit.

10-24. Why is it important in a pulse mode circuit either to control the pulse width of each input pulse or to use a double-rank circuit? Explain.

10-25. Design a pulse mode circuit with a Moore output Z that occurs after the last pulse in the input sequence $X1$ $X2$ or after any pulse that occurs on input $X2$.
(*a*) Draw a state diagram for the design.
(*b*) Obtain the flow map.
(*c*) Use an implication table to see if the flow map can be reduced.
(*d*) Obtain the composite Karnaugh map for the next state and external outputs.
(*e*) Obtain the composite Karnaugh map for the T excitation inputs for the master rank for a double-rank design.
(*f*) Write the design equations for the design.
(*g*) Draw a double-rank implementation for the pulse mode circuit.

10-26. Repeat 10-25 for a pulse mode circuit with two inputs $X1$ and $X2$, a Moore output $Z1$, and a Mealy output $Z2$. The Moore output occurs either after the fourth consecutive pulse on input $X2$ or after any pulse that occurs on input $X1$. The Mealy output is coincident with the second consecutive pulse on input $X2$.

10-27. Analyze the double-rank pulse mode circuit in Fig. P10-27 by obtaining its state diagram.

10-28. Design the double-rank pulse mode circuit represented by the state diagram in Fig. P10-28. Use positive edge-triggered T flip-flops for the master rank and D flip-flops for the slave rank.

10-29. Design the double-rank pulse mode circuit represented by the state diagram in Fig. P10-29. Use positive edge-triggered T flip-flops for the master rank and D flip-flops for the slave rank. For proper circuit operation does the design require:

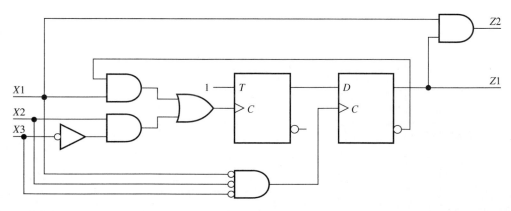

FIGURE P10-27

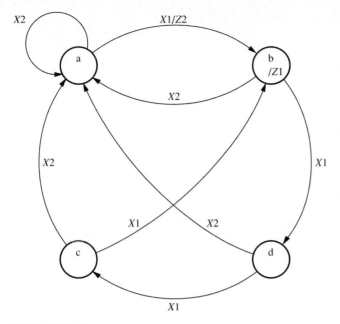

FIGURE P10-28

        (*a*) Illegal state recovery? Explain.
        (*b*) A race-free state assignment? Explain.
        (*c*) Logic hazard-free equations? Explain.
        (*d*) Essential hazard removal? Explain.
10-30. Draw the double-rank pulse mode asynchronous circuit implementation for the combinational lock in Example 10-6 using positive edge-triggered T flip-flops for the master rank and D flip-flops for the slave rank.

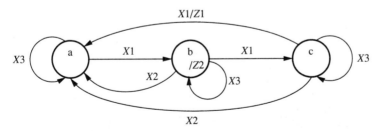

FIGURE P10-29

Overview of IEEE Standard 91-1984

Explanation of Logic Symbols

F.A. Mann
Semiconductor Group

TEXAS
INSTRUMENTS

Contents

List of Tables

List of Illustrations

1.0 INTRODUCTION

The International Electrotechnical Commission (IEC) has been developing a very powerful symbolic language that can show the relationship of each input of a digital logic circuit to each output without showing explicitly the internal logic. At the heart of the system is dependency notation, which will be explained in Section 4.

The system was introduced in the USA in a rudimentary form in IEEE/ANSI Standard Y32.14-1973. Lacking at that time a complete development of dependency notation, it offered little more than a substitution of rectangular shapes for the familiar distinctive shapes for representing the basic functions of AND, OR, negation, etc. This is no longer the case.

Internationally, Working Group 2 of IEC Technical Committee TC-3 has prepared a new document (Publication 617-12) that consolidates the original work started in the mid 1960's and published in 1972 (Publication 117-15) and the amendments and supplements that have followed. Similarly for the USA, IEEE Committee SCC 11.9 has revised the publication IEEE Std 91/ANSI Y32.14. Now numbered simply IEEE Std 91-1984, the IEEE standard contains all of the IEC work that has been approved, and also a small amount of material still under international consideration. Texas Instruments is participating in the work of both organizations and this document introduces new logic symbols in accordance with the new standards. When changes are made as the standards develop, future editions will take those changes into account.

The following explanation of the new symbolic language is necessarily brief and greatly condensed from what the standards publications will contain. This is not intended to be sufficient for those people who will be developing symbols for new devices. It is primarily intended to make possible the understanding of the symbols used in various data books and the comparison of the symbols with logic diagrams, functional block diagrams, and/or function tables will further help that understanding.

2.0 SYMBOL COMPOSITION

A symbol comprises an outline or a combination of outlines together with one or more qualifying symbols. The shape of the symbols is not significant. As shown in Figure 1, general qualifying symbols are used to tell exactly what logical operation is performed by the elements. Table I shows general qualifying symbols defined in the new standards. Input lines are placed on the left and output lines are placed on the right. When an exception is made to that convention, the direction of signal flow is indicated by an arrow as shown in Figure 11.

All outputs of a single, unsubdivided element always have identical internal logic states determined by the function of the element except when otherwise indicated by an associated qualifying symbol or label inside the element.

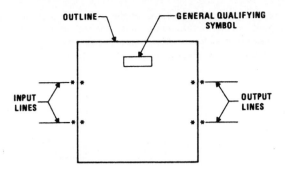

*Possible positions for qualifying symbols relating to inputs and outputs

Figure 1. Symbol Composition

The outlines of elements may be abutted or embedded in which case the following conventions apply. There is no logic connection between the elements when the line common to their outlines is in the direction of signal flow. There is at least one logic connection between the elements when the line common to their outlines is perpendicular to the direction of signal flow. The number of logic connections between elements will be clarified by the use of qualifying symbols and this is discussed further under that topic. If no indications are shown on either side of the common line, it is assumed there is only one connection.

When a circuit has one or more inputs that are common to more than one element of the circuit, the common-control block may be used. This is the only distinctively shaped outline used in the IEC system. Figure 2 shows that unless otherwise qualified by dependency notation, an input to the common-control block is an input to each of the elements below the common-control block.

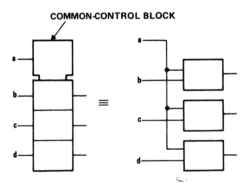

Figure 2. Common-Control Block

A common output depending on all elements of the array can be shown as the output of a common-output element. Its distinctive visual feature is the double line at its top. In addition the common-output element may have other inputs as shown in Figure 3. The function of the common-output element must be shown by use of a general qualifying symbol.

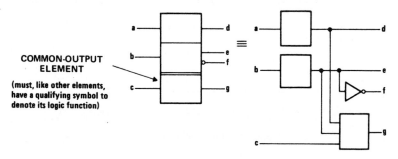

COMMON-OUTPUT ELEMENT

(must, like other elements, have a qualifying symbol to denote its logic function)

Figure 3. Common-Output Element

3.0   QUALIFYING SYMBOLS

3.1   General Qualifying Symbols

Table I shows general qualifying symbols defined by IEEE Standard 91. These characters are placed near the top center or the geometric center of a symbol or symbol element to define the basic function of the device represented by the symbol or of the element.

3.2   Qualifying Symbols for Inputs and Outputs

Qualifying symbols for inputs and outputs are shown in Table II and will be familiar to most users with the possible exception of the logic polarity and analog signal indicators. The older logic negation indicator means that the external 0 state produces the internal 1 state. The internal 1 state means the active state. Logic negation may be used in pure logic diagrams; in order to tie the external 1 and 0 logic states to the levels H (high) and L (low), a statement of whether positive logic (1 = H, 0 = L) or negative logic (1 = L, 0 = H) is being used is required or must be assumed. Logic polarity indicators eliminate the need for calling out the logic convention and are used in various data books in the symbology for actual devices. The presence of the triangular polarity indicator indicates that the L logic level will produce the internal 1 state (the active state) or that, in the case of an output, the internal 1 state will produce the external L level. Note how the active direction of transition for a dynamic input is indicated in positive logic, negative logic, and with polarity indication.

The internal connections between logic elements abutted together in a symbol may be indicated by the symbols shown in Table II. Each logic connection may be shown by the presence of qualifying symbols at one or both sides of the common line and if confusion can arise about the numbers of connections, use can be made of one of the internal connection symbols.

Table I. General Qualifying Symbols

SYMBOL	DESCRIPTION	CMOS EXAMPLE	TTL EXAMPLE
&	AND gate or function.	'HC00	SN7400
≥ 1	OR gate or function. The symbol was chosen to indicate that at least one active input is needed to activate the output.	'HC02	SN7402
$= 1$	Exclusive OR. One and only one input must be active to activate the output.	'HC86	SN7486
=	Logic identity. All inputs must stand at the same state.	'HC86	SN74180
2k	An even number of inputs must be active.	'HC280	SN74180
2k + 1	An odd number of inputs must be active.	'HC86	SN74ALS86
1	The one input must be active.	'HC04	SN7404
$\triangleright$ or $\triangleleft$	A buffer or element with more than usual output capability (symbol is oriented in the direction of signal flow).	'HC240	SN74S436
∏	Schmitt trigger; element with hysteresis.	'HC132	SN74LS18
X/Y	Coder, code converter (DEC/BCD, BIN/OUT, BIN/7-SEG, etc.).	'HC42	SN74LS347
MUX	Multiplexer/data selector.	'HC151	SN74150
DMUX or DX	Demultiplexer.	'HC138	SN74138
Σ	Adder.	'HC283	SN74LS385
P−Q	Subtracter.	*	SN74LS385
CPG	Look-ahead carry generator	'HC182	SN74182
π	Multiplier.	*	SN74LS384
COMP	Magnitude comparator.	'HC85	SN74LS682
ALU	Arithmetic logic unit.	'HC181	SN74LS381
⊓	Retriggerable monostable.	'HC123	SN74LS422
1⊓	Nonretriggerable monostable (one-shot)	'HC221	SN74121
G	Astable element. Showing waveform is optional.	*	SN74LS320
!G	Synchronously starting astable.	*	SN74LS624
G!	Astable element that stops with a completed pulse.	*	*
SRGm	Shift register. m = number of bits.	'HC164	SN74LS595
CTRm	Counter. m = number of bits; cycle length = 2^m.	'HC590	SN54LS590
CTR DIVm	Counter with cycle length = m.	'HC160	SN74LS668
RCTRm	Asynchronous (ripple-carry) counter; cycle length = 2^m.	'HC4020	*
ROM	Read-only memory.	*	SN74187
RAM	Random-access read/write memory.	'HC189	SN74170
FIFO	First-in, first-out memory.	*	SN74LS222
I = 0	Element powers up cleared to 0 state.	*	SN74AS877
I = 1	Element powers up set to 1 state.	'HC7022	SN74AS877
Φ	Highly complex function; "gray box" symbol with limited detail shown under special rules.	*	SN74LS608

*Not all of the general qualifying symbols have been used in TI's CMOS and TTL data books, but they are included here for the sake of completeness.

Table II. Qualifying Symbols for Inputs and Outputs

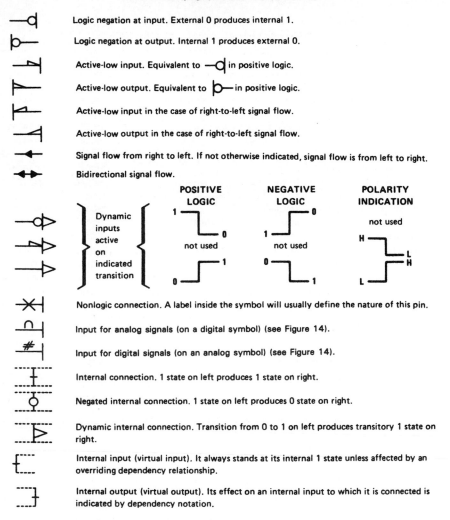

Logic negation at input. External 0 produces internal 1.

Logic negation at output. Internal 1 produces external 0.

Active-low input. Equivalent to ─◁ in positive logic.

Active-low output. Equivalent to ▷─ in positive logic.

Active-low input in the case of right-to-left signal flow.

Active-low output in the case of right-to-left signal flow.

Signal flow from right to left. If not otherwise indicated, signal flow is from left to right.

Bidirectional signal flow.

	POSITIVE LOGIC	NEGATIVE LOGIC	POLARITY INDICATION
Dynamic inputs active on indicated transition			not used

Nonlogic connection. A label inside the symbol will usually define the nature of this pin.

Input for analog signals (on a digital symbol) (see Figure 14).

Input for digital signals (on an analog symbol) (see Figure 14).

Internal connection. 1 state on left produces 1 state on right.

Negated internal connection. 1 state on left produces 0 state on right.

Dynamic internal connection. Transition from 0 to 1 on left produces transitory 1 state on right.

Internal input (virtual input). It always stands at its internal 1 state unless affected by an overriding dependency relationship.

Internal output (virtual output). Its effect on an internal input to which it is connected is indicated by dependency notation.

The internal (virtual) input is an input originating somewhere else in the circuit and is not connected directly to a terminal. The internal (virtual) output is likewise not connected directly to a terminal. The application of internal inputs and outputs requires an understanding of dependency notation, which is explained in Section 4.

Table III. Symbols Inside the Outline

Postponed output (of a pulse-triggered flip-flop). The output changes when input initiating change (e.g., a C input) returns to its initial external state or level. See § 5.

Bi-threshold input (input with hysteresis).

N-P-N open-collector or similar output that can supply a relatively low-impedance L level when not turned off. Requires external pull-up. Capable of positive-logic wired-AND connection.

Passive-pull-up output is similar to N-P-N open-collector output but is supplemented with a built-in passive pull-up.

N-P-N open-emitter or similar output that can supply a relatively low-impedance H level when not turned off. Requires external pull-down. Capable of positive-logic wired-OR connection.

Passive-pull-down output is similar to N-P-N open-emitter output but is supplemented with a built-in passive pull-down.

3-state output.

Output with more than usual output capability (symbol is oriented in the direction of signal flow).

EN Enable input
   When at its internal 1-state, all outputs are enabled.
   When at its internal 0-state, open-collector and open-emitter outputs are off, three-state outputs are at normally defined internal logic states and at external high-impedance state, and all other outputs (e.g., totem-poles) are at the internal 0-state.

J, K, R, S, T Usual meanings associated with flip-flops (e.g., R = reset, T = toggle)

D Data input to a storage element equivalent to:

$-m$ $\leftarrow m$ Shift right (left) inputs, m = 1, 2, 3, etc. If m = 1, it is usually not shown.

$+m$ $-m$ Counting up (down) inputs, m = 1, 2, 3, etc. If m = 1, it is usually not shown.

Binary grouping. m is highest power of 2.

CT = 15 The contents-setting input, when active, causes the content of a register to take on the indicated value.

CT = 9 The content output is active if the content of the register is as indicated.

Input line grouping . . . indicates two or more terminals used to implement a single logic input.

e.g., The paired expander inputs of SN7450.

"1" Fixed-state output always stands at its internal 1 state. For example, see SN74185.

In an array of elements, if the same general qualifying symbol and the same qualifying symbols associated with inputs and outputs would appear inside each of the elements of the array, these qualifying symbols are usually shown only in the first element. This is done to reduce clutter and to save time in recognition. Similarly, large identical elements that are subdivided into smaller elements may each be represented by an unsubdivided outline. The SN54HC242 or SN54LS440 symbol illustrates this principle.

3.3 Symbols Inside the Outline

Table III shows some symbols used inside the outline. Note particularly that open-collector (open-drain), open-emitter (open-source), and three-state outputs have distinctive symbols. Also note that an EN input affects all of the outputs of the circuit and has no effect on inputs. When an enable input affects only certain outputs and/or affects one or more inputs, a form of dependency notation will indicate this (see 4.9). The effects of the EN input on the various types of outputs are shown.

It is particularly important to note that a D input is always the data input of a storage element. At its internal 1 state, the D input sets the storage element to its 1 state, and at its internal 0 state it resets the storage element to its 0 state.

The binary grouping symbol will be explained more fully in Section 8. Binary-weighted inputs are arranged in order and the binary weights of the least-significant and the most-significant lines are indicated by numbers. In this document weights of input and output lines will be represented by powers of two usually only when the binary grouping symbol is used, otherwise decimal numbers will be used. The grouped inputs generate an internal number on which a mathematical function can be performed or that can be an identifying number for dependency notation (Figure 28). A frequent use is in addresses for memories.

Reversed in direction, the binary grouping symbol can be used with outputs. The concept is analogous to that for the inputs and the weighted outputs will indicate the internal number assumed to be developed within the circuit.

Other symbols are used inside the outlines in accordance with the IEC/IEEE standards but are not shown here. Generally these are associated with arithmetic operations and are self-explanatory.

When nonstandardized information is shown inside an outline, it is usually enclosed in square brackets [like these].

4.0 DEPENDENCY NOTATION

4.1 General Explanation

Dependency notation is the powerful tool that sets the IEC symbols apart from previous systems and makes compact, meaningful, symbols possible. It provides the means of denoting the relationship between inputs, outputs, or inputs and outputs without actually showing all the elements and interconnections involved. The information provided by dependency notation supplements that provided by the qualifying symbols for an element's function.

In the convention for the dependency notation, use will be made of the terms "affecting" and "affected." In cases where it is not evident which inputs must be considered as being the affecting or the affected ones (e.g., if they stand in an AND relationship), the choice may be made in any convenient way.

So far, eleven types of dependency have been defined and all of these are used in various TI data books. X dependency is used mainly with CMOS circuits. They are listed below in the order in which they are presented and are summarized in Table IV following 4.12.

Section	Dependency Type or Other Subject
4.2	G, AND
4.3	General Rules for Dependency Notation
4.4	V, OR
4.5	N, Negate (Exclusive-OR)
4.6	Z, Interconnection
4.7	X, Transmission
4.8	C, Control
4.9	S, Set and R, Reset
4.10	EN, Enable
4.11	M, Mode
4.12	A, Address

4.2 G (AND) Dependency

A common relationship between two signals is to have them ANDed together. This has traditionally been shown by explicitly drawing an AND gate with the signals connected to the inputs of the gate. The 1972 IEC publication and the 1973 IEEE/ANSI standard showed several ways to show this AND relationship using dependency notation. While ten other forms of dependency have since been defined, the ways to invoke AND dependency are now reduced to one.

In Figure 4 input **b** is ANDed with input **a** and the complement of **b** is ANDed with **c**. The letter G has been chosen to indicate AND relationships and is placed at input **b**, inside the symbol. A number considered appropriate by the symbol designer (1 has been used here) is placed after the letter G and also at each affected input. Note the bar over the 1 at input **c**.

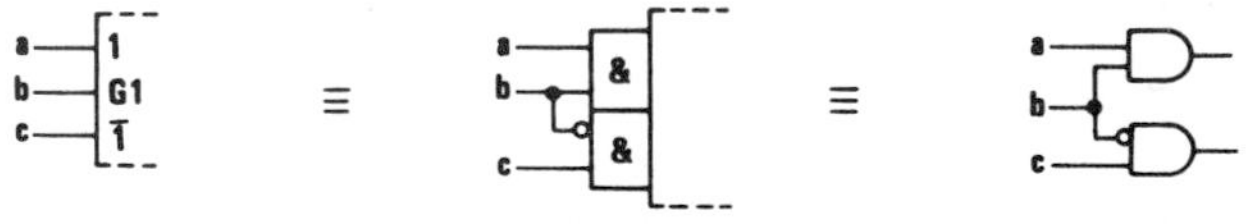

Figure 4. G Dependency Between Inputs

In Figure 5, output **b** affects input **a** with an AND relationship. The lower example shows that it is the internal logic state of **b**, unaffected by the negation sign, that is ANDed. Figure 6 shows input **a** to be ANDed with a dynamic input **b**.

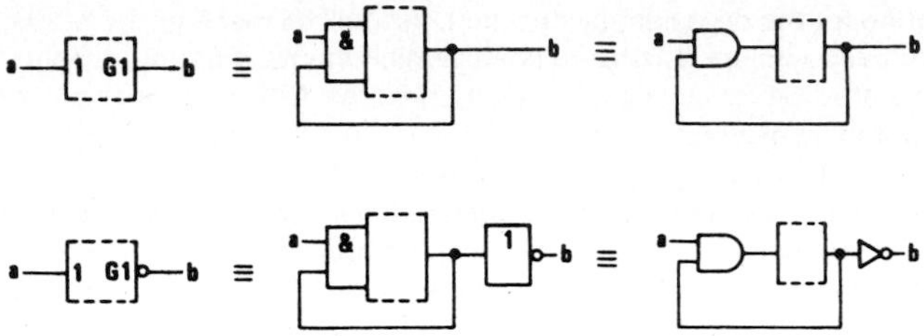

Figure 5. G Dependency Between Outputs and INputs

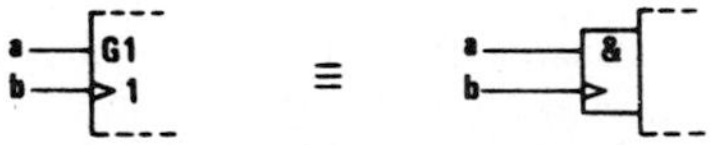

Figure 6. G Dependency with a Dynamic Input

The rules for G dependency can be summarized thus:

> When a Gm input or output (m is a number) stands at its internal 1 state, all inputs and outputs affected by Gm stand at their normally defined internal logic states. When the Gm input or output stands at its 0 state, all inputs and outputs affected by Gm stand at their internal 0 states.

4.3 Conventions for the Application of Dependency Notation in General

The rules for applying dependency relationships in general follow the same pattern as was illustrated for G dependency.

Application of dependency notation is accomplished by:

1) labeling the input (or output) *affecting* other inputs or outputs with the letter symbol indicating the relationship involved (e.g., G for AND) followed by an identifying number, appropriately chosen, and
2) labeling each input or output *affected* by that affecting input (or output) with that same number.

If it is the complement of the internal logic state of the affecting input or output that does the affecting, then a bar is placed over the identifying numbers at the affected inputs or outputs (Figure 4).

If two affecting inputs or outputs have the same letter and same identifying number, they stand in an OR relationship to each other (Figure 7).

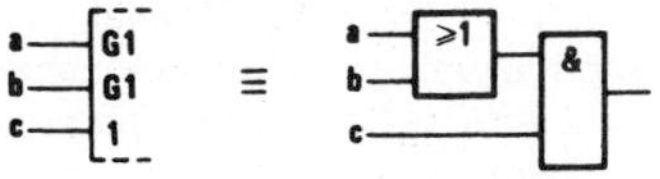

Figure 7. ORed Affecting Inputs

If the affected input or output requires a label to denote its function (e.g., "D"), this label will be *prefixed* by the identifying number of the affecting input (Figure 15).

If an input or output is affected by more than one affecting input, the identifying numbers of each of the affecting inputs will appear in the label of the affected one, separated by commas. The normal reading order of these numbers is the same as the sequence of the affecting relationships (Figure 15).

If the labels denoting the functions of affected inputs or outputs must be numbers (e.g., outputs of a coder), the identifying numbers to be associated with both affecting inputs and affected inputs or outputs will be replaced by another character selected to avoid ambiguity, e.g., Greek letters (Figure 8).

Figure 8. Substitution for Numbers

4.4 V (OR) Dependency

The symbol denoting OR dependency is the letter V (Figure 9).

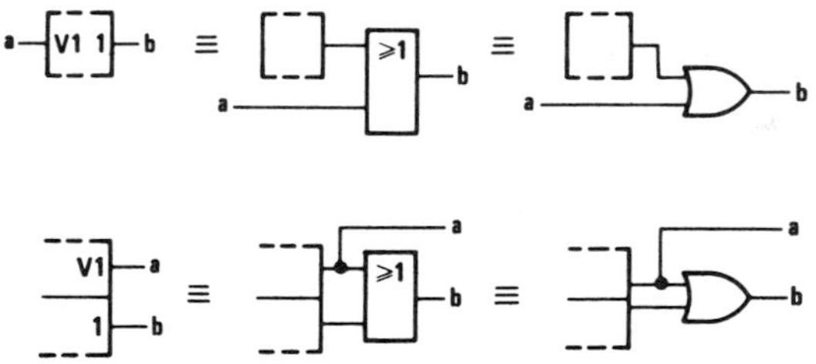

Figure 9. V (OR) Dependency

When a Vm input or output stands at its internal 1 state, all inputs and outputs affected by Vm stand at their internal 1 states. When the Vm input or output stands at its internal 0 state, all inputs and outputs affected by Vm stand at their normally defined internal logic states.

4.5 N (Negate) (Exclusive-OR) Dependency

The symbol denoting negate dependency is the letter N (Figure 10). Each input or output affected by an Nm input or output stands in an Exclusive-OR relationship with the Nm input or output.

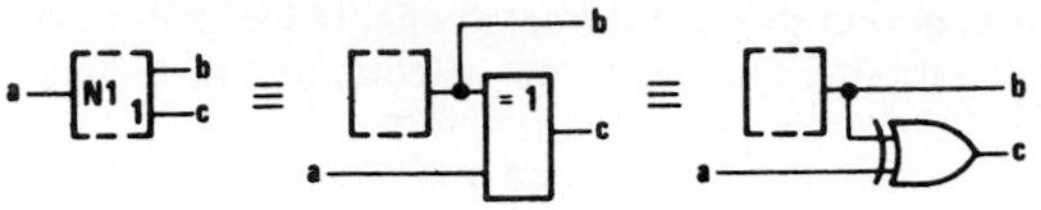

If a = 0, then c = b
If a = 1, then c = b̄

Figure 10. N (Negate) (Exclusive-OR) Dependency

When an Nm input or output stands at its internal 1 state, the internal logic state of each input and each output affected by Nm is the complement of what it would otherwise be. When an Nm input or output stands at its internal 0 state, all inputs and outputs affected by Nm stand at their normally defined internal logic states.

4.6 Z (Interconnection) Dependency

The symbol denoting interconnection dependency is the letter Z.

Interconnection dependency is used to indicate the existence of internal logic connections between inputs, outputs, internal inputs, and/or internal outputs.

The internal logic state of an input or output affected by a Zm input or output will be the same as the internal logic state of the Zm input or output, unless modified by additional dependency notation (Figure 11).

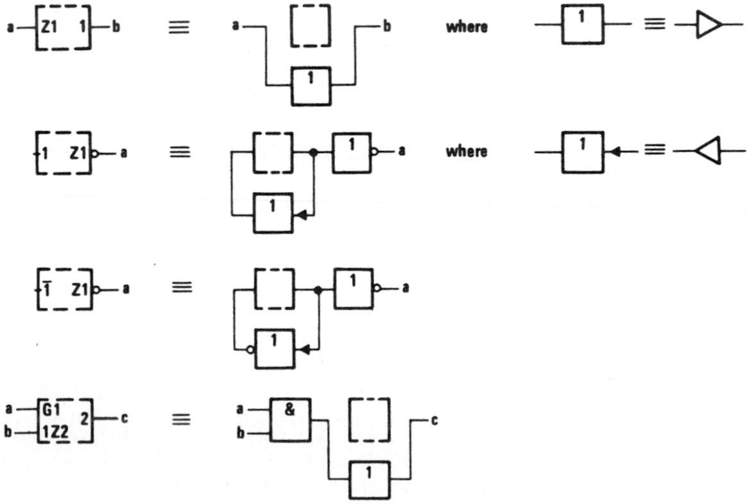

Figure 11. Z (Interconnection) Dependency

4.7   X (Transmission) Dependency

The symbol denoting transmission dependency is the letter X.

Transmission dependency is used to indicate controlled bidirectional connections between affected input/output ports (Figure 12).

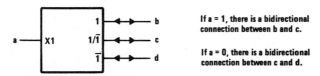

If a = 1, there is a bidirectional
connection between b and c.

If a = 0, there is a bidirectional
connection between c and d.

Figure 12. X (Transmission) Dependency

When an Xm input or output stands at its internal 1 state, all input-output ports affected by this Xm input or output are bidirectionally connected together and stand at the same internal logic state or analog signal level. When an Xm input or output stands at its internal O state, the connection associated with this set of dependency notation does not exist.

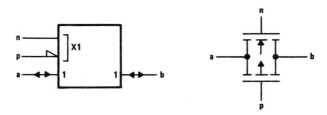

Figure 13. CMOS Transmission Gate Symbol and Schematic

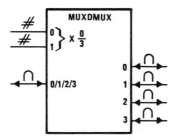

Figure 14. Analog Data Selector (Multiplexer/Demultiplexer)

Although the transmission paths represented by X dependency are inherently bidirectinal, use is not always made of this property. This is analogous to a piece of wire, which may be constrained to carry current in only one direction. If this is the case in a particular application, then the directional arrows shown in Figures 12, 13, and 14 would be omitted.

4.8 C (Control) Dependency

The symbol denoting control dependency is the letter C.

Control inputs are usually used to enable or disable the data (D, J, K, R, or S) inputs of storage elements. They may take on their internal 1 states (be active) either statically or dynamically. In the latter case the dynamic input symbol is used as shown in the third example of Figure 15.

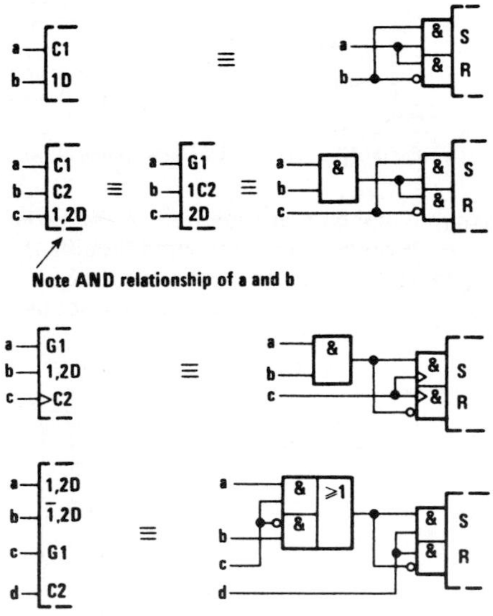
C1
1D
&
S
R
C1
C2
1,2D
G1
1C2
2D
Note AND relationship of a and b
G1
1,2D
C2
1,2D
1̄,2D
G1
C2
≥1

Input c selects which of a or b is stored when d goes low.

Figure 15. C (Control) Dependency

When a C*m* input or output stands at its internal 1 state, the inputs affected by C*m* have their normally defined effect on the function of the element, i.e., these inputs are enabled. When a C*m* input or output stands at its internal 0 state, the inputs affected by C*m* are disabled and have no effect on the function of the element.

4.9 S (Set) and R (Reset) Dependencies

The symbol denoting set dependency is the letter S. The symbol denoting reset dependency is the letter R.

Set and reset dependencies are used if it is necessary to specify the effect of the combination $R = S = 1$ on a bistable element. Case 1 in Figure 16 does not use S or R dependency.

When an Sm input is at its internal 1 state, outputs affected by the Sm input will react, regardless of the state of an R input, as they normally would react to the combination $S = 1, R = 0$. See cases 2, 4, and 5 in Figure 16.

When an Rm input is at its internal 1 state, outputs affected by the Rm input will react, regardless of the state of an S input, as they normally would react to the combination $S = 0, R = 1$. See cases 3, 4, and 5 in Figure 16.

When an Sm or Rm input is at its internal 0 state, it has no effect.

Note that the noncomplementary output patterns in cases 4 and 5 are only pseudo stable. The simultaneous return of the inputs to $S = R = 0$ produces an unforeseeable stable and complementary output pattern.

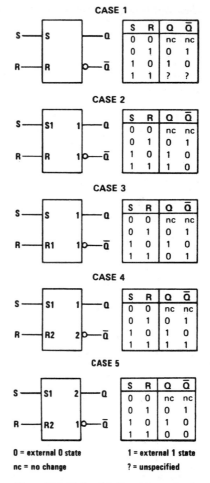

CASE 1

S	R	Q	Q̄
0	0	nc	nc
0	1	0	1
1	0	1	0
1	1	?	?

CASE 2

S	R	Q	Q̄
0	0	nc	nc
0	1	0	1
1	0	1	0
1	1	1	0

CASE 3

S	R	Q	Q̄
0	0	nc	nc
0	1	0	1
1	0	1	0
1	1	0	1

CASE 4

S	R	Q	Q̄
0	0	nc	nc
0	1	0	1
1	0	1	0
1	1	1	1

CASE 5

S	R	Q	Q̄
0	0	nc	nc
0	1	0	1
1	0	1	0
1	1	0	0

0 = external 0 state   1 = external 1 state

nc = no change   ? = unspecified

Figure 16. S (Set) and R (Reset) Dependencies

4.10 EN (Enable) Dependency

The symbol denoting enable dependency is the combination of letters EN.

An ENm input has the same effect on outputs as an EN input, see 3.1, but it affects only those outputs labeled with the identifying number m. It also affects those inputs labeled with the identifying number m. By contrast, an EN input affects all outputs and no inputs. The effect of an ENm input on an affected input is identical to that of a Cm input (Figure 17).

When an EN*m* input stands at its internal 1 state, the inputs affected by EN*m* have their normally defined effect on the function of the element and the outputs affected by this input stand at their normally defined internal logic states, i.e., these inputs and outputs are enabled.

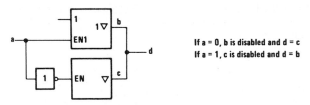

If a = 0, b is disabled and d = c
If a = 1, c is disabled and d = b

Figure 17. EN (Enable) Dependency

When an EN*m* input stands at its internal 0 state, the inputs affected by EN*m* are disabled and have no effect on the function of the element, and the outputs affected by EN*m* are also disabled. Open-collector outputs are turned off, three-state outputs stand at their normally defined internal logic states but externally exhibit high impedance, and all other outputs (e.g., totem-pole outputs) stand at their internal 0 states.

4.11 M (MODE) Dependency

The symbol denoting mode dependency is the letter M.

Mode dependency is used to indicate that the effects of particular inputs and outputs of an element depend on the mode in which the element is operating.

If an input or output has the same effect in different modes of operation, the identifying numbers of the relevant affecting M*m* inputs will appear in the label of that affected input or output between parentheses and separated by solidi (Figure 22).

4.11.1 M Dependency Affecting Inputs

M dependency affects inputs the same as C dependency. When an M*m* input or M*m* output stands at its internal 1 state, the inputs affected by this M*m* input or M*m* output have their normally defined effect on the function of the element, i.e., the inputs are enabled.

When an M*m* input or M*m* output stands at its internal 0 state, the inputs affected by this M*m* input or M*m* output have no effect on the function of the element. When an affected input has several sets of labels separated by solidi (e.g., C4/2→/3+), any set in which the identifying number of the M*m* input or M*m* output appears has no effect and is to be ignored. This represents disabling of some of the functions of a multifunction input.

The circuit in Figure 18 has two inputs, **b** and **c**, that control which one of four modes (0, 1, 2, or 3) will exist at any time. Inputs **d, e,** and **f** are D inputs subject to dynamic control (clocking) by the **a** input. The numbers 1 and 2 are in the series chosen to indicate the modes so inputs **e** and **f** are only enabled in mode 1 (for parallel loading) and input **d** is only enabled in mode 2 (for serial loading). Note that input **a** has three functions. It is the clock for entering data. In mode 2, it causes right shifting of data, which means a shift away from the control block. In mode 3, it causes the contents of the register to be incremented by one count.

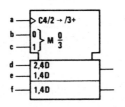

Note that all operations are synchronous.

In MODE 0 (b = 0, c = 0), the outputs remain at their existing states as none of the inputs has an effect.

In MODE 1 (b = 1, c = 0), parallel loading takes place thru inputs e and f.

In MODE 2 (b = 0, c = 1), shifting down and serial loading thru input d take place.

In MODE 3 (b = c = 1), counting up by increment of 1 per clock pulse takes place.

Figure 18. M (Mode) Dependency Affecting Inputs

4.11.2 M Dependency Affecting Outputs

When an M*m* input or M*m* output stands at its internal 1 state, the affected outputs stand at their normally defined internal logic states, i.e., the outputs are enabled.

When an M*m* input or M*m* output stands at its internal 0 state, at each affected output any set of labels containing the identifying number of that M*m* input or M*m* output has no effect and is to be ignored. When an output has several different sets of labels separated by solidi (e.g., 2,4/3,5), only those sets in which the identifying number of this M*m* input or M*m* output appears are to be ignored.

Figure 19 shows a symbol for a device whose output can behave like either a 3-state output or an open-collector output depending on the signal applied to input **a**. Mode 1 exists when input **a** stands at its internal 1 state and, in that case, the three-state symbol applies and the open-element symbol has no effect. When **a** = 0, mode 1 does not exist so the three-state symbol has no effect and the open-element symbol applies.

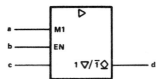

Figure 19. Type of Output Determined by Mode

In Figure 20, if input **a** stands at its internal 1 state establishing mode 1, output **b** will stand at its internal 1 state only when the content of the register equals 9. Since output **b** is located in the common-control block with no defined function outside of mode 1, the state of this output outside of mode 1 is not defined by the symbol.

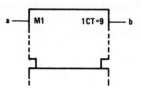

Figure 20. An Output of the Common-Control Block

In Figure 21, if input **a** stands at its internal 1 state establishing mode 1, output **b** will stand at its internal 1 state only when the content of the register equals 15. If input **a** stands at its internal 0 state, output **b** will stand at its internal 1 state only when the content of the register equals 0.

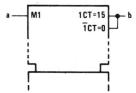

Figure 21. Determining and Output's Function

In Figure 22 inputs **a** and **b** are binary weighted to generate the numbers 0, 1, 2, or 3. This determines which one of the four modes exists.

Figure 22. Dependent Relationships Affected by Mode

At output **e** the label set causing negation (if **c** = 1) is effective only in modes 2 and 3. In modes 0 and 1 this output stands at its normally defined state as if it had no labels. At output **f** the label set has effect when the mode is not 0 so output **e** is negated (if **c** = 1) in modes 1, 2, and 3. In mode 0 the label set has no effect so the output stands at its normally defined state. In this example $\overline{0},4$ is equivalent to (1/2/3)4. At output **g** there are two label sets. The first set, causing negation (if **c** = 1), is effective only in mode 2. The second set, subjecting **g** to AND dependency on **d**, has effect only in mode 3.

Note that in mode 0 none of the dependency relationships has any effect on the outputs, so **e**, **f**, and **g** will all stand at the same state.

4.12 A (Address) Dependency

The symbol denoting address dependency is the letter A.

Address dependency provides a clear representation of those elements, particularly memories, that use address control inputs to select specified sections of a multidimensional arrays. Such a section of a memory array is usually called a word. The purpose of address dependency is to allow a symbolic presentation of the entire array. An input of the array shown at a particular element of this general section is common to the corresponding elements of all selected sections of the array. An output of the array shown at a particular element of this general section is the result of the OR function of the outputs of the corresponding elements of selected sections.

Inputs that are not affected by any affecting address input have their normally defined effect on all sections of the array, whereas inputs affected by an address input have their normally defined effect only on the section selected by that address input.

An affecting address input is labeled with the letter A followed by an identifying number that corresponds with the address of the particular section of the array selected by this input. Within the general section presented by the symbol, inputs and outputs affected by an A*m* input are labeled with the letter A, which stands for the identifying numbers, i.e., the addresses, of the particular sections.

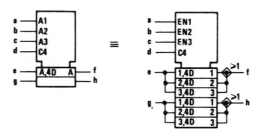

Figure 23. A (Address) Dependency

Figure 23 shows a 3-word by 2-bit memory having a separate address line for each word and uses EN dependency to explain the operation. To select word 1, input **a** is taken to its 1 state, which establishes mode 1. Data can now be clocked into the inputs marked "1,4D." Unless words 2 and 3 are also selected, data cannot be clocked in at the inputs marked "2,4D" and "3,4D." The outputs will be the OR functions of the selected outputs, i.e., only those enabled by the active EN functions.

The identifying numbers of affecting address inputs correspond with the addresses of the sections selected by these inputs. They need not necessarily differ from those of other affecting dependency-inputs (e.g., G, V, N, . . .), because in the general section presented by the symbol they are replaced by the letter A.

If there are several sets of affecting A*m* inputs for the purpose of independent and possibly simultaneous access to sections of the array, then the letter A is modified to 1A, 2A, ... Because they have access to the same sections of the array, these sets of A inputs may have the same identifying numbers. The symbols for 'HC170 or SN74LS170 make use of this.

Figure 24 is another illustration of the concept.

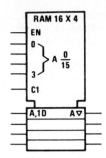

Figure 24. Array of 16 Sections of Four Transparent Latches with 3-State Outputs Comprising a 16-Word X 4-Bit Random-Access Memory

Table IV. Summary of Dependency Notation

TYPE OF DEPENDENCY	LETTER SYMBOL*	AFFECTING INPUT AT ITS 1-STATE	AFFECTING INPUT AT ITS 0-STATE
Address	A	Permits action (address selected)	Prevents action (address not selected)
Control	C	Permits action	Prevents action
Enable	EN	Permits action	Prevents action of inputs ◇outputs off ▽outputs at external high impedance, no change in internal logic state Other outputs at internal 0 state
AND	G	Permits action	Imposes 0 state
Mode	M	Permits action (mode selected)	Prevents action (mode not selected)
Negate (Ex-OR)	N	Complements state	No effect
Reset	R	Affected output reacts as it would to S = 0, R = 1	No effect
Set	S	Affected output reacts as it would to S = 1, R = 0	No effect
OR	V	Imposes 1 state	Permits action
Transmission	X	Bidirectional connection exists	Bidirectional connection does not exist
Interconnection	Z	Imposes 1 state	Imposes 0 state

*These letter symbols appear at the AFFECTING input (or output) and are followed by a number. Each input (or output) AFFECTED by that input is labeled with that same number. When the labels EN, R, and S appear at inputs without the following numbers, the descriptions above do not apply. The action of these inputs is described under "Symbols Inside the Outline," see 3.3.

5.0 BISTABLE ELEMENTS

The dynamic input symbol, the postponed output symbol, and dependency notation provide the tools to differentiate four main types of bistable elements and make synchronous and asynchronous inputs easily recognizable (Figure 25). The first column shows the essential distinguishing features; the other columns show examples.

Transparent latches have a level-operated control input. The D input is active as long as the C input is at its internal 1 state. The outputs respond immediately. Edge-triggered elements accept data from D, J, K, R, or S inputs on the active transition of C. Pulse-triggered elements require the setup of data before the start of the control pulse; the C input is considered static since the data must be maintained as long as C is at its 1 state. The output is postponed until C returns to its 0 state. The data-lock-out element is similar to the pulse-triggered version except that the C input is considered dynamic in that shortly after C goes through its active transition, the data inputs are disabled and data does not have to be held. However, the output is still postponed until the C input returns to its initial external level.

Notice that synchronous inputs can be readily recognized by their dependency labels (1D, 1J, 1K, 1S, 1R) compared to the asynchronous inputs (S, R), which are not dependent on the C inputs.

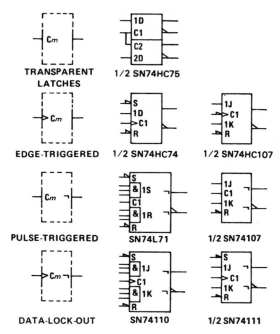

Figure 25. Four Types of Bistable Circuits

6.0 CODERS

The general symbol for a coder or code converter is shown in Figure 26. X and Y may be replaced by appropriate indications of the code used to represent the information at the inputs and at the outputs, respectively.

Figure 26. Coder General Symbol

Indication of code conversion is based on the following rule:

> Depending on the input code, the internal logic states of the inputs determine an internal value. This value is reproduced by the internal logic states of the outputs, depending on the output code.

The indication of the relationships between the internal logic states of the inputs and the internal value is accomplished by:

1) labeling the inputs with numbers. In this case the internal value equals the sum of the weights associated with those inputs that stand at their internal 1-state, or by
2) replacing X by an appropriate indication of the input code and labeling the inputs with characters that refer to this code.

The relationships between the internal value and the internal logic states of the outputs are indicated by:

1) labeling each output with a list of numbers representing those internal values that lead to the internal 1-state of that output. These numbers shall be separated by solidi as in Figure 27. This labeling may also be applied when Y is replaced by a letter denoting a type of dependency (see Section 7). If a continuous range of internal values produces the internal 1 state of an output, this can be indicated by two numbers that are inclusively the beginning and the end of the range, with these two numbers separated by three dots (e.g., 4 . . . 9 = 4/5/6/7/8/9) or by
2) replacing Y by an appropriate indiction of the output code and labeling the outputs with characters that refer to this code as in Figure 28.

Alternatively, the general symbol may be used together with an appropriate reference to a table in which the relationship between the inputs and outputs is indicated. This is a recommended way to symbolize a PROM after it has been programmed.

FUNCTION TABLE

INPUTS			OUTPUTS			
c	b	a	g	f	e	d
0	0	0	0	0	0	0
0	0	1	0	0	0	1
0	1	0	0	0	1	0
0	1	1	0	1	1	0
1	0	0	0	1	0	1
1	0	1	0	0	0	0
1	1	0	0	0	0	0
1	1	1	1	0	0	0

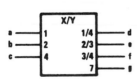

Figure 27. An X/Y Code Converter

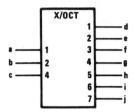

FUNCTION TABLE

INPUTS			OUTPUTS						
c	b	a	j	i	h	g	f	e	d
0	0	0	0	0	0	0	0	0	0
0	0	1	0	0	0	0	0	0	1
0	1	0	0	0	0	0	0	1	0
0	1	1	0	0	0	0	1	0	0
1	0	0	0	0	0	1	0	0	0
1	0	1	0	0	1	0	0	0	0
1	1	0	0	1	0	0	0	0	0
1	1	1	1	0	0	0	0	0	0

Figure 28. An X/Octal Code Converter

7.0 USE OF A CODER TO PRODUCE AFFECTING INPUTS

It often occurs that a set of affecting inputs for dependency notation is produced by decoding the signals on certain inputs to an element. In such a case use can be made of the symbol for a coder as an embedded symbol (Figure 29).

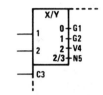

Figure 29. Producing Various Types of Dependencies

If all affecting inputs produced by a coder are of the same type and their identifying numbers shown at the outputs of the coder, Y (in the qualifying symbol X/Y) may be replaced by the letter denoting the type of dependency. The indications of the affecting inputs should then be omitted (Figure 30).

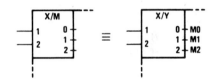

Figure 30. Producing One Type of Dependency

8.0 USE OF BINARY GROUPING TO PRODUCE AFFECTING INPUTS

If all affecting inputs produced by a coder are of the same type and have consecutive identifying numbers not necessarily corresponding with the numbers that would have been shown at the outputs of the coder, use can be made of the binary grouping symbol. k external lines effectively generate 2^k internal inputs. The bracket is followed by the letter denoting the type of dependency followed by m1/m2. The m1 is to be replaced by the smallest identifying number and the m2 by the largest one, as shown in Figure 31.

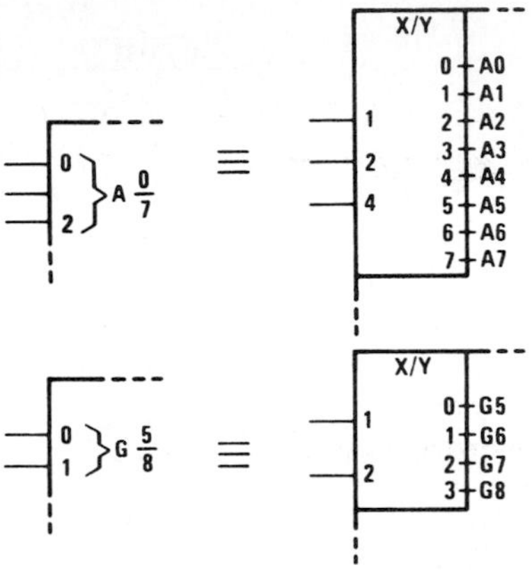

Figure 31. Use of the Binary Grouping Symbol

9.0 SEQUENCE OF INPUT LABELS

If an input having a single functional effect is affected by other inputs, the qualifying symbol (if there is any) for that functional effect is preceded by the labels corresponding to the affecting inputs. The left-to-right order of these preceding labels is the order in which the effects or modifications must be applied. The affected input has no functional effect on the element if the logic state of any one of the affecting inputs, considered separately, would cause the affected input to have no effect, regardless of the logic states of other affecting inputs.

If an input has several different functional effects or has several different sets of affecting inputs, depending on the mode of action, the input may be shown as often as required. However, there are cases in which this method of presentation is not advantageous. In those cases the input may be shown once with the different sets of labels separated by solidi (Figure 32). No meaning is attached to the order of these sets of labels. If one of the functional effects of an input is that of an unlabeled input to the element, a solidus will precede the first set of labels shown.

If all inputs of a combinational element are disabled (caused to have no effect on the function of the element), the internal logic states of the outputs of the element are not specified by the symbol. If all inputs of a sequential element are disabled, the content of this element is not changed and the outputs remain at their existing internal logic states.

Labels may be factored using algebraic techniques (Figure 33).

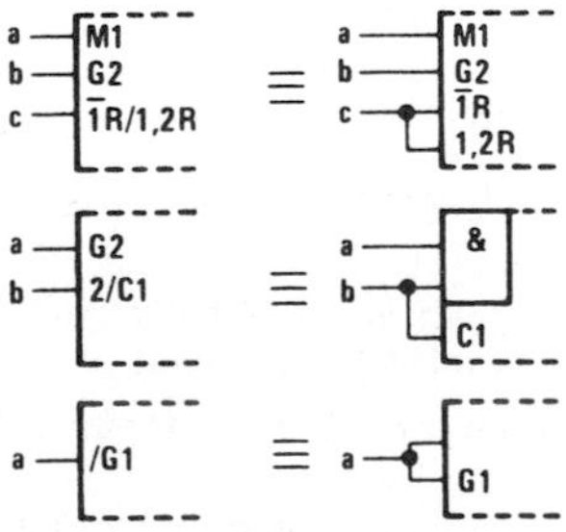

Figure 32. Input Labels

```
    ┌ ─ ─ ─ ┐         ┌ ─ ─ ─ ┐
────┤ (1/2)D │   ≡   ──┤ 1D/2D │
    └ ─ ─ ─ ┘         └ ─ ─ ─ ┘
```

```
    ┌ ─ ─ ─ ─ ┐             ┌ ─ ─ ─ ─ ─ ─ ─ ┐
────▷ 1,2,3,4(5+/6−) │  ≡  ──▷ 1,2,3,4,5+/1,2,3,4,6− │
    └ ─ ─ ─ ─ ┘             └ ─ ─ ─ ─ ─ ─ ─ ┘
```

Figure 33. Factoring Input Labels

10.0 SEQUENCE OF OUTPUT LABELS

If an output has a number of different labels, regardless of whether they are identifying numbers of affecting inputs or outputs or not, these labels are shown in the following order:

1) If the postponed output symbol has to be shown, this comes first, if necessary preceded by the indications of the inputs to which it must be applied
2) Followed by the labels indicating modifications of the internal logic state of the output, such that the left-to-right order of these labels corresponds with the order in which their effects must be applied
3) Followed by the label indicating the effect of the output on inputs and other outputs of the element.

Symbols for open-circuit or three-state outputs, where applicable, are placed just inside the outside boundary of the symbol adjacent to the output line (Figure 34).

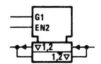

Figure 34. Placement of 3-State Symbols

If an output needs several different sets of labels that represent alternative functions (e.g., depending on the mode of action), these sets may be shown on different output lines that must be connected outside the outline. However, there are cases in which this method of presentation is not advantageous. In those cases the output may be shown once with the different sets of labels separated by solidi (Figure 35).

Two adjacent identifying numbers of affecting inputs in a set of labels that are not already separated by a nonnumeric character should be separated by a comma.

If a set of labels of an output not containing a solidus contains the identifying number of an affecting M*m* input standing at its internal 0 state, this set of labels has no effect on that output.

Labels may be factored using algebraic techniques (Figure 36).

```
    ┌ ─ ─ ─ ─ ─ ─ ┐              ┌ ─ ─ ─ ─ ─ ┐
a──┤ M1  1CT=9/1CT=15 ├─b    ≡   a──┤ M1   1CT=9 ├─b
    └ ─ ─ ─ ─ ─ ─ ┘                 │     1CT=15 │
                                    └ ─ ─ ─ ─ ─ ┘
```

```
    ┌ ─ ─ ─ ─ ─ ─ ┐              ┌ ─ ─ ─ ─ ─ ┐
a──┤ M1  1CT=9/1CT=15▷─b    ≡   a──┤ M1   1CT=9 ▷─b
    └ ─ ─ ─ ─ ─ ─ ┘                 │     1CT=15 ▷
                                    └ ─ ─ ─ ─ ─ ┘
```

Figure 35. Output Labels

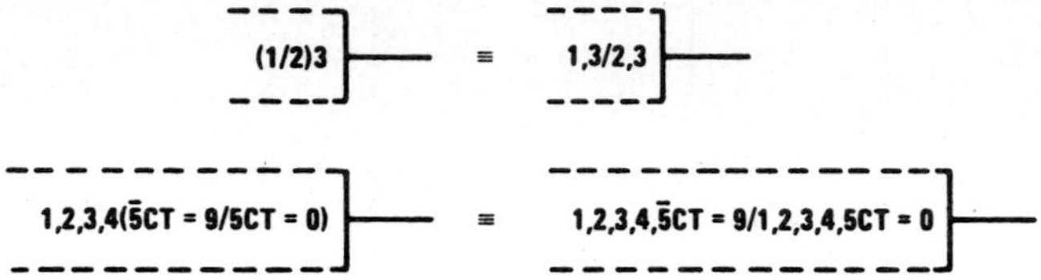

Figure 36. Factoring Output Labels

F.A. Mann received his Bachelor of Science degree from the United States Naval Academy, Annapolis, MD in 1953 and his Master's degree in Engineering Administration from Southern Methodist University, Dallas, TX in 1970. He joined Texas Instruments in 1957 and has worked as a Manufacturing Engineer, Product Marketing Engineer, and Semiconductor Data Sheet Manager. Currently, he serves as Technical Advisor for data sheets and other technical documentation for semiconductors. He has served on many JEDEC committees since 1963 and has been a USA delegate to the technical committee for semiconductor devices of the International Electrotechnical Commission since 1970. He is a member of the IEEE committees for Logic Symbols and for Logic Diagrams, and participates in meetings of an international working group on logic symbols.

Mr. Mann can be reached at (214) 995-2867 in Dallas.

APPENDIX
B

DIGITAL
LOGIC
DEVICES

DIGITAL LOGIC PRODUCTS

TI's digital logic family offers everything from standard bipolar devices to the latest VLSI products. Thanks to TI-proprietary technologies such as IMPACT™ and EPIC™, many TI logic devices feature substantially faster operating speeds and power savings beyond comparable products.

For instance, IMPACT (Implanted Advanced Composed Technology) creates 2-μm features, producing dramatic decreases in device size, length of signal paths and sidewall capacitance. This technology and a derivative, IMPACT-X (which utilizes trench isolation), have made it possible to create high-performance products such as: TI's 8-bit 'AS888 processor slice and 32-bit processor (see Section 4, subsection on High Performance VLSI Processors); 10-ns programmable array logic (PAL®) devices; and memory management products (see Section 4, subsection on Cache Tags, and DRAM controllers). The 1-μm EPIC™ process, Enhanced Performance Implanted CMOS, is bringing an Advanced CMOS Logic (ACL) family to the forefront of the industry. Three times faster than its standard counterparts, the ACL family includes more than 100 of the most popular 54/74 logic functions.

Along with these premiere products, TI's digital family now includes new BiCMOS bus interface devices (SN74BCT') which combine the best of bipolar and CMOS technologies. These devices can reduce system power consumption as much as 25% while maintaining advanced speed and output drive. All TI digital logic products have passed a rigorous quality and reliability program, making them prime candidates for ship-to-stock and just-in-time programs. The reader is urged to utilize the 1988 Master Selection Guide as a quick reference to TI's entire digital logic family.

The reader should refer to the 'Alphanumeric Index' and the 'Functional Index' in Section 1 and to the order forms at the back of the Guide for additional information on technical documentation.

Contents   Page

Contents	**Page**

658 *Digital Logic Devices* *Appendix B*

Contents

GATES

Positive-NAND Gates

DESCRIPTION	TYPE	TECHNOLOGY									DOCUMENT
		STD TTL	LS	S	ALS	AS	F	HC	AC	ACT	
8-Input	'30	•	•	•							SDLD001
					A	•					SDAD001B
								•			SCLD001A
							•				SDFD001
	'11030								•	•	SCAD001
13-Input	'133				•						SDAD001B
								•			SCLD001A
				•							SDLD001
12-Input	'134			•							SDLD001
Dual 2-Input	'8003				•						SDAD001B
Dual 4-Input	'13	•	•								SDLD001
	'20	•	•	•							SDLD001
					A	•					SDAD001B
								•			SCLD001A
							•				SDFD001
	'40				A						SDAD001B
		•	•	•							SDLD001
	'1020				A						SDAD001B
	'11020								•	•	SCAD001
Triple 3-Input	'10	•	•	•							SDLD001
					•	•					SDAD001B
								•			SCLD001A
							•				SDFD001
	'1010				A						SDAD001B
	'11010								•	•	SCAD001
Quad 2-Input	'00				A	•					SDAD001B
		•	•	•							SDLD001
								•			SCLD001A
							•				SDFD001
	'26	•	•								SDLD001
	'37	•	•	•							SDLD001
					A						SDAD001B
	'38	•	•	•							SDLD001
					A						SDAD001B
	'39	•									SDLD001
	'132	•	•	•							SDLD001
								•			SCLD001A
	'1000				A	A					SDAD001B
	'11000								•	•	SCAD001
Hex 2-Input	'804				A	B					SDAD001B
								•			SCLD001A
	'1804				A	•					SDAD001B

How to read Digital Logic Products selection tables:
The following symbols are common to all selection tables on pages 659 to 689.
- • = Product available in technology indicated
- ▲ = New Product planned in technology indicated
- A = "A" suffix version available in technology indicated
- B = "B" suffix version available in technology indicated

Positive-NAND Gate with Open-Collector Outputs

DESCRIPTION	TYPE	TECHNOLOGY									DOCUMENT
		STD TTL	LS	S	ALS	AS	F	HC	AC	ACT	
Dual 4-Input	'22	•		•							SDLD001
					B						SDAD001B
Triple 3-Input	'12	•	•								SDLD001
					A						SDAD001B
Quad 2-Input	'01	•	•								SDLD001
					•						SDAD001B
								•			SCAD001
	'03	•	•	•							SDLD001
					B						SDAD001B
								•			SCLD001A
	'1003				A						SDAD001B

Positive-AND Gate with Open-Collector Outputs

DESCRIPTION	TYPE	TECHNOLOGY									DOCUMENT
		STD TTL	LS	S	ALS	AS	F	HC	AC	ACT	
Triple 3-Input	'15		•	•							SDLD001
					A						SDAD001B
Quad 2-Input	'09	•	•	•							SDLD001
					•						SDAD001B
								•			SCLD001A
							•				SDFD001
	'7001							•			SCLD001A

Positive-AND Gates

DESCRIPTION	TYPE	TECHNOLOGY									DOCUMENT
		STD TTL	LS	S	ALS	AS	F	HC	AC	ACT	
Dual 4-Input	'21				A	•					SDAD001B
			•								SDLD001
								•			SCLD001A
							•				SDFD001
	'11021								•	•	SCAD001
Triple 3-Input	'11		•	•							SDLD001
					A	•					SDAD001B
								•			SCLD001A
							•				SDFD001
	'1011				A						SDAD001B
	'11011								•	•	SCAD001
Quad 2-Input	'08	•	•	•							SDLD001
					•	•					SDAD001B
								•			SCLD001A
							•				SDFD001
	'1008				A	A					SDAD001B
	'11008								•	•	SCAD001

See "How to read Digital Logic Products selection tables" on page 659.

Positive-OR Gates

DESCRIPTION	TYPE	TECHNOLOGY									DOCUMENT
		STD TTL	LS	S	ALS	AS	F	HC	AC	ACT	
Triple 3-Input	'4075							•			SCLD001A
Quad 2-Input	'32	•	•	•							SDLD001
					•	•					SDAD001B
								•			SCLD001A
							•				SDFD001
	'1032				A	A					SDAD001B
	'11032								•	•	SCAD001
Quad 2-Input	'7032							•			SCLD001A
Hex 2-Input	'832				A	B					SDAD001B
								•			SCLD001A
	'1832				A	•					SDAD001B

Positive-NOR Gates

DESCRIPTION	TYPE	TECHNOLOGY									DOCUMENT
		STD TTL	LS	S	ALS	AS	F	HC	AC	ACT	
Dual 4-Input with Strobe	'25	•									SDLD001
Dual 4-Input	'4002							•			SCLD001A
Dual 5-Input	'260			•							SDLD001
Triple 3-Input	'27	•	•								SDLD001
					•	•					SDAD001B
								•			SCLD001A
							•				SDFD001
	'11027								•	•	SCAD001
Quad 2-Input	'02	•	•	•							SDLD001
					•	•					SDAD001B
								•			SCLD001A
							•				SDFD001
	'28	•	•								SDLD001
					A						SDAD001B
	'33	•	•								SDLD001
					A						SDAD001B
	'36							•			SCLD001A
							•				SDFD001
	'1002				A						SDAD001B
	'1036					A					SDAD001B
	'7002							•			SCLD001A
	'11002								•	•	SCAD001
Hex 2-Input	'805				A	B					SDAD001B
								•			SCLD001A
	'1805				A	•					SDAD001B

See "How to read Digital Logic Products selection tables" on page 659.

Positive-OR/NOR Gates

DESCRIPTION	TYPE	TECHNOLOGY									DOCUMENT
		STD TTL	LS	S	ALS	AS	F	HC	AC	ACT	
8-Input	'4078							A			SCLD001A

Exclusive OR/NOR Gates

DESCRIPTION	TYPE	TECHNOLOGY						DOCUMENT
		STD TTL	LS	S	ALS	AS	HC	
Quad 2-Input Exclusive OR Gates with Totem-Pole Outputs	'86	•	A	•				SDLD001
					•			SDAD001B
							•	SCLD001A
	'386						•	SCLD001A
Quad 2-Input Exclusive OR Gates with Open-Collector Outputs	'136	•	•					SDLD001
					•			SDAD001B
						▲		SDAD001B
Quad 2-Input Exclusive-NOR Gates	'266		•					SDLD001
							•	SCLD001A
	'810				•	▲		SDAD001B
	'7266						•	SCLD001A
Quad 2-Input Exclusive-NOR Gates with Open-Collector Outputs	'811				•	▲		SDAD001B
Quad Exclusive OR/NOR Gates	'135			•				SDLD001

AND-NOR Gates

DESCRIPTION	TYPE	TECHNOLOGY							DOCUMENT
		STD TTL	LS	S	ALS	AS	HC	HCT	
2-Wide 4-Input	'55		•						SDLD001
4-Wide 4-2-3-2 Input	'64			•					SDLD001
4-Wide 2-2-3-2 Input	'54	•	•						SDLD001
Dual 2-Wide 2-Input	'51	•	•	•					SDLD001
							•		SCLD001A

AND-NOR Gates with Open-Collector Outputs

DESCRIPTION	TYPE	TECHNOLOGY							DOCUMENT
		STD TTL	LS	S	ALS	AS	HC	HCT	
4-Wide 4-2-3-2-Input	'65			•					SDLD001

See "How to read Digital Logic Products selection tables" on page 659.

Expandable Gates

DESCRIPTION	TYPE	TECHNOLOGY							DOCUMENT
		STD TTL	LS	S	ALS	AS	HC	HCT	
Dual 2-Wide AND-OR-Invert	'50	●							SDLD001
Dual 4-Input Positive-NOR with Strobe	'23	●							SDLD001

Multifunction Gates and Elements

DESCRIPTION	TYPE	TECHNOLOGY							DOCUMENT
		STD TTL	LS	S	ALS	AS	HC	HCT	
Inverter,3-/4-Input NAND/NOR Combination	'7006						●		SCLD001A
6-Section NAND Invert,NOR	'7008						●		SCLD001A
Quadruple Complementary Output Logic Element	'265	●							SDLD001

Delay Elements

DESCRIPTION	TYPE	TECHNOLOGY							DOCUMENT
		STD TTL	LS	S	ALS	AS	HC	HCT	
Inverting and Noninverting Elements 2-Input NAND-Buffer	'31		●						SDLD001

See "How to read Digital Logic Products selection tables" on page 659.

INVERTERS/NONINVERTING BUFFERS

Hex Inverters/Noninverters

DESCRIPTION	TYPE	TECHNOLOGY										DOCUMENT
		STD TTL	LS	S	ALS	AS	F	HC	AC	ACT	HCU	
Hex Inverters	'04	●	●	●								SDLD001
					B	●						SDAD001B
								●				SCLD001A
											●	SCLD001A
							●					SDFD001
	'11004								●	●		SCAD001
	'05	●	●	●								SDLD001
					A							SDAD001B
								●				SCLD001A
	'06	●										SDLD001
	'14	●	●									SDLD001
								●				SCLD001A
	'16	●										SDLD001
	'19		●									SDLD001
	'1004				●	A						SDAD001B
	'1005				●							SDAD001B
Hex Noninverter	'34				●	●						SDAD001B
	'11034								●	●		SCAD001

See "How to read Digital Logic Products selection tables" on page 659.

DRIVER AND BUS TRANSCEIVERS

Hex Drivers

DESCRIPTION	TYPE	TECHNOLOGY							DOCUMENT
		STD TTL	LS	S	ALS	AS	HC	HCT	
Hex 2-Input Driver	'808				A	B			SDAD001B
								•	SCLD001A
	'1808				A	•			SDAD001B
Hex Driver	'07	•							SDLD001
	'17	•							SDLD001
	'35				A				SDAD001B
	'1034				•	A			SDAD001B
	'1035				•				SDAD001B
Noninverting Hex Buffers/ Drivers	'365	A	A						SDLD001
							•		SCLD001A
	'366	A	A						SDLD001
							•		SCLD001A
	'367	A	A						SDLD001
							•		SCLD001A
	'368	A	A						SDLD001
							•		SCLD001A

Drivers with Open-Collector Outputs

DESCRIPTION	TYPE	TECHNOLOGY							DOCUMENT
		STD TTL	LS	S	ALS	AS	HC	HCT	
Noninverting Octal Buffers, Drivers	'757					•			SDAD001B
	'760				▲	•			SDAD001B
Inverting Octal Buffers, Drivers	'756				•	•			SDAD001B
	'763				•	•			SDAD001B
Inverting and Noninverting Octal Buffers, Drivers	'762					•			SDAD001B

See "How to read Digital Logic Products selection tables" on page 659.

Bus Transceivers with Open Collector Outputs

DESCRIPTION	TYPE	TECHNOLOGY							DOCUMENT
		STD TTL	LS	S	ALS	AS	F	HC	
Noninverting Quad Transceivers	'759					•			SDAD001B
Inverting Quad Transceivers	'758				•	•			SDAD001B
12 mA/24 mA/ 40 mA Sink Transceivers	'615				•				SDAD001B
	'621				A	•			SDAD001B
							•		SDFD001
	'639				A	•			SDAD001B
	'641				A	•			SDAD001B
			•						SDLD001
12 mA/24 mA/ 48 mA Sink Inverting Output Transceivers	'614				•				SDAD001B
	'622				A	•			SDAD001B
							•		SDFD001
	'638				A	•			SDAD001B
	'642				A	•			SDAD001B
			•						SDLD001
	'653				•				SDAD001B
12 mA/24 mA/ 48 mA Sink, True and Inverting Output Transceivers	'644				A	•			SDAD001B
			•						SDLD001
Registered With Multiplexed 12 mA/24 mA/ 48 mA True Output Transceivers	'647				•				SDAD001B
	'654				•				SDVD001
Registered with Multiplexed 12 mA/24 mA/ 48 mA Inverting Output Transceivers	'649				•				SDVD001

See "How to read Digital Logic Products selection tables" on page 659.

Drivers with 3-State Outputs

DESCRIPTION	TYPE	STD TTL	LS	S	ALS	AS	F	HC	HCT	AC	ACT	BCT	DOCUMENT
Quad Buffers/ Drivers with Independent Output Controls	'125	•	A										SDLD001
								•					SCLD001A
	'126	•	A										SDLD001
								•					SCLD001A
Noninverting Octal Buffers/ Drivers	'241		•	•									SDLD001
					A	•							SDAD001B
								•	•				SCLD001A
							•						SDFD001
												▲	TBA
	'11241										▲		SCAD001
	'244		•	•									SDLD001
					A	•							SDAD001B
								•	•				SCLD001A
							•						SDFD001
												▲	TBA
	'11244									•	•		SCAD001
	'465		•										SDLD001
					A								SDAD001B
	'467				A								SDAD001B
	'541		•										SDLD001
					•								SDAD001B
								•					SCAD001
									•				SCLD001A
	'1244				A								SDAD001B
Inverting Octal Buffers/ Drivers	'231				•	•							SDAD001B
	'240		•	•									SDLD001
					A	•							SDAD001B
								•	•				SCLD001A
							•						SDFD001
												▲	TBA
	'11240									•			SCAD001
											•		SCAD001
	'466		•										SDLD001
					A								SDAD001B
	'468				A								SDAD001B
	'540		•										SDLD001
					•								SDAD001B
								•					SCAD001
									•				SCAD001
	'1240				•								SDAD001B

See "How to read Digital Logic Products selection tables" on page 659.

Bus Transceivers with 3-State Outputs (Continued)

DESCRIPTION	TYPE	TECHNOLOGY										DOCUMENT
		LS	S	ALS	AS	F	HC	HCT	AC	ACT	BCT	
12 mA/24 mA/ 48 mA Sink, True Output Transceivers	'623			A	●							SDAD003
		●										SDLD001
							●	●				SCLD001A
						●						SDFD001
	'645						●					SCLD001A
								●				SCLD001A
				A	●							SDAD001B
		●										SDLD001
	'654			●								SDAD001B
	'1640			A								SDAD001B
	'1645			A								SDAD001B
	'11623								▲	▲		SCAD001
Universal Transceiver/ Port Controllers	'852				▲							SDVD001
	'856				▲							SDVD001
	'877				▲							SDVD001

Line Drivers/ Bus Transceivers/ MOS Drivers

DESCRIPTION	TYPE	TECHNOLOGY							DOCUMENT
		STD TTL	LS	S	ALS	AS	HC	BCT	
Bus Transceivers	'2242				●				SDAD001B
	'2620					●			SDAD001B
	'2623					●			SDAD001B
	'2640					●			SDAD001B
	'2645					●			SDAD001B
Line Drivers	'2240				●				SDAD001B
	'2240							▲	TBA
	'2241							▲	TBA
	'2244				▲				TBA
	'2244							▲	TBA
	'2540				●				SDAD001B
	'2541				●				SDAD001B

Line Drivers

DESCRIPTION	TYPE	TECHNOLOGY							DOCUMENT
		STD TTL	LS	S	ALS	AS	HC	HCT	
Octal Buffers AND/Line Drivers with Input Pull-up Resistors	'746				●				SDAD001B
	'747				●				SDAD001B
Octal/Line Drivers/with 3-State Output	'2540				●				SDAD001B
	'2541				●				SDAD001B

See "How to read Digital Logic Products selection tables" on page 659.

Bus Transceivers with 3-State Outputs (Continued)

DESCRIPTION	TYPE	TECHNOLOGY										DOCUMENT
		LS	S	ALS	AS	F	HC	HCT	AC	ACT	BCT	
Octal Transceivers	'643						•					SCAD001
								•				SCAD001
				A	•							SDAD001B
	'11643								▲	▲		SCAD001
	'1245			A								SDAD003
											▲	TBA
Octal Bus Transceivers with Registers	'543					▲						SDFD001
	'544					▲						SDFD001
	'646						•	•				SCLD001
				•	•							SDAD001B
		•										SDLD001
	'648						•	•				SCLD001A
				•	•							SDVD001
		•										SDLD001
	'651						•	•				SCLD001A
				•	•							SDVD001
		•										SDLD001
	'652						•	•				SCLD001A
				•	•							SDVD001
		•										SDLD001
	'11646								▲	▲		SCAD001
	'11648								▲	▲		SCAD001
	'11651								▲	▲		SCAD001
	'11652								▲	▲		SCAD001
8-/9-Bit Bus Transceivers with Parity Checker/ Generator	'658						•	•				SCLD001A
	'659						•	•				SCLD001A
	'664						•	•				SCLD001A
	'665						•	•				SCLD001A
	'29833			•								SDVD001
											▲	TBA
	'29834			•								SDVD001
											▲	TBA
	'29853			•								SDVD001
											▲	TBA
	'29854			•								SDVD001
											▲	TBA
Noninverting 9-Bit Transceivers	'29863			•								SDVD001
											•	SCLS055
Inverting 9-Bit Transceivers	'29864			•								SDVD001
											▲	TBA
Noninverting 10-Bit Transceivers	'29861			•								SDVD001
											•	SCLS056
Inverting 10-Bit Transceivers	'29862			•								SDVD001
											▲	TBA

See "How to read Digital Logic Products selection tables" on page 659.

Drivers with 3-State Outputs (Continued)

DESCRIPTION	TYPE	TECHNOLOGY											DOCUMENT
		STD TTL	LS	S	ALS	AS	F	HC	HCT	AC	ACT	BCT	
Inverting and Noninverting Octal Buffers/ Drivers	'230					•							SDAD001B
Noninverting 10-Bit Buffers/ Drivers	'2827											•	SCLS051
	'29827				•								SDVD001
												•	SCLS052
Inverting 10-Bit Buffers/ Drivers	'2828											•	SCLS051
	'29828				•								SDVD001
												•	SCLS052

Bus Transceivers with 3-State Outputs

DESCRIPTION	TYPE	TECHNOLOGY										DOCUMENT
		LS	S	ALS	AS	F	HC	HCT	AC	ACT	BCT	
Noninverting Quad Transceivers	'243	•										SDLD001
				A	•							SDAD001B
							•	•				SCLD001A
						•						SDFD001
Inverting Quad Transceivers	'242	•										SDLD001
				B	•							SDAD001B
							•	•				SCLD001A
						•						SDFD001
	'1242			•								SDAD001B
Quad Transceivers	'442	•										SDLD001
Octal Transceivers		•										SDLD001
	'245			A	•							SDAD001B
							•					SCLD001A
									•			SCAD001
						•						SDFD001
											▲	TBA
	'11245								▲	▲		SCAD001
	'620						•	•				SCLD001A
				A	•							SDAD003
						•						SDFD001
	'11620								▲	▲		SCAD001
	'640						•					SCAD001
								•				SCAD001
				A	•							SDAD001B
		•										SDLD001
	'11640								▲	▲		SCAD001

See "How to read Digital Logic Products selection tables" on page 659.

50-Ohm/75-Ohm Line Drivers

DESCRIPTION	TYPE	TECHNOLOGY							DOCUMENT
		STD TTL	LS	S	ALS	AS	HC	HCT	
Quad 2-Input Positive-NOR	'128	●							SDLD001
Dual 4-Input Positive-NAND	'140			●					SDLD001
Hex 2-Input Positive-NAND	'804				A	B			SDAD001B
							●		SCLD001A
	'1804				A	●			SDAD001B
Hex 2-Input Positive-NOR	'805				A	B			SDAD001B
							●		SCLD001A
	'1805				A	●			SDAD001B
Hex 2-Input Positive-AND	'808				A	B			SDAD001B
							●		SCLD001A
	'1808				A	●			SDAD001B
Hex 2-Input Positive-OR	'832				A	B			SDAD001B
							●		SCLD001A
	'1832				A	●			SDD001B

Multifunction Drivers

DESCRIPTION	TYPE	TECHNOLOGY							DOCUMENT
		STD TTL	LS	S	ALS	AS	HC	HCT	
Dual Pulse Synchronizers/ Drivers	'120	●							SDLD001

See "How to read Digital Logic Products selection tables" on page 659.

FLIP-FLOP

Dual and Single Flip-Flops

DESCRIPTION	TYPE	TECHNOLOGY									DOCUMENT
		STD TTL	LS	S	ALS	AS	HC	AC	ACT	F	
Dual J-K Edge Triggered	'73	•	A								SDLD001
							•				SCLD001A
	'76	•	A								SDLD001
							•				SCLD001A
	'78		A								SDLD001
							•				SCLD001A
	'107	•	A								SDLD001
							•				SCLD001A
	'109	•	A								SDLD001
					A	•					SDAD001B
							•				SCLD001A
										•	SDFD001
	'112		A	A							SDLD001
					A						SDAD001B
							•				SCLD001A
										•	SDFD001
	'113		A	A							SDLD001
					A						SDAD001B
							•				SCLD001A
										•	SDFD001
	'114		A	A							SDLD001
					A						SDAD001B
							•				SCLD001A
										•	SDFD001
	'11109							•	•		SCAD001
Single J-K Edge Triggered	'70	•									SDLD001
Dual D-Type	'74	•	A	•							SDLD001
					A	•					SDAD001B
							•				SCLD001A
										•	SDFD001
	'11074							•	•		SCAD001
Dual D-Type with 2-Input NAND/NOR Gates	'7074						•				SCLD001A
	'7075						•				SCLD001A
	'7076						•				SCLD001A
Dual 4-Bit D-Type Edge-Triggered	'874				B	•					SDAD001B
	'876				A	•					SDAD001B
	'878				A	•					SDAD001B
	'879				A	•					SDAD001B

See "How to read Digital Logic Products selection tables" on page 659.

Quad and Hex Flip-Flops

DESCRIPTION	OUTPUTS	NO. OF FF's	TYPE	TECHNOLOGY							DOCUMENT
				STD TTL	LS	S	ALS	AS	HC	F	
D-Type	Q,Q̄	4	'175	•	•	•					SDLD001
							•	•			SDAD001B
									•		SCLD001A
										•	SDFD001
			'379		•						SDLD001
									•		SCLD001A
										•	SDFD001
	Q	6	'174	•	•	•					SDLD001
							•	•			SDAD001B
									•		SCLD001A
										•	SDFD001
			'378		•						SDLD001
									•		SCLD001A
										•	SDFD001
J-K	Q	4	'276	•							SDLD001
			'279	•	A						SDLD001
			'376	•							SDLD001

Octal, 9-Bit, and 10-Bit D-Type Flip-Flops

DESCRIPTION	NO. OF BITS	OUTPUTS	TYPE	TECHNOLOGY											DOCUMENT
				STD TTL	LS	S	ALS	AS	HC	HCT	AC	ACT	BCT	F	
True Data	Octal	3-State	'374		•	•									SDLD001
							•	•							SDAD001B
									•	•					SCLD001A
													▲		TBA
														•	SDFD001
			'574				A	•							SDAD001B
									•	•					SCLD001A
														•	SDFD001
			'11374								▲	▲			SCAD001
True Data with Clear	Octal	2-State	'273	•	•										SDLD001
						•									SDAD001B
									•						SCLD001A
														•	SDFD001
		3-State	'575				A	•							SDAD001B
			'874				A	•							SDAD001B
			'878				A	•							SDAD001B
True with Enable	Octal	2-State	'377		•										SDLD001
									•						SCLD001A
														•	SDFD001

See "How to read Digital Logic Products selection tables" on page 659.

Octal, 9-Bit, and 10-Bit D-Type Flip-Flops (Continued)

DESCRIPTION	NO. OF BITS	OUTPUTS	TYPE	TECHNOLOGY											DOCUMENT
				STD TTL	LS	S	ALS	AS	HC	HCT	AC	ACT	BCT	F	
Inverting	Octal	3-State	'534				●	●							SDAD001B
										●					SCLD001A
											●				SCLD001A
													▲		TBA
														●	SDFD001
			'564				A								SDAD001B
									●	●					SCLD001A
														●	SDFD001
			'576				A	●							SDAD001B
			'826					●							SDVD001
			'11534								▲	▲			SCAD001
Inverting with Clear	Octal	3-State	'577				A	●							SDAD001B
			'879				A	●							SDAD001B
Inverting with Preset	Octal	3-State	'876				A	●							SDAD001B
True	Octal	3-State	'825					●							SDVD001
	9-Bit	3-State	'823					●							SDVD001
			'1823					●							SDAS126
Inverting	9-Bit	3-State	'824					●							SDVD001
True	10-Bit	3-State	'821					●							SDVD001
			'1821					●							SDAS131
Inverting	10-Bit	3-State	'822					●							SDVD001

See "How to read Digital Logic Products selection tables" on page 659.

LATCHES AND MULTIVIBRATORS

Quad Latches with 2-State Outputs

DESCRIPTION	TYPE	TECHNOLOGY							DOCUMENT
		STD TTL	LS	S	ALS	AS	HC	HCT	
Bistable	'75	●	●						SDLD001
								●	SCLD001A
	'375		●						SDLD001
								●	SCLD001A
S-R	'279	●	A						SDLD001

Monostable Multivibrators

DESCRIPTION	TYPE	TECHNOLOGY							DOCUMENT
		STD TTL	LS	S	ALS	AS	HC	HCT	
Single	'121	●							SDLD001
	'122	●	●						SDLD001
	'130	●							SDLD001
Dual	'123	●	●						SDLD001
	'221	●	●						SDLD001
	'423		●						SDLD001

D-Type Octal, 9-Bit, and 10-Bit Read-Back Latches

DESCRIPTION	NO. OF BITS	TYPE	TECHNOLOGY							DOCUMENT
			STD TTL	LS	S	ALS	AS	HC	HCT	
Edge-Triggered Inverting and Noninverting	Octal	'996				●				SDVD001
Transparent True	Octal	'990				●				SDVD001
	9-Bit	'992				●				SDVD001
	10-Bit	'994				●				SDVD001
Transparent Noninverting	Octal	'991				●				SDVD001
	9-Bit	'993				●				SDVD001
	10-Bit	'995				●				SDVD001
Transparent with Clear True Outputs	Octal	'666				●				SDVD001
Transparent with Clear Inverting Outputs	Octal	'667				●				SDVD001

See "How to read Digital Logic Products selection tables" on page 659.

Octal, 9-Bit, and 10-Bit Latches

DESCRIPTION	NO. OF BITS	OUTPUT	TYPE	STD TTL	LS	S	ALS	AS	HC	HCT	AC	ACT	BCT	F	DOCUMENT
Transparent	Octal	3-State	'373		•	•									SDLD001
							•	•							SDAD001B
									•	•					SCLD001A
													•		TBA
														•	SDFD001
			'573				B	•							SDAD001B
									•	•					SCLD001A
														•	SDFD001
			'11373								•	•			SCAD001
Dual 4-Bit Transparent	Octal	2-State	'116	•											SDLD001
		3-State	'873				B	•							SDAD001B
Inverting Transparent	Octal	3-State	'533				•	•							SDAD001B
									•	•					SCLD001A
													•		TBA
														•	SDFD001
			'11533								•	•			SCAD001
			'563				A								SDAD001B
									•	•					SCLD001A
														•	SDFD001
			'580				A	•							SDAD001B
Dual 4-Bit Inverting Transparent	Octal	3-State	'880				A	•							SDAD001B
2-Input Multiplexed	Octal	3-State	'604						•						SCLD001A
		OC	'607		•										SDLD001
Addressable	Octal	2-State	'259	•	B										SDLD001
								•							SDAD001B
									•						SCLD001A
		Q Only	'4724						•						SCLD001A
True	10-Bit	3-State	'841				•	•							SDVD001
			'1841					•							SDAS130
True	9-Bit	3-State	'843				•	•							SDVD001
			'1843					•							SDAS127
True	Octal	3-State	'845				•	•							SDVD001
Inverting	10-Bit	3-State	'842				•	•							SDVD001
	9-Bit	3-State	'844				•	•							SDVD001
	Octal	3-State	'846				•	•							SDVD001

See "How to read Digital Logic Products selection tables" on page 659.

REGISTERS

Shift Registers

DESCRIPTION	NO. OF BITS	MODES				TYPE	TECHNOLOGY							DOCUMENT
		S-	S	L	H		STD TTL	LS	S	ALS	AS	HC	F	
Sign Protected		X		X	X	'322		A						SDLD001
Parallel-In Parallel-Out Bidirectional	4	X	X	X	X	'194	•	A	•					SDLD001
											•			SDAD001B
												•		SCLD001A
	8	X	X	X	X	'198	•							SDLD001
		X	X	X	X	'299		•	•					SDLD001
										•				SDAD001B
												•		SCLD001A
													•	SDFD001
		X	X	X	X	'323		•						SDLD001
										•				SDAD001B
												•		SCLD001A
													•	SDFD001
Parallel-In Parallel-Out	4	X		X		'95	A	B						SDLD001
											•			SDAD001B
		X		X		'195	•	•	•					SDLD001
												•		SCLD001A
		X		X		'295		B						SDLD001
		X		X		'395		A						SDLD001
	5	X		X		'96	•	•						SDLD001
	8	X		X	X	'199	•							SDLD001
Serial-In Parallel-Out	8	X				'164	•	•						SDLD001
												•		SCLD001A
Parallel-In Serial-Out	8	X		X	X	'165	•	A						SDLD001
												•		SCLD001A
		X		X	X	'166	•	A						SDLD001
												•		SCLD001A
	16	X		X	X	'674		•						SDLD001
Serial-In Serial-Out	8	X				'91		•						SDLD001

NOTE: Modes; S- = S-R, S = S-L, L = Load, H = Hold

See "How to read Digital Logic Products selection tables" on page 659.

Shift Registers with Latches

DESCRIPTION	NO. OF BITS	OUTPUT	TYPE	TECHNOLOGY						DOCUMENT
				STD TTL	LS	S	ALS	AS	HC	
Parallel-In, Parallel-Out with Output Latches	4	3-State	'671		●					SDLD001
			'672		●					SDLD001
Serial-In Parallel-Out with Output Latches	8	Buffered	'594		●					SDLD001
									●	SCLD001A
		3-State	'595		●					SDLD001
									●	SCLD001A
		OC	'599		●					SDLD001
	16	2-State	'673		●					SDLD001
Parallel-In, Serial-Out with Input Latches	8	2-State	'597		●					SDLD001
Parallel I/O Ports with Input Latches Multiplexed Serial Inputs	8	3-State	'598		●					SDLD001

Sign-Protected Registers

DESCRIPTION	NO. OF BITS	MODES				TYPE	TECHNOLOGY						DOCUMENT
		S-	S	L	H		STD TTL	LS	S	ALS	AS	HC	
Sign-Protected Registers	8	X		X	X	'322	A						SDLD001

Register Files

DESCRIPTION	OUTPUT	TYPE	TECHNOLOGY						DOCUMENT
			STD TTL	LS	S	ALS	AS	HC	
Dual 16 Words × 4 Bits	3-State	'870				▲	●		SDVD001
		'871				▲	●		SDVD001
4 Words × 4 Bits	OC	'170	●	●					SDLD001
	3-State	'670		●					SDLD001
8 Words × 2 Bits	3-State	'172	●						SDLD001
64 Words × 40 Bits	3-State	'8834					▲		TBA

See "How to read Digital Logic Products selection tables" on page 659.

Other Registers

DESCRIPTION	TYPE	TECHNOLOGY							DOCUMENT
		STD TTL	LS	S	ALS	AS	HC	BCT	
Quadruple Multiplexers with Storage	'298	●	●						SDLD001
						●			SDAD001B
							●		SCLD001A
8-Bit Universal Shift Registers	'299		●	●					SDLD001
					●				SDAD001B
Quadruple Bus Buffer Register	'173	●	A						SDLD001
							●		SCLD001A
Data Selector/ Multiplexer/ Register	'356		●						SDLD001
							●		SCLD001A
Dual-Rank 8-Bit Shift Register	'963				▲				SDVD001
	'964				▲				SDVD001
8-Bit Diagnostic/ Pipeline Register	'819				●			●	SDAS105
	'29818				●				SDAS105
								▲	TBA

See "How to read Digital Logic Products selection tables" on page 659.

COUNTERS

Synchronous Counters — Positive-Edge Triggered

DESCRIPTION	PARALLEL LOAD	TYPE	TECHNOLOGY							DOCUMENT
			STD TTL	LS	S	ALS	AS	HC	F	
Decade	Sync	'160	●	A						SDLD001
						B	●			SDAD001B
								●		SCLD001A
									●	SDFD001
		'162		A	●					SDLD001
						B	●			SDAD001B
								●		SCLD001A
									●	SDFD001
		'560				A				SDAD001B
		'692						▲		SCLD001A
Decade Up/Down	Sync	'168				B	●			SDAD001B
									●	SDFD001
	Async	'190	●	●						SDLD001
						●				SDAD001B
								●		SCLD001A
		'192	●	●						SDLD001
						●				SDAD001B
								●		SCLD001A
	Sync	'568				A				SDAD001B
									●	SDFD001
		'696						▲		SCLD001A
4-Bit Binary	Sync	'161	●	A						SDLD001
						B	●			SDAD001B
								●		SCLD001A
									●	SDFD001
		'163	●	A	●					SDLD001
						B	●			SDAD001B
								●		SCLD001A
									●	SDFD001
		'561				A				SDAD001B
		'669		●						SDLD001
		'691						▲		SCLD001A
		'693						▲		SCLD001A
8-Bit Binary	Sync	'8161					●			SDAS116
		'8163					●			SDAS104

See "How to read Digital Logic Products selection tables" on page 659.

Synchronous Counters — Positive-Edge Triggered (Continued)

DESCRIPTION	PARALLEL LOAD	TYPE	TECHNOLOGY							DOCUMENT
			STD TTL	LS	S	ALS	AS	HC	F	
4-Bit Binary Up/Down	Async	'191	●	●						SDLD001
						●				SDAD001B
								●		SCLD001A
		'193	●	●						SDLD001
						●				SDAD001B
								●		SCLD001A
	Sync	'169		B	●					SDLD001
						B	●			SDAD001B
									●	SDFD001
		'569				A				SDAD001B
									●	SDFD001
		'697		●						SDLD001
		'699		●						SDLD001
8-Bit Up/Down	Sync	'8169				●				SDAS117
	Async CLR	'867				●	●			SDVD001
	Sync CLR	'869				●	●			SDVD001
Divide-by-10 Johnson Counter		'4017						●		SCLD001A
Divide-by-8 Johnson Counter		'7022						●		SCLD001A

Asynchronous Counters (Ripple Clock) — Negative-Edge Triggered

DESCRIPTION	PARALLEL LOAD	TYPE	TECHNOLOGY						DOCUMENT
			STD TTL	LS	S	ALS	AS	HC	
Decade	Set-to-9	'90		●					SDLD001
	Yes	'176	●						SDLD001
	Yes	'196	●	●	●				SDLD001
	Set-to-9	'290		●					SDLD001
4-Bit Binary	None	'93	A	●					SDLD001
	Yes	'177	●						SDLD001
	Yes	'197	●	●	●				SDLD001
	None	'293	●	●					SDLD001
Divide-by-12 Dual Decade	None	'92	A	●					SDLD001
		'390	●	●					SDLD001
								●	SCLD001A
	Set-to-9	'490						●	SCLD001A
Dual 4-Bit Binary	None	'393	●	●					SDLD001
								●	SCLD001A
7-Bit Binary		'4024						●	SCLD001A
12-Bit Binary		'4040						●	SCLD001A
14-Bit Binary		'4020						●	SCLD001A
		'4060						●	SCLD001A
		'4061						●	SCLD001A

See "How to read Digital Logic Products selection tables" on page 659.

8-Bit Binary Counters with Registers

DESCRIPTION	PARALLEL LOAD	TYPE	TECHNOLOGY						DOCUMENT
			STD TTL	LS	S	ALS	AS	HC	
Parallel Register Outputs	3-State	'590		•					SDLD001
								•	SCLD001A
	OC	'591		•					SDLD001
Parallel Register Inputs	2-State	'592		•					SDLD001
Parallel I/O	3-State	'593		•					SDLD001

Frequency Dividers, Rate Multipliers

DESCRIPTION	TYPE	TECHNOLOGY						DOCUMENT
		STD TTL	LS	S	ALS	AS	HC	
60-Bit Binary Rate Multiplier	'97	•						SDLD001
Decade Rate Multiplier	'167	•						SDLD001
Programmable Frequency Dividers/ Digital Timers	'292		•					SDLD001
	'294		•					SDLD001

See "How to read Digital Logic Products selection tables" on page 659.

DECODERS, ENCODERS,DATA SELECTORS/MULTIPLEXERS AND SHIFTERS

Encoders/Data Selectors/Multiplexers

DESCRIPTION	OUTPUT	TYPE	TECHNOLOGY							DOCUMENT
			STD TTL	LS	S	ALS	AS	HC	F	
Quad 2-to-1	2-State	'157	●	●	●					SDLD001
						●	●			SDAD001B
								●		SCLD001A
									●	SDFD001
		'158		●	●					SDLD001
						●	●			SDAD001B
								●		SCLD001A
									●	SDFD001
		'298	●	●						SDLD001
							●			SDAD001B
								●		SCLD001A
		'399		●						SDLD001
	3-State	'257		B	●					SDLD001
						●				SDAD001B
							●			SDAD001B
								●		SCLD001A
									●	SDFD001
		'258		B	●					SDLD001
						●	●			SDAD001B
								●		SCLD001A
									●	SDFD001
Octal 2-to-1 with Storage	3-State	'604		●						SDLD001
								●		SCLD001A
	OC	'605		●						SDLD001
	3-State	'606		●						SDLD001
	OC	'607		●						SDLD001
Dual 4-to-1	2-State	'153	●	●	●					SDLD001
						●	●			SDAD001B
								●		SCLD001A
									●	SDFD001
	3-State	'253		●	●					SDLD001
						●	●			SDAD001B
								●		SCLD001A
									●	SDFD001
	2-State	'352		●						SDLD001
						●	●			SDAD001B
								●		SCLD001A
									●	SDFD001
	3-State	'353		●						SDLD001
						●	●			SDAD001B
								●		SCLD001A
									●	SDFD001

See "How to read Digital Logic Products selection tables" on page 659.

Encoders/Data Selectors/Multiplexers (Continued)

DESCRIPTION	OUTPUT	TYPE	TECHNOLOGY							DOCUMENT
			STD TTL	LS	S	ALS	AS	HC	F	
Hex 2-to-1 Universal Multiplexer	3-State	'857				●	●			SDAD001B
8-to-1	2-State	'151	A	●	●					SDLD001
						●	●			SDAD001B
								●		SCLD001A
									●	SDFD001
		'152						●		SCLD001A
	3-State	'251	●	●	●					SDLD001
						●				SDAD001B
								●		SCLD001A
									●	SDFD001
		'354		●						SDLD001
								●		SCLD001A
16-to-1	2-State	'150	●							SDLD001
	3-State	'250					●			SDVD001
		'850					●			SDVD001
		'851					●			SDVD001
Full BCD	2-State	'147	●	●						SDLD001
								●		SCLD001A
Cascadable Octal	2-State	'148	●	●						SDLD001
								●		SCLD001A
	3-State	'348		●						SDLD001

See "How to read Digital Logic Products selection tables" on page 659.

Decoders/Demultiplexers

DESCRIPTION	OUTPUT	TYPE	TECHNOLOGY								DOCUMENT
			STD TTL	LS	S	ALS	AS	HC	HCT	F	
Dual 2-to-4	2-State	'239						•			SCLD001A
		'139		A	•						SDLD001
						•					SDAD001A
								•			SCLD001A
		'155	•	A							SDLD001
	OC	'156	•	•							SDLD001
						•					SDAD001B
3-to-8	2-State	'138		•	•						SDLD001
						•	•				SDAD001B
								•	•		SCLD001A
										•	SDFD001
3-to-8	2-State	'237						•	•		SCLD001A
		'238						•	•		SCLD001A
		'131				•	•				SDAD001B
		'137		•							SDLD001
						•	•				SDAD001B
								•	•		SCLD001A
4-to-10 BCD-to-Decimal	2-State	'42	A	•							SDLD001
								•			SCLD001A
4-to-16	3-State	'154	•								SDLD001
						•					SDAD001B
								•			SCLD001A
	OC	'159	•								SDLD001
	2-State	'4514						•			SCLD001A
		'4515						•			SCLD001A

Shifters

DESCRIPTION	OUTPUT	TYPE	TECHNOLOGY								DOCUMENT
			STD TTL	LS	S	ALS	AS	HC	HCT	F	
4-Bit Shifter	3-State	'350								•	SDFD001
Parallel 16-Bit Multimode Barrel Shifter	3-State	'897					A				SDVD001
32-Bit Barrel Shifter	3-State	'8838					•				SDVD001
32-Bit Shuffle/Exchange	3-State	'8839					•				SDVD001

Open-Collector Display Decoders/Drivers

DESCRIPTION	OFF-STATE OUTPUT VOLTAGE	TYPE	TECHNOLOGY							DOCUMENT
			STD TTL	LS	S	ALS	AS	HC	HCT	
BCD-to-Decimal	30 V	'45	•							SDLD001
	15 V	'145	•	•						SDLD001
BCD-to-7 Segment	30 V	'46	A							SDLD001
	15 V	'47	A	•						SDLD001
		'247		•						SDLD001

See "How to read Digital Logic Products selection tables" on page 659.

Open Collector Display Decoder/Drivers with Counters/Latch

DESCRIPTION	TYPE	TECHNOLOGY							DOCUMENT
		STD TTL	L	S	ALS	AS	HC	HCT	
BCD Counter/4-Bit Latch/BCD-to-7-Segment Decoder/ LAD Driver	'143	•							SDLD001

Voltage-Controlled Oscillators

NO. OF VCOs	COMP'L Z_{OUT}	ENABLE	RANGE OUTPUT	R_{EXT}	fmax MHz	TYPE	TECHNOLOGY		DOCUMENT
							LS	S	
Single	No	No	No	No	70	'321	•		SDLD001
Single	Yes	Yes	Yes	No	20	'624	•		SDLD001
Single	Yes	Yes	Yes	Yes	20	'628	•		SDLD001
Dual	No	Yes	Yes	No	60	'124		•	SDLD001
Dual	Yes	No	No	No	20	'625	•		SDLD001
Dual	No	Yes	Yes	No	20	'629	•		SDLD001

Memory/Microprocessor Controllers

DESCRIPTION		TYPE	TECHNOLOGY					DOCUMENT
			LS	ALS	AS	HCT	TMS	
System Controllers, Universal OR for '888		'890			•			SDBS002
Memory Mappers	3-State	'612	•					SDVD001
	OC	'613	•					SDVD001
Memory Mappers with Output Latches	3-State	'610	•					SDVD001
	OC	'611	•					SDVD001
Dynamic Memory Controllers	16K, 64K, 256K	'2967		•				SDVD001
		'2968		•				SDVD001
		'4500					A	SDVD001
		'4502				B		SDVD001
	16K, 64K, 256K, 1 Meg	'6301		•				SDVD001
		'6302		•				SDVD001

Digital Loops

DESCRIPTION	TYPE	TECHNOLOGY							DOCUMENT
		STD TTL	LS	S	ALS	AS	HC	HCT	
Digital Phase-Lock Loop	'297		•						SDLD001

See "How to read Digital Logic Products selection tables" on page 659.

COMPARATORS AND ERROR DETECTION CIRCUITS

4-Bit Comparators

P=Q	P>Q	P̄>Q̄	OUTPUT	ENABLE	TYPE	STD TTL	LS	S	HC	DOCUMENT
DESCRIPTION						**TECHNOLOGY**				
Yes	Yes	No	2-State	No	'85	•	•	•		SDLD001
									•	SCLD001A

8-Bit Identity Comparators

INPUT	P=Q	P̄=Q̄	P>Q	P̄>Q̄	P<Q	OUTPUT	ENABLE	TYPE	LS	ALS	AS	HC	AC	ACT	F	DOCUMENT
DESCRIPTION								**TYPE**	**TECHNOLOGY**							**DOCUMENT**
20KΩ Pull-up	Yes	No	No	No	No	OC	Yes	'518		•						SDAD001B
															•	SDFD001
	No	Yes	No	No	No	2-State	Yes	'520		•						SDAD001B
															•	SDFD001
								'11520					•	•		SCAD001
	No	Yes	No	No	No	OC	Yes	'522		•						SDAD001B
	No	Yes	No	Yes	No	2-State	No	'682	•							SDLD001
												•				SCLD001A
Standard	Yes	No	No	No	No	OC	Yes	'519		•						SDAD001B
															•	SDFD001
	No	Yes	No	No	No	2-State	Yes	'521		•						SDAD001B
															•	SDFD001
								'11521					•	•		SCAD001
	No	Yes	No	Yes	No	2-State	No	'684	•							SDLD001
												•				SCLD001A
	No	Yes	No	No	No	2-State	Yes	'688		•						SDAD001A
									•							SDLD001
												•				SCLD001A
	No	Yes	No	No	No	OC	Yes	'689	•							SDLD001
										•						SDAD001A
Latched P	No	No	Yes	No	Yes	2-State	Yes	'885				•				SDVD001
Latched P and Q	Yes	No	Yes	No	Yes	Latched	Yes	'866				•				SDAD001A

Other Identity Comparators

DESCRIPTION	TYPE	LS	ALS	AS	HC	DOCUMENT
		TECHNOLOGY				
6-Bit Identity Comparator Controlling a 2-to-4-Bit Decoder	'29806		•			SDVD001
9-Bit Identity Comparator	'29809		•			

See "How to read Digital Logic Products selection tables" on page 659.

Address Comparators

DESCRIPTION	OUTPUT ENABLE	LATCHED ENABLE	TYPE	TECHNOLOGY					DOCUMENT
				S	ALS	AS	HC	HCT	
16-Bit	Yes		'677		•				SDAD001B
							•		SCLD001A
		Yes	'678		•				SDAD001B
							•		SCLD001A
12-Bit	Yes		'679		•				SDAD001B
							•		SCLD001A
		Yes	'680		•				SDAD001B
							•		SCLD001A

Parity Generators/Checkers, Error Detection and Correction Circuits

DESCRIPTION		NO. OF BITS	TYPE	TECHNOLOGY							DOCUMENT
				STD TTL	LS	S	ALS	AS	HC	F	
Odd/Even Parity Generators/Checkers		9	'180	•							SDLD001
									•		SCLD001A
		9	'280		•	•					SDLD001
									•		SCLD001A
							•	•			SDVD001
										•	SDFD001
		9	'286					•			SDAD001B
Error Detection and Correction Circuits	3-State	16	'616				•				SDVD001
	3-State	16	'630		•						SDLD001
	OC	16	'631		•						SDLD001
	3-State	32	'632				B	•			SDVD001
	OC	32	'633				▲				SDVD001
	3-State	32	'634				A	▲			SDVD001
	OC	32	'635				▲				SDVD001

Fuse-Programmable Comparators

DESCRIPTION	TYPE	TECHNOLOGY							DOCUMENT
		STD TTL	LS	S	ALS	AS	HC	HCT	
16-Bit Identity Comparator	'526				•				SDAD001B
12-Bit Identity Comparator	'528				•				SDAD001B
12-Bit Identity Comparator Controlling a 2-to-4 Bit Decoder	'812				•				SDVD001
8-Bit Identity Comparator and 4-Bit Comparator	'527				•				SDAD001B

See "How to read Digital Logic Products selection tables" on page 659.

ARITHMETIC CIRCUITS AND PROCESSOR ELEMENTS

Parallel Binary Adders

DESCRIPTION	TYPE	TECHNOLOGY							DOCUMENT
		STD TTL	LS	S	ALS	AS	HC	F	
4-Bit	'83	•	•						SDLD001
	'283	•	•	•					SDLD001
							•		SCLD001A
								•	SDFD001

Accumulators, Arithmetic Logic Units, Look-Ahead Carry Generators

DESCRIPTION		TYPE	TECHNOLOGY									DOCUMENT
			STD TTL	LS	S	ALS	AS	HCT	AC	ACT	F	
4-Bit Parallel Binary Accumulators		'681		•								SDLD001
16-Bit by 16-Bit Multiplier/Accumulators		'1010					•					SDVD001
										•		SCSS003B
4-Bit Arithmetic Logic Units:		'181		•	•							SDLD001
							B					SDAD001B
Function Generators		'11181							•	•		SCAD001
		'1181					•					SDVD001
		'381	A	•								SDLD001
											•	SDFD001
		'881					A					SDVD001
		'11881								•		SCAD001
4-Bit Arithmetic Logic Unit with Ripple Carry		'382		•							•	SDLD001
Look Ahead Carry Generators	16-Bit	'182			•							SDLD001
							▲					SDAD001B
		'282					▲					SDAD001B
	32-Bit	'882					A					SDAD001B
		'11882							•	•		SCAD001

See "How to read Digital Logic Products selection tables" on page 659.

PROGRAMMABLE LOGIC ARRAYS

Standard High-Speed PAL® Circuits (ALS)

TYPE	INPUTS	OUTPUTS		NO. OF PINS	PACKAGES	DOCUMENT
		NO.	TYPE			
PAL16L8A	16	8	Active Low	20	FK,FN,J,N	SDZD001B
PAL16R4A	16	4	Registered			
PAL16R6A	16	6	Registered			
PAL16R8A	16	8	Registered			
PAL16R6A-2	16	8	Active Low			
PAL16R4A-2	16	4	Registered			
PAL16R6A-2	16	6	Registered			
PAL16R8A-2	16	8	Registered			
PAL20L8A	20	8	Active Low	24	FK,FN,JT,NT	SDZD001B
PAL20R4A	20	4	Registered			
PAL20R6A	20	6	Registered			
PAL20R8A	20	8	Registered			
PAL20L8A-2	20	8	Active Low			
PAL20R4A-2	20	4	Registered			
PAL20R6A-2	20	6	Registered			
PAL20R8A-2	20	8	Registered			

High Performance PAL® Circuits (ALS)

TYPE	INPUTS	OUTPUTS		NO. OF PINS	PACKAGES	DOCUMENT
		NO.	TYPE			
TIBPAL16L8-10	16	8	Active High	20	FK,FN,J,N	SDZD001B
TIBPAL16R4-10	16	4	Registered			
TIBPAL16R6-10	16	6	Registered			
TIBPAL16R8-10	16	8	Registered			
TIBPAL16L8-12	16	8	Active High			
TIBPAL16R4-12	16	4	Registered			
TIBPAL16R6-12	16	6	Registered			
TIBPAL16R8-12	16	8	Registered			
TIBPAL16H8-15	16	8	Active High	20	FK,FN,J,N	TBA
TIBPAL16HD8-15	16	8	Active High			
TIBPAL16L8-15	16	8	Active Low	20	FK,FN,J,N	SDZD001B
TIBPAL16LD8-15	16	8	Active Low	20	FK,FN,J,N	TBA
TIBPAL16R4-15	16	4	Registered	20	FK,FN,J,N	SDZD001B
TIBPAL16R6-15	16	6	Registered			
TIBPAL16R8-15	16	8	Registered			
TIBPAL16H8-25	16	8	Active High	20	FK,FN,J,N	TBA
TIBPAL16HD8-25	16	8	Active High			
TIBPAL16L8-25	16	8	Active Low	20	FK,FN,J,N	SDZD001B
TIBPAL16LD8-25	16	8	Active Low	20	FK,FN,J,N	TBA
TIBPAL16R4-25	16	4	Registered	20	FK,FN,J,N	SDZD001B
TIBPAL16R6-25	16	6	Registered			
TIBPAL16R8-25	16	8	Registered			
TIBPAL16L8-30	16	8	Active Low			
TIBPAL16R4-30	16	4	Registered			
TIBPAL16R6-30	16	6	Registered			
TIBPAL16R8-30	16	8	Registered			
TIBPAL20L8-15	20	8	Active Low	24	FK,FN,JT,NT	SDZD001B
TIBPAL20R4-15	20	4	Registered			
TIBPAL20R6-15	20	6	Registered			
TIBPAL20R8-15	20	8	Registered			
TIBPAL20L8-25	20	8	Active Low			
TIBPAL20R4-25	20	4	Registered			
TIBPAL20R6-25	20	6	Registered			
TIBPAL20R8-25	20	8	Registered			
TIBPAL20L10-20	20	10	Active Low			
TIBPAL20X4-20	20	4	Registered			
TIBPAL20X8-20	20	8	Registered			
TIBPAL20X10-20	20	10	Registered			
TIBPAL20L10-30	20	10	Active Low	24	FK,FN,JT,NT	TBA
TIBPAL20X4-30	20	4	Registered			
TIBPAL20X8-30	20	8	Registered			
TIBPAL20X10-30	20	10	Registered			

High Performance PAL® Circuits (ALS) (Continued)

TYPE	INPUTS	OUTPUTS		NO. OF PINS	PACKAGES	DOCUMENT
		NO.	TYPE			
TIBPALR19L8	19	8	Active Low	24	FK,FN,JT,NT	SDZD001B
TIBPALR19R4	19	4	Registered			
TIBPALR19R6	19	6	Registered			
TIBPALR19R8	19	8	Registered			
TIBPALT19L8	19	8	Active Low			
TIBPALT19R4	19	4	Registered			
TIBPALT19R6	19	6	Registered			
TIBPALT19R8	19	8	Registered			

High Performance CMOS PAL® Circuits

TYPE	INPUTS	OUTPUTS		NO. OF PINS	PACKAGES	DOCUMENT
		NO.	TYPE			
TICPAL16L8-55	16	8	Active High	20	JL,N	TBA
TICPAL16R4-55	16	4	Registered			
TICPAL16R6-55	16	6	Registered			
TICPAL16R8-55	16	8	Registered			

High Performance IMPACT™ Programmable Array Logic

TYPE	INPUTS	OUTPUTS		NO. OF PINS	PACKAGES	DOCUMENT
		NO.	TYPE			
TIBPAL22V10	12 Inputs or 11 Inputs with CLK	10	I/O	24	NT,FN	SDPS015
TIBPAL22V10A						
TIBPAL22VP10-20	12 Inputs or 11 Inputs with CLK	10	I/O	24	NT,FN	SDPS106

Field Programmable Logic Array (ALS)

TYPE	INPUTS	OUTPUTS		NO. OF PINS	ARRAY	PACKAGES	DOCUMENT
		NO.	TYPE				
TIFPLA839	14	6	3-State	24	14 × 32 × 6	FK,FN,N,NT	SDZD001A
TIFPLA840	14	6	OC				
TIB82S167B	14	6	3-State	24	14 × 48 × 6		
82S167A	14	6	3-State				
TIB82S105B	16	8	3-State	28	14 × 48 × 6	FK,FN,JD,N	
82S105A	16	8	3-State				

APPENDIX
C

SELECTED DATA SHEETS

PAL/PLD Device Menu

DEVICE NAME	INPUTS	OUTPUTS	PRODUCT TERMS/OUTPUT	SPEED (t_{PD} in ns)	STANDBY I_{CC} (mA)	DATA SHEET PAGE NO.
PAL8L14A	8	14	1	25	90	5-141
PAL6L16A	6	16	1	25	90	5-141
PAL10H8	10	8	2	35	90	5-56
PAL12H6	12	6	2,4	35	90	5-56
PAL14H4	14	4	4	35	90	5-56
PAL16H2	16	2	8	35	90	5-56
PAL10L8	10	8	2	35	90	5-56
PAL12L6	12	6	2,4	35	90	5-56
PAL14L4	14	4	4	35	90	5-56
PAL16L2	16	2	8	35	90	5-56
PAL16C1	16	2	16	40	90	5-56
PAL16L8D	16*	8	7	10	180	5-29
AmPAL16L8D				10	180	5-183
PAL16L8B				15	180	5-31
AmPAL16L8B				15	180	5-197
PALC16L8Z-25				25	0.1	5-50
PALC16L8Q-25				25	45	5-33
PAL16L8B-2				25	90	5-35
AmPAL16L8AL				25	90	5-197
PAL16L8A				25	180	5-37
AmPAL16L8A				25	155	5-197
PAL16L8B-4				35	55	5-39
AmPAL16L8Q				35	45	5-197
PAL16L8A-2				35	90	5-41
AmPAL16L8L				35	80	5-197
AmPAL16L8				35	155	5-197
PAL16L8A–4				55	50	5-43
PAL16P8A	16*	8	7	25/30**	180	5-17
AmPAL18P8B	18*	8	8	15	180	5-202
AmPAL18P8AL				25	90	5-202
AmPAL18P8A				25	180	5-202
AmPAL18P8Q				35	55	5-202
AmPAL18P8L				35	90	5-202
PAL12L10	12	10	2	40	100	5-147
PAL14L8	14	8	2,4	40	100	5-147
PAL16L6	16	6	2,4	40	100	5-147
PAL18L4	18	4	4,6	40	100	5-147
PAL20L2	20	2	8	40	100	5-147
PAL20C1	20	2	16	40	100	5-147
PAL20L8B	20*	8	7	15	210	5-125
PAL20L8B-2				25	105	5-126
PAL20L8A				25	210	5-128
PALC20L8Z-35				35	0.1	5-133
PAL20L8A-2				35	105	5-130
PALC20L8Z-45				45	0.1	5-133

Table 1. Simple Combinatorial PAL Devices

PAL/PLD Device Menu

DEVICE NAME	INPUTS	OUTPUTS	PRODUCT TERMS/OUTPUT	SPEED (t_{PD} in ns)	STANDBY I_{cc} (mA)	DATA SHEET PAGE NO.
AmPAL20L10B	20*	10	3	15	210	5-306
AmPAL20L10-20				20	165	5-306
AmPAL20L10AL				25	105	5-306
PAL20L10A				30	165	5-113
PAL20S10	20*	10	0–16††	35	240	5-103
AmPAL22P10B	22*	10	8	15	210	5-306
AmPAL22P10AL				25	105	5-306
AmPAL22P10A				25	210	5-306
AmPAL22XP10-20	22*	10	2/6†	20	210	5-286
AmPAL22XP10-30L				30	105	5-286
AmPAL22XP10-30				30	180	5-286
AmPAL22XP10-40L				40	105	5-286

* Includes feedback † Has an exclusive-OR gate
** Depending on polarity †† Product term steering

Table 1. Simple Combinatorial PAL Devices (Cont'd.)

DEVICE NAME	INPUTS	OUTPUTS	FLIP-FLOPS	PRODUCT TERMS/OUTPUT	SPEED (f_{MAX} in MHz)	STANDBY I_{cc} (mA)	DATA SHEET PAGE NO.
PAL16R8D	16*	8	8	8	55	180	5-29
AmPAL16R8D					55	180	5-183
PAL16R8B					37	180	5-31
AmPAL16R8B					40	180	5-197
PALC16R8Z-25					28.5	0.1	5-50
PALC16R8Q-25					28.5	45	5-33
PAL16R8B-2					25	90	5-35
AmPAL16R8AL					28.5	90	5-197
PAL16R8A					25	180	5-37
AmPAL16R8A					28.5	155	5-197
PAL16R8B-4					16	55	5-39
AmPAL16R8Q					18	45	5-197
PAL16R8A-2					16	90	5-41
AmPAL16R8L					18	80	5-197
AmPAL16R8					18	155	5-197
PAL16R8A-4					11	50	5-43
PAL16R6D	16*	8	6	8	55	180	5-29
AmPAL16R6D					55	180	5-183
PAL16R6B					37	180	5-31
AmPAL16R6B					40	180	5-197
PALC16R6Z-25					28.5	0.1	5-50
PALC16R6Q-25					28.5	45	5-33
PAL16R6B-2					25	90	5-35
AmPAL16R6AL					28.5	90	5-197
PAL16R6A					25	180	5-37

Table 2. Simple Registered PAL Devices

PAL/PLD Device Menu

DEVICE NAME	INPUTS	OUTPUTS	FLIP-FLOPS	PRODUCT TERMS/OUTPUT	SPEED (f_{MAX} in MHz)	STANDBY I_{cc} (mA)	DATA SHEET PAGE NO.
AmPAL16R6A					28.5	180	5-197
PAL16R6B-4					16	55	5-39
AmPAL16R6Q					18	45	5-197
PAL16R6A-2					16	90	5-41
AmPAL16R6L					18	90	5-197
AmPAL16R6					18	180	5-197
PAL16R6A-4					11	50	5-43
PAL16R4D	16*	8	4	8	55	180	5-29
AmPAL16R4D					55	180	5-183
PAL16R4B					37	180	5-31
AmPAL16R4B					40	180	5-197
PALC16R4Z-25					28.5	0.1	5-50
PALC16R4Q-25					28.5	45	5-33
PAL16R4B-2					25	90	5-35
AmPAL16R4AL					28.5	90	5-197
PAL16R4A					25	180	5-37
AmPAL16R4A					28.5	180	5-197
PAL16R4B-4					16	55	5-39
AmPAL16R4Q					18	45	5-197
PAL16R4A-2					16	90	5-41
AmPAL16R4L					18	90	5-197
AmPAL16R4					18	180	5-197
PAL16R4A-4					11	50	5-43
PAL16X4	16*	8	4	8†	14	225	5-51
PAL16RP8A	16*	8	8	8	25**	180	5-17
PAL16RP6A	16*	8	6	8	25**	180	5-17
PAL16RP4A	16*	8	4	8	25**	180	5-17
PAL20R8B	20*	8	8	8	37	210	5-125
PAL20R8B-2					25	105	5-126
PAL20R8A					25	210	5-128
PALC20R8Z-35					20	0.1	5-133
PAL20R8A-2					16	105	5-130
PALC20R8Z-45					15.3	0.1	5-133
PAL20R6B	20*	8	6	8	37	210	5-125
PAL20R6B-2					25	105	5-126
PAL20R6A					25	210	5-128
PALC20R6Z-35					20	0.1	5-133
PAL20R6A-2					16	105	5-130
PALC20R6Z-45					15.3	0.1	5-133
PAL20R4B	20*	8	4	8	37	210	5-125
PAL20R4B-2					25	105	5-126
PAL20R4A					25	210	5-128
PALC20R4Z-35					20	0.1	5-133
PAL20R4A-2					16	105	5-130
PALC20R4Z-45					15.3	0.1	5-133

Table 2. Simple Registered PAL Devices (Cont'd.)

PAL/PLD Device Menu

DEVICE NAME	INPUTS	OUTPUTS	FLIP-FLOPS	PRODUCT TERMS/OUTPUT	SPEED (f$_{MAX}$ in MHz)	STANDBY I$_{cc}$ (mA)	DATA SHEET PAGE NO.
PAL20RS10	20*	10	10	0–16††	20	240	5-103
PAL20RS8	20*	10	8	0–16††	20	240	5-103
PAL20RS4	20*	10	4	0–16††	20	240	5-103
AmPAL22V10-15	22*	10	0–10§	8–16§§	40	180	5-249
PALC22V10H-25					33.3	90	5-79
AmPAL22V10A					28.5	180	5-260
PALC22V10H-35					20	90	5-79
AmPAL22V10					18	180	5-260
AmPAL20RP10B	22*	10	10	8	37	210	5-306
AmPAL20RP10AL					25	105	5-306
AmPAL20RP10A					25	210	5-306
AmPAL20RP8B	22*	10	8	8	37	210	5-306
AmPAL20RP8AL					25	105	5-306
AmPAL20RP8A					25	210	5-306
AmPAL20RP6B	22*	10	6	8	37	210	5-306
AmPAL20RP6AL					25	105	5-306
AmPAL20RP6A					25	210	5-306
AmPAL20RP4B	22*	10	4	8	37	210	5-306
AmPAL20RP4AL					25	105	5-306
AmPAL20RP4A					25	210	5-306
PAL32R16	32*	16	16§	0–16††	16	280	5-158

* Includes feedback
** With polarity fuse intact
† Has an exclusive-OR gate
†† Product term steering
§ Flip-flops can be bypassed
§§ Has varied product term distribution

Table 2. Simple Registered PAL Devices (Cont'd.)

PAL/PLD Device Menu

DEVICE NAME	INPUTS	OUTPUTS	FLIP-FLOPS	FLIP-FLOP TYPES	PRODUCT TERMS/OUTPUT	SPEED (f_{MAX} in MHz)	STAND BY I_{cc} (mA)	DATASHEET PAGE NO.
PAL20X10A	20*	10	10	D,T,JK,SR	2/2†	22.2	180	5-113
PAL20X8A	20*	10	8	D,T,JK,SR	2/2†	22.2	180	5-113
PAL20X4A	20*	10	4	D,T,JK,SR	2/2†	22.2	180	5-113
AmPAL20XRP10-20	22*	10	10	D,T,JK,SR	2/6†	30.3	210	5-286
AmPAL20XRP10-30L						22.2	105	5-286
AmPAL20XRP10-30						22.2	180	5-286
AmPAL20XRP10-40L						14.3	105	5-286
AmPAL20XRP8-20	22*	10	8	D,T,JK,SR	2/6, 8†	30.3	210	5-286
AmPAL20XRP8-30L						22.2	105	5-286
AmPAL20XRP8-30						22.2	180	5-286
AmPAL20XRP8-40L						14.3	105	5-286
AmPAL20XRP6-20	22*	10	6	D,T,JK,SR	2/6, 8†	30.3	210	5-286
AmPAL20XRP6-30L						22.2	105	5-286
AmPAL20XRP6-30						22.2	180	5-286
AmPAL20XRP6-40L						14.3	105	5-286
AmPAL20XRP4-20	22*	10	4	D,T,JK,SR	2/6, 8†	30.3	210	5-286
AmPAL20XRP4-30L						22.2	105	5-286
AmPAL20XRP4-30						22.2	180	5-286
AmPAL20XRP4-40L						14.3	105	5-286
PAL22RX8A	22*	8	8§	D,T,JK,SR	1/8†	28.5	210	5-87
AmPAL23S8-20	23*	8	14§	D,B◊	6–12§§	33	200	5-169
AmPAL23S8-25						28.5	200	5-169
AmPALC29M16-35	29*	16	16§	D,B,L◊	8–16§§	20	120	5-231
AmPALC29M16-45						15	120	5-231
PAL32VX10A	32*	10	10§	D,T,JK,SR, B◊	1/8–16†	25	180	5-70
PAL32VX10						22.2	180	5-70

* Includes feedback   §§ Has varied product term distribution
† Has an exclusive-OR gate   ◊ B=flip-flops are or can be buried; L=latched outputs possible
§ Some flip-flops can be bypassed

Table 3. State Machine PAL Devices

DEVICE NAME	INPUTS	OUTPUTS	PRODUCT TERMS/OUTPUT	SPEED (t_{PD} in ns)	STANDBY I_{cc} (mA)	DATA SHEET PAGE NO.
PAL16RA8	16*	8	4	30**	170	5-11
PAL20RA10-20	20*	10	4	20**	200	5-95
PAL20RA10				30**	200	5-97
AmPALC29MA16-35	29*	16	4–12††	35	120	5-209
AmPALC29MA16-45				45	120	5-209

* Includes feedback
** With polarity fuse intact
†† Has product term steering

Table 4. Asynchronous PAL Devices

PAL/PLD Device Menu

DEVICE NAME	INPUTS	OUTPUTS	FLIP-FLOPS	PRODUCT TERMS/OUTPUT	SPEED (t_{PD} or f_{MAX})	I_{EE} (mA)	DATA SHEET PAGE NO.
PAL10H20P8	20*	8	0	0–8††	6 ns	210	5-386
PAL10H20G8	20*	8	8§	0–8††	6 ns	225	5-382
PAL10H20EV/EG8	20*	8	8§	8–12§§	125 MHz	220	5-381

* Includes feedback   § Flip-flops can be bypassed
†† Has product term steering   §§ Has varied product term distribution

Table 5. 10KH-Compatible PAL Devices

DEVICE NAME	INPUTS	OUTPUTS	FLIP-FLOPS	PRODUCT TERMS/OUTPUT	SPEED (t_{PD} or f_{MAX})	I_{EE} (mA)	DATA SHEET PAGE NO.
PAL10020EV/EG8	20*	8	8§	8–12§§	125 MHz	220	5-381

* Includes feedback   § Flip-flops can be bypassed
§§ Has varied product term distribution

Table 6. 100K-Compatible PAL Devices

DEVICE NAME	INPUTS	OUTPUTS	STATES (MAX)	BRANCHES PER STATE	SPEED (f_{MAX} in MHz)	I_{cc} (mA)	DATA SHEET PAGE NO.
PMS14R21A	8	8	128	4	30	210	5-315
PMS14R21					25	210	5-315
PLS105-37	16	8	<64*	*	37	200	5-331
PLS167-33	14	6	<128*	*	33	200	5-331
PLS168-33	12	8	<1028*	*	33	200	5-331
Am29PL141	6	16	64	2	20	450	5-339
Am2971	6	14	N/A	N/A	85	310	5-365

* Depends highly on state diagram topology. May be limited by number of product terms or number of flip-flops.

Table 7. Programmable Sequencers

DEVICE NAME	I/O PINS	CLBs	SPEED (INTERNAL TOGGLE f_{MAX} in MHz)	STANDBY I_{cc} (mA)	DATA SHEET PAGE NO.
M2064-70	58	64	70	5	5-483
M2064-50			50	5	5-483
M2064-33			33	5	5-483
M2018-50	74	100	50	5	5-483
M2018-33			33	5	5-483

Table 8. LCA Devices

PAL16RP8A Series   16P8A, 16RP8A 16RP6A, 16RP4A

Features/Benefits

- Programmable polarity
- High speed at 25 ns tPD
- Register preload
- Power-up reset
- Security fuse

Description

The PAL16RP8A Series is equivalent to the PAL16R8A Series, with the addition of programmable polarity. With programmable polarity unused, these devices are equivalent to the PAL16R8A Series.

Variable Input/Output Pin Ratio

The registered devices have eight dedicated input lines, and each combinatorial output is an I/O pin. The combinatorial device has ten dedicated input lines, and only six of the eight combinatorial outputs are I/O pins. Buffers for device inputs have complementary outputs to provide user-programmable input signal polarity. Unused input pins should be tied directly to VCC or GND.

Programmable Three-State Outputs

Each output has a three-state output buffer with programmable three-state control. On combinatorial outputs, a product term controls the buffer, allowing enable and disable to be a function of any combination of device inputs or output feedback. The output provides a bidirectional I/O pin in the combinatorial configuration, and may be configured as a dedicated input if the buffer is always disabled.

Registers with Feedback

Registered outputs are provided for data storage and synchronization. Registers are composed of D-type flip-flops which are loaded on the low-to-high transition of the clock input.

PAL16RP8A Series

Ordering Information

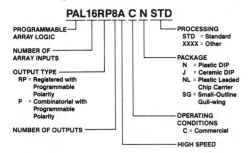

PAL16RP8A C N STD

PROGRAMMABLE ARRAY LOGIC

NUMBER OF ARRAY INPUTS

OUTPUT TYPE
RP = Registered with Programmable Polarity
P = Combinatorial with Programmable Polarity

NUMBER OF OUTPUTS

PROCESSING
STD = Standard
XXXX = Other

PACKAGE
N = Plastic DIP
J = Ceramic DIP
NL = Plastic Leaded Chip Carrier
SG = Small-Outline Gull-wing

OPERATING CONDITIONS
C = Commercial

HIGH SPEED

Polarity

Each of these devices offers programmable polarity on each output. If the polarity fuse is unused, the output is active low. If the polarity fuse is programmed, the output is inverted to active high.

Preload and Power-Up Reset

Each device also offers register preload for device testability. The registers can be preloaded from the outputs by using supervoltages in order to simplify functional testing. This series also offers Power-Up Reset, whereby the registers power up to a logic LOW, setting the active-low outputs to a logic HIGH.

Performance

Performance varies according to the use of the programmable polarity. Active low outputs have a tPD of 25 ns, while active high outputs have a tPD of 30 ns due to the extra inversion. All devices consume 180 mA maximum ICC.

Packages

The commercial PAL16RP8A Series is available in the plastic DIP (N), ceramic DIP (J), plastic leaded chip carrier (NL), and small outline (SG) packages.

	ARRAY INPUTS	OUTPUTS		t_{PD}* (ns)	I_{CC} (mA)
		COMBINATORIAL	REGISTERED		
PAL16P8A	16	8	0	25/30	180
PAL16RP8A	16	0	8	25/30	180
PAL16RP6A	16	6	2	25/30	180
PAL16RP4A	16	4	4	25/30	180

*25 ns active low, 30 ns active high

10236A
JANUARY 1988

PAL16RP8A Series
16P8A, 16RP8A, 16RP6A, 16RP4A

DIP/SO Pinouts

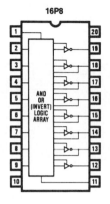

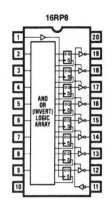

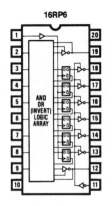

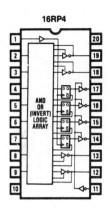

16P8

16RP8

16RP6

16RP4

PLCC Pinouts

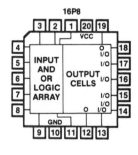

16P8

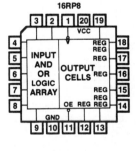

16RP8

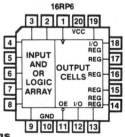

16RP6

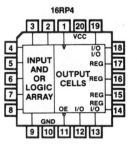

16RP4

Package Drawings

PAL16RP8A Series
16P8A, 16RP8A, 16RP6A, 16RP4A

Absolute Maximum Ratings

	Operating	Programming
Supply voltage V_{CC}	−0.5 V to 7.0 V	−0.5 V to 12.0 V
Input voltage	−1.5 V to 5.5 V	−1.0 V to 22.0 V
Off-state output voltage	5.5 V	12.0 V
Storage temperature		−65°C to +150°C

Operating Conditions

SYMBOL	PARAMETER			COMMERCIAL[1] MIN TYP MAX			UNIT
V_{CC}	Supply voltage			4.75	5	5.25	V
t_w	Width of clock	Low		20	14		ns
		High		10	6		
t_{su}	Set up time from input or feedback to clock	16RP8A 16RP6A 16RP4A	Polarity fuse intact	25	15		ns
			Polarity fuse programmed	30	20		
t_h	Hold time			0	−10		ns
T_A	Operating free-air temperature			0		75	°C

Electrical Characteristics Over Operating Conditions

SYMBOL	PARAMETER	TEST CONDITIONS		MIN	TYP	MAX	UNIT
V_{IL}[2]	Low-level input voltage					0.8	V
V_{IH}[2]	High-level input voltage			2			V
V_{IC}	Input clamp voltage	V_{CC} = MIN	I_I = −18 mA		−0.8	−1.5	V
I_{IL}[3]	Low-level input current	V_{CC} = MAX	V_I = 0.4 V		−0.02	−0.25	mA
I_{IH}[3]	High-level input current	V_{CC} = MAX	V_I = 2.4 V			25	μA
I_I	Maximum input current	V_{CC} = MAX	V_I = 5.5 V			100	μA
V_{OL}	Low-level output voltage	V_{CC} = MIN	I_{OL} = 24 mA		0.3	0.5	V
V_{OH}	High-level output voltage	V_{CC} = MIN	I_{OH} = −3.2 mA	2.4	2.8		V
I_{OZL}[3]	Off-state output current	V_{CC} = MAX	V_O = 0.4 V			−100	μA
I_{OZH}[3]			V_O = 2.4 V			100	μA
I_{OS}[4]	Output short-circuit current	V_{CC} = 5 V	V_O = 0 V	−30	−70	−130	mA
I_{CC}	Supply current	V_{CC} = MAX			120	180	mA

1. The PAL16RP8A Series is designed to operate over the full military operating conditions. For availability and specifications, contact Monolithic Memories.
2. These are absolute voltages with respect to the ground pin on the device and include all overshoots due to system and/or tester noise. Do not attempt to test these values without suitable equipment.
3. I/O pin leakage is the worst case of I_{IL} and I_{OZL} (or I_{IH} and I_{OZH}).
4. No more than one output should be shorted at a time, and duration of the short circuit should not exceed one second.

PAL16RP8A Series
16P8A, 16RP8A, 16RP6A, 16RP4A

Switching Characteristics Over Operating Conditions

SYMBOL	PARAMETER		TEST CONDITIONS	MIN	TYP	MAX	UNIT
t_{PD}	Input or feedback to output 16P8A, 16RP6A, 16RP4A	Polarity fuse intact			15	25	ns
		Polarity fuse programmed			20	30	
t_{CLK}	Clock to output				10	15	ns
t_{CF}	Clock to feedback				8	10	ns
t_{PZX}	Pin 11 to output enable except 16P8A		$R_1 = 200\ \Omega$ $R_2 = 390\ K\Omega$		10	20	ns
t_{PXZ}	Pin 11 to output disable except 16P8A				11	20	ns
t_{EA}	Input to output enable	16P8A, 16RP6A, 16RP4A			10	25	ns
t_{ER}	Input to output disable	16P8A, 16RP6A, 16RP4A			13	25	ns
f_{MAX}	Maximum frequency 16RP8A, 16RP6A, 16RP4A	External — Polarity fuse intact		25	40		MHz
		External — Polarity fuse programmed		22	33		
		Internal — Polarity fuse intact		28.5	43		
		Internal — Polarity fuse programmed		25	35		
		No feedback		33	50		

Switching Test Load
(refer to page 5-164)

Programmers/Development Systems
(refer to Programmer Reference Guide, page 3-81)

Register Preload Waveform
(refer to page 5-164)

Power-Up Reset Waveform
(refer to page 5-164)

Schematic of Inputs and Outputs
(refer to page 5-164)

PAL16RP8A Series
16P8A, 16RP8A, 16RP6A, 16RP4A

Switching Waveforms

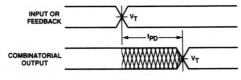

Combinatorial Output

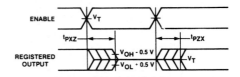

Pin 11 to Output Disable/Enable

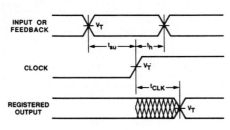

Registered Output

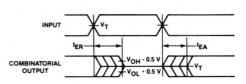

Input to Output Disable/Enable

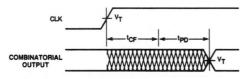

Clock to Feedback to Combinatorial Output (see path below)

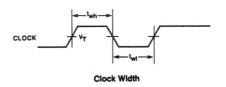

Clock Width

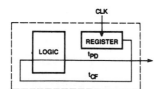

Key to Timing Diagrams

WAVEFORM	INPUTS	OUTPUTS
	DON'T CARE; CHANGE PERMITTED	CHANGING; STATE UNKNOWN
	NOT APPLICABLE	CENTER LINE IS HIGH IMPEDANCE STATE
	MUST BE STEADY	WILL BE STEADY

Notes:
1. VT=1.5V
2. Input pulse amplitude 0 V to 3.0 V
3. Input rise and fall times 2-5 ns typical

Logic Diagram 16P8A

PAL16RP8A Series
16P8A, 16RP8A, 16RP6A, 16RP4A

Logic Diagram   16RP8A

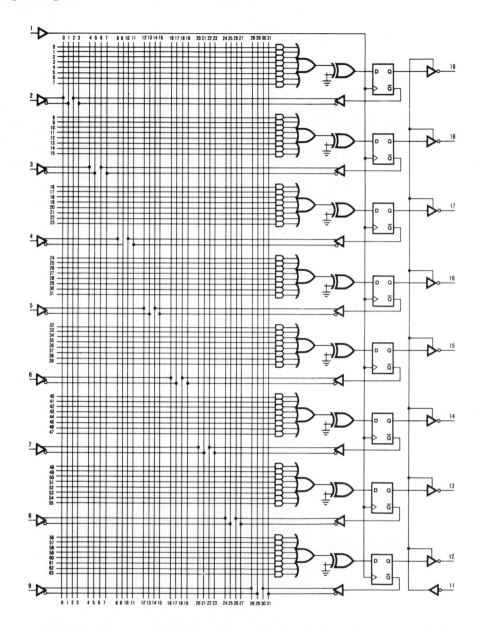

PAL16RP8A Series
16P8A, 16RP8A, 16RP6A, 16RP4A

Logic Diagram   16RP6A

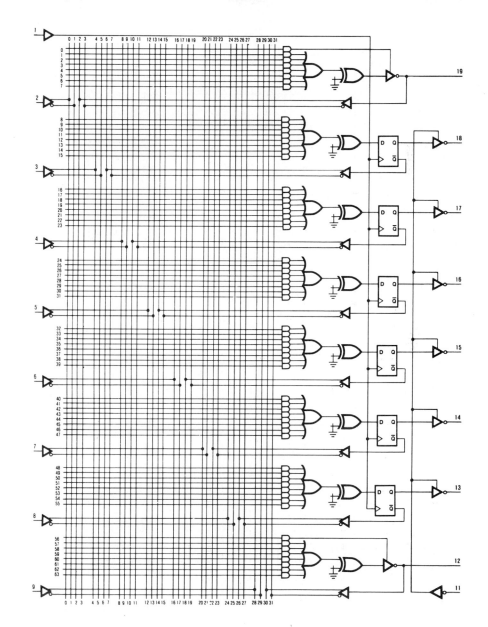

PAL16RP8A Series
16P8A, 16RP8A, 16RP6A, 16RP4A

Logic Diagram   16RP4A

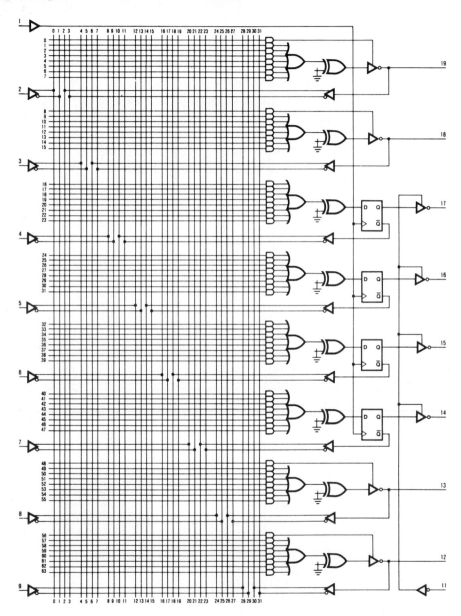

Combinatorial PAL10H8 Series

10H8, 12H6, 14H4, 16H2 16C1
10L8, 12L6, 14L4, 16L2

Features/Benefits

- Combinatorial architectures
- Active high or active low options
- Security fuse

Ordering Information

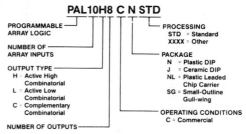

PAL10H8 C N STD

PROGRAMMABLE ARRAY LOGIC

NUMBER OF ARRAY INPUTS

OUTPUT TYPE
H = Active High Combinatorial
L = Active Low Combinatorial
C = Complementary Combinatorial

NUMBER OF OUTPUTS

PROCESSING
STD = Standard
XXXX = Other

PACKAGE
N = Plastic DIP
J = Ceramic DIP
NL = Plastic Leaded Chip Carrier
SG = Small-Outline Gull-wing

OPERATING CONDITIONS
C = Commercial

	INPUTS	OUTPUTS	POLARITY	t_{PD} (ns)	I_{CC} (mA)
PAL10H8	10	8	HIGH	35	90
PAL12H6	12	6	HIGH	35	90
PAL14H4	14	4	HIGH	35	90
PAL16H2	16	2	HIGH	35	90
PAL16C1	16	2	BOTH	40	90
PAL10L8	10	8	LOW	35	90
PAL12L6	12	6	LOW	35	90
PAL14L4	14	4	LOW	35	90
PAL16L2	16	2	LOW	35	90

Description

The PAL10H8 Series is made up of nine combinatorial 20-pin PAL devices. They implement simple combinatorial logic, with no feedback. Each has sixteen product terms total, divided among the outputs, with two to sixteen product terms per output.

Polarity

Both active high and active low versions are available for each architecture. The 16C1 offers both polarities of its single output.

Performance

The standard series has a propagation delay (tpd) of 35 nanoseconds (ns), except for the 16C1 at 40 ns. Standard supply current is 90 milliamps (mA).

Packages

The commercial PAL10H8 Series is available in the plastic DIP (N), ceramic DIP (J), plastic leaded chip carrier (NL), and small outline (SG) packages.

Combinatorial PAL10H8 Series
10H8, 12H6, 14H4, 16H2, 16C1, 10L8, 12L6, 14L4, 16L2

DIP/SO Pinouts

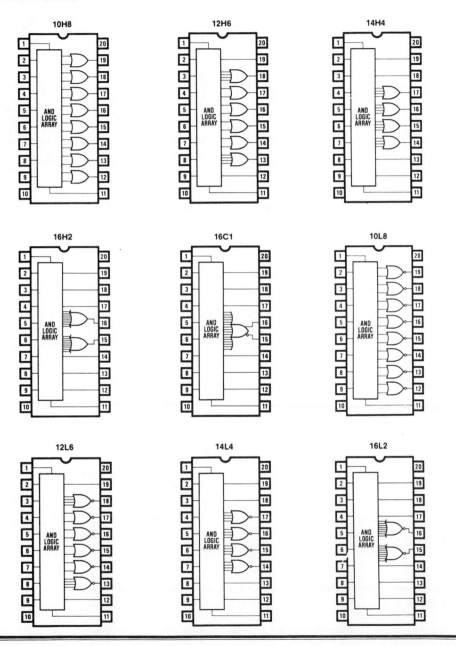

Combinatorial PAL10H8 Series
10H8, 12H6, 14H4, 16H2, 16C1, 10L8, 12L6, 14L4, 16L2

PLCC Pinouts

10H8
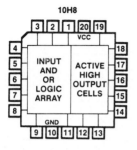

12H6
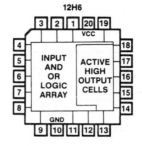

14H4
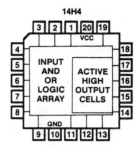

10L8
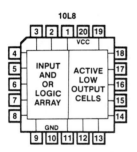

12L6
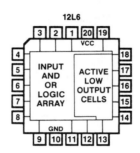

14L4
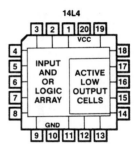

16H2
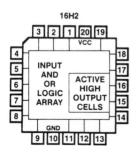

16L2
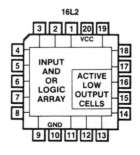

16C1
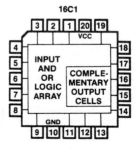

Package Drawings

Combinatorial PAL10H8 Series
10H8, 12H6, 14H4, 16H2, 16C1, 10L8, 12L6, 14L4, 16L2

Absolute Maximum Ratings

	Operating	Programming
Supply voltage V_{CC}	–0.5 V to 7.0 V	–0.5 V to 12.0 V
Input voltage	–1.5 V to 5.5 V	–1.0 V to 22.0 V
Off-state output voltage	5.5 V	12.0 V
Storage temperature		–65°C to +150°C

Operating Conditions

SYMBOL	PARAMETER	MIN	TYP	MAX	UNIT
V_{CC}	Supply voltage	4.75	5	5.25	V
T_A	Operating free-air temperature	0	25	75	°C

Electrical Characteristics Over Operating Conditions

SYMBOL	PARAMETER	TEST CONDITIONS		MIN	TYP	MAX	UNIT
V_{IL}[1]	Low-level input voltage					0.8	V
V_{IH}[1]	High-level input voltage			2			V
V_{IC}	Input clamp voltage	V_{CC} = MIN	I_I = –18 mA		–0.8	–1.5	V
I_{IL}	Low-level input current	V_{CC} = MAX	V_I = 0.4 V		–0.02	–0.25	mA
I_{IH}	High-level input current	V_{CC} = MAX	V_I = 2.4 V			25	μA
I_I	Maximum input current	V_{CC} = MAX	V_I = 5.5 V			100	μA
V_{OL}	Low-level output voltage	V_{CC} = MIN	I_{OL} = 8 mA		0.3	0.5	V
V_{OH}	High-level output voltage	V_{CC} = MIN	I_{OH} = –3.2 mA	2.4	2.8		V
I_{OS}[2]	Output short-circuit current	V_{CC} = 5 V	V_O = 0 V	–30	–70	–130	mA
I_{CC}	Supply current	V_{CC} = MAX			55	90	mA

Switching Characteristics Over Operating Conditions

SYMBOL	PARAMETER		TEST CONDITIONS	MIN	TYP	MAX	UNIT
t_{PD}	Input or feedback to output	Except 16C1	R_1 = 560 Ω R_2 = 1.1 kΩ		25	35	ns
		16C1			25	40	

1. These are absolute values with respect to the ground pin on the device and include all overshoots due to system and/or tester noise. Do not attempt to test these values without suitable equipment.
2. No more than one output should be shorted at a time, and duration of the short circuit should not exceed one second.

Combinatorial PAL10H8 Series
10H8, 12H6, 14H4, 16H2, 16C1, 10L8, 12L6, 14L4, 16L2

Switching Waveforms

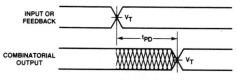

Combinatorial Output

Notes:
1. VT = 1.5 V.
2. Input pulse amplitude 0 V to 3.0 V.
3. Input rise and fall times 2-5 ns typical.

Switching Test Load
(refer to page 5-164)

Programmers/Development Systems
(refer to Programmer Reference Guide, page 3-81)

Schematic of Inputs and Outputs
(refer to page 5-164)

Combinatorial PAL10H8 Series
10H8, 12H6, 14H4, 16H2, 16C1, 10L8, 12L6, 14L4, 16L2

Logic Diagram   **10H8**

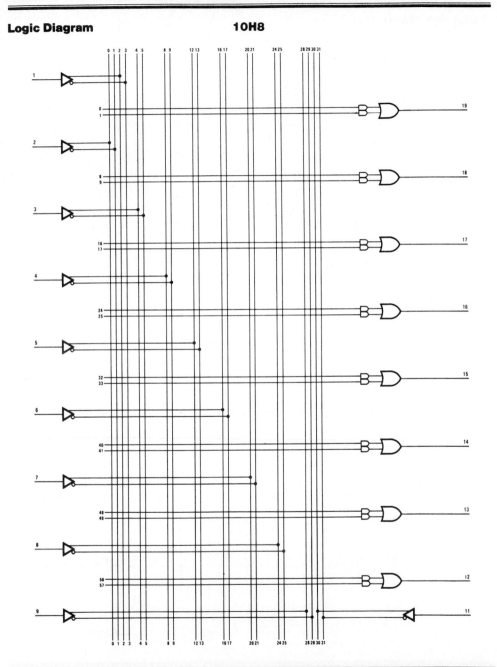

Combinatorial PAL10H8 Series
10H8, 12H6, 14H4, 16H2, 16C1, 10L8, 12L6, 14L4, 16L2

Logic Diagram

12H6

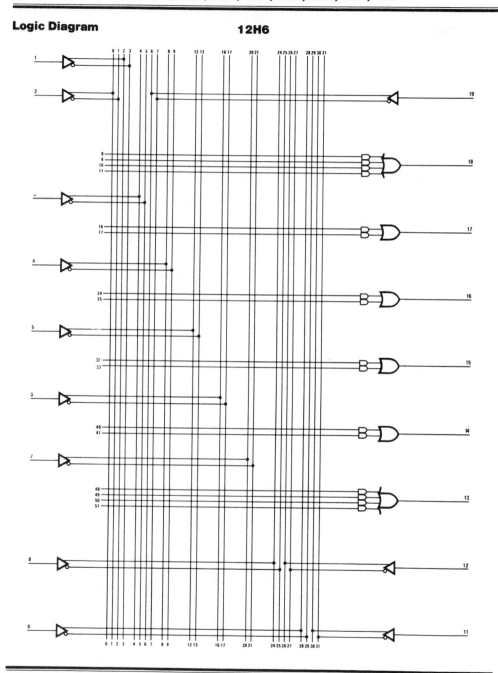

Combinatorial PAL10H8 Series
10H8, 12H6, 14H4, 16H2, 16C1, 10L8, 12L6, 14L4, 16L2

Logic Diagram   **14H4**

Combinatorial PAL10H8 Series
10H8, 12H6, 14H4, 16H2, 16C1, 10L8, 12L6, 14L4, 16L2

Logic Diagram   **10L8**

Combinatorial PAL10H8 Series
10H8, 12H6, 14H4, 16H2, 16C1, 10L8, 12L6, 14L4, 16L2

Logic Diagram   **12L6**

Combinatorial PAL10H8 Series
10H8, 12H6, 14H4, 16H2, 16C1, 10L8, 12L6, 14L4, 16L2

Logic Diagram   **16H2**

Combinatorial PAL10H8 Series
10H8, 12H6, 14H4, 16H2, 16C1, 10L8, 12L6, 14L4, 16L2

Logic Diagram   **16C1**

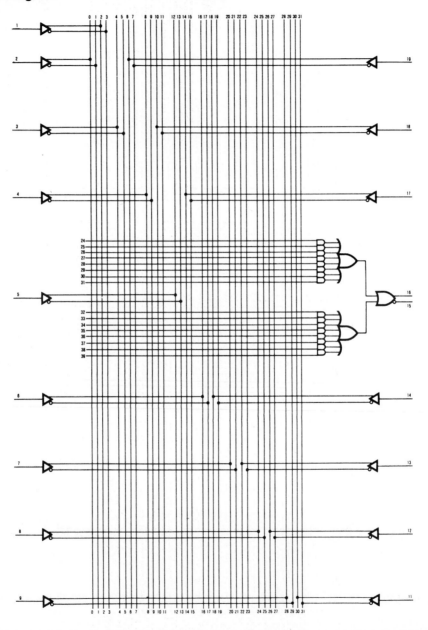

Combinatorial PAL10H8 Series
10H8, 12H6, 14H4, 16H2, 16C1, 10L8, 12L6, 14L4, 16L2

Logic Diagram   14L4

Combinatorial PAL10H8 Series
10H8, 12H6, 14H4, 16H2, 16C1, 10L8, 12L6, 14L4, 16L2

Logic Diagram   **16L2**

APPENDIX
D

DIGITAL
CIRCUIT
TYPES

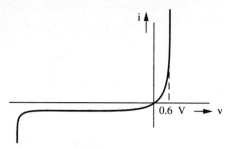

(*a*) Diode characteristic.

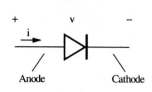

(*b*) Diode symbol.

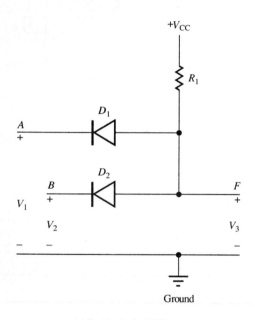

(*c*) Positive logic AND gate.

Voltage truth table		
V_1	V_2	V_3
0 V	0 V	0.6 V
0 V	3 V	0.6 V
3 V	0 V	0.6 V
3 V	3 V	3.6 V

Logic truth table		
A	B	F
0	0	0
0	1	0
1	0	0
1	1	1

Logic 0 = 0 V to 0.6 V
Logic 1 = 3 V to 3.6 V

(*d*) Truth tables.

FIGURE D-1
Diode Logic (DL).

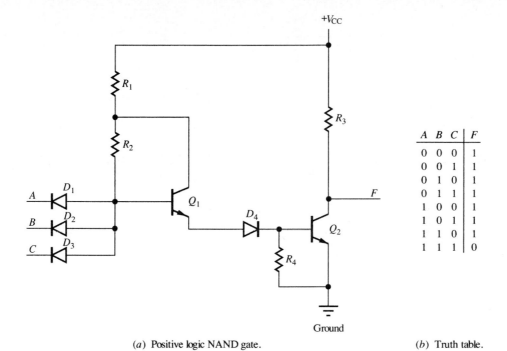

A	B	C	F
0	0	0	1
0	0	1	1
0	1	0	1
0	1	1	1
1	0	0	1
1	0	1	1
1	1	0	1
1	1	1	0

(a) Positive logic NAND gate.   (b) Truth table.

FIGURE D-2
Diode-Transistor Logic (DTL).

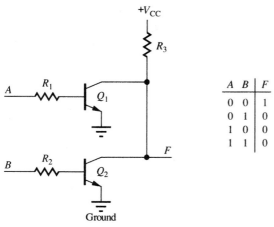

A	B	F
0	0	1
0	1	0
1	0	0
1	1	0

(a) Positive logic NOR gate.   (b) Truth table.

FIGURE D-3
Resistor-Transitor Logic (RTL).

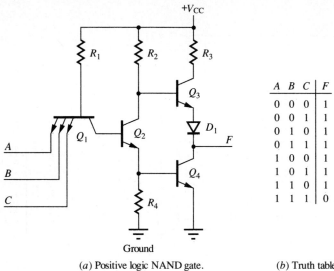

A	B	C	F
0	0	0	1
0	0	1	1
0	1	0	1
0	1	1	1
1	0	0	1
1	0	1	1
1	1	0	1
1	1	1	0

(*a*) Positive logic NAND gate.   (*b*) Truth table.

FIGURE D-4
Transistor-Transistor Logic (TTL).

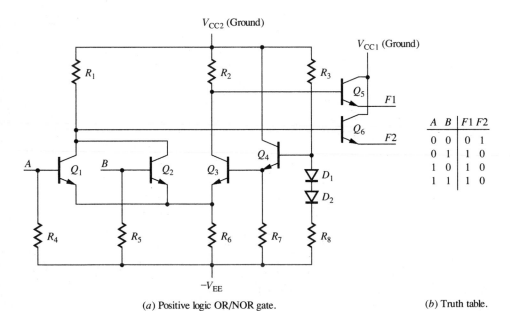

A	B	F1	F2
0	0	0	1
0	1	1	0
1	0	1	0
1	1	1	0

(*a*) Positive logic OR/NOR gate.   (*b*) Truth table.

FIGURE D-5
Emitter-Coupled Logic (ECL).

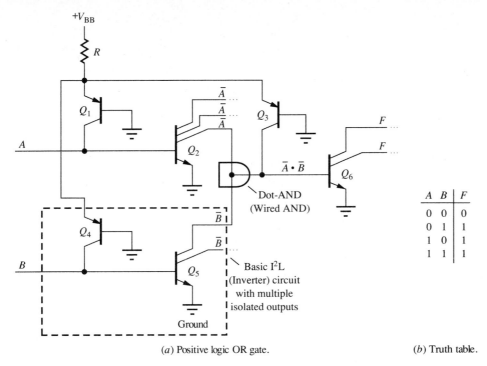

A	B	F
0	0	0
0	1	1
1	0	1
1	1	1

(*a*) Positive logic OR gate.   (*b*) Truth table.

FIGURE D-6
Integrated-Injection Logic (I^2L).

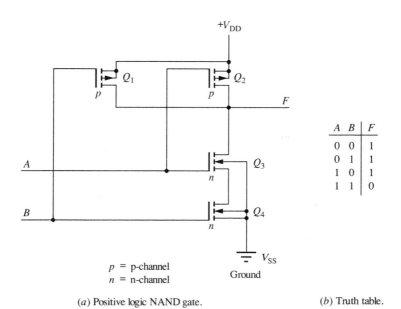

A	B	F
0	0	1
0	1	1
1	0	1
1	1	0

p = p-channel
n = n-channel

(*a*) Positive logic NAND gate.   (*b*) Truth table.

FIGURE D-7
Complementary Metal-Oxide Semiconductor (CMOS) Logic.

INDEX

INDEX

Excitation input signals, 444, 449
Excitation tables, for bistable devices, 494–497
Exclusive NOR element, 239, 241–246
  implementing logic functions using, 364–373
Exclusive NOR gate, maximum length shift counter and, 551
Exclusive OR element, 239, 241–246
  binary to gray code converter using, 97
  implementing logic functions using, 364–373
  logic symbols for, 97
  with programmable polarity fuse, 394
Exclusive OR function, binary to gray code conversion, 95–97
Exclusive OR gate, 97
  maximum length shift counter and, 551, 553, 555
Expansion Theorem (Shannon), 331–334, 347
External logic state, 195, 196

F

Fanout, 344
Field-programmable devices, 375
Finite state machine, 452